Student Solutions Manual to Accompany

Atkins, Jones, and Laverman's

CHEMICAL PRINCIPLES
The Quest for Insight

Seventh Edition

Christina Johnson
University of California, San Diego

Laurence Lavelle
University of California, Los Angeles

Yinfa Ma
Missouri University of Science and Technology

D0166089

ISBN-13: 978-1-319-01756-9
ISBN-10: 1-319-01756-8

Printed in the United States of America

Second Printing

Macmillan Learning
One New York Plaza
New York, NY 10004-1562

www.macmillanhighered.com

CONTENTS

PREFACE

This Student Solutions Manual accompanies the textbook *Chemical Principles: The Quest for Insight,* seventh edition, by Peter Atkins, Loretta Jones, and Leroy Laverman —an authoritative and thorough introduction to chemistry for students anticipating careers in science or engineering disciplines. We have followed the order of topics in the textbook Focus by Focus. The parallel between the symbols, concepts, and style of this supplement and the textbook enable the reader to easily move back and forth between the two.

The Student Solutions Manual contains solutions and answers to the odd-numbered exercises in the textbook, including the Fundamentals and Major Techniques sections. To display intermediate results we have disregarded rules concerning significant figures, but have adhered to them in reporting the final answers. All values used in performing the calculation herein can be found in either the Focus to which the problem is found, or in Appendix 2 of the textbook. In exercises with multiple parts, the properly rounded values are used for subsequent calculations. Student answers may differ slightly if only the final values are rounded, a method recommended by some instructors and the text, but not readily implemented in a printed manual.

ACKNOWLEDGEMENTS

The Solutions Manual authors, Laurence Lavelle, Yinfa Ma, and Christina Johnson, would like to thank Valerie Keller for her invaluable assistance in ensuring the accuracy of this text, Maria Vlasak for the copyediting, and Jodi Isman for guiding the text through production. We'd also like to thank Heidi Bamatter for project managing this text with attentiveness.

FUNDAMENTALS

A.1   (a) law; (b) hypothesis; (c) hypothesis; (d) hypothesis; (e) hypothesis

A.3   (a) chemical;   (b) physical;   (c) physical

A.5   The temperature of the injured camper, the evaporation and condensation of water are physical properties. The ignition of propane is a chemical change.

A.7   (a) physical;   (b) chemical;   (c) chemical

A.9   (a) intensive;   (b) intensive;   (c) intensive;   (d) extensive

A.11   (a) $1000 \text{ grain} \times \left(\dfrac{1 \text{ kilograin}}{1,000 \text{ grain}} \right) = 1 \text{ kilograin}$

(b) $0.01 \text{ batman} \times \left(\dfrac{100 \text{ centibatman}}{1 \text{ batman}} \right) = 1 \text{ centibatman}$

(c) $1 \times 10^6 \text{ mutchkin} \times \left(\dfrac{1 \text{ megamutchkin}}{10^6 \text{ mutchkin}} \right) = 1 \text{ megamutchkin}$

A.13 $1.00 \text{ cup} \times \left(\dfrac{1 \text{ pint}}{2 \text{ cups}} \right) \times \left(\dfrac{1 \text{ quart}}{2 \text{ pints}} \right) \times \left(\dfrac{0.946 \text{ L}}{1 \text{ quart}} \right) \times \left(\dfrac{1000 \text{ mL}}{1 \text{ L}} \right) = 236 \text{ mL}$

A.15   (a) $\lambda = 5.4 \times 10^2 \ \mu\text{m} \times \left[\dfrac{10^6 \text{ pm}}{1 \ \mu\text{m}} \right] = 5.4 \times 10^8 \text{ pm}$ $<$ (b) $= 1.3 \times 10^9 \text{ pm}$

A.17 $d = \dfrac{m}{V}$

$$= \left(\frac{11.23\ g}{2.93\ mL - 2.34\ mL} \right) \left(\frac{1\ mL}{1\ cm^3} \right)$$

$$= 19.0\ g \cdot cm^{-3}$$

A.19 $d = \dfrac{m}{V}$, rearranging gives $V = \dfrac{m}{d}$

$$= \left(\frac{0.750\ carat}{3.51\ g \cdot cm^{-3}} \right) \left(\frac{200\ mg}{1\ carat} \right) \left(\frac{1\ g}{1000\ mg} \right)$$

$$= 0.0427\ cm^3$$

A.21 $105.50\ g - 43.50\ g = 62.00\ g = m_{H_2O}$

$$d = \frac{m}{V} \qquad V = \frac{m}{d} = \frac{m_{H_2O}}{d_{H_2O}} = \frac{62.00\ g}{0.9999\ g \times cm^{-3}} = 62.00\ cm^3$$

$$d_{liquid} = \left(\frac{96.75\ g - 43.50\ g}{62.00\ cm^3} \right) = 0.8589\ g \times cm^{-3}$$

A.23 $d = \dfrac{m}{V}$, rearrange gives $V = \dfrac{m}{d}$

$$V = \frac{20.0\ g}{2.70\ g \cdot cm^{-3}} = 7.41\ cm^3$$

Since the area is one centimeter, the thickness must be 7.41 cm

A.25 0.3423; keep four significant figures

A.27 The result should have three significant figures because the number 3.25 has the least significant figures. The results should be 0.989.

A.29 (a) $4.82\ nm \times \left(\dfrac{1000\ pm}{1\ nm} \right) = 4.82 \times 10^3\ pm$

(b) $\left(\dfrac{1.83\ mL}{min} \right) \times \left(\dfrac{1\ cm^3}{1\ mL} \right) \times \left(\dfrac{1000\ mm^3}{1\ cm^3} \right) \times \left(\dfrac{1\ min}{60\ s} \right) = 30.5\ mm^3/s$

(c) $1.88 \text{ ng} \times (\dfrac{1\text{g}}{10^9\text{ng}}) \times (\dfrac{1\text{kg}}{10^3\text{g}}) = 1.88 \times 10^{-12} \text{ kg}$

(d) $(\dfrac{2.66\text{g}}{\text{cm}^3}) \times (\dfrac{1\text{kg}}{1000\text{g}}) \times (\dfrac{10^6 \text{cm}^3}{1\text{m}^3}) = 2.66 \times 10^3 \text{ kg/m}^3$

(e) $(\dfrac{0.044\text{g}}{\text{L}}) \times (\dfrac{1000\text{mg}}{1\text{g}}) \times (\dfrac{1\text{L}}{1000\text{cm}^3}) = 0.044 \text{ mg/cm}^3$

A.31 (a) $d = \dfrac{m}{V}$

$$= \left(\dfrac{0.213\text{g}}{1.100\text{ cm}\times0.531\text{ cm}\times0.212\text{ cm}} \right) = \left(\dfrac{0.213 \text{ g}}{0.1238 \text{ cm}^3} \right)$$

$$= 1.72 \text{g} \cdot \text{cm}^{-3}$$

This determination is more precise because the volume is not limited to two significant figures as it is in part (b)

(b) $d = \dfrac{m}{V}$

$$= \left(\dfrac{41.003\text{g} - 39.753\text{g}}{20.37\text{mL} - 19.65\text{mL}} \right)$$

$$= \left(\dfrac{1.250\text{g}}{0.72\text{mL}} \right) \left(\dfrac{1\text{mL}}{1\text{cm}^3} \right)$$

$$= 1.7 \text{g} \cdot \text{cm}^{-3}$$

A.33 (a) Formula: $^{\circ}\text{X} = 50 + 2 \times {}^{\circ}\text{C}$

(b) $^{\circ}\text{X} = 50 + 2 \times 22 ^{\circ}\text{C} = 94 ^{\circ}\text{X}$

A.35 $E_K = \dfrac{1}{2}mv^2$

$$= \dfrac{1}{2}(4.2 \text{ kg})(14 \text{ km} \cdot \text{h}^{-1})^2 \left(\dfrac{1\text{ h}}{3600 \text{ s}} \right)^2 \left(\dfrac{1000 \text{ m}}{1 \text{ km}} \right)^2$$

$$= 32 \text{ kg} \cdot \text{m}^2 \cdot \text{s}^{-2}$$

$$= 32 \text{ J}$$

A.37 $m = 2.8$ metric tons, $v_i = 100 \text{ km} \cdot \text{hr}^{-1}$, $v_f = 50 \text{ km} \cdot \text{hr}^{-1}$

$$E_K = \frac{1}{2}mv^2$$

$$E_{K(init)} = \frac{1}{2}(2.8 \text{ metric tons})\left(\frac{10^3 \text{ kg}}{1 \text{ metric ton}}\right)$$

$$\left[\left(\frac{100 \text{ km}}{1 \text{ hr}}\right)\left(\frac{1 \text{ hr}}{60 \text{ min.}}\right)\left(\frac{1 \text{ min.}}{60 \text{ sec.}}\right)\right]^2$$

$$= 1.08 \text{ kg} \cdot \text{km}^2 \cdot \text{s}^{-2} = 1.08 \times 10^6 \text{ kg} \cdot \text{m}^2 \cdot \text{s}^{-2} = 1{,}080 \text{ kJ}$$

$$E_{K(final)} = \frac{1}{2}(2.8 \text{ metric tons})\left(\frac{10^3 \text{ kg}}{1 \text{ metric ton}}\right)$$

$$\left[\left(\frac{50 \text{ km}}{1 \text{ hr}}\right)\left(\frac{1 \text{ hr}}{60 \text{ min.}}\right)\left(\frac{1 \text{ min.}}{60 \text{ sec.}}\right)\right]^2$$

$$= 0.27 \text{ kg} \cdot \text{km}^2 \cdot \text{s}^{-2} = 0.27 \times 10^6 \text{ kg} \cdot \text{m}^2 \cdot \text{s}^{-2} = 270 \text{ kJ}$$

$$E_{K(init)} - E_{K(final)} = (1{,}080 - 270) \text{ kJ} = 810 \text{ kJ} = 8.1 \times 10^2 \text{ kJ}$$

This amount of energy could have been recovered, neglecting friction and other losses, or used to drive the vehicle up a hill.

$$E_P = mgh \quad g = 9.81 \text{ ms}^{-2}$$

Setting potential energy equal to $8.1 \times 10^2 \text{ kJ} = 8.1 \times 10^5 \text{ kg m}^2 \text{ s}^{-2}$ and solving for height gives

$$h = \frac{E_P}{mg} = \left(\frac{8.1 \times 10^5 \text{ kg} \cdot \text{m}^2 \cdot \text{s}^{-2}}{(2800 \text{ kg})(9.81 \text{ ms}^{-2})}\right) = 29 \text{ m}$$

A.39 $E_P = mgh$

$$= (40.0 \text{ g})(9.81 \text{ m} \cdot \text{s}^{-2})(0.50 \text{ m})\left(\frac{1 \text{ kg}}{1000 \text{ g}}\right)$$

$$= 0.20 \text{ kg} \cdot \text{m}^2 \cdot \text{s}^{-2} \text{ for one raise of a fork.}$$

For 30 raises, $(30)(0.20 \text{ kg} \cdot \text{m}^2 \cdot \text{s}^{-2}) = 6.0 \text{ J}$

A.41   Since $R = R_E + h$,

$$E_p = \Delta E_P = \frac{-Gm_E m}{R_E + h} - \left(\frac{-Gm_E m}{R_E} \right)$$

$$= \frac{-Gm_E m}{R_E} \left(\frac{1}{1 + \dfrac{h}{R_E}} - 1 \right) = \frac{Gm_E m}{R_E} \left(1 - \frac{1}{1 + \dfrac{h}{R_E}} \right)$$

If $x = \dfrac{h}{R_E}$, then $\dfrac{1}{1 + \dfrac{h}{R_E}} = 1 - \dfrac{h}{R_E}$ from the expansion in x.

Then

$$E_p = \frac{Gm_E m}{R_E} \left(1 - \left(1 - \frac{h}{R_E} \right) \right) = \frac{Gm_E mh}{R_E^2} = mgh$$

where $g = \dfrac{Gm_E}{R_E^2}$

B.1   number of beryllium atoms $= \dfrac{\text{mass of sample}}{\text{mass of one atom}}$

$$= \left(\frac{0.210 \text{ g}}{1.50 \times 10^{-26} \text{ kg} \cdot \text{atom}^{-1}} \right) \left(\frac{1 \text{ kg}}{1000 \text{ g}} \right)$$

$$= 1.40 \times 10^{22} \text{ atoms}$$

B.3   (a) 5p, 6n, 5e; (b) 5p, 5n, 5e; (c) 15p, 16n, 15e; (d) 92p, 146n, 92e

B.5   (a) ^{194}Ir ; (b) ^{22}Ne ; (c) ^{51}V

B.7

Element	Symbol	Protons	Neutrons	Electrons	Mass number
Chlorine	^{36}Cl	17	19	17	36
Zinc	^{65}Zn	30	35	30	65
Calcium	^{40}Ca	20	20	20	40
Lanthanum	^{137}La	57	80	57	137

B.9 (a) they all have the same mass; (b) they have differing numbers of protons, neutrons, and electrons

B.11 (a) Mass fraction due to the neutrons:

$(56-26)(1.675\times10^{-27})/[(30\times1.675\times10^{-27})+(26\times1.673\times10^{-27})]$

$= 0.5360;$

(b) Mass fraction due to the protons:

$(26)(1.673\times10^{-27})/[(30\times1.675\times10^{-27})+(26\times1.673\times10^{-27})]$

$= 0.4640;$

(c) Mass fraction due to the electrons: $(26 \times \dfrac{1}{1840})/56 = 2.523 \times 10^{-4}$

(can be ignored);

(d) $1000. \text{ kg} \times 0.5360 = 536.0 \text{ kg}$

B.13 (a) Scandium is a Group 3 metal. (b) Strontium is a Group 2 metal. (c) Sulfur is a Group 16 nonmetal. (d) Antimony is a Group 15 metalloid.

B.15 (a) Sr, metal; (b) Xe, nonmetal; (c) Si, metalloid

B.17 (a) Alkali metal: none; (b) Transition metals: cadmium; (c) lanthanoid: cerium

B.19 (a) d block; (b) p block; (c) d block; (d) s block; (e) p block; (f) d block

B.21 (a) Pb; group 14; period 6; metal; (b) Cs; group 1; period 6; metal.

C.1 Container (a) holds a mixture (one is single compound and another is single element); Container (b) holds a single element.

C.3 The chemical formula of xanthophyll is $C_{40}H_{56}O_2$.

C.5 (a) $C_3H_7O_2N$; (b) C_2H_7N

C.7 (a) Cesium is a metal in Group 1; it will form Cs^+ ions. (b) Iodine is a nonmetal in Group 17/VII and will form I^- ions. (c) Selenium is a Group 16/VI nonmetal and will form Se^{2-} ions. (d) Calcium is a Group 2 metal and will form Ca^{2+} ions.

C.9 (a) $^{10}Be^{2+}$ has 4 protons, 6 neutrons, and 2 electrons. (b) $^{17}O^{2-}$ has 8 protons, 9 neutrons, and 10 electrons. (c) $^{80}Br^-$ has 35 protons, 45 neutrons, and 36 electrons. (d) $^{75}As^{3-}$ has 33 protons, 42 neutrons, and 36 electrons.

C.11 (a) $^{19}F^-$; (b) $^{24}Mg^{2+}$; (c) $^{128}Te^{2-}$; (d) $^{86}Rb^+$

C.13 (a) Aluminum forms Al^{3+} ions; tellurium forms Te^{2-} ions. Two aluminum atoms produce a charge of $2 \times +3 = +6$. Three tellurium atoms produce a charge of $3 \times -2 = -6$. The formula for aluminum telluride is Al_2Te_3. (b) Magnesium forms Mg^{2+} ions and oxygen forms O^{2-} ions. A magnesium ion produces a charge of $+2$, which is required to balance the charge on one O^{2-} ion. The formula for magnesium oxide is MgO. (c) Sodium forms $+1$ ions; sulfur forms -2 ions. The formula for sodium sulfide is Na_2S. (d) Rubidium forms $+1$ ions and iodine forms -1 ions. One iodide ion is required to balance the charge of one rubidium ion, so the formula is RbI.

C.15 (a) Group III; (b) aluminum, Al

C.17 (a) Mass fraction due to the neutrons: $(48-22)/48 = 0.542$

(b) Mass fraction due to the protons: $22/48 = 0.458$

(c) Mass fraction due to the electrons: $(22 \times \dfrac{1}{1840})/48 = 2.49 \times 10^{-4}$

(can be ignored)

(d) $30.\,kg \times \left(\dfrac{22}{48}\right) = 14\,kg$

C.19 (a) Na_2HPO_3

(b) $(NH_4)_2CO_3$

(c) Since the copper ion has the same charge as one magnesium ion, the copper cation carries a charge of +2.

(d) Since three tin ions have the same charge as six potassium ions, each tin ion carries a charge of +2.

C.21 (a) HCl, molecular compound (in the gas phase); (b) S_8, element (molecular substance); (c) CoS, ionic compound; (d) Ar, element; (e) CS_2, molecular compound;   (f) $SrBr_2$, ionic compound

D.1 (a) Bromite ion; (b) HSO_3^-

D.3 (a) $MnCl_2$. Mn forms 2+ ions and chlorine forms -1 ions.

(b) $Ca_3(PO_4)_2$. Calcium forms +2 ions and the phosphate ion is PO_4^{3-}.

(c) $Al_2(SO_3)_3$. Aluminum forms +3 ions and the sulfite ion is SO_3^{2-}.

(d) Mg_3N_2. Magnesium forms 2+ ions and the nitride ion is N^{3-}.

D.5 (a) Phosphorus pentafluoride.

(b) Iodine trifluoride.

(c) Oxygen difluoride.

(d) Diboron tetrachloride.

(e) Cobalt (II) sulfate heptahydrate.

(f) Mercury (II) bromide.

(g) Iron(III) hydrogen phosphate (or ferric hydrogen phosphate; or iron(III) biphosphate).

(h) Tungsten(V) oxide.

(i) Osmium (III) bromide.

D.7 (a) calcium phosphate; (b) tin(IV) sulfide, stannic sulfide; (c) vanadium(V) oxide; (d) copper(I) oxide, cuprous oxide

D.9 (a) sulfur hexafluoride; (b) dinitrogen pentoxide; (c) nitrogen triiodide; (d) xenon tetrafluoride; (e) arsenic tribromide; (f) chlorine dioxide

D.11 (a) hydrochloric acid; (b) sulfuric acid; (c) nitric acid; (d) acetic acid; (e) sulfurous acid; (f) phosphoric acid

D.13 (a) $HClO_4$; (b) $HClO$; (c) HIO; (d) HF; (e) H_3PO_3; (f) HIO_4

D.15 (a) TiO_2; (b) $SiCl_4$; (c) CS_2; (d) SF_4; (e) Li_2S; (f) SbF_5; (g) N_2O_5; (h) IF_7

D.17 (a) ZnF_2 (b) $Ba(NO_3)_2$ (c) AgI (d) Li_3N (e) Cr_2S_3

D. 19 (a) $BaCl_2$; (b) Ionic

D.21 (a) sodium sulfite; (b) iron(III) oxide or ferric oxide; (c) iron(II) oxide or ferrous oxide; (d) magnesium hydroxide; (e) nickel(II) sulfate hexahydrate; (f) phosphorus pentachloride; (g) chromium(III) dihydrogen phosphate; (h) diarsenic trioxide; (i) ruthenium(II) chloride

D.23   (a) $CuCO_3$ copper (II) carbonate; (b) K_2SO_3 potassium sulfite; (c) LiCl, lithium chloride

D.25   (a) heptane;   (b) propane;   (c) pentane;   (d) butane

D.27   (a) cobalt(III) oxide monohydrate; $Co_2O_3 \cdot H_2O$; (b) cobalt(II) hydroxide; $Co(OH)_2$

D.29   E = Si; SiH_4, silicon tetrahydride; Na_4Si, sodium silicide

D.31   (a) lithium aluminum hydride, ionic (with a molecular anion); (b) sodium hydride, ionic

D.33   (a) selenic acid;   (b) sodium arsenate;   (c) calcium tellurite;   (d) barium arsenate;   (e) antimonic acid;   (f) nickel(III) selenate

D.35   (a) alcohol; (b) carboxylic acid; (c) haloalkane

E.1 $1.0 \text{ mole} \times \dfrac{6.022 \times 10^{23} \text{ Ag atom}}{1.0 \text{ mole Ag}} \times \dfrac{288 \text{pm}}{1 \text{ Ag atom}} \times \dfrac{1 \text{m}}{10^{12} \text{ pm}}$

$= 1.73 \times 10^{14} \text{ m} = 1.73 \times 10^{11} \text{ km}$

E.3   You will need to add three At atoms to balance the nine Ga atoms on the left pan because an At atom is three times as massive as a Ga atom.

E.5   (a) moles of people $= \dfrac{7.0 \times 10^9 \text{ people}}{6.022 \times 10^{23} \text{ people} \cdot \text{mol}^{-1}}$

$= 1.2 \times 10^{-14} \text{ mol}$

(b) $\text{time} = \dfrac{1\,\text{mol peas}}{1.2 \times 10^{-14}\,\text{mol} \cdot \text{s}^{-1}} = 8.3 \times 10^{13}\,\text{s}$

$(8.3 \times 10^{13}\,\text{s})\left(\dfrac{1\,\text{h}}{3600\,\text{s}}\right)\left(\dfrac{1\,\text{day}}{24\,\text{h}}\right)\left(\dfrac{1\,\text{yr}}{365\,\text{days}}\right) = 2.6 \times 10^{6}\,\text{years}$

E.7 $\text{moles of C} = (2.1 \times 10^{9}\,\text{C atoms}) \times \left(\dfrac{1\,\text{mole C}}{6.022 \times 10^{23}\,\text{C}}\right)$

$= 3.5 \times 10^{-15}\,\text{moles C}$

E.9 (a) $MgSO_4 \cdot 7\,H_2O$ formula mass $= 246.48\,\text{g} \cdot \text{mol}^{-1}$

$\text{atoms of O} = \left(\dfrac{5.15\,\text{g}}{246.48\,\text{g} \cdot \text{mol}^{-1}}\right)\left(\dfrac{11\,\text{mol O atoms}}{\text{mol } MgSO_4 \cdot 7\,H_2O}\right)$

$(6.022 \times 10^{23}\,\text{atoms} \cdot \text{mol}^{-1}) = 1.38 \times 10^{23}$

(b) $\text{formula units} = \left(\dfrac{5.15\,\text{g}}{246.48\,\text{g} \cdot \text{mol}^{-1}}\right)(6.022 \times 10^{23}\,\text{atoms} \cdot \text{mol}^{-1})$

$= 1.26 \times 10^{22}$

(c) $\text{moles of } H_2O = 7\left(\dfrac{5.15\,\text{g}}{246.48\,\text{g} \cdot \text{mol}^{-1}}\right) = 0.146\,\text{mol}$

E.11 (a) mass of average Li atom

$= \left(\dfrac{7.42}{100}\right)(9.988 \times 10^{-24}\,\text{g}) + \left(\dfrac{92.58}{100}\right)(1.165 \times 10^{-23}\,\text{g})$

$= 1.153 \times 10^{-23}\,\text{g} \cdot \text{atom}^{-1}$

$\text{molar mass} = (1.153 \times 10^{-23}\,\text{g} \cdot \text{atom}^{-1})(6.022 \times 10^{23}\,\text{atoms} \cdot \text{mol}^{-1})$

$= 6.94\,\text{g} \cdot \text{mol}^{-1}$

(b) mass of average Li atom

$$= \left(\frac{5.67}{100} \right)(9.988 \times 10^{-24} \text{ g}) + \left(\frac{100 - 5.67}{100} \right)(1.165 \times 10^{-23} \text{ g})$$

$$= 1.1556 \times 10^{-23} \text{ g} \cdot \text{atom}^{-1}$$

$$\text{molar mass} = (1.1556 \times 10^{-23} \text{ g} \cdot \text{atom}^{-1})(6.022 \times 10^{23} \text{ atoms} \cdot \text{mol}^{-1})$$

$$= 6.96 \text{ g} \cdot \text{mol}^{-1}$$

E.13　The percentage $^{10}\text{B} = 100 - \text{percentage } ^{11}\text{B}$

$$\text{molar mass} = \left(\frac{\% \, ^{10}\text{B}}{100\%} \right)(\text{mass } ^{10}\text{B}) + \left(\frac{\% \, ^{11}\text{B}}{100\%} \right)(\text{mass } ^{11}\text{B})$$

$$= \left(\frac{100\% - \% \, ^{11}\text{B}}{100\%} \right)(\text{mass } ^{10}\text{B}) + \left(\frac{\% \, ^{11}\text{B}}{100\%} \right)(\text{mass } ^{11}\text{B})$$

Rearranging gives

$$\% \, ^{11}\text{B} = \frac{100 \cdot \text{molar mass} - 100 \cdot \text{mass } ^{10}\text{B}}{\text{mass } ^{11}\text{B} - \text{mass } ^{10}\text{B}}$$

$$= \frac{100(10.81 \text{ g} \cdot \text{mol}^{-1}) - 100(10.013 \text{ g} \cdot \text{mol}^{-1})}{11.093 \text{ g} \cdot \text{mol}^{-1} - 10.013 \text{ g} \cdot \text{mol}^{-1}}$$

$$= 73.8 \, \%$$

$$\% \, ^{10}\text{B} = 26.2 \, \%$$

E.15　The molar mass of the metal $= 74.10 \text{ g} \cdot \text{mol}^{-1} - 34.02 \text{ g} \cdot \text{mol}^{-1}$

$= 40.08 \text{ g} \cdot \text{mol}^{-1}$ (Ca); The molar mass of calcium sulfide (CaS)

$= 72.15 \text{ g} \cdot \text{mol}^{-1}$

E.17　(a)　$\dfrac{75 \text{ g}}{114.82 \text{ g} \cdot \text{mol}^{-1}} = 0.65 \text{ mol In}$

$\dfrac{80 \text{ g}}{127.60 \text{ g} \cdot \text{mol}^{-1}} = 0.63 \text{ mol Te}$

75 g of indium contains more moles of atoms than 80 g of tellurium

(b)　$\dfrac{15.0 \text{ g}}{30.97 \text{ g} \cdot \text{mol}^{-1}} = 0.484 \text{ mol P}$

$\dfrac{15.0 \text{ g}}{32.07 \text{ g} \cdot \text{mol}^{-1}} = 0.468 \text{ mol S}$

15.0 g of P has slightly more atoms than 15.0 g of S.

(c) Because the two samples have the same number of atoms, they will have the same number of moles, which is given by

$$\left(\frac{7.36 \times 10^{27} \text{ atoms}}{6.022 \times 10^{23} \text{ atoms} \times \text{mol}^{-1}}\right) = 1.22 \times 10^4 \text{ moles}$$

E.19 (a) molar mass of D_2O: $(2.014 \text{ g/mol}) \times 2 + 15.999 \text{ g/mol} = 20.027 \text{ g/mol}$

(b) Assuming the volume occupied by a D_2O molecule is the same as that of H_2O, the density will be proportional to the molar mass, and the

density of D_2O is: $d = \dfrac{\text{molar mass of } D_2O}{\text{molar mass of } H_2O} \times \text{density of water}$

$$= \left(\frac{20.027 \text{ g} \cdot \text{mol}^{-1}}{18.016 \text{ g} \cdot \text{mol}^{-1}}\right) \times 1.00 \text{ g} \cdot \text{cm}^{-3} = 1.11 \text{ g} \cdot \text{cm}^{-3}$$

(c) The volume of the spherical tank $V = \dfrac{4}{3}\pi r^3 = \dfrac{4}{3}\pi \times (6 \text{ m})^3 = 905. \text{ m}^3$

$= 9.05 \times 10^8 \text{ cm}^3$; the volume calculated from the density data:

$$v = \frac{m}{d} = \frac{1.00 \times 10^9 \text{ g}}{1.11 \text{ g} \cdot \text{cm}^{-3}} = 9.01 \times 10^8 \text{ cm}^3 = 901 \text{ m}^3$$

The volume calculated from the density data is slightly smaller than real tank volume.

(d) The reported D_2O mass was not accurate.

(e) The assumption in part (b) is reasonable because the nuclei are so small and the volume difference between a D_2O molecule and a H_2O can be ignored.

E.21 (a) molar mass of $Al_2O_3 = 101.96 \text{ g} \cdot \text{mol}^{-1}$

$$n_{Al_2O_3} = \frac{10.0 \text{ g}}{101.96 \text{ g} \cdot \text{mol}^{-1}} = 0.0981 \text{ mol}$$

$$N_{Al_2O_3} = (0.0981 \text{ mol})(6.022 \times 10^{23} \text{ molecules} \cdot \text{mol}^{-1}) = 5.91 \times 10^{22} \text{ molecules}$$

(b) molar mass of HF $= 20.01 \text{ g} \cdot \text{mol}^{-1}$

$$n_{HF} = \frac{25.92 \times 10^{-3} \text{ g}}{20.01 \text{ g} \cdot \text{mol}^{-1}} = 1.30 \times 10^{-3} \text{ mol}$$

$$N_{HF} = (1.30 \times 10^{-3} \text{ mol})(6.022 \times 10^{23} \text{ } molecules \cdot \text{mol}^{-1})$$

$$= 7.83 \times 10^{20} \text{ molecules}$$

(c) molar mass of hydrogen peroxide = 34.02 g · mol⁻¹

$$n_{H_2O_2} = \frac{1.55 \times 10^{-3} \text{ g}}{34.02 \text{ g} \cdot \text{mol}^{-1}} = 4.56 \times 10^{-5} \text{ mol}$$

$$N_{H_2O_2} = (4.56 \times 10^{-5} \text{ mol})(6.022 \times 10^{23} \text{ molecules} \cdot \text{mol}^{-1})$$

$$= 2.75 \times 10^{19} \text{ molecules}$$

(d) molar mass of glucose = 180.15 g · mol⁻¹

$$n_{glucose} = \frac{1250 \text{ g}}{180.15 \text{ g} \cdot \text{mol}^{-1}} = 6.94 \text{ mol}$$

$$N_{glucose} = (6.94 \text{ mol})(6.022 \times 10^{23} \text{ molecules} \cdot \text{mol}^{-1}) = 4.18 \times 10^{24} \text{ molecules}$$

(e) molar mass of N atoms = 14.01 g · mol⁻¹

$$n_N = \frac{4.37 \text{ g}}{14.01 \text{ g} \cdot \text{mol}^{-1}} = 0.312 \text{ mol}$$

$$N_N = (0.312 \text{ mol})(6.022 \times 10^{23} \text{ atoms} \cdot \text{mol}^{-1}) = 1.88 \times 10^{23} \text{ atoms}$$

molar mass of N_2 molecules = 28.02 g · mol⁻¹

$$n_{N_2} = \frac{4.37 \text{ g}}{28.02 \text{ g} \cdot \text{mol}^{-1}} = 0.156 \text{ mol}$$

$$N_{N_2} = (0.156 \text{ mol})(6.022 \times 10^{23} \text{ molecules} \cdot \text{mol}^{-1}) = 9.39 \times 10^{22} \text{ molecules}$$

E.23   (a) molar mass of $CuBr_2$ = 223.35 g · mol⁻¹

$$n_{Cu^{2+}} = \left(\frac{3.00 \text{ g CuBr}_2}{223.35 \text{ g} \cdot \text{mol}^{-1} \text{ CuBr}_2} \right) \times \left(\frac{1 \text{ mol Cu}^{2+}}{1 \text{ mol CuBr}_2} \right) = 0.0134 \text{ mol Cu}^{2+}$$

(b) molar mass of SO_3 = 80.06 g · mol⁻¹

$$n_{SO_3} = \left(\frac{0.700 \text{ g SO}_3}{80.06 \text{ g} \cdot \text{mol}^{-1} \text{ SO}_3} \right) = 8.74 \times 10^{-3} \text{ mol SO}_3$$

(c) molar mass of $UF_6 = 352.03$ $g \cdot mol^{-1}$

$$n_F = 25.2 \text{ kg} \times \left(\frac{1000 \text{ g}}{1 \text{ kg}} \right) \times \left(\frac{1 \text{ mol } UF_6}{352.03 \text{ g } UF_6} \right) \times \left(\frac{6 \text{ mol } F^-}{1 \text{ mol } UF_6} \right) = 430 \text{ mol } F^-$$

(d) molar mass of $Na_2CO_3 \cdot 10H_2O = 286.21$ $g \cdot mol^{-1}$

$$n_{H_2O} = \left(\frac{2.00 \text{ g } Na_2CO_3 \cdot 10H_2O}{286.21 \text{ g} \cdot mol^{-1} \text{ } Na_2CO_3 \cdot 10H_2O} \right) \times \left(\frac{10 \text{ mol } H_2O}{1 \text{ mol } Na_2CO_3 \cdot 10H_2O} \right)$$

$$= 0.0699 \text{ mol } H_2O$$

E.25 (a) number of formula units $= (0.750 \text{ mol})(6.022 \times 10^{23}$ formula

units $\cdot mol^{-1}) = 4,52 \times 10^{23}$ formula units

(b) molar mass of $Ag_2SO_4 = 311.80$ $g \cdot mol^{-1}$

$$\left(\frac{2.39 \times 10^{20} \text{ formula units}}{6.022 \times 10^{23} \text{ formula units} \cdot mol^{-1}} \right) (311.80 \text{ g} \cdot mol^{-1}) \left(\frac{1000 \text{ mg}}{1 \text{ g}} \right)$$

$$= 124 \text{ mg}$$

(c) molar mass of $NaHCO_2 = 68.01$ $g \cdot mol^{-1}$

$$\left(\frac{3.429 \text{ g}}{68.01 \text{ g} \cdot mol^{-1}} \right) (6.022 \times 10^{23} \text{ formula units} \cdot mol^{-1})$$

$$= 3.036 \times 10^{22} \text{ formula units}$$

E.27 (a) molar mass of $H_2O = 18.02$ $g \cdot mol^{-1}$

$$\left(\frac{18.02 \text{ g} \cdot mol^{-1}}{6.022 \times 10^{23} \text{ molecules} \cdot mol^{-1}} \right) = 2.992 \times 10^{-23} \text{ g} \cdot \text{molecule}^{-1}$$

(b) $N_{H_2O} = \left(\dfrac{1000 \text{ g}}{18.02 \text{ g} \cdot mol^{-1}} \right) (6.022 \times 10^{23} \text{ molecules} \cdot mol^{-1})$

$$= 3.34 \times 10^{25} \text{ molecules}$$

E.29 (a) molar mass of $CuCl_2 \cdot 4H_2O = 206.53$ $g \cdot mol^{-1}$

$$n = \left(\frac{8.61 \text{ g CuCl}_2 \cdot 4\text{H}_2\text{O}}{206.53 \text{ g} \cdot \text{mol}^{-1} \text{ CuCl}_2 \cdot 4\text{H}_2\text{O}} \right) = 0.0417 \text{ mol CuCl}_2 \cdot 4\text{H}_2\text{O}$$

(b) Because there are two mol Cl^- per mole of compound, the number of moles will be twice the amount in part (a), 0.0834 mol Cl^-

(c) There are four mole of H_2O for each mole compound. Therefore,

Number of water molecules = (0.0417×4) mole $H_2O \times (6.022 \times 10^{23}$

Molecules $\cdot$ mol^{-1}) $= 1.00 \times 10^{23}$ H_2O molecules

(d) fraction of mass due to O $= \dfrac{4(16.00 \text{g} \cdot \text{mol}^{-1})}{206.53 \text{ g} \cdot \text{mol}^{-1}} = 0.3099$

E.31   (a) $\left(\dfrac{10 \text{ mol H}_2\text{O}}{1 \text{ mol hydrated cpd}} \right)\left(\dfrac{18.01 \text{ g H}_2\text{O}}{1 \text{ mol H}_2\text{O}} \right)\left(\dfrac{1 \text{ mol hydrated cpd}}{286.148 \text{ g hydrated cpd}} \right) \times 100$

= 62.94% H_2O

Therefore 1.6 kg out of 2.5 kg was water.

$$\frac{?\$}{\text{L H}_2\text{O}} = \left(\frac{\$175.00}{2.5 \text{ kg hydrated cpd}} \right)\left(\frac{2.5 \text{ kg hydrated cpd}}{1.6 \text{ kg H}_2\text{O}} \right)\left(\frac{1 \text{ kg H}_2\text{O}}{1 \text{ L H}_2\text{O}} \right)$$

= \$109.28 per L H_2O

(b) Since the anhydrous compound costs \$195.00 for 2.5kg, or

\$78.00/kg. In the hydrous compound, only 2.5 − 1.6 = 0.9 kg is

Na_2CO_3, a fair price would have been $\left(\dfrac{\$78.00}{\text{kg}} \right) \times 0.9 = \$70.20$.

E.33   The molar mass of $Ca_5(PO_4)_3OH$ is 502.31 g/mol;

The molar mass of $Ca_5(PO_4)_3F$ is 504.30 g/mol;

% increase in mass $= \left(\dfrac{504.30 - 502.31}{504.30} \right) \times 100 = 0.39\%$

F.1   (a) $C_{10}H_{16}O$

(b) 10×12.01 g $= 120.1$ g $\times \left(\dfrac{100}{152.2264}\right) = 78.90\%$

16×1.0079 g $= 16.1264$ g $\times \left(\dfrac{100}{152.2264}\right) = 10.59\%$

1×16.00 g $= 16.00$ g $\times \left(\dfrac{100}{152.2264}\right) = 10.51\%$

F.3   (a) HNO_3; (b) O (oxygen)

F.5 $C_7H_{15}NO_3$

$$7 \times 12.01 \text{ g} = 84.07 \text{ g} \times \left(\frac{100}{161.20 \text{ g}}\right) = 52.15\%$$

$$15 \times 1.0079 \text{ g} = 15.1185 \text{ g} \times \left(\frac{100}{161.20 \text{ g}}\right) = 9.3787\%$$

$$1 \times 14.01 \text{ g} = 14.01 \text{ g} \times \left(\frac{100}{161.20 \text{ g}}\right) = 8.691\%$$

$$3 \times 16.00 \text{ g} = \underline{48.00 \text{ g}} \times \left(\frac{100}{161.20 \text{ g}}\right) = \underline{29.78\%}$$

$$161.20 \text{ g} \qquad\qquad 100.00\%$$

F.7   (a) M_2O 88.8% M   For 100 g of compound, 88.8 g is M, 11.2 g is O.

$$\frac{? \text{ g } M_2O}{\text{mole } M_2O} = \left(\frac{100 \text{ g } M_2O}{11.2 \text{ g O}}\right)\left(\frac{16.00 \text{ g O}}{1 \text{ mol O}}\right)\left(\frac{1 \text{ mol O}}{1 \text{ mol } M_2O}\right)$$

$$= 143 \text{ g} \cdot \text{mol}^{-1} \, M_2O$$

Therefore, $143 - 16 = 127$ g/mol are due to M. Since there are two moles of M per mole of M_2O, the molar mass of M $= 63.4$ g/mol. That molar mass matches Cu.

(b) copper(I) oxide

F.9   For 100 g of compound,

$$\text{moles of C} = \frac{63.15\ \text{g}}{12.011\ \text{g}\cdot\text{mol}^{-1}} = 5.258\ \text{mol C}$$

$$\text{moles of H} = \frac{5.30\ \text{g}}{1.0079\ \text{g}\cdot\text{mol}^{-1}} = 5.26\ \text{mol H}$$

$$\text{moles of O} = \frac{31.55\ \text{g}}{16.00\ \text{g}\cdot\text{mol}^{-1}} = 1.972\ \text{mol O}$$

Dividing each number by 1.972 gives a ratio of 1 O : 2.67 C : 2.67 H.

The formula is $C_8H_8O_3$

F.11   (a) For 100 g of compound,

$$\text{moles of Na} = \frac{32.79\ \text{g}}{22.99\ \text{g}\cdot\text{mol}^{-1}} = 1.426\ \text{mol}$$

$$\text{moles of Al} = \frac{13.02\ \text{g}}{26.98\ \text{g}\cdot\text{mol}^{-1}} = 0.4826\ \text{mol}$$

$$\text{moles of F} = \frac{54.19\ \text{g}}{19.00\ \text{g}\cdot\text{mol}^{-1}} = 2.852\ \text{mol}$$

Dividing each number by 0.4826 gives a ratio of 1 Al : 2.95 Na : 5.91 F.

The formula is Na_3AlF_6.

(b) For 100 g of compound,

$$\text{moles of K} = \frac{31.91\ \text{g}}{39.10\ \text{g}\cdot\text{mol}^{-1}} = 0.8161\ \text{mol}$$

$$\text{moles of Cl} = \frac{28.93\ \text{g}}{35.45\ \text{g}\cdot\text{mol}^{-1}} = 0.8161\ \text{mol}$$

mass of O is obtained by difference:

$$\text{moles of O} = \frac{100\ \text{g} - 31.91\ \text{g} - 28.93\ \text{g}}{16.00\ \text{g}\cdot\text{mol}^{-1}} = 2.448\ \text{mol}$$

Dividing each number by 0.8161 gives a ratio of 1.00 K : 1 Cl : 3.00 O.

The formula is $KClO_3$.

(c) For 100 g of compound,

$$\text{moles of N} = \frac{12.2\ \text{g}}{14.01\ \text{g}\cdot\text{mol}^{-1}} = 0.871\ \text{mol}$$

$$\text{moles of H} = \frac{5.26\,g}{1.0079\,g \cdot mol^{-1}} = 5.22\,mol$$

$$\text{moles of P} = \frac{26.9\,g}{30.97\,g \cdot mol^{-1}} = 0.869\,mol$$

$$\text{moles of O} = \frac{55.6\,g}{16.00\,g \cdot mol^{-1}} = 3.475\,mol$$

Dividing each number by 0.869 gives a ratio of 1.00 N : 6.01 H : 1.00 P : 4.00 O. The formula is NH_6PO_4 or $[NH_4][H_2PO_4]$, ammonium dihydrogen phosphate.

F.13   (a) moles of P $= \dfrac{4.14\,g}{30.97\,g \cdot mol^{-1}} = 0.134\,mol$

$$\text{moles of Cl} = \frac{27.8\,g - 4.14\,g}{35.45\,g \cdot mol^{-1}} = 0.667\,mol$$

Dividing each number by 0.134 mol gives a ratio of 4.98 Cl : 1 P. The formula is PCl_5.

(b) Name is phosphorus pentachloride.

F.15   For 100 g of compound,

$$\text{moles of C} = \left(\frac{67.49\,g}{12.01\,g \cdot mol^{-1}} \right) = 5.619\,mol$$

$$\text{moles of H} = \left(\frac{4.60\,g}{1.008\,g \cdot mol^{-1}} \right) = 4.563\,mol$$

$$\text{moles of Cl} = \left(\frac{12.45\,g}{35.45\,g \cdot mol^{-1}} \right) = 0.3512\,mol$$

$$\text{moles of N} = \left(\frac{9.84\,g}{14.01\,g \cdot mol^{-1}} \right) = 0.7024\,mol$$

$$\text{moles of O} = \left(\frac{5.62\,g}{16.00\,g \cdot mol^{-1}} \right) = 0.3512\,mol$$

Dividing each number by 0.3512 mol gives a ratio of 16.00 C : 13.00 H :

1.00 Cl : 2.00N : 1.00 O. The formula is $C_{16}H_{13}ClN_2O$

F.17 For 100 g of the osmium carbonyl compound,

$$\text{moles of C} = \frac{15.89 \text{ g}}{12.01 \text{ g} \cdot \text{mol}^{-1}} = 1.323 \text{ mol}$$

$$\text{moles of O} = \frac{21.18 \text{ g}}{16.00 \text{ g} \cdot \text{mol}^{-1}} = 1.324 \text{ mol}$$

$$\text{moles of Os} = \frac{62.93 \text{ g}}{190.2 \text{ g} \cdot \text{mol}^{-1}} = 0.3309 \text{ mol}$$

Dividing each number by 0.3309 mol gives a ratio of 4.00 C : 4.00 O :

1.00 Os.

(a) The empirical formula is OsC_4O_4.

(b) The formula mass of OsC_4O_4 is 302.24 g · mol^{-1}. The molar mass is

907 g · mol^{-1} which is three times the formula mass, so the molecular

formula is $Os_3C_{12}O_{12}$.

F.19 For 100 g of caffeine,

$$\text{moles of C} = \frac{49.48 \text{ g}}{12.01 \text{ g} \cdot \text{mol}^{-1}} = 4.12 \text{ mol}$$

$$\text{moles of H} = \frac{5.19 \text{ g}}{1.0079 \text{ g} \cdot \text{mol}^{-1}} = 5.15 \text{ mol}$$

$$\text{moles of N} = \frac{28.85 \text{ g}}{14.01 \text{ g} \cdot \text{mol}^{-1}} = 2.059 \text{ mol}$$

$$\text{moles of O} = \frac{16.48 \text{ g}}{16.00 \text{ g} \cdot \text{mol}^{-1}} = 1.03 \text{ mol}$$

Dividing each number by 1.03 mol gives a ratio of 4.00 C : 5.00 H : 2.00

N : 1.00 O. The formula is $C_4H_5N_2O$ with a molar formula mass of

$97.10 \text{ g} \cdot \text{mol}^{-1}$. Because the molecular molar mass is twice this value, the actual formula will be $C_8H_{10}N_4O_2$.

F.21 For the 1.78 mg sample of didemnin-A,

$$\text{mmoles of C} = \frac{1.11 \text{ mg}}{12.01 \text{ g} \cdot \text{mol}^{-1}} = 0.0924 \text{ mmol}$$

$$\text{mmoles of H} = \frac{0.148 \text{ mg}}{1.0079 \text{ g} \cdot \text{mol}^{-1}} = 0.147 \text{ mmol}$$

$$\text{mmoles of N} = \frac{0.159 \text{ mg}}{14.01 \text{ g} \cdot \text{mol}^{-1}} = 0.0113 \text{ mmol}$$

$$\text{mmoles of O} = \frac{0.363 \text{ mg}}{16.00 \text{ g} \cdot \text{mol}^{-1}} = 0.0227 \text{ mmol}$$

Dividing by 0.0113 mmol gives 8.18 C : 13.0 H : 1.00 N : 2.00 O.

Multiplying by 6 gives the formula $C_{49}H_{78}N_6O_{12}$ with a molar mass of $942 \text{ g} \cdot \text{mol}^{-1}$.

F.23 Calculate the mass percent carbon for each fuel from its formula.

ethene, C_2H_4 $\quad \dfrac{2(12.01)}{2(12.01)+4(1.0079)} \times 100 = 85.63\% \text{ C}$

propanol, C_3H_7OH $\quad \dfrac{3(12.01)}{3(12.01)+8(1.0079)+16.00} \times 100 = 59.96\% \text{ C}$

heptane, C_7H_{16} $\quad \dfrac{7(12.01)}{7(12.01)+16(1.0079)} \times 100 = 83.91\% \text{ C}$

ethene (85.63%) > heptane (83.91%) > propanol (59.96%)

F.25 (a) empirical formula: C_2H_3Cl, molecular formula: $C_4H_6Cl_2$
(b) empirical formula: CH_4N, molecular formula: $C_2H_8N_2$

F.27 This problem requires that we relate unknowns to each other appropriately by writing a balanced chemical equation and using other information in the problem.

$$x\,NaNO_3 + y\,Na_2SO_4 \rightarrow (x+2y)\,Na^+ + x\,NO_3^- + y\,SO_4^{2-}$$

$$\frac{1.61\text{ g Na}^+}{22.99\text{ g}\cdot\text{mol}^{-1}\text{Na}^+} = 0.07003\text{ mol Na}^+ = x+2y$$

$$5.37\text{ g total} - 1.61\text{ g Na}^+ = 3.76\text{ g} = (62.01\text{ g}\cdot\text{mol}^{-1})x + (96.07\text{ g}\cdot\text{mol}^{-1})y$$

Rearrange and substitute:

$$3.76 = 62.01(0.07003 - 2y) + 96.07y$$

$$0.06065 = 0.07003 - 2y + 1.548y$$

$$0.009385 = 0.4516y$$

$$y = 0.02078\text{ moles sulfate},\ x = 0.02847\text{ moles nitrate}$$

Therefore, the mass of sodium nitrate in the mixture was

$$0.02847\text{ mol}\left(\frac{85.00\text{ g NaNO}_3}{1\text{ mol}}\right) = 2.42\text{ g NaNO}_3$$

$$\frac{2.42\text{ g NaNO}_3}{5.37\text{ g total}} \times 100 = 45.1\%\text{ NaNO}_3$$

G.1   (a) False. They can only be separated by chemical methods.

(b) False. The composition of a solution is fixed.

(c) False. The properties of a compound are unlike those of the elements that compose it.

G.3   (a) heterogeneous, decanting;   (b) heterogeneous, dissolving followed by filtration and distillation;   (c) homogeneous, distillation

G.5   (a) molarity of $Na_2CO_3 = \dfrac{2.111\text{ g}}{(105.99\text{ g}\cdot\text{mol}^{-1})(0.2500\text{ L})} = 0.07967$

M Na₂CO₃

$$V = \frac{(2.15\times10^{-3}\text{ mol Na}^+)(1\text{ mol Na}_2CO_2)}{(0.079\,67\text{ mol}\cdot\text{L}^{-1}\text{ Na}_2CO_3)(2\text{ mol Na}^+)} = 1.35\times10^{-2}\text{ L or 13.5 mL}$$

(b) $$V = \frac{(4.98 \times 10^{-3} \text{ mol CO}_3^{2-})(1 \text{ mol Na}_2\text{CO}_2)}{(0.079\,67 \text{ mol} \cdot \text{L}^{-1} \text{ Na}_2\text{CO}_3)(1 \text{ mol CO}_3^{2-})}$$

$$= 6.25 \times 10^{-2} \text{ L or } 62.5 \text{ mL}$$

(c) $$V = \frac{(50.0 \times 10^{-3} \text{ g Na}_2\text{CO}_3)}{(105.99 \text{ g} \cdot \text{mol}^{-1})(0.079\,67 \text{ mol} \cdot \text{L}^{-1} \text{ Na}_2\text{CO}_3)}$$

$$= 5.92 \times 10^{-3} \text{ L or } 5.92 \text{ mL}$$

G.7 mass of KNO_3 = 510. g × 5.45% = 27.8 g KNO_3

mass of H_2O = 510. g − 27.8 g = 482.2 g H_2O

Measuring 482.2 g H_2O on a balance and pure it into a beaker. Then you weigh 27.8 g of KNO_3 and mix it with the water until it totally dissolve.

G.9 mass of $AgNO_3$ = $(0.179 \text{ mol} \cdot \text{L}^{-1})(0.5000 \text{ L})(169.88 \text{ g} \cdot \text{mol}^{-1}) = 15.2$ g

G.11 volume of glucose = $(4.50 \times 10^{-3} \text{ mol}) \left(\frac{1 \text{ L}}{0.278 \text{ mol}}\right) = 1.62 \times 10^{-2} \text{ L} = 16.2 \text{ mL}$

G.13 concentration of diluted NH_4NO_3 solution = 0.050 M;

moles of NH_4NO_3 each plant receive = 5.0×10^{-3} mole; since each NH_4NO_3 contains 2 N atoms, the moles of N atoms each plant receive = 1.0×10^{-2} mole

G.15 (a) V(0.778 mol·L^{-1}) = (0.1500 L)(0.0234 mol·L^{-1})

V = 4,51×10^{-3} L or 4.51 mL

(b) The concentration desired is one-fifth of the starting NaOH solution, so the stockroom attendant will need to add four volumes of water to one volume of the 2.5 mol·L^{-1} solution. To prepare 60.0 mL of solution, divide 60.0 by 5; so 12.0 mL of 2.5 mol·L^{-1} NaOH solution are added to 48.0 mL of water. (See the solution to G.11).

G.17   (a)   mass of $CuSO_4 = (0.20 \; mol \cdot L^{-1})(0.250 \; L)(159.60 \; g \cdot mol^{-1})$

$= 8.0 \; g$

(b)   mass of $CuSO_4 \cdot 5 \, H_2O = (0.20 \; mol \cdot L^{-1})(0.250 \; L)(249.68 \; g \cdot mol^{-1})$

$= 12 \; g$

G.19   (a)   molarity $= \dfrac{(1.345 \; mol \cdot L^{-1})(0.012 \; 56 \; L)}{(0.2500 \; L)} = 0.06757 \; mol \cdot L^{-1}$

(b)   molarity $= \dfrac{(0.366 \; mol \cdot L^{-1})(0.025 \; 00 \; L)}{(0.125 \; 00 \; L)} = 0.0732 \; mol \cdot L^{-1}$

G.21   (a) Total moles of K^+:

$$\left(\frac{0.500 \, g \, KCl}{74.55 \, g \cdot mol^{-1}} \right) + \left(\frac{0.500 \, g \, K_2S}{110.26 \, g \cdot mol^{-1}} \right) \left(\frac{2 \, mol \, K^+}{1 \, mol \, K_2S} \right) +$$

$$\left(\frac{0.500 \, g \, K_3PO_4}{212.27 \, g \cdot mol^{-1}} \right) \left(\frac{3 \, mol \, K^+}{1 \, mol \, K_3PO_4} \right)$$

$$= 6.71 \times 10^{-3} \; mol + 9.07 \times 10^{-3} \; mol + 7.07 \times 10^{-3} \; mol$$

$$= 2.29 \times 10^{-2} \; mol$$

Molarity of $K^+ = \left(\dfrac{2.29 \times 10^{-2} \; mol \, K^+}{0.500 \, L} \right) = 4.58 \times 10^{-2} \; M$

(b) molarity of S^{2-}:

$$\left(\frac{0.500 \, g \, K_2S}{110.26 \, g \cdot mol^{-1}} \right) \left(\frac{1 \, mol \, S^{2-}}{1 \, mol \, K_2S} \right) \left(\frac{1}{0.500 \, L} \right) = 9.07 \times 10^{-3} \; M$$

G.23   The total molar concentration of Cl^-:

$$\left[\left(\frac{0.50 \, g \, NaCl}{58.44 \, g \cdot mol^{-1}} \right) + \left(\frac{0.30 \, g \, KCl}{74.55 \, g \cdot mol^{-1}} \right) \right] \times \left(\frac{1}{0.100 \, L} \right) = 0.13 \; M \; Cl^-$$

G.25   We can show that fewer than 1 molecule of X would be left after only 70 doublings.

$$10 \text{ mL} \times \frac{0.10 \text{ mol}}{1000 \text{ mL}} \times \frac{6.02214 \times 10^{23} \text{ molecules}}{1 \text{ mol}} = 6.0 \times 10^{20} \text{ molecules are}$$

in the first 10 mL aliquot of the solution. In order to find the number of times the volume must be doubled to get to one molecule, we can solve for the number of times this amount of molecules must be cut in half until it equals 1.

$$6.0 \times 10^{20} \text{ molecules} \times \left(\frac{1}{2}\right)^n = 1 \text{ molecule}$$

$$\log(6.0 \times 10^{20}) + n\log\left(\frac{1}{2}\right) = \log(1)$$

$$20.8 + n(-0.301) = 0$$

$$20.8 = 0.301n$$

$$n = 69$$

The other 21 additional doublings involve solutions with no X remaining. There can be no health benefits if there are no molecules of the active substance, X, left in the solution.

G.27 Find the volume of concentrated HCl solution that is equivalent to 10.0 L of the dilute HCl solution with respect to the number of moles of solute present in each volume.

? mL con. HCl(aq)

$$= 10.0 \text{ L dilute HCl(aq)} \left(\frac{0.7436 \text{ mol HCl}}{1 \text{ L dilute HCl(aq)}}\right)\left(\frac{36.46 \text{ g HCl}}{1 \text{ mol HCl}}\right)$$

$$\times \left(\frac{100 \text{ g con. HCl(aq)}}{37.50 \text{ g HCl}}\right)\left(\frac{1 \text{ cm}^3}{1.205 \text{ g}}\right)$$

$$= 600 \text{ mL con. HCl}$$

Therefore 600. mL of concentrated HCl(aq) must be diluted up to a final volume of 10.0 L by adding water in order to form 0.7436 M HCl(aq).

G.29 The level of instrument can detect:

$$\left(\frac{1.00\times10^{-8}\ \text{mol Pb}}{1.00\ \text{L solution}}\right)\left(\frac{207.2\ \text{g Pb}}{\text{mol Pb}}\right)\left(\frac{1\ \text{L solution}}{1000\ \text{mL solution}}\right)$$

$$\left(\frac{1.00\ \text{mL solution}}{1.00\ \text{g solution}}\right)\times 1.00\times10^{9}\ \text{solution} = 2.07\ \text{ppb} < 10\ \text{ppb}$$

The instrument can detect lower than 10 ppb.   It is satisfactory.

H.1   (a) You cannot add a different compound or element to the chemical equation that is not produced (or involved) in the chemical reaction. In this case, O atom is not produced by the defined reaction.

(b) $2\,Cu\ +\ SO_2\ \rightarrow\ 2\,CuO\ +\ S$

H.3 $2\,SiH_4\ +\ 4\,H_2O\ \rightarrow\ 2\,SiO_2\ +\ 8\,H_2$

H.5   (a) $NaBH_4(s)\ +\ 2\,H_2O(l)\ \rightarrow\ NaBO_2(aq)\ +\ 4\,H_2(g)$

(b) $Mg(N_3)_2(s)\ +\ 2\,H_2O(l)\ \rightarrow\ Mg(OH)_2(aq)\ +\ 2\,HN_3(aq)$

(c) $2\,NaCl(aq)\ +\ SO_3(g)\ +\ H_2O(l)\ \rightarrow\ Na_2SO_4(aq)\ +\ 2\,HCl(aq)$

(d) $4\,Fe_2P(s)\ +\ 18\,S(s)\ \rightarrow\ P_4S_{10}(s)\ +\ 8\,FeS(s)$

H.7   (a) $Ca(s)\ +\ 2\,H_2O(l)\ \rightarrow\ H_2(g)\ +\ Ca(OH)_2(aq)$

(b) $Na_2O(s)\ +\ H_2O(l)\ \rightarrow\ 2\,NaOH(aq)$

(c) $3\,Mg(s)\ +\ N_2(g)\ \rightarrow\ Mg_3N_2(s)$

(d) $4\,NH_3(g)\ +\ 7\,O_2(g)\ \rightarrow\ 6\,H_2O(g)\ +\ 4\,NO_2(g)$

H.9   (a) $3\,Pb(NO_3)_2(aq)\ +\ 2\,Na_3PO_4(aq)\ \rightarrow\ Pb_3(PO_4)_2(s)\ +\ 6\,NaNO_3(aq)$

(b) $Ag_2CO_3(aq)\ +\ 2\,NaBr(aq)\ \rightarrow\ 2\,AgBr(s)\ +\ Na_2CO_3(aq)$

H.11   (I) $3\,Fe_2O_3(s)\ +\ CO(g)\ \rightarrow\ 2\,Fe_3O_4(s)\ +\ CO_2(g)$

(II) $Fe_3O_4(s)\ +\ 4\,CO(g)\ \rightarrow\ 3\,Fe(s)\ +\ 4\,CO_2(g)$

H.13 (I) $N_2(g) + O_2(g) \rightarrow 2\,NO(g)$

(II) $2\,NO(g) + O_2(g) \rightarrow 2\,NO_2(g)$

H.15 $4\,HF(aq) + SiO_2(s) \rightarrow SiF_4(aq) + 2\,H_2O(l)$

H.17 $C_7H_{16}(l) + 11\,O_2(g) \rightarrow 7\,CO_2(g) + 8\,H_2O(g)$

H.19 $C_{14}H_{18}N_2O_5(s) + 16\,O_2(g) \rightarrow 14\,CO_2(g) + 9\,H_2O(l) + N_2(g)$

H.21 $2\,C_{10}H_{15}N(s)\ +\ 26\,O_2(g)\ \rightarrow\ 19\,CO_2(g)\ +\ 13\,H_2O(l)\ +\ CH_4N_2O(aq)$

H.23 (I) $H_2S(g) + 2\,NaOH(s) \rightarrow Na_2S(aq) + 2\,H_2O(l)$

(II) $4\,H_2S(g) + Na_2S(alc) \rightarrow Na_2S_5(alc) + 4\,H_2(g)$

(III) $2\,Na_2S_5(alc) + 9\,O_2(g) + 10\,H_2O(l) \rightarrow 2\,Na_2S_2O_3 \cdot 5\,H_2O(s) + 6\,SO_2(g)$

H.25 (a) We can find the empirical formulas from the percent compositions.

First oxide:

P $43.64\text{ g} \div 30.97\text{ g/mol} = 1.409\text{ mol}$

O $56.36\text{ g} \div 16.00\text{ g/mol} = 3.523\text{ mol}$

$\dfrac{3.523\text{ mol}}{1.409\text{ mol}} = 2.500$ 1:2.5 or 2:5 P_2O_5

Second oxide:

P $56.34\text{ g} \div 30.97\text{ g/mol} = 1.819\text{ mol}$

O $43.66\text{ g} \div 16.00\text{ g/mol} = 2.729\text{ mol}$

$\dfrac{2.729\text{ mol}}{1.819\text{ mol}} = 1.500$ 1:1.5 or 2:3 P_2O_3

These empirical formulas could be named diphosphorus pentoxide and diphosphorus trioxide. The names according to the Stock system are given below with the formulas of the actual compounds.

(b) P_4O_{10} (phosphorus (V) oxide), P_4O_6 (phosphorus (III) oxide);

Since the molar masses of the empirical formulas are 142 g/mol and 110 g/mol respectively, and these masses are both half as big as the molar masses of the compounds, both molecular formulas are twice the empirical formulas.

(c) $P_4(s) + 3\, O_2(g) \rightarrow P_4O_6(s)$; $P_4(s) + 5\, O_2(g) \rightarrow P_4O_{10}(s)$

I.1 The picture would show a precipitate, $CaSO_4(s)$, at the bottom of the flask. Sodium and chloride ions, $NaCl(aq)$, would remain throughout the solution.

I.3 (a) CH_3OH, nonelectrolyte; (b) $BaCl_2$, strong electrolyte; (c) KF, strong electrolyte

I.5 (a) $3\, BaBr_2(aq) + 2\, Li_3PO_4(aq) \rightarrow Ba_3(PO_4)_2(s) + 6\, LiBr(aq)$

$3\, Ba^{2+}(aq) + 6\, Br^-(aq) + 6\, Li^+(aq) + 2\, PO_4^{3-}(aq) \rightarrow$

$Ba_3(PO_4)_2(s) + 6\, Li^+(aq) + 6\, Br^-(aq)$

net ionic equation: $3\, Ba^{2+}(aq) + 2\, PO_4^{3-}(aq) \rightarrow Ba_3(PO_4)_2(s)$

(b) $2\, NH_4Cl(aq) + Hg_2(NO_3)_2(aq) \rightarrow 2\, NH_4NO_3(aq) + Hg_2Cl_2(s)$

$2\, NH_4^+(aq) + 2\, Cl^-(aq) + Hg_2^{2+}(aq) + 2\, NO_3^-(aq) \rightarrow$

$Hg_2Cl_2(s) + 2\, NH_4^+(aq) + 2\, NO_3^-(aq)$

net ionic equation: $Hg_2^{2+}(aq) + 2\, Cl^-(aq) \rightarrow Hg_2Cl_2(s)$

(c) $2\, Co(NO_3)_3(aq) + 3\, Ca(OH)_2(aq) \rightarrow 2\, Co(OH)_3(s) + 3\, Ca(NO_3)_2(aq)$

$2\, Co^{3+}(aq) + 6\, NO_3^-(aq) + 3\, Ca^{2+}(aq) + 6\, OH^-(aq) \rightarrow$

$2\, Co(OH)_3(s) + 3\, Ca^{2+}(aq) + 6\, NO_3^-(aq)$

Net ionic equation: $Co^{3+}(aq) + 3\, OH^-(aq) \rightarrow Co(OH)_3(s)$

I.7 (a) soluble; (b) slightly soluble; (c) insoluble; (d) insoluble

I.9 (a) $Na^+(aq)$ and $I^-(aq)$;

(b) $Ag^+(aq)$ and $CO_3^{2-}(aq)$, Ag_2CO_3 is insoluble. The very small amount that does go into solution will be present as Ag^+ and CO_3^{2-} ions.

(c) $NH_4^+(aq)$ and $PO_4^{3-}(aq)$ (d) $Fe^{2+}(aq)$ and $SO_4^{2-}(aq)$

I.11 (a) $Fe(OH)_3$, precipitate; (b) Ag_2CO_3, precipitate forms; (c) No precipitate will form because all possible products are soluble in water.

I.13 (a) net ionic equation: $Fe^{2+}(aq) + S^{2-}(aq) \rightarrow FeS(s)$;

spectator ions: Na^+, Cl^-

(b) net ionic equation: $Pb^{2+}(aq) + 2I^-(aq) \rightarrow PbI_2(s)$;

spectator ions: K^+, NO_3^-

(c) net ionic equation: $Ca^{2+}(aq) + SO_4^{2-}(aq) \rightarrow CaSO_4(s)$;

spectator ions: NO_3^-, K^+

(d) net ionic equation: $Pb^{2+}(aq) + CrO_4^{2-}(aq) \rightarrow PbCrO_4(s)$;

spectator ions: Na^+, NO_3^-

(e) net ionic equation: $Hg_2^{2+}(aq) + SO_4^{2-}(aq) \rightarrow Hg_2SO_4(s)$;

spectator ions: K^+, NO_3^-

I.15 (a) overall equation: $(NH_4)_2CrO_4(aq) + BaCl_2(aq)$
$\rightarrow BaCrO_4(s) + 2NH_4Cl(aq)$

complete ionic equation:
$2NH_4^+(aq) + CrO_4^{2-}(aq) + Ba^{2+}(aq) + 2Cl^-(aq)$
$\rightarrow BaCrO_4(s) + 2NH_4^+(aq) + 2Cl^-(aq)$

net ionic equation: $Ba^{2+}(aq) + CrO_4^{2-}(aq) \rightarrow BaCrO_4(s)$;

spectator ions: NH_4^+, Cl^-

(b) $CuSO_4(aq) + Na_2S(aq) \rightarrow CuS(s) + Na_2SO_4(aq)$

complete ionic equation:

$Cu^{2+}(aq) + SO_4^{2-}(aq) + 2 Na^+(aq) + S^{2-}(aq)$

$\rightarrow CuS(s) + 2 Na^+(aq) + SO_4^{2-}(aq)$

net ionic equation: $Cu^{2+}(aq) + S^{2-}(aq) \rightarrow CuS(s)$;

spectator ions: Na^+, SO_4^{2-}

(c) $3 FeCl_2(aq) + 2 (NH_4)_3PO_4(aq) \rightarrow Fe_3(PO_4)_2(s) + 6 NH_4Cl(aq)$

complete ionic equation:

$3 Fe^{2+}(aq) + 6 Cl^-(aq) + 6 NH_4^+(aq) + 2 PO_4^{3-}(aq)$

$\rightarrow Fe_3(PO_4)_2(s) + 6 NH_4^+(aq) + 6 Cl^-(aq)$

net ionic equation: $3 Fe^{2+}(aq) + 2 PO_4^{3-}(aq) \rightarrow Fe_3(PO_4)_2(s)$;

spectator ions: Cl^-, NH_4^+

(d) $K_2C_2O_4(aq) + Ca(NO_3)_2(aq) \rightarrow CaC_2O_4(s) + 2 KNO_3(aq)$

complete ionic equation:

$2 K^+(aq) + C_2O_4^{2-}(aq) + Ca^{2+}(aq) + 2 NO_3^-(aq)$

$\rightarrow CaC_2O_4(s) + 2 K^+(aq) + 2 NO_3^-(aq)$

net ionic equation: $Ca^{2+}(aq) + C_2O_4^{2-}(aq) \rightarrow CaC_2O_4(s)$;

spectator ions: K^+, NO_3^-

(e) $NiSO_4(aq) + Ba(NO_3)_2(aq) \rightarrow Ni(NO_3)_2(aq) + BaSO_4(s)$

complete ionic equation:

$Ni^{2+}(aq) + SO_4^{2-}(aq) + Ba^{2+}(aq) + 2 NO_3^-(aq)$

$\rightarrow Ni^{2+}(aq) + 2 NO_3^-(aq) + BaSO_4(s)$

net ionic equation: $Ba^{2+}(aq) + SO_4^{2-}(aq) \rightarrow BaSO_4(s)$;

spectator ions: Ni^{2+}, NO_3^-

I.17 (a) $AgNO_3$ and Na_2CrO_4

(b) $CaCl_2$ and Na_2CO_3

(c) $Cd(ClO_4)_2$ and $(NH_4)_2S$

I.19 You should use the reagent to only react with one of the ions and form precipitate, so that two ions can be separated.

(a) Diluted H_2SO_4 solution will be used as reagent.

$$Pb^{2+}(aq) \; + \; SO_4^{2-}(aq) \; \rightarrow \; PbSO_4(s)$$

(b) H_2S solution will also be used as reagent.

$$Mg^{2+}(aq) \; + \; S^{2-}(aq) \; \rightarrow \; MgS(s)$$

I.21 (a) $2\,Ag^+(aq) + SO_4^{2-}(aq) \rightarrow Ag_2SO_4(s)$

(b) $Hg^{2+}(aq) + S^{2-}(aq) \rightarrow HgS(s)$

(c) $3\,Ca^{2+}(aq) + 2\,PO_4^{3-}(aq) \rightarrow Ca_3(PO_4)_2(s)$

(d) $AgNO_3$ and Na_2SO_4; $Na^+,\,NO_3^-$;

$Hg(CH_3CO_2)_2$ and Li_2S; $Li^+,\,CH_3CO_2^-$

$CaCl_2$ and K_3PO_4; $K^+,\,Cl^-$

I.23 white ppt. = $AgCl(s)$, Ag^+; no ppt. with H_2SO_4, no Ca^{2+};

black ppt. = ZnS, Zn^{2+}

I.25 (a) $2\,NaOH(aq) + Cu(NO_3)_2(aq) \; \rightarrow \; Cu(OH)_2(s) + 2\,NaNO_3(aq)$

complete ionic equation:

$2\,Na^+(aq) \; + \; 2\,OH^-(aq) + Cu^{2+}(aq) + 2\,NO_3^-(aq)$

$\rightarrow Cu(OH)_2(s) + 2\,Na^+(aq) + 2\,NO_3^-(aq)$

net ionic equation: $Cu^{2+}(aq) + 2\,OH^-(aq) \rightarrow Cu(OH)_2(s)$

(b) $M_{Na^+} = \left(\dfrac{40.0\ mL \times 0.100\ M}{50.0\ mL} \right) = 0.0800\ M$

I.27 (a) $Ag^+(aq) + I^-(aq) \rightarrow AgI(s)$

(b) Concentration of Ag =

$$(0.356 \text{ g AgI}) \times (\frac{1 \text{ mole AgI}}{234.77 \text{ g AgI}}) / (0.1500 \text{L}) = 1.01 \times 10^{-2} \text{M}$$

J.1 (a) base; (b) acid; (c) base; (d) acid; (e) base

J.3 The solution is acidic with weak conductivity. It must be CH_3COOH.

J.5 (a) overall equation: $HF(aq) + NaOH(aq) \rightarrow NaF(aq) + H_2O(l)$

complete ionic equation:

$HF(aq) + Na^+(aq) + OH^-(aq) \rightarrow Na^+(aq) + F^-(aq) + H_2O(l)$

net ionic equation: $HF(aq) + OH^-(aq) \rightarrow F^-(aq) + H_2O(l)$

(b) overall equation: $(CH_3)_3N(aq) + HNO_3(aq) \rightarrow (CH_3)_3NHNO_3(aq)$

complete ionic equation:
$(CH_3)_3N(aq) + H^+(aq) + NO_3^-(aq)$
$\rightarrow (CH_3)_3NH^+(aq) + NO_3^-(aq)$

net ionic equation: $(CH_3)_3N(aq) + H^+(aq) \rightarrow (CH_3)_3NH^+(aq)$

(c) overall equation: $LiOH(aq) + HI(aq) \rightarrow LiI(aq) + H_2O(l)$

complete ionic equation:

$Li^+(aq) + OH^-(aq) + H^+(aq) + I^-(aq) \rightarrow Li^+(aq) + I^-(aq) + H_2O(l)$

net ionic equation: $OH^-(aq) + H^+(aq) \rightarrow 2 H_2O(l)$

J.7 (a) $HBr(aq) + KOH(aq) \rightarrow KBr(aq) + H_2O(l)$

(b) $Zn(OH)_2(aq) + 2 HNO_2(aq) \rightarrow Zn(NO_2)_2(aq) + 2 H_2O(l)$

(c) $Ca(OH)_2(aq) + 2 HCN(aq) \rightarrow Ca(CN)_2(aq) + 2 H_2O(l)$

(d) $3 KOH(aq) + H_3PO_4(aq) \rightarrow K_3PO_4(aq) + 3 H_2O(l)$

J.9   (a) CH₃CO₂K, potassium acetate;

$$CH_3CO_2H(aq) + K^+(aq) + OH^-(aq) \rightarrow K^+(aq) + CH_3CO_2^-(aq) + H_2O(l)$$

(b) (NH₄)₃PO₄ , ammonium phosphate;

$$3\,NH_3(aq) + 3\,H^+ + PO_4^{3-}(aq) \rightarrow 3\,NH_4^+(aq) + PO_4^{3-}(aq)$$

(c) Ca(BrO₂)₂, calcium bromite;

$$Ca^{2+}(aq) + 2\,OH^-(aq) + 2\,HBrO_2(aq)$$
$$\rightarrow 2\,H_2O(l) + Ca^{2+}(aq) + 2\,BrO_2^-(aq)$$

(d) Na₂S, sodium sulfide;

$$2\,Na^+(aq) + 2OH^-(aq) + H_2S\,(aq)$$
$$\rightarrow 2\,H_2O(l) + 2\,Na^+(aq) + S^{2-}(aq)$$

J.11   Image (b) best represents a solution of hydrochloric acid.

J.13   (a) acid: $H_3O^+(aq)$; base: $CH_3NH_2(aq)$; (b) acid: CH_3COOH; base: CH_3NH_2; (c) acid: $HI(aq)$; base: $CaO(s)$

J.15   Since X turns litmus red and conducts electricity poorly, it is a weak acid. We can find the empirical formula from the percent composition.

C   26.68 g ÷ 12.01 g/mol = 2.221 mol

H   2.239 g ÷ 1.0079 g/mol = 2.221 mol

O   71.081 g ÷ 16.00 g/mol = 4.443 mol

So the subscripts are 1:1:2 on the empirical formula.

(a) CHO_2

(b) Since the molar mass of the empirical formula is 45.0 g·mol⁻¹ while the   molar mass of X is 90.0 g·mol⁻¹, the molecular formula is twice the empirical formula or $C_2H_2O_4$.

(c) The weak acid whose formula matches the one given in part (b) is oxalic acid.

$$(COOH)_2(aq) + 2\,NaOH(aq) \rightarrow Na_2C_2O_4(aq) + 2\,H_2O(l)$$

net ionic equation: $(COOH)_2(aq) + 2\,OH^- \rightarrow C_2O_4^{2-}(aq) + 2\,H_2O(l)$

J.17   (a) $C_6H_5O^-$ (aq) $+ H_2O(l) \rightarrow C_6H_5OH$ (aq) $+ OH^-$(aq)

(b) ClO^- (aq) $+ H_2O(l) \rightarrow HClO(aq) + OH^-$ (aq)

(c) $C_5H_5NH^+$ (aq) $+ H_2O(l) \rightarrow C_5H_5N(aq) + H_3O^+$(aq)

(d) NH_4^+ (aq) $+ H_2O(l) \rightarrow NH_3(aq) + H_3O^+$ (aq)

J.19   (a) ? M $C_6H_5NH_3^+$

$$= \left(\frac{40.0 \text{ g } C_6H_5NH_3Cl}{210.0 \text{ mL}} \right) \left(\frac{1000 \text{ mL}}{L} \right) \left(\frac{1 \text{ mol } C_6H_5NH_3Cl}{129.45 \text{ g } C_6H_5NH_3Cl} \right)$$

$$= 1.47 \text{ M } C_6H_5NH_3Cl = 1.47 \text{ M } C_6H_5NH_3^+$$

(b) $C_6H_5NH_3^+ + H_2O(l) \rightarrow C_6H_5NH_2 (aq) + H_3O^+$
     acid   base   conjugate   conjugate
                                     base   acid

J.21   (a) AsO_4^{3-}(aq) $+ H_2O(l) \rightarrow HAsO_4^{2-}$(aq) $+ OH^-$(aq)

$HAsO_4^{2-}$(aq) $+ H_2O(l) \rightarrow H_2AsO_4^-$(aq) $+ OH^-$(aq)

$H_2AsO_4^-$(aq) $+ H_2O(l) \rightarrow H_3AsO_4(aq) + OH^-$(aq)

In each equation, H_2O is the acid.

(b) ? mol Na^+

$$= 35.0 \text{ g } Na_3AsO_4 \left(\frac{1 \text{ mol } Na_3AsO_4}{207.89 \text{ g } Na_3AsO_4} \right) \times \left(\frac{3 \text{ mol } Na^+}{1 \text{ mol } Na_3AsO_4} \right)$$

$$= 0.505 \text{ mol } Na^+$$

J.23   (a) $CO_2(g) + H_2O(l) \rightarrow H_2CO_3(aq)$ (carbonic acid)

(b) $SO_3(g) + H_2O(l) \rightarrow H_2SO_4(aq)$ (sulfuric acid)

K.1   (a) +2;   (b) +2;   (c) +6;   (d) +4;   (e) +1

K.3   (a) +4;   (b) +4;   (c) −2;   (d) +5;   (e) +1;   (f) 0

K.5 Fe^{2+} ions and copper will form. Iron is a reducing agent, so it reduces

Cu^{2+} to copper metal. The iron is then oxidized to Fe^{2+}.

K.7 (a) Methanol CH_3OH (aq) is oxidized to formic acid (the carbon atom goes from an oxidation number of +2 to +4). The O_2 (g) is reduced to O^{2-} present in water.

(b) Mo is reduced from +5 to +4, while *some* sulfur (that which ends up as S(s)) is oxidized from –2 to 0. The sulfur present in MoS_2 (s) remains in the –2 oxidation state.

(c) Tl^+ is both oxidized and reduced. The product Tl(s) is a reduction of Tl^+ (from +1 to 0) while the Tl^{3+} is produced via an oxidation of Tl^+. A reaction in which a single substance is both oxidized and reduced is known as a *disproportionation reaction*.

K.9 (a) oxidizing agent: H^+ in HCl(aq); reducing agent: Zn(s)

(b) oxidizing agent: SO_2 (g); reducing agent: H_2S(g)

(c) oxidizing agent: B_2O_3 (s); reducing agent: Mg(s)

K.11 $CO_2(g) + 4 H_2(g) \rightarrow CH_4(g) + 2 H_2O(l)$
oxidation-reduction reaction;
CO_2 is oxidizing reagent; H_2 is reducing reagent

K.13 (a) $2 NO_2 (g) + O_3 (g) \rightarrow N_2O_5 (g) + O_2 (g)$

(b) $S_8 (s) + 16 Na(s) \rightarrow 8 Na_2S(s)$

(c) $2 Cr^{2+} (aq) + Sn^{4+} (aq) \rightarrow 2 Cr^{3+} (aq) + Sn^{2+} (aq)$

(d) $2 As(s) + 3 Cl_2 (g) \rightarrow 2 AsCl_3 (l)$

K.15 (a) oxidizing agent: WO_3 (s); reducing agent: H_2 (g)

(b) oxidizing agent: HCl reducing agent: Mg(s)

(c) oxidizing agent: $SnO_2(s)$; reducing agent: $C(s)$

(d) oxidizing agent: $N_2O_4(g)$; reducing agent: $N_2H_4(g)$

K.17 (a) $Cl_2(g) \ + \ H_2O(l) \rightarrow HClO(aq) \ + \ HCl(aq)$

oxidizing agent: $Cl_2(g)$; reducing agent: $Cl_2(g)$

(b) $4\,NaClO_3(aq) + \ 2\,SO_2(g) + 2\,H_2SO_4(aq, dilute)$

$\rightarrow 4\,NaHSO_4(aq) \ + \ 4\,ClO_2(g)$

oxidizing agent: $NaClO_3(aq)$; reducing agent: $SO_2(g)$

(c) $2\,CuI(aq) \rightarrow \ 2\,Cu(s) \ + \ I_2(s)$

oxidizing agent: $CuI(aq)$; reducing agent: $CuI(aq)$

K.19 (a) $Mg(s) + Cu^{2+}(aq) \rightarrow Mg^{2+}(aq) + Cu(s)$

(b) $Fe^{2+}(aq) + Ce^{4+}(aq) \rightarrow Fe^{3+}(aq) + Ce^{3+}(aq)$

(c) $H_2(g) + Cl_2(g) \rightarrow 2\,HCl(g)$

(d) $4\,Fe(s) + 3\,O_2(g) \rightarrow 2\,Fe_2O_3(s)$

K.21 (a) $-\frac{1}{2}$; (b) -1; (c) -1; (d) -1; (e) $-\frac{1}{3}$

K.23 (a) $ClO_4^- \rightarrow ClO_2$, Cl goes from +7 to +4; need a reducing agent

(b) $SO_4^{2-} \rightarrow SO_2$, S goes from +6 to +4; need a reducing agent

K.25 (a) redox reaction: oxidizing agent: $I_2O_5(s)$; reducing agent: $CO(g)$

(b) redox reaction: oxidizing agent: $I_2(aq)$; reducing agent: $S_2O_3^{2-}(aq)$

(c) precipitation reaction: $Ag^+(aq) + Br^-(aq) \rightarrow AgBr(s)$

(d) redox reaction: oxidizing agent: $UF_4(g)$; reducing agent: $Mg(s)$

L.1 0.050 mole Br_2 will be obtained

L.3 $Li_3N(s) \ + 2\,H_2(g) \ \rightarrow \ LiNH_2(s) + \ 2\,LiH(s)$

(a) moles of H_2 needed to react with 1.5 mg Li_3N:

$$= 1.5 \times 10^{-3} \text{ g } Li_3N \left(\frac{1 \text{ mol } Li_3N}{34.83 \text{ g } Li_3N} \right) \left(\frac{2 \text{ mol } H_2}{1 \text{ mol } Li_3N} \right) = 8.6 \times 10^{-5} \text{ mol } H_2$$

(b) mass of Li_3N to produce 0.650 mol LiH:

$$= 0.650 \text{ mol LiH} \left(\frac{1 \text{ mol } Li_3N}{2 \text{ mol LiH}} \right) \left(\frac{34.83 \text{ g } Li_3N}{1 \text{ mol } Li_3N} \right) = 11.3 \text{ g } Li_3N$$

L.5 $6 NH_4ClO_4(s) + 10 Al(s) \rightarrow 5 Al_2O_3(s) + 3 N_2(g) + 6 HCl(g) + 9 H_2O(g)$

(a) $(1.325 \text{ kg } NH_4ClO_4) \left(\frac{1 \text{ mol } NH_4ClO_4}{117.49 \text{ g } NH_4ClO_4} \right) \left(\frac{1000 \text{ g}}{1 \text{ kg}} \right)$

$$\left(\frac{10 \text{ mol Al}}{6 \text{ mol } NH_4ClO_4} \right) \left(\frac{26.98 \text{ g Al}}{1 \text{ mol Al}} \right) = 507.1 \text{ g Al}$$

(b) $3500 \text{ kg Al} \left(\frac{1000 \text{ g Al}}{1 \text{ kg Al}} \right) \left(\frac{1 \text{ mol Al}}{26.98 \text{ g Al}} \right)$

$$\times \left(\frac{5 \text{ mol } Al_2O_3}{10 \text{ mol Al}} \right) \left(\frac{101.96 \text{ g } Al_2O_3}{1 \text{ mol } Al_2O_3} \right)$$

$$= 6.613 \times 10^6 \text{ g } Al_2O_3 \text{ or } 6.613 \times 10^3 \text{ kg } Al_2O_3$$

L.7 $2 C_{57}H_{110}O_6(s) + 163 O_2(g) \rightarrow 114 CO_2(g) + 110 H_2O(l)$

(a) $(454 \text{ g fat}) \left(\frac{1 \text{ mol fat}}{891.44 \text{ g fat}} \right) \left(\frac{110 \text{ mol } H_2O}{2 \text{ mol fat}} \right) \left(\frac{18.02 \text{ g } H_2O}{1 \text{ mol } H_2O} \right)$

$$= 505 \text{ g } H_2O$$

(b) $(454 \text{ g fat}) \left(\frac{1 \text{ mol fat}}{891.44 \text{ g}} \right) \left(\frac{163 \text{ mol } O_2}{2 \text{ mol fat}} \right) \left(\frac{32.00 \text{ g } O_2}{1 \text{ mol } O_2} \right)$

$$= 1.33 \times 10^3 \text{ g } O_2$$

L.9 $2 C_8H_{18}(l) + 25 O_2(g) \rightarrow 16 CO_2(g) + 18 H_2O(l)$

$d = 0.79 \text{ g} \cdot \text{mL}^{-1}$, density of gasoline

$$(3.8 \text{ L gas}) \left(\frac{1000 \text{ mL}}{1 \text{ L}} \right) \left(\frac{0.79 \text{ g gas}}{1 \text{ mL}} \right) \left(\frac{1 \text{ mol gas}}{114.22 \text{ g gas}} \right) \left(\frac{18 \text{ mol H}_2\text{O}}{2 \text{ mol C}_8\text{H}_{18}} \right)$$

$$\left(\frac{18.02 \text{ g H}_2\text{O}}{1 \text{ mol H}_2\text{O}} \right) = 4.3 \times 10^3 \text{ g H}_2\text{O or } 4.3 \text{ kg H}_2\text{O}$$

L.11 $CaCO_3(s) + 2 \text{ HCl(aq)} \rightarrow CaCl_2(aq) + H_2O(l) + CO_2(g)$

$Mg(OH)_2(s) + 2 \text{ HCl(aq)} \rightarrow MgCl_2(aq) + 2 \text{ H}_2\text{O(l)}$

The mass of HCl can be neutralized by 400 mg $CaCO_3$:

400 mg $CaCO_3$

$$\left(\frac{1 \text{ g}}{1000 \text{ mg}} \right) \left(\frac{1 \text{ mol CaCO}_3}{99.087 \text{ g CaCO}_3} \right) \left(\frac{2 \text{ mol HCl}}{1 \text{ mol CaCO}_3} \right) \left(\frac{36.461 \text{ g HCl}}{1 \text{ mol HCl}} \right)$$

$= 0.294 \text{ g HCl}$

The mass of HCl can be neutralized by 150. mg $Mg(OH)_2$:

150. mg $Mg(OH)_2$ $\times$

$$\left(\frac{1 \text{ g}}{1000 \text{ mg}} \right) \left(\frac{1 \text{ mol Mg(OH)}_2}{58.319 \text{ g Mg(OH)}_2} \right) \left(\frac{2 \text{ mol HCl}}{1 \text{ mol Mg(OH)}_2} \right) \left(\frac{36.461 \text{ g HCl}}{1 \text{ mol HCl}} \right)$$

$= 0.188 \text{ g HCl}$

Total mass of HCl can be neutralized: $0.294 \text{ g} + 0.188 \text{ g} = 0.482 \text{ g HCl}$

L.13   The molar reaction ratio of $Ca(OH)_2$ to HNO_3 is 1:2

Molarity of $Ca(OH)_2 = \dfrac{12.15 \text{ mL} \times 0.144 \text{ M}}{25.00 \text{ mL}} \left(\dfrac{1}{2} \right) = 3.50 \times 10^{-2} \text{ M}$

L.15   (a) $HCl + NaOH \rightarrow NaCl + H_2O$

$$17.40 \text{ mL} \left(\frac{0.234 \text{ mol HCl}}{1000 \text{ mL}} \right) \left(\frac{1 \text{ mol NaOH}}{1 \text{ mol HCl}} \right) = 0.00407 \text{ mol}$$

concentration of $NaOH = \dfrac{0.00407 \text{ mol}}{15.00 \times 10^{-3} \text{ L}} = 0.271 \text{ mol} \cdot \text{L}^{-1}$

(b) $(0.271 \text{ mol} \cdot L^{-1})(0.01500 \text{ L})(40.00 \text{ g} \cdot \text{mol}^{-1}) = 0.163 \text{ g NaOH}$

L.17 (a) $Ba(OH)_2(aq) + 2 HNO_3(aq) \rightarrow Ba(NO_3)_2(aq) + 2 H_2O(l)$

$$\frac{\text{mol HNO}_3}{\text{L HNO}_3(aq)} = (11.56 \text{ mL Ba(OH)}_2(aq)) \left(\frac{9.670 \text{ g Ba(OH)}_2}{250. \text{ mL Ba(OH)}_2(aq)}\right)$$

$$\times \frac{\left(\dfrac{1 \text{ mol Ba(OH)}_2}{171.36 \text{ g Ba(OH)}_2}\right)\left(\dfrac{2 \text{ mol HNO}_3}{1 \text{ mol Ba(OH)}_2}\right)}{(25.0 \text{ mL HNO}_3(aq))\left(\dfrac{1 \text{ L}}{1000 \text{ mL}}\right)}$$

$$= 0.209 \text{ mol} \cdot L^{-1}$$

(b) mass of HNO_3 in solution:

$$\left(\frac{0.209 \text{ mol}}{1000 \text{ mL}}\right)(25.0 \text{ mL})\left(\frac{63.02 \text{ g HNO}_3}{1 \text{ mol HNO}_3}\right) = 0.329 \text{ g}$$

L.19 $HX(aq) + NaOH(aq) \rightarrow NaX(aq) + H_2O(l)$

$$(68.8 \text{ mL})\left(\frac{0.750 \text{ mol NaOH}}{1000 \text{ mL NaOH}}\right) = 0.0516 \text{ mol NaOH}$$

3.25 g HX corresponds to 0.0516 mol NaOH used

$$\frac{3.25 \text{ g}}{0.0516 \text{ mol}} = 63.0 \text{ g} \cdot \text{mol}^{-1} = \text{molar mass of acid}$$

L.21 $NaI(aq) + AgNO_3(aq) \rightarrow AgI(s) + NaNO_3(aq)$

$$M \text{ of AgNO}_3 = \left(\frac{1.76 \text{ g AgI}}{234.77 \text{ g} \times \text{mol}^{-1}}\right)\left(\frac{1 \text{ mol AgNO}_3}{1 \text{ mol AgI}}\right)\left(\frac{1}{0.0500 \text{ L}}\right) = 0.150 \text{ M}$$

L.23 (a) $Na_2CO_3(aq) + 2 HCl(aq) \rightarrow 2 NaCl(aq) + H_2CO_3(aq)$

(b) First find the concentration of the diluted acid.

$$? \text{ M HCl(aq) dilute} = \left(\frac{0.832 \text{ g } Na_2CO_3}{0.100 \text{ L base solution}} \right) \left(\frac{1 \text{ mol } Na_2CO_3}{105.99 \text{ g } Na_2CO_3} \right)$$

$$\times \left(\frac{0.025 \text{ L base solution}}{0.031\ 25 \text{ L acid solution}} \right)$$

$$\times \left(\frac{2 \text{ mol HCl}}{1 \text{ mol } Na_2CO_3} \right) = 0.126 \text{ M HCl(aq) dilute}$$

The original HCl solution is 100 times more concentrated than the solution used for titration (diluted 10.00 mL to 1000 mL), so the original concentration of the HCl solution is $12.6 \text{ mol} \cdot L^{-1}$.

L.25   Mass of copper =

$$(0.0244 \text{ L} \times 0.0010 \text{ M } I_3^-)(\frac{2 \text{ mole Cu}}{1 \text{ mole } I_3^-})(\frac{63.55 \text{ g Cu}}{1 \text{ mole Cu}}) = 3.1 \times 10^{-3} \text{ g Cu}$$

Percent of copper = $\dfrac{3.1 \times 10^{-3} \text{ g}}{1.10} \times 100 = 0.28\%$

L.27 $I_3^- + SnCl_x + (y-x)Cl^- \rightarrow 3\ I^- + SnCl_y$

The information given can be used to find the molar mass of the reactant in order to identify it.

$$25.00 \text{ mL} \left(\frac{0.120 \text{ mol } I_3^-}{1000 \text{ mL}} \right) = 3.00 \times 10^{-3} \text{ mol } I_3^-$$

$$30.00 \text{ mL} \left(\frac{19.0 \text{ g tin chloride}}{1000 \text{ mL}} \right) = 0.570 \text{ g tin chloride}$$

If the reaction is 1:1 then the # moles of I_3^- is the same as the number of moles of $SnCl_x$. In that case, the molar mass of the tin chloride reactant is

$\dfrac{0.570 \text{ g}}{3.00 \times 10^{-3} \text{ mol}} = 190. \text{ g} \cdot \text{mol}^{-1}$. This molar mass matches that of $SnCl_2$,

$189.61 \text{ g} \cdot \text{mol}^{-1}$. Tin (II) chloride also has the correct mass percent tin.

$$\frac{118.71 \text{ g} \cdot \text{mol}^{-1} \text{ Sn}}{189.61 \text{ g} \cdot \text{mol}^{-1} \text{ SnCl}_2} \times 100 = 62.6\%$$

Since the product compound is oxidized relative to the reactant, we can expect it to be Sn(IV). The net ionic equation for the reaction is $I_3^- + Sn^{2+} \rightarrow 3\,I^- + Sn^{4+}$. Another way to write a balanced reaction would be $I_3^- + SnCl_2(aq) + 2\,Cl^- \rightarrow 3\,I^- + SnCl_4(aq)$.

L.29 (a) $S_2O_3^{2-}$ is both oxidized and reduced.

(b) Find the number of grams of thiosulfate ion in 10.1 mL of solution.

$$g\ S_2O_3^{2-} = 10.1\ \text{mL}\ HSO_3^-(aq)\left(\frac{1.45\ g\ HSO_3^-(aq)}{1\ \text{mL}\ HSO_3^-(aq)}\right)\left(\frac{55.0\ g\ HSO_3^-}{100\ g\ HSO_3^-(aq)}\right)$$

$$\times\left(\frac{1\ \text{mol}\ HSO_3^-}{81.0\ g\ HSO_3^-}\right)\left(\frac{1\ \text{mol}\ S_2O_3^{2-}}{1\ \text{mol}\ HSO_3^-}\right)\left(\frac{112.0\ g\ S_2O_3^{2-}}{1\ \text{mol}\ S_2O_3^{2-}}\right)$$

$$= 11.1\ g\ S_2O_3^{2-}\ \text{present initially}$$

L.31 $XCl_4 + 2\,NH_3 \rightarrow XCl_2(NH_3)_2 + Cl_2$

The reactant and product that contain X are in a 1:1 ratio, so 3.571 g of the reactant is equivalent to 3.180 g of the product. The molar mass of the reactant is $x + 4(35.453\ \text{g/mol})$ while that of the product is $x + 2(35.453\ \text{g/mol}) + 2(14.01\ \text{g/mol}) + 6(1.0079\ \text{g/mol})$. Therefore we can set up the following proportion in order to solve for x in g/mol:

$$\frac{x+141.8}{x+104.97} = \frac{3.571}{3.180}$$

$$1.1230x - x = 23.923$$

$$x = 194.6\ \text{g}\cdot\text{mol}^{-1},\ \text{or Pt}$$

L.33 The number of moles of product is

$2.27\ \text{g} \div 208.23\ \text{g}\cdot\text{mol}^{-1} = 0.0109$ moles $BaCl_2$. An equivalent number of moles is represented by 3.25 g of $BaBr_x$, so its molar mass is

$3.25\ \text{g} \div 0.0109$ moles $= 298\ \text{g}\cdot\text{mol}^{-1}\ BaBr_x$. Since 137.33 g is attributable to Ba, 161 g must be Br. Each Br has a mass of 80.4 g/mol so there must be two moles of Br for each mole of Ba in the reactant.

$$x = 2, \ BaBr_2 + Cl_2 \rightarrow BaCl_2 + Br_2$$

L.35 First equation must be balanced as following:

$$Fe + Br_2 \rightarrow FeBr_2$$

$$3 \ FeBr_2 + Br_2 \rightarrow Fe_3Br_8$$

$$Fe_3Br_8 + 4 \ Na_2CO_3 \rightarrow 8 \ NaBr + 4 \ CO_2 + Fe_3O_4$$

The masses of Fe are needed to produce 2.5 t of NaBr:

$$2.50 \ t \times \left(\frac{1000 \ kg}{1 \ t}\right)\left(\frac{1000 \ g}{1 \ kg}\right)\left(\frac{1 \ mol \ NaBr}{102.9 \ g \ NaBr}\right)\left(\frac{1 \ mol \ Fe_3Br_8}{8 \ mol \ NaBr}\right)$$

$$\times \left(\frac{3 \ mol \ FeBr_2}{1 \ mol \ Fe_3Br_8}\right)\left(\frac{1 \ mol \ Fe}{1 \ mol \ FeBr_2}\right)\left(\frac{55.84 \ g \ Fe}{1 \ mol \ Fe}\right) = 5.09 \times 10^5 \ g \ Fe$$

$$= 509 \ kg \ Fe$$

L.37 (a) $M_1V_1 = M_2V_2$ is used for calculation of dilution problems:

$$V_1 = \left(\frac{1.00 \ L \times 0.50 \ M}{16.0 \ M}\right) = 0.031 \ L = 31 \ mL$$

Pipette 31 mL of 16 M HNO_3 into a 1.00 L volumetric flask that contains about 800 mL of H_2O. Dilute to the mark with H_2O. Shake the flask to mix the solution thoroughly.

(b) $V_{NaOH} = \left(\frac{100 \ mL \times 0.50 \ M}{0.20 \ M}\right) = 2.5 \times 10^2 \ mL$

L.39 Total mass of tin oxide is: 28.35g − 26.45 g = 1.90 g

(a) $1.50 \ g \ Sn \times \left(\frac{1 \ mol \ Sn}{118.71 \ g \ Sn}\right) = 1.264 \times 10^{-2} \ mol \ Sn$

Moles of O: $(1.90 - 1.50)g \times \left(\frac{1 \ mol \ O}{16.00 \ g \ O}\right) = 0.025 \ mol \ O$

Mole ratio of Sn:O is 1:2

Empirical formula: SnO_2

(b) tin(IV) oxide

L.41   (a) It will not affect the reported KOH concentration. The reaction is mole-to-mole interaction (HCl/KOH) (as long as the volume of KOH is accurately measured).

(b) It will make the reported KOH concentration too high.

(c) It will make the reported KOH concentration too high.

(d) It will make the reported KOH concentration too high.

M.1   Theoretical yield of $N_2H_4 =$

$$(35.0 \text{ g NH}_3)(\frac{1 \text{ mole NH}_3}{17.031 \text{ g NH}_3})(\frac{1 \text{ mole N}_2\text{H}_4}{2 \text{ mole NH}_3})(\frac{32.046 \text{ g N}_2\text{H}_4}{1 \text{ mole N}_2\text{H}_4}) = 32.9 \text{ g N}_2\text{H}_4$$

Percentage yield of $N_2H_4 = \frac{25.2}{32.9} \times 100 = 76.6\%$

M.3 $CaCO_3(s) \rightarrow CaO(s) + CO_2(g)$

theoretical yield:

$$(42.73 \text{ g CaCO}_3)\left(\frac{1 \text{ mol CaCO}_3}{100.09 \text{ g CaCO}_3}\right)\left(\frac{1 \text{ mol CO}_2}{1 \text{ mol CaCO}_3}\right)\left(\frac{44.01 \text{ g CO}_2}{1 \text{ mol CO}_2}\right)$$

$= 18.79 \text{ g CO}_2$

actual yield:

$\frac{17.5 \text{ g}}{18.79 \text{ g}} \times 100\% = 93.1\%$ yield

M.5   (a) BrF_3; (b) 12 moles of ClO_2F and 2 moles of Br_2 will be produced; 1 mole of BrF_3 will remain

M.7   (a) $B_2O_3(s) + 3 \text{ Mg}(s) \rightarrow 3 \text{ MgO}(s) + 2 \text{ B}(s)$

(b) mass of B produced from 125 kg of B_2O_3:

$$125 \times 10^3 \text{g B}_2\text{O}_3\left(\frac{1 \text{ mol B}_2\text{O}_3}{69.619 \text{ g B}_2\text{O}_3}\right)\left(\frac{2 \text{ mol B}}{1 \text{ mol B}_2\text{O}_3}\right)\left(\frac{10.811 \text{ g B}}{1 \text{ mol B}}\right)$$

$= 3.88 \times 10^4 \text{ g B}$

mass of B produced from 125 kg of Mg:

$$125 \times 10^3 \text{g Mg} \left(\frac{1 \text{ mol Mg}}{24.305 \text{ g Mg}} \right) \left(\frac{2 \text{ mol B}}{3 \text{ mol Mg}} \right) \left(\frac{10.811 \text{ g B}}{1 \text{ mol B}} \right)$$

$$= 3.71 \times 10^4 \text{ g B}$$

Mg is limiting. 3.71×10^4 g B can be produced.

M.9 (a) $Cu^{2+}(aq) + 2\,OH^-(aq) \rightarrow Cu(OH)_2(s)$

(b) $2.00 \text{ g NaOH} \times \left(\dfrac{1 \text{ mol NaOH}}{40.0 \text{ g NaOH}} \right) = 0.0500 \text{ mol NaOH}$

$(0.0800 \text{ L})(0.500 \text{ M}) = 0.0400 \text{ mol Cu(NO}_3)_2$

Moles of NaOH required to react with 0.04 mol $Cu(NO_3)_2$:

$$0.0400 \text{ mol Cu(NO}_3)_2 \times \left(\frac{2 \text{ mol NaOH}}{1 \text{ mol Cu(NO}_3)_2} \right) = 0.0800 \text{ mol NaOH}$$

0.0800 mol > 0.0500 mol; therefore, NaOH is limiting reagent.

The mass of $Cu(OH)_2(s)$:

$$0.0500 \text{ mol NaOH} \times \left(\frac{1 \text{ mol Cu(OH)}_2}{2 \text{ mol NaOH}} \right) \times \left(\frac{97.57 \text{ g Cu(OH)}_2}{1 \text{ mol Cu(OH)}_2} \right) = 2.44 \text{ g}$$

M.11 (a) $P_4(s) + 3\,O_2(g) \rightarrow P_4O_6(s)$

$P_4O_6(s) + 2\,O_2(g) \rightarrow P_4O_{10}(s)$

In the first reaction, 5.77 g P_4 uses

$$(5.77 \text{ g P}_4) \left(\frac{1 \text{ mol P}_4}{123.88 \text{ g P}_4} \right) \left(\frac{3 \text{ mol O}_2}{1 \text{ mol P}_4} \right) \left(\frac{32.00 \text{ g O}_2}{1 \text{ mol O}_2} \right) = 4.47 \text{ g O}_2 \text{ (g)}$$

excess $O_2 = 5.77 \text{ g} - 4.47 \text{ g O}_2 = 1.30 \text{ g O}_2$

In the second reaction, 5.77 g P_4 uses

$$\left(\frac{5.77 \text{ g P}_4}{123.88 \text{ g} \cdot \text{mol}^{-1} \text{ P}_4} \right) \left(\frac{1 \text{ mol P}_4O_6}{1 \text{ mol P}_4} \right) \left(\frac{2 \text{ mol O}_2}{1 \text{ mol P}_4O_6} \right) \left(\frac{32.00 \text{ g O}_2}{1 \text{ mol O}_2} \right)$$

$$= 2.98 \text{ g O}_2$$

limiting reagent: O_2

(b) $\left(\dfrac{1.30\ \text{g}\ O_2}{32.00\ \text{g}\cdot\text{mol}^{-1}\ O_2}\right)\left(\dfrac{1\ \text{mol}\ P_4O_{10}}{2\ \text{mol}\ O_2}\right)\left(\dfrac{283.88\ \text{g}\ P_4O_{10}}{1\ \text{mol}\ P_4O_{10}}\right)=5.77\ \text{g}\ P_4O_{10}$

(c) $\left(\dfrac{1.30\ \text{g}\ O_2}{32.00\ \text{g}\cdot\text{mol}^{-1}\ O_2}\right)\left(\dfrac{1\ \text{mol}\ P_4O_6}{2\ \text{mol}\ O_2}\right)\left(\dfrac{219.88\ \text{g}\ P_4O_6}{1\ \text{mol}\ P_4O_6}\right)$

$=4.47\ \text{g}\ P_4O_6$ used

In the first reaction, $5.77\ \text{g}\ P_4$ produces

$\left(\dfrac{5.77\ \text{g}\ P_4}{123.88\ \text{g}\cdot\text{mol}^{-1}}\right)\left(\dfrac{219.88\ \text{g}\ P_4O_6}{1\ \text{mol}\ P_4O_6}\right)\left(\dfrac{1\ \text{mol}\ P_4O_6}{1\ \text{mol}\ P_4}\right)=10.2\ \text{g}\ P_4O_6$

excess reagent: $10.2\ \text{g}-4.47\ \text{g}=5.7\ \text{g}\ P_4O_6$

M.13 $\quad C_xH_yCl_z\ +\ (x+\dfrac{y}{4})\,O_2\ \rightarrow\ x\,CO_2\ +\ \dfrac{y}{2}\,H_2O\ +\ \dfrac{z}{2}\,Cl_2$

$\dfrac{1.52\ \text{g}}{360.88\ \text{g}\cdot\text{mol}^{-1}}=4.21\times10^{-3}$ mol Arochlor yields $\dfrac{2.224\ \text{g}}{44.0\ \text{g}\cdot\text{mol}^{-1}}$

$=5.055\times10^{-2}\ \text{mol}\ CO_2$

$\dfrac{2.53\ \text{g}}{360.88\ \text{g}\cdot\text{mol}^{-1}}=7.01\times10^{-3}$ mol Arochlor yields $\dfrac{0.2530\ \text{g}}{18.01\ \text{g}\cdot\text{mol}^{-1}}$

$=1.405\times10^{-2}\ \text{mol}\ H_2O$

Therefore, $x=\dfrac{5.055\times10^{-2}}{4.21\times10^{-3}}=12.0$ and $y=\dfrac{1.405\times10^{-2}}{7.01\times10^{-3}}=2.00$

$12.011x+1.0079y+35.453z=360.88$

$12.011(12.0)+1.0079(2.00)+35.453z=360.88$

$35.453z=214.7$

$z=6.06$

Since the number or Cl atoms per Arochlor 1254 molecule must be a whole number, the number of chlorine atoms is 6.

M.15 (a) $2\ Al(s)\ +\ 3\ Cl_2(g)\ \rightarrow\ 2\ AlCl_3(s)$

(b) Mass of $AlCl_3$ produced based on 255 g Al:

$$255. \text{ g Al} \left(\frac{1 \text{ mol Al}}{26.982 \text{ g Al}} \right) \left(\frac{2 \text{ mol AlCl}_3}{2 \text{ mol Al}} \right) \left(\frac{133.341 \text{ g AlCl}_3}{1 \text{ mol AlCl}_3} \right)$$

$$= 1.26 \times 10^3 \text{ g AlCl}_3$$

Mass of $AlCl_3$ produced based on 535. g Cl_2:

$$535. \text{ g Cl}_2 \left(\frac{1 \text{ mol Cl}_2}{70.906 \text{ g Cl}_2} \right) \left(\frac{2 \text{ mol AlCl}_3}{3 \text{ mol Cl}_2} \right) \left(\frac{133.341 \text{ g AlCl}_3}{1 \text{ mol AlCl}_3} \right)$$

$$= 671. \text{ g AlCl}_3$$

Cl_2 is limiting. Maximum amount of $AlCl_3$ can be produced = 671 g

(c) %yield $= \dfrac{300}{671} \times 100 = 44.7\%$

M.17 HA + XOH $\rightarrow$ H$_2$O + XA

Moles of HA: $\left(\dfrac{2.45 \text{ g HA}}{231 \text{ g} \cdot \text{mol}^{-1}} \right) = 0.0106 \text{ mol}$

Moles of XOH: $\left(\dfrac{1.50 \text{ g XOH}}{125 \text{ g} \cdot \text{mol}^{-1}} \right) = 0.0120 \text{ mol}$

Molar mass of XA $= (231 - 1.0) + (125 - 17) = 338 \text{ g} \cdot \text{mol}^{-1}$

Mole of XA produced: $\left(\dfrac{2.91 \text{ g XA}}{338 \text{ g} \cdot \text{mol}^{-1}} \right) = 8.61 \times 10^{-3} \text{ mol}$

Theoretical yield: (based on the limiting reagent, HA)

$$0.0106 \text{ mol HA} \times \left(\frac{1 \text{ mol XA}}{1 \text{ mole HA}} \right) = 0.0106 \text{ mol XA}$$

Percentage yield: $\left(\dfrac{8.61 \times 10^{-3} \text{ mol XA}}{0.0106 \text{ mole XA}} \right) \times 100\% = 81.2\%$

M.19 $(0.682 \text{ g CO}_2) \left(\dfrac{1 \text{ mol CO}_2}{44.01 \text{ g CO}_2} \right) \left(\dfrac{1 \text{ mol C}}{1 \text{ mol CO}_2} \right) = 0.0155 \text{ mol C}$

$(0.0155 \text{ mol C})(12.01 \text{ g} \cdot \text{mol}^{-1} \text{ C}) = 0.186 \text{ g C}$

$(0.174 \text{ g H}_2\text{O})\left(\dfrac{1 \text{ mol H}_2\text{O}}{18.02 \text{ g H}_2\text{O}}\right)\left(\dfrac{2 \text{ mol H}}{1 \text{ mol H}_2\text{O}}\right) = 0.0193 \text{ mol H}$

$(0.0193 \text{ mol H})(1.0079 \text{ g} \cdot \text{mol}^{-1} \text{ H}) = 0.0195 \text{ g H}$

$(0.110 \text{ g N}_2)\left(\dfrac{1 \text{ mol N}_2}{28.02 \text{ g N}_2}\right)\left(\dfrac{2 \text{ mol N}}{1 \text{ mol N}_2}\right) = 0.007\,85 \text{ mol N}$

$(0.007\,85 \text{ mol N})(14.01 \text{ g} \cdot \text{mol}^{-1} \text{ N}) = 0.110 \text{ g N}$

mass of $\text{O} = 0.376 \text{ g} - (0.186 \text{ g} + 0.0195 \text{ g} + 0.110 \text{ g}) = 0.060 \text{ g O}$

$\dfrac{0.060 \text{ g O}}{16.00 \text{ g O}} = 0.0038 \text{ mol O}$

Dividing each amount by 0.0038 gives C : H : N : O ratios

$= 4.1 : 5.1 : 2.1 : 1$. The empirical formula is $C_4H_5N_2O$.

The molecular mass of caffeine is $194 \text{ g} \cdot \text{mol}^{-1}$. Its empirical mass is

$97.10 \text{ g} \cdot \text{mol}^{-1}$.

molecular formula $= 2 \times$ empirical formula $= C_8H_{10}N_4O_2$

$2\,C_8H_{10}N_4O_2(s) + 19\,O_2(g) \rightarrow 16\,CO_2(g) + 10\,H_2O(l) + 4\,N_2(g)$

M.21 Calculate the mass percentage composition; doing so eases the comparison of data from multiple analyses.

$2.20 \text{ g CO}_2 \times \left(\dfrac{1 \text{ mol CO}_2}{44.01 \text{ g CO}_2}\right) \times \left(\dfrac{1 \text{ mol C}}{1 \text{ mol CO}_2}\right) \times \left(\dfrac{12.01 \text{ g C}}{1 \text{ mol C}}\right) = 0.600 \text{ g C}$

$\%\text{C} = \dfrac{0.600 \text{ g C}}{1.35 \text{ g unknown}} \times 100\% = 44.5\% \text{ C}$

$0.901 \text{ g H}_2\text{O} \times \left(\dfrac{1 \text{ mol H}_2\text{O}}{18.02 \text{ g H}_2\text{O}}\right) \times \left(\dfrac{2 \text{ mol H}}{1 \text{ mol H}_2\text{O}}\right) \times \left(\dfrac{1.0079 \text{ g H}}{1 \text{ mol H}}\right) = 0.101 \text{ g H}$

$\%\text{H} = \dfrac{0.101 \text{ g H}}{1.35 \text{ g unknown}} \times 100\% = 7.48\% \text{ H}$

$0.130 \text{ g N}_2 \times \left(\dfrac{1 \text{ mol N}_2}{28.02 \text{ g N}_2}\right) \times \left(\dfrac{2 \text{ mol N}}{1 \text{ mol N}_2}\right) \times \left(\dfrac{14.01 \text{ g N}}{1 \text{ mol N}}\right) = 0.130 \text{ g N}$

$\%\text{N} = \dfrac{0.130 \text{ g N}}{0.500 \text{ g unknown}} \times 100\% = 26.0\% \text{ N}$

Oxygen must be present in the compound because the percentages of C, H, and N only account for 78.0% of its composition. Combustion analysis does not generate data directly for oxygen; we calculate it by difference.

$$\%O = 100 - 78.0 = 22\% \text{ O}$$

To find the empirical formula, assume a sample size of 100 g and find the mole ratios.

$$44.5 \text{ g C} \times \left(\frac{1 \text{ mol C}}{12.01 \text{ g C}} \right) = 3.71 \text{ mol} \times \left(\frac{1}{1.38 \text{ mol}} \right) = 2.69 \times 3 = 8.07$$

$$7.48 \text{ g H} \times \left(\frac{1 \text{ mol H}}{1.0079 \text{ g H}} \right) = 7.42 \text{ mol} \times \left(\frac{1}{1.38 \text{ mol}} \right) = 5.38 \times 3 = 16.1$$

$$26.0 \text{ g N} \times \left(\frac{1 \text{ mol N}}{14.01 \text{ g N}} \right) = 1.86 \text{ mol} \times \left(\frac{1}{1.38 \text{ mol}} \right) = 1.35 \times 3 = 4.05$$

$$22.0 \text{ g O} \times \left(\frac{1 \text{ mol O}}{16.00 \text{ g O}} \right) = 1.38 \text{ mol} \times \left(\frac{1}{1.38 \text{ mol}} \right) = 1.00 \times 3 = 3.00$$

The empirical formula is $C_8H_{16}N_4O_3$.

M.23 $3 \text{ Ca(NO}_3)_2 \text{(aq)} + 2 \text{ H}_3\text{PO}_4 \text{(aq)} \rightarrow \text{Ca}_3(\text{PO}_4)_2 \text{(s)} + 6 \text{ HNO}_3 \text{(aq)}$

(a) The solid is calcium phosphate, $Ca_3(PO_4)_2$.

(b) $(206 \text{ g Ca(NO}_3)_2) \left(\dfrac{1 \text{ mol Ca(NO}_3)_2}{164.10 \text{ g Ca(NO}_3)_2} \right) \left(\dfrac{2 \text{ mol H}_3\text{PO}_4}{3 \text{ mol Ca(NO}_3)_2} \right)$

$\left(\dfrac{97.99 \text{ g H}_3\text{PO}_4}{1 \text{ mol H}_3\text{PO}_4} \right) = 82.01 \text{ g H}_3\text{PO}_4$

Therefore $Ca(NO_3)_2$ is the limiting reagent.

$(206 \text{ g Ca(NO}_3)_2) \left(\dfrac{1 \text{ mol Ca(NO}_3)_2}{164.10 \text{ g Ca(NO}_3)_2} \right) \left(\dfrac{1 \text{ mol Ca}_3(\text{PO}_4)_2}{3 \text{ mol Ca(NO}_3)_2} \right)$

$\left(\dfrac{310.18 \text{ g Ca}_3(\text{PO}_4)_2}{1 \text{ mol Ca}_3(\text{PO}_4)_2} \right) = 130. \text{ g Ca}_3(\text{PO}_4)_2$

M.25 If the 2-naphthol $(144.16 \text{ g} \cdot \text{mol}^{-1})$ were pure, it would give the following combustion analysis:

$$\%C = \frac{10(12.01 \text{ g} \cdot \text{mol}^{-1})}{(144.16 \text{ g} \cdot \text{mol}^{-1} \text{ naphthol})} \times 100\% = 83.31\% \text{ C}$$

$$\%H = \frac{8(1.0079 \text{ g} \cdot \text{mol}^{-1})}{(144.16 \text{ g} \cdot \text{mol}^{-1} \text{ naphthol})} \times 100\% = 5.59\% \text{ H}$$

The observed percentages are low as is expected for a sample contaminated with a substance that contains no C or H. Because the sample does not contain C or H, the percent purity can be easily obtained by

$$\%\text{purity (based on C)} = \frac{\% \text{ found}}{\% \text{ theoretical}} = \frac{77.48\% \text{ mixture}}{83.31\% \text{ pure naphthol}} \times 100\%$$

$$= 93.00\%$$

$$\%\text{purity (based on H)} = \frac{\% \text{ found}}{\% \text{ theoretical}} = \frac{5.20\% \text{ mixture}}{5.59\% \text{ pure naphthol}} \times 100\%$$

$$= 93.0\%$$

M.27 (a) $C_xH_yCl_z + (x + \frac{y}{4} - \frac{z}{2}) O_2 \rightarrow x \, CO_2 + \frac{y}{2} H_2O$

$$\frac{2.492 \text{ g CO}_2}{44.0 \text{ g} \cdot \text{mol}^{-1}} = 5.664 \times 10^{-2} \text{ mol CO}_2 = \text{mol C}$$

$$\frac{0.6495 \text{ g H}_2\text{O}}{18.01 \text{ g} \cdot \text{mol}^{-1}} = 3.608 \times 10^{-2} \text{ mol H}_2\text{O} = 7.216 \times 10^{-2} \text{ mol H}$$

$$(5.664 \times 10^{-2} \text{ mol C}) \left(\frac{12.011 \text{ g}}{\text{mol}} \right) = 0.6803 \text{ g C}$$

$$(7.216 \times 10^{-2} \text{ mol H}) \left(\frac{1.0079 \text{ g}}{\text{mol}} \right) = 0.07273 \text{ g H}$$

Mass of oxygen: 1.000 g compound $- (0.6803 \text{ g} + 0.07273 \text{ g}) = 0.2470$ g

$0.2470 \text{ g O} \div 16.00 \text{ g} \cdot \text{mol}^{-1} = 1.544 \times 10^{-2} \text{ mol O}$

$$\frac{5.664 \times 10^{-2}}{1.544 \times 10^{-2}} = 3.67 \qquad \frac{7.216 \times 10^{-2}}{1.544 \times 10^{-2}} = 4.67$$

The mole ratio of C:H:O is 3.67:4.67:1 or 11:14:3, so the empirical formula is $C_{11}H_{14}O_3$.

(b) The molar mass of the empirical formula is 194 g/mol, which is half of 388.46 g/mol. Therefore, the molecular formula of the compound is $C_{22}H_{28}O_6$.

FOCUS 1

ATOMS

1A.1 (a) Radiation may pass through a metal foil.   (b) All light (electromagnetic radiation) travels at the same speed; the slower speed supports the particle model.   (c) This observation supports the radiation model.   (d) This observation supports the particle model; electromagnetic radiation has no mass and no charge.

1A.3 All of these can be determined using $E = h\nu$ and $c = \nu\lambda$.

(a) No; speed is constant (c).   (b) No; $\nu \propto \lambda^{-1}$.   (c) Yes. The electrical field corresponds to the amplitude; as the frequency decreases the waves broaden and the extent of the change (the slope of the wave) decreases.   (d) No; $E \propto \nu$.

1A.5 microwaves < visible light < ultraviolet light < x-rays < γ-rays

1A.7 (a) $\lambda = \dfrac{c}{\upsilon} = \dfrac{2.998\times10^8 \text{ m}\cdot\text{s}^{-1}}{7.1\times10^{14} \text{ s}^{-1}} = 4.2\times10^{-7}\text{ m} = 420\text{ nm}$

(b) $\lambda = \dfrac{c}{\upsilon} = \dfrac{2.998\times10^8 \text{ m}\cdot\text{s}^{-1}}{2.0\times10^{18} \text{ s}^{-1}} = 1.5\times10^{-10}\text{ m} = 150\text{ pm}$

1A.9 All of these can be determined using $E = h\nu$ and $c = \nu\lambda$. For example, in the first entry frequency is given, so:

$\lambda = \dfrac{c}{\upsilon} = \dfrac{2.998\times10^8 \text{ m}\cdot\text{s}^{-1}}{8.7\times10^{14} \text{ s}^{-1}} = 3.4\times10^{-7}\text{ m} = 340\text{ nm};$

and $E = h\nu = (6.626\times10^{-34}\text{ J}\cdot\text{s})(8.7\times10^{14}\text{ s}^{-1}) = 5.8\times10^{-19}\text{ J}$

Frequency (2 s.f.)	Wavelength (2 s.f.)	Energy of photon (2 s.f.)	Event
8.7×10^{14} Hz	340 nm	5.8×10^{-19} J	Suntan
5.0×10^{14} Hz	600 nm	3.3×10^{-19} J	Reading
300 MHz	1 m	2×10^{-25} J	Microwave popcorn
1.2×10^{17} Hz	2.5 nm	7.9×10^{-17} J	Dental x-ray

1A.11 In each of these series, the principal quantum number for the lower energy level involved is the same for each absorption line. Thus, for the Lyman series, the lower energy level is $n = 1$; for the Balmer series, $n = 2$; for Paschen series, $n = 3$; and for the Brackett series, $n = 4$.

1A.13 (a) The Rydberg equation gives ν when $\mathfrak{R} = 3.29 \times 10^{15}$ s^{-1}, from which one can calculate λ from the relationship $c = \nu\lambda$.

$$\nu = \mathfrak{R}\left(\frac{1}{n_1^2} - \frac{1}{n_2^2}\right)$$

and $c = \nu\lambda = 2.997\ 92 \times 10^8\ m \cdot s^{-1}$

$$c = \mathfrak{R}\left(\frac{1}{n_1^2} - \frac{1}{n_2^2}\right)\lambda$$

$$2.99\ 792 \times 10^8\ m \cdot s^{-1} = (3.29 \times 10^{15}\ s^{-1})\left(\frac{1}{1} - \frac{1}{4}\right)\lambda$$

$$\lambda = 1.21 \times 10^{-7}\ m = 121\ nm$$

(b) Lyman series

(c) This absorption lies in the ultraviolet region.

1A.15 The ultraviolet spectrum of atomic hydrogen (known as the Lyman series) has $n_1 = 1$; so, for a line at 102.6 nm:

$$\upsilon = \frac{c}{\lambda} = \frac{2.998 \times 10^8\ m \cdot s^{-1}}{102.6 \times 10^{-9}\ m} = 2.922 \times 10^{15}\ s^{-1};$$

using the Rydberg equation we can now solve for n_2 :

$$v = \Re\left(\frac{1}{n_1^2} - \frac{1}{n_2^2}\right); \quad \frac{1}{n_2^2} = \frac{1}{1^2} - \frac{v}{\Re}$$

$$\frac{1}{n_2^2} = 1 - \frac{2.922 \times 10^{15} \text{ s}^{-1}}{3.29 \times 10^{15} \text{ s}^{-1}} = 0.112$$

$$n_2^2 = 9; \quad n_2 = 3$$

The transition is $n_1 = 1$ to $n_2 = 3$.

1A.17 For hydrogenlike one-electron ions, we use the Z-dependent Rydberg relation with the relationship $c = \lambda v$ to determine the transition wavelength. For He^+, $Z = 2$.

$$v = Z^2 \Re\left(\frac{1}{n_1^2} - \frac{1}{n_2^2}\right) = \left(2^2\right)\left(3.29 \times 10^{15} \text{ s}^{-1}\right)\left(\frac{1}{1^2} - \frac{1}{2^2}\right) = 9.87 \times 10^{15} \text{ Hz}$$

$$\lambda = \frac{c}{v} = \frac{2.997\,92 \times 10^8 \text{ m} \cdot \text{s}^{-1}}{9.87 \times 10^{15} \text{ s}^{-1}} = 3.04 \times 10^{-8} \text{ m} = 30.4 \text{ nm}$$

1B.1 (a) False. The total intensity is proportional to T^4. (Stefan-Boltzmann Law)   (b) True.   (c) False. Photons of radio-frequency radiation are lower in energy than photons of ultraviolet radiation.

1B.3 The photoelectric effect (d) best supports the idea that EMR has the properties of particles. To explain the photoelectric effect, Einstein proposed that EMR consists of particles (or packets) of energy called photons, each photon having a fixed energy equal to hv.

1B.5 The energy is first converted from eV to joules:

$$E = \left(140.511 \times 10^3 \text{ eV}\right)\left(1.6022 \times 10^{-19} \text{ J} \cdot \text{eV}^{-1}\right) = 2.2513 \times 10^{-14} \text{ J}$$

From $E = hv$ and $c = v\lambda$ we can write

$$\lambda = \frac{hc}{E}$$

$$= \frac{(6.626\,08 \times 10^{-34}\ \text{J} \cdot \text{s})\,(2.997\,92 \times 10^{8}\ \text{m} \cdot \text{s}^{-1})}{2.2513 \times 10^{-14}\ \text{J}}$$

$$= 8.8237 \times 10^{-12}\ \text{m or } 8.8237\ \text{pm}$$

1B.7 (a) From $c = \nu\lambda$ and $E = h\nu$, we can write

$$E = hc\lambda^{-1}$$

$$= (6.626\,08 \times 10^{-34}\ \text{J s})\,(2.997\,92 \times 10^{8})\,(589 \times 10^{-9}\ \text{m})^{-1}$$

$$= 3.37 \times 10^{-19}\ \text{J}$$

(b) $E = \left(\dfrac{5.00 \times 10^{-3}\ \text{g Na}}{22.99\ \text{g} \cdot \text{mol}^{-1}\ \text{Na}} \right) (6.022 \times 10^{23}\ \text{atoms} \cdot \text{mol}^{-1})$

$$\times\ (3.37 \times 10^{-19}\ \text{J} \cdot \text{atom}^{-1})$$

$$= 44.1\ \text{J}$$

(c) $E = (6.022 \times 10^{23}\ \text{atoms} \cdot \text{mol}^{-1})(3.37 \times 10^{-19}\ \text{J} \cdot \text{atom}^{-1})$

$$= 2.03 \times 10^{5}\ \text{J or } 203\ \text{kJ}$$

1B.9 $32\ \text{W} = 32\ \text{J} \cdot \text{sec}^{-1}$, so in 2 seconds 64 J will be emitted.

For violet light ($\lambda = 420\ \text{nm} = 420 \times 10^{-9}\ \text{m}$) the energy per photon is:

$$E = hc\lambda^{-1}$$

$$= (6.626\,08 \times 10^{-34}\ \text{J} \cdot \text{s})\,(2.997\,92 \times 10^{8})\,(420 \times 10^{-9}\ \text{m})^{-1}$$

$$= 4.7 \times 10^{-19}\ \text{J} \cdot \text{photon}^{-1}$$

$$\text{number of photons} = (64\ \text{J})\,(4.7 \times 10^{-19}\ \text{J} \cdot \text{photon}^{-1})^{-1}$$

$$= 1.4 \times 10^{20}\ \text{photons}$$

$$\text{moles of photons} = (1.4 \times 10^{20}\ \text{photons}) \left(\frac{1\ \text{mol}}{6.022 \times 10^{23}\ \text{photons}} \right)$$

$$= 2.3 \times 10^{-4}\ \text{mol photons}$$

1B.11 Using Wien's law: $T\lambda_{max} = \text{constant} = 2.88 \times 10^{-3} \text{ K} \cdot \text{m}$.

$\lambda_{max} = 850 \text{ nm} = 8.5 \times 10^{-7} \text{ m}$,

then $T = \dfrac{2.88 \times 10^{-3} \text{ K} \cdot \text{m}}{8.5 \times 10^{-7} \text{ m}} = 3400 \text{ K}$

1B.13 Wien's law states that $T\lambda_{max} = \text{constant} = 2.88 \times 10^{-3} \text{ K} \cdot \text{m}$.

If $T/\text{K} = 1540°\text{C} + 273°\text{C} = 1813 \text{ K}$, then $\lambda_{max} = \dfrac{2.88 \times 10^{-3} \text{ K} \cdot \text{m}}{1813 \text{ K}}$

$\lambda_{max} = 1.59 \times 10^{-6} \text{ m, or } 1590 \text{ nm}$

This wavelength falls in the infrared region.

1B.15 (a) Use the de Broglie relationship, $\lambda = hp^{-1} = h(mv)^{-1}$.

$m_e = (9.109\,39 \times 10^{-28} \text{ g})(1 \text{ kg}/1000 \text{ g}) = 9.109\,39 \times 10^{-31} \text{ kg}$

$(3.6 \times 10^3 \text{ km} \cdot \text{s}^{-1})(1000 \text{ m} \cdot \text{km}^{-1}) = 3.6 \times 10^6 \text{ m} \cdot \text{s}^{-1}$

$\lambda = h(mv)^{-1}$

$= \dfrac{6.626\,08 \times 10^{-34} \text{ J} \cdot \text{s}}{(9.109\,39 \times 10^{-31} \text{ kg})(3.6 \times 10^6 \text{ m} \cdot \text{s}^{-1})}$

$= 2.0 \times 10^{-10} \text{ m}$

(b) $E = hv$

$= (6.626\,08 \times 10^{-34} \text{ J} \cdot \text{s})(2.50 \times 10^{16} \text{ s}^{-1})$

$= 1.66 \times 10^{-17} \text{ J}$

(c) The photon needs to contain enough energy to eject the electron from the surface as well as to cause it to move at $3.6 \times 10^3 \text{ km} \cdot \text{s}^{-1}$. The energy involved is the kinetic energy of the electron, which equals $\frac{1}{2}mv^2$.

$$E_{photon} = 1.66 \times 10^{-17} \text{ J} + \frac{1}{2} mv^2$$

$$= 1.66 \times 10^{-17} \text{ J} + \frac{1}{2}(9.109\ 39 \times 10^{-31} \text{ kg})\ (3.6 \times 10^6 \text{ m} \cdot \text{s}^{-1})^2$$

$$= 1.66 \times 10^{-17} \text{ J} + 5.9 \times 10^{-18} \text{ J}$$

$$= 2.25 \times 10^{-17} \text{ J}$$

But we are asked for the wavelength of the photon, which we can get from

$E = h\nu$ and $c = \nu\lambda$ or $E = hc\lambda^{-1}$.

$$2.25 \times 10^{-17} \text{ kg} \cdot m^2 \cdot s^{-2} = (6.626\ 08 \times 10^{-34} \text{ kg} \cdot m^2 \cdot s^{-1})$$

$$\times (2.997\ 92 \times 10^8 \text{ m} \cdot s^{-1})\lambda^{-1}$$

$$\lambda = 8.8 \times 10^{-9} \text{ m}$$

$$= 8.8 \text{ nm}$$

(d) 8.8 nm is in the x-ray/gamma ray region.

1B.17 The de Broglie relationship, $\lambda = hp^{-1} = h(mv)^{-1}$, states that the wavelength of a particle is inversely proportional to its mass. Therefore, the heavier person (80 kg) should have a shorter wavelength than the lighter person (60 kg) if they are moving at the same speed.

1B.19 The wavelength for both a proton and a neutron can be calculated using the de Broglie relationship; thus:

$$\lambda_{proton} = h(m_{proton}v)^{-1}$$

$$= \frac{6.626\ 08 \times 10^{-34} \text{ J} \cdot \text{s}}{(1.673 \times 10^{-27} \text{ kg})\ (2.75 \times 10^5 \text{ m} \cdot \text{s}^{-1})}$$

$$= 1.44 \times 10^{-12} \text{ m} = 1.44 \text{ pm}.$$

$$\lambda_{neutron} = h(m_{neutron}v)^{-1}$$

$$= \frac{6.626\ 08 \times 10^{-34} \text{ J} \cdot \text{s}}{(1.675 \times 10^{-27} \text{ kg})\ (2.75 \times 10^5 \text{ m} \cdot \text{s}^{-1})}$$

$$= 1.44 \times 10^{-12} \text{ m} = 1.44 \text{ pm}.$$

The wavelength of a proton and a neutron are identical to 3 significant figures.

1B.21 To answer this question, we need to convert the quantities to a consistent set of units, in this case, SI units.

(5.15 ounce) (28.3 g · ounce^{-1}) (1 kg/1000 g) = 0.146 kg

$$\left(\frac{92 \text{ mi}}{h}\right)\left(\frac{1 \text{ h}}{3600 \text{ s}}\right)\left(\frac{1 \text{ km}}{0.6214 \text{ mi}}\right)\left(\frac{1000 \text{ m}}{1 \text{ km}}\right) = 41 \text{ m} \cdot \text{s}^{-1}$$

Use the de Broglie relationship.

$$\lambda = hp^{-1} = h(mv)^{-1}$$
$$= h(mv)^{-1}$$
$$= \frac{6.626 \, 08 \times 10^{-34} \text{ J} \cdot \text{s}}{(0.146 \text{ kg}) (0.041 \text{ km} \cdot \text{s}^{-1})}$$
$$= \frac{6.626 \, 08 \times 10^{-34} \text{ kg} \cdot \text{m}^2 \cdot \text{s}^{-1}}{(0.146 \text{ kg}) (41 \text{ m} \cdot \text{s}^{-1})}$$
$$= 1.1 \times 10^{-34} \text{ m}$$

1B.23 From the de Broglie relationship, $p = h\lambda^{-1}$ or $h = mv\lambda$, we can calculate the velocity of the neutron:

$$v = \frac{h}{m\lambda}$$
$$= \frac{(6.626 \, 08 \times 10^{-34} \text{ kg} \cdot \text{m}^2 \cdot \text{s}^{-1})}{(1.674 \, 93 \times 10^{-27} \text{ kg}) (100 \times 10^{-12} \text{ m})} \quad \text{(remember that } 1 \text{ J} = 1 \text{ kg} \cdot \text{m}^2 \cdot \text{s}^{-2})$$
$$= 3.96 \times 10^3 \text{ m} \cdot \text{s}^{-1}$$

1B.25 The uncertainty principle states that $\Delta p \Delta x = \frac{1}{2}\hbar$; so, for an electron

$\Delta p = m_e \Delta v$, then $m_e \Delta v \Delta x = \frac{1}{2}\hbar$ and $\Delta v = \frac{1}{2}\frac{\hbar}{m_e \Delta x}$; if we assume that the

uncertainty in the position of the electron is equal to the diameter of the lead atom $\Delta x = 350 \text{ pm} = 350 \times 10^{-12} \text{ m}$.

Remembering that $1 \text{ J} = 1 \text{ kg} \cdot \text{m}^2 \cdot \text{s}^{-2}$, gives

$$\hbar = \left(1.054 \quad 457 \times 10^{-34} J \cdot s\right)\left(\frac{1 \, kg \cdot m^2 \cdot s^{-2}}{J}\right) = 1.054 \quad 457 \times 10^{-34} \, kg \cdot m^2 \cdot s^{-1}$$

Then

$$\Delta v = \frac{1}{2} \left(\frac{1.054\ 457 \times 10^{-34}\ \text{kg} \cdot \text{m}^2 \cdot \text{s}^{-1}}{(9.109\ 38 \times 10^{-31}\text{kg})(350 \times 10^{-12}\ \text{m})} \right)$$

$$\Delta v = 1.65 \times 10^5\ \text{m} \cdot \text{s}^{-1}$$

1B.27 The minimum uncertainty in the position of the bowling ball can be calculated using the uncertainty principle:

$$\Delta x = \frac{1}{2} \left(\frac{\hbar}{m \Delta v} \right) = \frac{1}{2} \left(\frac{1.054\ 457 \times 10^{-34}\ \text{kg} \cdot \text{m}^2 \cdot \text{s}^{-1}}{(8.00\ \text{kg})(5.0\ \text{m} \cdot \text{s}^{-1})} \right)$$

$$\Delta x = 1.3 \times 10^{-36}\ \text{m}$$

1C.1 (a) For movement between energy levels separated by a difference of 1 in principal quantum number, the expression is

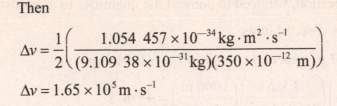

$$\Delta E = E_{n+1} - E_n = \frac{(n+1)^2 h^2}{8mL^2} - \frac{n^2 h^2}{8mL^2} = \frac{(2n+1)h^2}{8mL^2}$$

For $n = 4$ and $n + 1 = 5$, $\Delta E = \dfrac{9h^2}{8mL^2}$

Then $\lambda_{5,4} = \dfrac{hc}{E} = \dfrac{8mhcL^2}{9h^2} = \dfrac{8mcL^2}{9h}$

For an electron in a 150. pm box, the expression becomes

$$\lambda_{5,4} = \frac{8(9.109\ 39 \times 10^{-31}\ \text{kg})\,(2.997\ 92 \times 10^8\ \text{m} \cdot \text{s}^{-1})\,(150 \times 10^{-12}\ \text{m})^2}{9(6.626\ 08 \times 10^{-34}\ \text{J} \cdot \text{s})}$$

$$= 8.24 \times 10^{-9}\ \text{m} = 8.24\ \text{nm}.$$

(b) Do the same as in (a):

$n = 3$ and $n + 1 = 4$, $\Delta E = \dfrac{7h^2}{8mL^2}$

$$\lambda_{4,3} = \frac{hc}{E} = \frac{8mhcL^2}{7h^2} = \frac{8mcL^2}{7h}$$

For an electron in a 150. pm box, the expression becomes

$$\lambda_{4,3} = \frac{8(9.109\,39 \times 10^{-31}\ \text{kg})\,(2.997\,92 \times 10^{8}\ \text{m} \cdot \text{s}^{-1})\,(150 \times 10^{-12}\ \text{m})^{2}}{7(6.626\,08 \times 10^{-34}\ \text{J} \cdot \text{s})}$$

$$= 1.06 \times 10^{-8}\ \text{m} = 10.6\ \text{nm}.$$

1C.3   Yes, there are degenerate levels. The first three cases of degenerate levels are:

$n_1 = 1,\ n_2 = 2$ is degenerate with $n_1 = 2,\ n_2 = 1$

$n_1 = 1,\ n_2 = 3$ is degenerate with $n_1 = 3,\ n_2 = 1$

$n_1 = 2,\ n_2 = 3$ is degenerate with $n_1 = 3,\ n_2 = 2$

1C.5   (a) Refer to the plot below for parts (a) thru (d); nodes are where the wavefunction is zero:

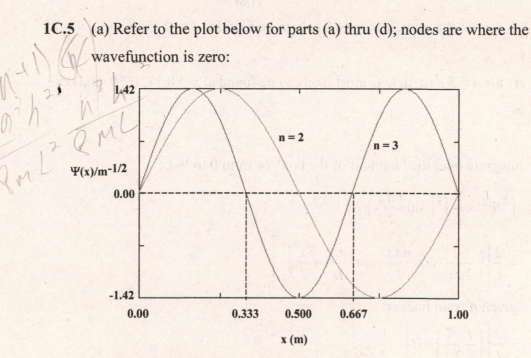

(b) for $n = 2$ there is one node at $x = 0.500$ m.

(c) for $n = 3$ there are two nodes, one at $x = 0.333$ and 0.667 m.

(d) the number of nodes is equal to $n - 1$

Refer to the plot below for parts (e) and (f):

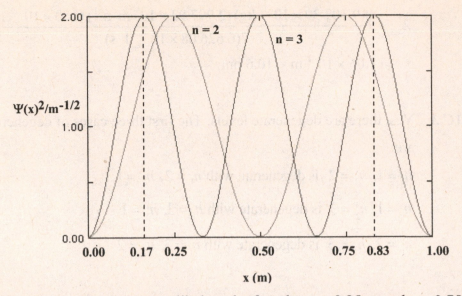

(e) for $n = 2$ a particle is most likely to be found at $x = 0.25$ m and $x = 0.75$ m.

(f) for $n = 3$ a particle is most likely to be found at $x = 0.17$, 0.50, and 0.83 m.

1C.7 Integrate over the "left half of the box" or from 0 to ½ L:

$$\int_0^{\frac{L}{2}} \Psi^2 = \frac{2}{L} \int_0^{\frac{L}{2}} \left(\sin \frac{n\pi x}{L} \right)^2 dx$$

$$= \frac{2}{L} \left[\left(\frac{-1}{2n\pi} \cdot \cos \frac{n\pi x}{L} \cdot \sin \frac{n\pi x}{L} + \frac{x}{2} \right) \Big|_0^{\frac{L}{2}} \right]$$

given n is an integer:

$$= \frac{2}{L} \left[\left(\frac{L/2}{2} \right) - 0 \right] = \frac{1}{2}$$

1D.1 (a) Energy will increase (energy is a function of n). (b) n increases (from $n = 1$ to $n = 2$). (c) l increases (from $l = 0$ for s to $l = 1$ for p). (d) radius increases (radius is a function of n).

1D.3　The equation derived in Example 1D.1 can be used:

$$\frac{\psi^2(r=0.55a_0,\theta,\phi)}{\psi^2(0,\theta,\phi)}=\frac{\dfrac{e^{-2(0.55a_0)/a_0}}{\pi a_0^3}}{\left(\dfrac{1}{\pi a_0^3}\right)}=0.33$$

1D.5　To show that three p-orbitals taken together are spherically symmetric, sum the three probability distributions (the wavefunctions squared) and show that the magnitude of the sum is not a function of θ or ϕ.

$$p_x=R(r)C\sin\theta\cos\phi$$
$$p_y=R(r)C\sin\theta\sin\phi$$
$$p_z=R(r)C\cos\theta$$

where $C=\left(\dfrac{3}{4\pi}\right)^{\frac{1}{2}}$

Squaring the three wavefunctions and summing them:

$$R(r)^2C^2\sin^2\theta\cos^2\phi+R(r)^2C^2\sin^2\theta\sin^2\phi+R(r)^2C^2\cos^2\theta$$
$$=R(r)^2C^2\left(\sin^2\theta\cos^2\phi+\sin^2\theta\sin^2\phi+\cos^2\theta\right)$$
$$=R(r)^2C^2\left(\sin^2\theta\left(\cos^2\phi+\sin^2\phi\right)+\cos^2\theta\right)$$

Using the identity $\cos^2 x+\sin^2 x=1$ this becomes

$$R(r)^2C^2\left(\sin^2\theta+\cos^2\theta\right)=R(r)^2C^2$$

With one electron in each p-orbital, the electron distribution is not a fuction of θ or ϕ and is, therefore, spherically symmetric.

1D.7　(a) The probability (P) of finding an electron within a sphere of radius a_o may be determined by integrating the appropriate wavefunction squared from 0 to a_o :

$$P=\frac{4}{a_o^2}\int_0^{a_o}r^2\exp\left(-\frac{2r}{a_o}\right)dr$$

This integral is easier to evaluate if we allow the following change of variables:

$$z = \frac{2r}{a_o} \quad \therefore \ z = 2 \text{ when } r = a_o,\ z = 0 \text{ when } r = 0, \text{ and } dr = \left(\frac{a_o}{2}\right) dz$$

$$P = \frac{1}{2} \int_0^2 z^2 \exp(-z)dz = -\frac{1}{2}(z^2 + 2z + 2)\exp(-z)\bigg|_0^2$$

$$= -\frac{1}{2}\Big[\big((4+4+2)\exp(-2)\big) - 2\Big]$$

$$= 0.323 \text{ or } 32.3\%$$

(b) Following the answer developed in (a) changing the integration limits to 0 to 2 a_o:

$$z = 4 \text{ when } r = 2a_o,\ z = 0 \text{ when } r = 0, \text{ and } dr = \left(\frac{a_o}{2}\right)dz$$

$$P = \frac{1}{2}\int_0^4 z^2 \exp(-z)dz = -\frac{1}{2}(z^2 + 2z + 2)\exp(-z)\bigg|_0^4$$

$$= -\frac{1}{2}\Big[\big(26\exp(-4)\big) - 4\Big]$$

$$= 0.761 \text{ or } 76.1\%$$

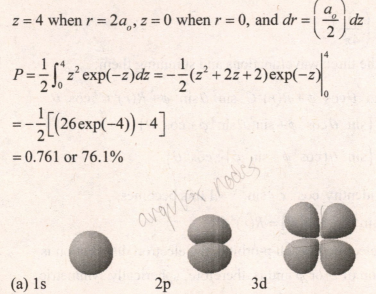

angular nodes

1D.9   (a) 1s   2p   3d

(b) A node is a region in space where the wavefunction ψ passes through 0.   (c) The simplest s-orbital has zero nodes, the simplest p-orbital has one nodal plane, and the simplest d-orbital has two nodal planes.   (d) Given the increase in number of nodes, an f-orbital would be expected to have three nodal planes.

1D.11 (a) one orbital;   (b) five orbitals;   (c) three orbitals;   (d) seven orbitals

1D.13 (a) seven values: 0, 1, 2, 3, 4, 5, 6;   (b) five values; $-2, -1, 0, 1, 2$;   (c) three values: $-1, 0, 1$;   (d) four subshells: $4s, 4p, 4d$, and $4f$

1D.15 (a) $n = 6; l = 1$;   (b) $n = 3; l = 2$;   (c) $n = 2; l = 1$;   (d) $n = 5; l = 3$

1D.17 (a) $-1, 0, +1$;   (b) $-2, -1, 0, +1, +2$;   (c) $-1, 0, +1$;
(d) $-3, -2, -1, 0, +1, +2, +3$.

1D.19 (a) three orbitals;   (b) five orbitals;   (c) one orbital;   (d) seven orbitals

1D.21 (a) 5d, five;   (b) 1s, one;   (c) 6f, seven;   (d) 2p, three

1D.23 (a) 3;   (b) 1;   (c) 4;   (d) 1

1D.25 (a) cannot exist;   (b) exists;   (c) cannot exist;   (d) exists

1E.1 (a) Energy increases; (b) n increases; (c) l increases; (d) radius increases. All of these are the same for a hydrogen atom. (Exercise 1D.1)

1E.3 (a) The total Coulomb potential energy $V(r)$ is the sum of the individual coulombic attractions and repulsions. There will be one attraction between the nucleus and each electron plus a repulsive term to represent the interaction between each pair of electrons. For lithium, there are three protons in the nucleus and three electrons. Each attractive Coulomb potential will be equal to

$$\frac{(-e)(+3e)}{4\pi\varepsilon_0 r} = \frac{-3e^2}{4\pi\varepsilon_0 r}$$

where $-e$ is the charge on the electron and $+3e$ is the charge on the nucleus, ε_0 is the vacuum permittivity, and r is the distance from the electron to the nucleus. The total attractive potential will thus be

$$\left(\frac{-3e^2}{4\pi\varepsilon_0 r_1}\right)+\left(\frac{-3e^2}{4\pi\varepsilon_0 r_2}\right)+\left(\frac{-3e^2}{4\pi\varepsilon_0 r_3}\right)=\left(\frac{-3e^2}{4\pi\varepsilon_0}\right)\left(\frac{1}{r_1}+\frac{1}{r_2}+\frac{1}{r_3}\right)$$

The repulsive terms will have the form

$$\frac{(-e)(-e)}{4\pi\varepsilon_0 r_{ab}}=\frac{e^2}{4\pi\varepsilon_0 r_{ab}}$$

where r_{ab} represents the distance between two electrons a and b. The total repulsive term will thus be

$$\frac{e^2}{4\pi\varepsilon_0 r_{12}}+\frac{e^2}{4\pi\varepsilon_0 r_{13}}+\frac{e^2}{4\pi\varepsilon_0 r_{23}}=\frac{e^2}{4\pi\varepsilon_0}\left(\frac{1}{r_{12}}+\frac{1}{r_{13}}+\frac{1}{r_{23}}\right)$$

This gives

$$V(r)=\left(\frac{-3e^2}{4\pi\varepsilon_0}\right)\left(\frac{1}{r_1}+\frac{1}{r_2}+\frac{1}{r_3}\right)+\frac{e^2}{4\pi\varepsilon_0}\left(\frac{1}{r_{12}}+\frac{1}{r_{13}}+\frac{1}{r_{23}}\right)$$

(b) The first term represents the coulombic attractions between the nucleus and each electron, and the second term represents the coulombic repulsions between each pair of electrons.

1E.5   (a) False. Z_{eff} is considerably affected by the total number of electrons present in the atom because the electrons in the lower energy orbitals will "shield" the electrons in the higher energy orbitals from the nucleus. This effect arises because the e-e repulsions tend to offset the attraction of the electron to the nucleus.   (b) True.   (c) False. The electrons are increasingly less able to penetrate to the nucleus as l increases.   (d) True.

1E.7   Only (d) is the configuration expected for a ground-state atom; the others all represent excited-state configurations.

1E.9 (a) This configuration is possible. (b) This configuration is not possible because $l = 0$ here, so m_l must also equal 0. (c) This configuration is not possible because the maximum value l can have is $n - 1$; $n = 4$, so $l_{max} = 3$.

1E.11 (a) sodium $[Ne] 3s^1$

       (b) silicon $[Ne] 3s^2 3p^2$

       (c) chlorine $[Ne] 3s^2 3p^5$

       (d) rubidium $[Kr] 5s^1$

1E.13 (a) silver $[Kr] 4d^{10} 5s^1$

       (b) beryllium $[He] 2s^2$

       (c) antimony $[Kr] 4d^{10} 5s^2 5p^3$

       (d) gallium $[Ar] 3d^{10} 4s^2 4p^1$

       (e) tungsten $[Xe] 4f^{14} 5d^4 6s^2$

       (f) iodine $[Kr] 4d^{10} 5s^2 5p^5$

1E.15 (a) tellurium; (b) vanadium; (c) carbon; (d) thorium

1E.17 (a) 4p; (b) 4s; (c) 6s; (d) 6s

1E.19 (a) 5; (b) 11; (c) 5; (d) 20

1E.21 (a) 3; (b) 2; (c) 3; (d) 2

1E.23

Element	Electron Configuration	Unpaired Electrons
Ga	$[Ar]\,3d^{10}\,4s^2\,4p^1$	1
Ge	$[Ar]\,3d^{10}\,4s^2\,4p^2$	2
As	$[Ar]\,3d^{10}\,4s^2\,4p^3$	3
Se	$[Ar]\,3d^{10}\,4s^2\,4p^4$	2
Br	$[Ar]\,3d^{10}\,4s^2\,4p^5$	1

1E.25　(a) ns^1;　(b) ns^2np^3;　(c) $(n-1)d^3ns^2$;　(d) $(n-1)d^{10}ns^1$

1F.1　(a) silicon (118 pm) > sulfur (104 pm) > chlorine (99 pm);

(b) titanium (147 pm) > chromium (129 pm) > cobalt (125 pm);

(c) mercury (155 pm) > cadmium (152 pm) > zinc (137 pm);

(d) bismuth (182 pm) > antimony (141 pm) > phosphorus (110 pm)

1F.3　$P^{3-} > S^{2-} > Cl^-$

1F.5　(a) Ca;　(b) Na;　(c) Na

1F.7　(a) oxygen (1310 kJ · mol^{-1}) > selenium (941 kJ · mol^{-1}) > tellurium (870 kJ · mol^{-1}); ionization energies generally decrease as one goes down a group.　(b) gold (890 kJ · mol^{-1}) > osmium (840 kJ · mol^{-1}) > tantalum (761 kJ· mol^{-1}); ionization energies generally decrease as one goes from right to left in the periodic table.　(c) lead (716 kJ · mol^{-1}) > barium (502 kJ · mol^{-1}) > cesium (376 kJ · mol^{-1}); ionization energies generally decrease as one goes from right to left in the periodic table.

1F.9   The first ionization energy for sulfur and phosphorus atoms are nearly the same, despite sulfur having a larger Z_{eff}; this is due to greater electron-electron repulsions in S, making the energy of the outermost electrons higher than predicted. Once the first electron is removed, the Z_{eff} becomes the predominant factor and the remaining electrons are held tighter due to the smaller size of the S^+ ion as compared with the P^+ ion; this is reflected in the much greater second ionization energy of sulfur as compared with phosphorus.

1F.11   (a) iodine;   (b) they are equal;   (c) sulfur;   (d) they are equal

1F.13   (a) The inert-pair effect is the term used to describe the fact that heavy (period 5 and greater) p-block elements have a tendency to form ions that are two units lower in charge than that expected based on their group number.   (b)   The inert-pair effect is presumed to only be observed for heavy elements because of the poor shielding ability of the d electrons in these elements, enhancing the ability of the s electrons to penetrate to the nucleus and therefore be bound tighter than expected.

1F.15   (a) A diagonal relationship is a similarity in chemical properties between an element in the periodic table and one lying one period lower and one group to the right. (b) It is caused by the similarity in size of the ions. The lower-right element in the pair would generally be larger because it lies in a higher period, but it also will have a higher oxidation state, which will cause the ion to be smaller. (c) For example, Al^{3+} and Ge^{4+} compounds show the diagonal relationship, as do Li^+ and Mg^{2+}.

1F.17   Only (b) Li and Mg exhibit a diagonal relationship.

1F.19 The ionization energies of the s-block metals are considerably lower, thus making it easier for them to lose electrons in chemical reactions.

1F.21 (a) metal;   (b) nonmetal;   (c) metal;   (d) metalloid;   (e) metalloid; (f) metal

1.1 For the Balmer Series $n_1 = 2$, so the fifth line in the spectrum should be $n_2 = 7$. Using the Rydberg equation we get

$$v = \Re\left(\frac{1}{n_1^2} - \frac{1}{n_2^2}\right) = \left(3.29 \times 10^{15}\ s^{-1}\right)\left(\frac{1}{2^2} - \frac{1}{7^2}\right) = 7.55 \times 10^{14}\ s^{-1}$$

$$\lambda = \frac{c}{v} = \frac{2.99792 \times 10^8\ m \cdot s^{-1}}{7.55 \times 10^{14}\ s^{-1}} = 3.97 \times 10^{-7}\ m = 397\ nm$$

1.3 For light with a wavelength of 633 nm the energy per photon is:

$$E = hc\lambda^{-1}$$
$$= (6.626\,08 \times 10^{-34}\ J \cdot s)\,(2.997\,92 \times 10^8\ m \cdot s^{-1})\,(633 \times 10^{-9}\ m)^{-1}$$
$$= 3.14 \times 10^{-19}\ J \cdot photon^{-1}.$$

The total energy produced is

$$(3.14 \times 10^{-19}\ J \cdot photon^{-1})\,(2.4 \times 10^{21}\ photons) = 750\ J.$$

1.5 (a)

$$\int_0^L \left(\sin\frac{\pi \cdot x}{L}\right) \cdot \left(\sin\frac{2\pi \cdot x}{L}\right)\,dx$$

$$= \frac{L}{2\pi}\left(\sin\frac{\pi \cdot x}{L}\right) - \frac{L}{6\pi}\left(\sin\frac{3\pi \cdot x}{L}\right)\Bigg|_0^L = 0$$

(b) Below is a plot of the first two wavefunctions describing the one-dimensional particle-in-a-box and the product of these two wavefunctions. Notice that the area above zero in the product exactly cancels the area below zero, making the integral of the product zero. This happens whenever a wavefunction that is unaltered by a reflection through the center of the box (wavefunctions with odd n) is multiplied by a wavefunction that changes sign everywhere when reflected through the center of the box (wavefunctions with even n).

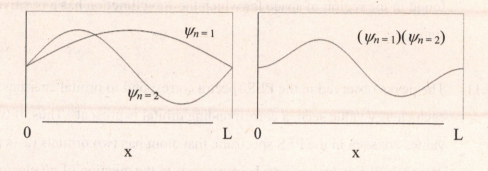

1.7   (a) Evaluating the integral for an $n = 1$ to $n = 3$ transition:

$$\int_0^L \sin\left(\frac{\pi x}{L}\right) \cdot x \cdot \sin\left(\frac{3\pi x}{L}\right)$$

$$= \frac{L^2}{8\pi^2}\left(\cos\left(\frac{2\pi x}{L}\right) + \frac{2\pi x}{L}\sin\frac{2\pi x}{L}\right) -$$

$$\frac{L^2}{32\pi^2}\left(\cos\left(\frac{4\pi x}{L}\right) + \frac{4\pi x}{L}\sin\frac{4\pi x}{L}\right)\Bigg|_0^L$$

$$= \frac{L^2}{8\pi^2}(1+0) - \frac{L^2}{32\pi^2}(1+0) - \left[\frac{L^2}{8\pi^2}(1+0) - \frac{L^2}{32\pi^2}(1+0)\right]$$

$$= 0$$

Because the integral is zero, one would not expect to observe a transition between the $n = 1$ and $n = 3$ states.

(b) Again, evaluating the integral:

$$\int_0^L \sin\left(\frac{\pi x}{L}\right) \cdot x \cdot \sin\left(\frac{2\pi x}{L}\right)$$

$$= \frac{L^2}{\pi^2}\left(\frac{1}{2}\left(\cos\left(\frac{\pi x}{L}\right) + \frac{\pi x}{L}\sin\left(\frac{\pi x}{L}\right)\right) - \frac{1}{18}\left(\cos\left(\frac{3\pi x}{L}\right) + \frac{3\pi x}{L}\sin\left(\frac{3\pi x}{L}\right)\right)\right)\Bigg|_0^L$$

Given the L^2 term, we see that the integral, and therefore I, will increase as the length of the box increases.

1.9   A $2p_x$ orbital has two lobes, one that has a wavefunction with a positive sign and one with a wavefunction with a negative sign. Thus there is a ½ or 0.50 probability that an electron excited to the $2p_x$-orbital would be found in the region of space for which the wavefunction has a positive sign.

1.11   The peaks observed in the PES spectra correspond to orbital energies; for each energy value seen, a corresponding orbital is present. Thus, if two values are seen in the PES spectrum, that atom has two orbitals (a 1s and a 2s); each PES value observed corresponds to the ejection of *all* electrons from that orbital. The PES value observed is approximately equal to the ionization energy of the first electron to be removed from that orbital; differences that are seen are due to the differences in how the various measurements are made. See Figure 1F.10 and Appendix 2 for the successive ionization energies of the elements.

(a) The observed values

75.7 eV (7.30 MJ×mol^{-1}) and 5.38 eV (0.519 MJ×mol^{-1}) correspond

respectively to the second (7300 kJ · mol^{-1}) and first (519 kJ · mol^{-1})

ionization energies of Li ($1s^2 2s^1$).

(b) The PES values observed

153 eV (14.8 MJ·mol^{-1}) and 9.33 eV (0.90 MJ·mol^{-1}) correspond

respectively to the third (14800 kJ · mol^{-1}) and first (900 kJ · mol^{-1})

ionization energies of Be ($1s^2 2s^2$).

1.13   A ground-state oxygen atom has four electrons in the p-orbitals. This configuration means that as one goes across the periodic table in Period 2, oxygen is the first element encountered in which the p-electrons must be paired. This added electron-electron repulsion energy causes the ionization potential to be lower.

1.15   molar volume $(cm^3 \cdot mol^{-1})$ = molar mass $(g \cdot mol^{-1})$/density $(g \cdot cm^{-3})$

Element	Molar vol.	Element	Molar vol.
Li	13	Na	24
Be	4.87	Mg	14.0
B	4.38	Al	9.99
C	5.29	Si	12.1
N	16	P	17.0
O	14.0	S	15.3
F	17.1	Cl	21.4
Ne	16.7	Ar	24.1

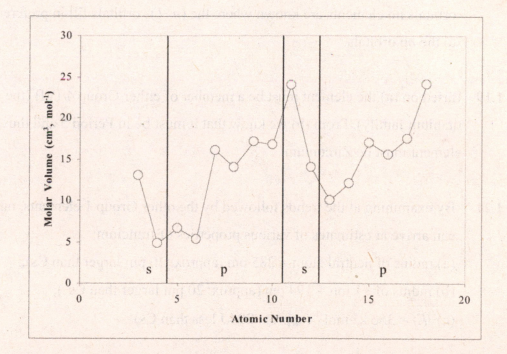

The molar volume roughly parallels atomic size (volume), which decreases as the s-sublevel begins to fill and subsequently increases as the p-sublevel fills (refer to the text discussion of periodic variation of atomic radii). In the above plot, this effect is most clearly seen in passing from Ne(10) to Na(11) and Mg(12), then to Al(13) and Si(14). Ne has a filled

2p-sublevel; the 3s-sublevel fills with Na and Mg; and the 3p-sublevel begins to fill with Al.

1.17   (a) In copper it is energetically favorable for an electron to be promoted from the 4s orbital to a 3d orbital, giving a completely filled 3d subshell. In the case of Cr, it is energetically favorable for an electron to be promoted from the 4s orbital to a 3d orbital to exactly ½ fill the 3d subshell.

(b) From Appendix 2C, the other elements for which anomalous electron configurations exist are Nb, Mo, Ru, Rh, Pd, Ag, Pt and Au. Of these, the explanation used for chromium and copper is valid for Mo, Pd, Ag and Au.

(c) Because the np orbitals are so much lower in energy than the $(n+1)$s orbitals no elements are known where the $(n+1)$s orbitals fill in preference to the np orbitals.

1.19   Based on (a) the element must be a member of either Group 4/IVB (the titanium family). From (b) we know that it must be in Period 5 and thus the element must be Zirconium.

1.21   By examining at the trends followed by the other Group 1 elements, one can arrive at estimates of various properties of francium:

(a) radius of neutral atom = 285 pm (approx. 20 pm larger than Cs);

(b) radius of +1 ion = 194 pm (approx. 20 pm larger than Cs^+);

(c) IE_1 = 356 kJ·mol^{-1} (approx. 20 kJ less than Cs)

1.23 $\underline{A}$ = Na; $\underline{B}$ = Cl; $\underline{C}$ = Na$^+$; $\underline{D}$ = Cl$^-$

The assignments can be made by looking at neutral atom and ionic radii (Figures 1F.4 and 1F.6)

1.25 (a) $1s^4 2s^1$; (b) +1; (c) One would expect the second inert gas in the other universe to still have the 1s, 2s, and 2p levels completely filled. Based on this the electron configuration of the neutral atom should be $1s^4 2s^4 2p^{12}$; therefore it should have $Z = 20$.

1.27 (a) $\lambda = \dfrac{c}{\nu} = \dfrac{2.997\ 92 \times 10^8\ \text{m} \cdot \text{s}^{-1}}{6.27 \times 10^{14}\ \text{s}^{-1}} = 4.78 \times 10^{-7}\ \text{m} = 478\ \text{nm}$

  (b) $\nu = \dfrac{c}{\lambda} = \dfrac{2.997\ 92 \times 10^8\ \text{m} \cdot \text{s}^{-1}}{421 \times 10^{-9}\ \text{m}} = 7.12 \times 10^{14}\ \text{s}^{-1}$

1.29 Neon ($Z = 10$) would have to lose a 2p electron to ionize; to accomplish this it would need to be exposed to radiation with sufficient energy to do so. Using the Rydberg equation (with $n_{lower} = 2$ and $n_{upper} = \infty$):

$$\Delta E = Z^2 h \mathfrak{R} \left(\frac{1}{n_{lower}^2} - \frac{1}{n_{upper}^2} \right)$$

$$\Delta E = 10^2 (6.626\ 08 \times 10^{-34}\ \text{J} \cdot \text{s})(3.29 \times 10^{15}\ \text{s}^{-1}) \left(\frac{1}{2^2} - \frac{1}{\infty^2} \right)$$

$$\Delta E = 5.45 \times 10^{-17}\ \text{J}$$

$$\lambda = hc\Delta E^{-1} = \frac{(6.626\ 08 \times 10^{-34}\ \text{J} \cdot \text{s})(2.997\ 92 \times 10^8\ \text{m} \cdot \text{s}^{-1})}{5.45 \times 10^{-17}\ \text{J}}$$

$$\lambda = 3.64 \times 10^{-9}\ \text{m} = 3.64\ \text{nm}.$$

The longest detectable wavelength would be 3.64 nm, which is in the x-ray region.

1.31 (a) An electron can be driven out of a metal only if a photon possessing a minimum energy equal to that of the work function for that metal strikes it. Here, to eject the electron from lithium we need

$$E_{work\ function} = (2.93\ \text{eV})(1.602\ 18 \times 10^{-19}\ \text{J} \cdot \text{eV}^{-1}) = 4.69 \times 10^{-19}\ \text{J}.$$

For the ruby-red laser (694 nm):

$$E_{694} = hc\lambda^{-1}$$

$$= (6.626\ 08 \times 10^{-34}\ \text{J} \cdot \text{s})\,(2.997\ 92 \times 10^{8}\ \text{m} \cdot \text{s}^{-1})\,(694 \times 10^{-9}\ \text{m})^{-1}$$

$$= 2.86 \times 10^{-19}\ \text{J}.$$

For the violet GaN laser (405 nm):

$$E_{405} = hc\lambda^{-1}$$

$$= (6.626\ 08 \times 10^{-34}\ \text{J} \cdot \text{s})\,(2.997\ 92 \times 10^{8}\ \text{m} \cdot \text{s}^{-1})\,(405 \times 10^{-9}\ \text{m})^{-1}$$

$$= 4.90 \times 10^{-19}\ \text{J}.$$

The violet GaN laser will provide enough energy to eject the electron.

(b) The kinetic energy (E_K) of the ejected electron can be determined by subtracting the work function from the energy supplied by the laser:

$$E_K = hc\lambda^{-1} - E_{work\ function}$$

$$= (4.90 \times 10^{-19}\ \text{J}) - (4.69 \times 10^{-19}\ \text{J})$$

$$= 2.10 \times 10^{-20}\ \text{J}.$$

1.33 The allowed energies of a particle of mass m in a one-dimensional box of length L are determined using $E_n = \dfrac{n^2 h^2}{8mL^2}$.

(a) if $L = 139\ \text{pm} = 1.39 \times 10^{-10}\ \text{m}$ for a $C-C$ bond, and $m_e = 9.109 \times 10^{-31}\ \text{kg}$, then

$$\Delta E = \frac{(6.626\ 08 \times 10^{-34}\ \text{J} \cdot \text{s})^2}{8(9.109 \times 10^{-31}\ \text{kg})(1.39 \times 10^{-10}\ \text{m})^2}(2^2 - 1^2)$$

$$= 9.35 \times 10^{-18}\ \text{J}$$

(b) This energy corresponds to a wavelength of 2.13×10^{-8} m or 21.3 nm. This falls within the x-ray region.

(c) If the chain is 10 carbons long, then there would be nine $C-C$ bonds. If a wavefunction extends over two adjacent carbon atoms then the minimum number of wavefunctions would be nine.

(d) For the chain of 10 carbons long, the $L = 1251\ \text{pm} = 1.251 \times 10^{-9}$ m (the length of nine $C-C$ bonds). For the $n = 5$ to $n = 6$ transition, we get:

$$\Delta E = \frac{(6.626\ 08 \times 10^{-34}\ \text{J}\cdot\text{s})^2}{8(9.109 \times 10^{-31}\ \text{kg})(1.251 \times 10^{-9}\ \text{m})^2}(6^2 - 5^2)$$

$$= 4.25 \times 10^{-19}\ \text{J}$$

(e) This energy corresponds to a wavelength of 4.69×10^{-7} m or 469 nm. This falls within the visible region.

(f) A wavelength of 696 nm corresponds to an energy of 2.85×10^{-19} J being required for the promotion of an electron from the $n = 6$ to $n = 7$ level. Rearranging the equation from part (a) to solve for L we get

$$L^2 = \frac{h^2}{8m_e\Delta E}(n_2^2 - n_1^2) = \frac{(6.626\ 08 \times 10^{-34}\ \text{J}\cdot\text{s})^2}{8(9.109 \times 10^{-31}\ \text{kg})(2.85 \times 10^{-19}\ \text{J})}(7^2 - 6^2)$$

$$= 2.75 \times 10^{-18}\ \text{m}^2$$

This will give $L = 1.658 \times 10^{-9}$ m $= 1658$ pm. Since each $C - C$ bond is 139 pm long, this length corresponds to the chain of carbon atoms having 12 $C - C$.

FOCUS 2
MOLECULES

2A.1 (a) 5; (b) 4; (c) 7; (d) 3

2A.3 (a) [Ar]; (b) $[Ar]3d^{10}4s^2$; (c) $[Kr]4d^5$; (d) $[Ar]3d^{10}4s^2$

2A.5 (a) $[Ar]3d^{10}$; (b) $[Xe]4f^{14}5d^{10}6s^2$; (c) $[Ar]3d^{10}$;
(d) $[Xe]4f^{14}5d^{10}$

2A.7 (a) $[Kr]4d^{10}5s^2$; same; In^+ and Sn^{2+} lose 5p valence electrons;
(b) none; (c) $[Kr]4d^{10}$; Pd

2A.9 (a) Co^{2+}; (b) Fe^{2+}; (c) Mo^{2+}; (d) Nb^{2+}

2A.11 (a) Co^{3+}; (b) Fe^{3+}; (c) Ru^{3+}; (d) Mo^{3+}

2A.13 (a) 4s; (b) 3p; (c) 3p; (d) 4s

2A.15 (a) −2; (b) −2; (c) +1; (d) +3; (e) +2

2A.17 (a) 3; (b) 6; (c) 6; (d) 2

2A.19 (a) $[Kr]4d^{10}5s^2$; no unpaired electrons; (b) $[Kr]4d^{10}$; no unpaired
electrons; (c) $[Xe]4f^{14}5d^4$; four unpaired electrons; (d) [Kr]; no
unpaired electrons; (e) $[Ar]3d^8$; two unpaired electrons

2A.21 (a) [Ar]; no unpaired electrons; (b) [Kr]$4d^{10}5s^2$; no unpaired

electrons; (c) [Xe]; no unpaired electrons; (d) [Kr]$4d^{10}$; no unpaired

electrons

2A.23 (a) Mg_3As_2; (b) In_2S_3; (c) AlH_3; (d) H_2Te; (e) BiF_3

2A. 25

(a) $:\!\ddot{C}l\!:^-$ Tl^{3+} $:\!\ddot{C}l\!:^-$ $:\!\ddot{C}l\!:^-$ (b) $:\!\ddot{S}\!:^{2-}$ Al^{3+} $:\!\ddot{S}\!:^{2-}$ Al^{3+} $:\!\ddot{S}\!:^{2-}$

(c) Ba^{2+} $:\!\ddot{O}\!:^{2-}$

2A.27 The coulombic attraction is directly proportional to the charge on each ion

(Equation 1) so the ions with the higher charges will give the greater

coulombic attraction. The answer is therefore (b) Ga^{3+}, O^{2-}.

2A.29 The Li^+ ion is smaller than the Rb^+ ion (58 vs 149 pm). Because the

lattice energy is related to the coulombic attraction between the ions, it

will be inversely proportional to the distance between the ions (see

Equation 2). Hence the larger rubidium ion will have the lower lattice

energy for a given anion.

2B.1

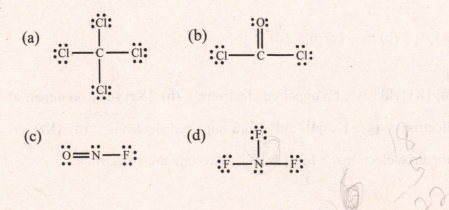

2B.3

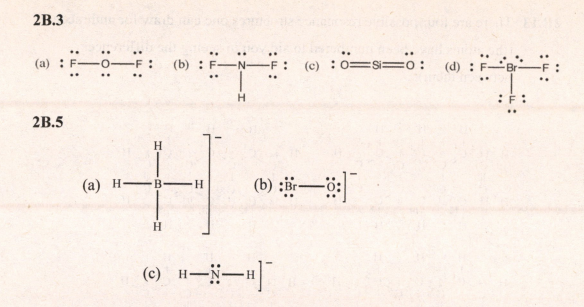

2B.5

2B.7 The structure has a total of 32 electrons; of these, 21 are accounted for by the chlorines (three Cl × seven valence electrons each); of the 11 electrons remaining, six come from the oxygen. This leaves five electrons unaccounted for; these must come from E. Therefore, E must be a member of the nitrogen family and since it is a third period element E must be phosphorous (P).

2B.9

2B.11

2B.13 There are four possible resonance structures one can draw for anthracene (the atoms have been numbered to aid you in seeing the differences between them):

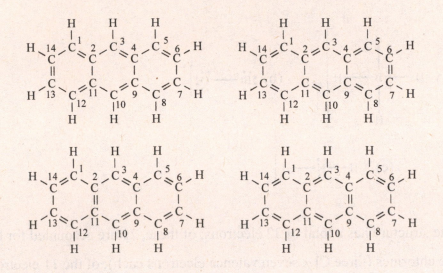

2B.15

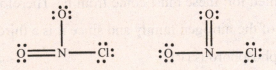

In both structures the N has a formal charge of +1 and the singly bound O has a formal charge of −1. All other atoms have formal charge of 0.

2B.17 The two resonance structures for cyclobutadiene are shown below (the carbons have been numbered for clarity):

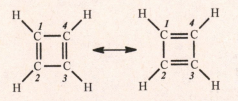

2B.19

(a) $\left[:N\!\equiv\!\overset{0}{}\overset{+1}{O}:\right]^+$

(b) $:\overset{0}{N}\!\equiv\!\overset{0}{N}:$

(c) $:\overset{-1}{C}\!\equiv\!\overset{+1}{O}:$

(d) $\left[:\overset{-1}{C}\!\equiv\!\overset{-1}{C}:\right]^{2-}$

(e) $\left[:\overset{-1}{C}\!\equiv\!\overset{0}{N}:\right]^-$

2B.21 Two possible structures for hypochlorous acid are:

$$H\!-\!\overset{+1}{\underset{..}{\overset{..}{Cl}}}\!-\!\overset{-1}{\underset{..}{\overset{..}{O}}}: \qquad H\!-\!\overset{0}{\underset{..}{\overset{..}{O}}}\!-\!\overset{0}{\underset{..}{\overset{..}{Cl}}}:$$

Based on formal charge, the structure on the right is the most likely.

2B.23

(a)

$\overset{0}{\underset{..}{\overset{..}{O}}}\!=\!\overset{0}{Cl}\!-\!\overset{0}{\underset{..}{\overset{..}{O}}}:$ with $\overset{0}{:\underset{..}{O}:}$ below and H

$\overset{-1}{:\underset{..}{\overset{..}{O}}}\!-\!\overset{+2}{Cl}\!-\!\overset{0}{\underset{..}{\overset{..}{O}}}:$ with $:\underset{-1}{\overset{..}{O}}:$ below and H

lower energy

(b)

$\overset{0}{\underset{..}{\overset{..}{O}}}\!=\!\overset{0}{C}\!=\!\overset{0}{\underset{..}{\overset{..}{S}}}$

$\overset{-1}{:\underset{..}{\overset{..}{O}}}\!-\!\overset{0}{C}\!\equiv\!\overset{+1}{S}:$

lower energy

(c)

$H\!-\!\overset{0}{C}\!\equiv\!\overset{0}{N}:$

$H\!-\!\overset{-1}{\underset{..}{C}}\!=\!\overset{+1}{N}:$

lower energy

2C.1 Radicals are species with an unpaired electron, therefore only (b) and (c) are radicals since they have an odd number of electrons. (a) and (d) have an even number of electrons allowing Lewis structures to be drawn with all electrons paired.

2C.3 (a) The periodate ion has one Lewis structure that obeys the octet rule:

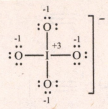

The formal charge of at I can be reduced to from +3 to 0 by including three double-bond contributions, thereby giving rise to four resonance forms.

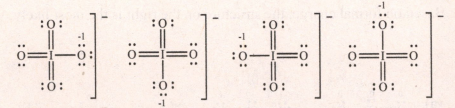

(b) The hydrogen phosphate ion has one Lewis structure that obeys the octet rule (the first structure shown below). The inclusion of one double bond to oxygen lowers the formal charge at P from +1 to 0. There are three resonance forms that include this contribution.

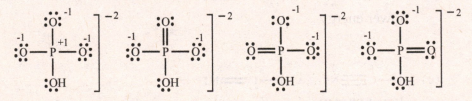

(c) There is one Lewis structure that obeys the octet rule shown below at the left. The formal charge at chlorine can be reduced to +1 by including one double bond contribution. The formal charge can be reduced to 0 if there are two double bond contributions. These contributions give rise to the following resonance structures.

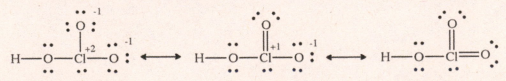

(d) The arsenate ion has one Lewis structure that obeys the octet rule.

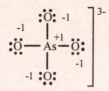

Just as in part (a), including one double bond to oxygen lowers the formal charge at As from +1 to 0. There are four resonance forms that include this contribution.

$$\begin{bmatrix} \ddots \overset{:O:}{\underset{||}{}} \\ :\overset{..}{O} - \overset{|}{As} - \overset{..}{O}: \,^{-1} \\ ^{-1} :\overset{..}{O}: \end{bmatrix}^{3-} \quad \begin{bmatrix} :\overset{..}{O}:^{-1} \\ \overset{..}{O} = \overset{|}{As} - \overset{..}{O}:^{-1} \\ ^{-1} :\overset{..}{O}: \end{bmatrix}^{3-} \quad \begin{bmatrix} :\overset{..}{O}:^{-1} \\ ^{-1}:\overset{..}{O} - \overset{|}{As} - \overset{..}{O}:^{-1} \\ || \\ :\overset{..}{O}: \end{bmatrix}^{3-} \quad \begin{bmatrix} :\overset{..}{O}:^{-1} \\ ^{-1}:\overset{..}{O} - \overset{|}{As} = \overset{..}{O} \\ ^{-1}:\overset{..}{O}: \end{bmatrix}^{3-}$$

2C.5   The Lewis structures are

(a) :C̈l — Ö:

(b) :C̈l — Ö — Ö — C̈l:

(c) Ö = N — Ö — C̈l: with :Ö: above N

Radicals are species with an unpaired electron, therefore only (a) is a radical.

2C.7

(a) $\left[:\overset{..}{\underset{..}{C}l} - \overset{..}{I} - \overset{..}{\underset{..}{C}l}: \right]^+$

I has 2 bonding pairs and 2 lone pairs

(b) $\left[\begin{array}{c} :\overset{..}{C}l: \\ :\overset{..}{\underset{..}{C}l} \diagdown \overset{|}{\underset{\diagup}{I}} \diagdown \\ :\overset{..}{\underset{..}{C}l}: \quad \overset{..}{\underset{..}{C}l}: \end{array} \right]^-$

I has 4 bonding pairs and 2 lone pairs

(c) :C̈l: on top, :C̈l — I with :C̈l: below

I has 3 bonding pairs and 2 lone pairs

(d) :C̈l: on top, :C̈l ∖ I — C̈l: with :C̈l: :C̈l: below

I has 5 bonding pairs and 1 lone pair

2C.9

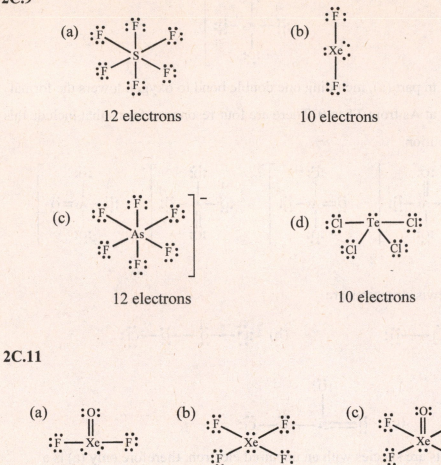

(a)   **12 electrons**

(b)   **10 electrons**

(c)   **12 electrons**

(d)   **10 electrons**

2C.11

(a)   **2 lone pair**

(b)   **2 lone pair**

(c)   **1 lone pair**

2C.13 (a) In $BeCl_2$, there are 4 electrons around the central beryllium:

(b) In ClO_2 there are an odd number of electrons around the central chlorine:

:O̤—C̤l—O̤:

2C.15

(a)

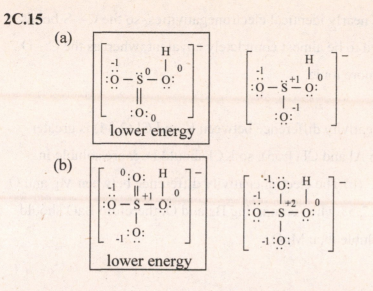

(b)

2C.17 (a) In the first structure, the formal charges at Xe and F are 0, whereas, in the second structure, Xe is −1, one F is 0, and the other F is +1. The first structure is favored on the basis of formal charges. (b) In the first structure, all of the atoms have formal charges of 0, whereas, in the second structure, one O atom has a formal charge of +1 and the other O has a formal charge of −1. The first structure is thus preferred.

2D.1 In (1.78) < Sn (1.96) < Sb (2.05) < Se (2.55)

2D.3 $BaBr_2$ would have bonds that are primarily ionic. The electronegativity difference is greater between Ba and Br than between B or Be and Br, making the Ba—Br bond more ionic.

2D.5 (a) The bond in HCl would be more ionic. The electronegativity difference is greater between H and Cl than between H and I, making the H—Cl bond more ionic.

(b) The bonds in CF_4 would be more ionic. The electronegativity difference is greater between C and F than between C and H, making the C—F bonds more ionic.

(c) C and S have nearly identical electronegativities, so the C—S bonds would be expected to be almost completely covalent, whereas the C—O bonds would be more ionic.

2D.7 (a) The electronegativity difference between K and Cl (2.34) is greater than that between Al and Cl (1.55), so KCl should be more soluble in water than $AlCl_3$. (b) The electronegativity difference between Mg and O is 2.13 while it is 2.55 when comparing Ba and O, therefore BaO should be more water soluble than MgO.

2D.9 $Rb^+ < Sr^{2+} < Be^{2+}$; smaller, more highly charged cations have greater polarizing power. The ionic radii are 149 pm, 116 pm, 27 pm, respectively.

2D.11 $O^{2-} < N^{3-} < Cl^- < Br^-$; the polarizability increases as the ion gets larger and less electronegative. The ionic radii for these species are 140 pm, 171 pm, 181 pm, 196 pm, respectively.

2D.13 (a) $CO_3^{2-} > CO_2 > CO$

CO_3^{2-} will have the longest C—O bond length. In CO there is a triple bond and in CO_2 the C—O bonds are double bonds. In carbonate, the bond is an average of three Lewis structures in which the bond is double in one form and single in two of the forms. We would thus expect the bond order to be approximately 1.3. Because the bond length is inversely related to the number of bonds between the atoms, we expect the bond length to be longest in carbonate.

(b) $SO_3^{2-} > SO_2 \sim SO_3$

Similar arguments can be used for these molecules as in part (a). In SO_2 and SO_3, the Lewis structures with the lowest formal charge at S have double bonds between S and each O. In the sulfite ion, however, there are three Lewis structures that have a zero formal charge at S. Each has one S—O double bond and two S—O single bonds. Because these S—O bonds would have a substantial amount of single bond character, they would be expected to be longer than those in SO_2 or SO_3. This is consistent with the experimental data that show the S—O bond lengths in SO_2 and SO_3 to be 143 pm, whereas those in SO_3^{2-} range from about 145 pm to 152 pm depending on the compound.

(c) $CH_3NH_2 > CH_2NH > HCN$

The C—N bond in HCN is a triple bond, in CH_2NH it is a double bond, and in CH_3NH_2 it is a single bond. The C—N bond in the last molecule would, therefore, be expected to be the longest.

2D.15 Of CF_4, CCl_4, and CBr_4, CF_4 is predicted to have the strongest C—X bond; it is the shortest of the three bonds. Note that electronegativity and polarity arguments would predict the C—F bond to be the weakest.

2D.17 (a) The C—O bonds in carbon dioxide are double bonds. The covalent radius for doubly bonded carbon is 67 pm and that of O is 60 pm. Thus we predict the C=O in CO_2 to be ca. 127 pm. The experimental bond length is 116.3 pm.

(b) The C—O bond is a double bond so it would be expected to be the same as in (a), 127 ppm. This is the experimentally found value. The C—N bonds are single bonds and so one might expect the bond distance to be the sum of the single bond C radius and the single bond N radius (77 plus 75 pm) which is 152 pm. However, because the C atom is involved in a multiple bond, its radius is actually smaller. The sum of that radius (67 pm) and the N single bond radius gives 142 pm, which is close to the experimental value of 133 pm.

(c) The O—Cl bond is a single bond so we expect the bond distance to be 74 pm + 98 pm or 172 pm. The experimental value is 169 pm.

(d) The N—Cl bond is a single bond so we predict the bond distance to be 75 pm + 98 pm or 173 pm. However, because the N atom is involved in a multiple bond with oxygen, its radius should actually be smaller. The sum of that radius (60 pm) and the Cl single bond radius gives a predicted bond length of 158 pm.

NOTE: Parts (b) and (d) of this question raise an interesting point concerning the radius to use for the A—X single bond in Y=A—X. The double bond that A forms with Y shrinks the A atom down to a smaller size, and even though the A—X link is a single bond, there is a strong argument for using the double-bond radius of A in predicting the A—X bond length. In many cases this assumption compares reasonably well with the experimental values seen, suggesting that the A atom has indeed been shrunk by its formation of a neighboring double bond. A caveat does warrant mention here: this assumption is based on a high degree of covalent character being present in the A—X bond; as the amount of ionic character increases, so does the bond length. This is reflected in the actual experimental value of 198 pm seen for the N—Cl bond. In fact. NOCl behaves as if it was actually O=N$^+$— C$^-$ in many of the reactions it undergoes, demonstrating that there exists a high degree of ionic character not accounted for in the simple model used here.

2D.19 (a) 77 pm + 58 pm = 135 pm;   (b) 111 pm + 58 pm = 169 pm;

(c) 141 pm + 58 pm = 199 pm. Bond distance increases with atomic size going down Group 14/IV.

2E.1   (a) Must have lone pairs;   (b) May have lone pairs.

2E.3

(a)   :N≡C—H   linear

(b)

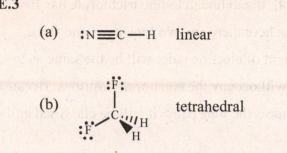

tetrahedral

2E.5   (a) The shape is angular; the electron pair on the central Cl atom results in a trigonal planar arrangement. (b) The bond angle will be slightly less than 120°.

2E.7   (a) The shape of the thionyl chloride molecule is trigonal pyramidal.

(b) The O—S—Cl angles are identical. The lone electron pair repels the bonded electron pairs equally thus, all O—S—Cl bond angles are compressed equally. (c) The expected bond angle is slightly less than 109.5°.

2E.9   (a) T-shaped;   (b) slightly less than 90°

2E.11

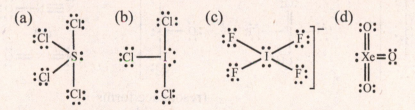

(a) The sulfur atom will have five pairs of electrons about it: one nonbonding pair and four bonding pairs to chlorine atoms. The arrangement of electron pairs will be trigonal bipyramidal; the nonbonding pair of electrons will prefer to lie in an equatorial position, because in that location the e-e repulsions will be lowest. The actual structure is described as a seesaw. AX_4E.

(b) Like the sulfur atom in (a), the iodine in iodine trichloride has five pairs of electrons about it, but here there are two lone pairs and three bonding pairs. The arrangement of electron pairs will be the same as in (a), and again the lone pairs will occupy the equatorial positions. Because the name of the molecule ignores the lone pairs, it will be classified as T-shaped. AX_3E_2.

(c) There are six pairs of electrons about the central iodine atom in IF_4^-. Of these, two are lone pairs and four are bonding pairs. The pairs will be placed about the central atom in an octahedral arrangement with the lone pairs opposite each other. This will minimize repulsions between them. The name given to the structure is square planar. AX_4E_2.

(d) In determining the shape of a molecule, double bonds count the same as single bonds. The XeO_3 structure has four "objects" about the central Xe atom: three bonds and one lone pair. These will be placed in a tetrahedral arrangement. Because the lone pair is ignored in naming the molecule, it will be classified as trigonal pyramidal. AX_3E.

2E.13   The Lewis structures are:

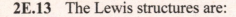

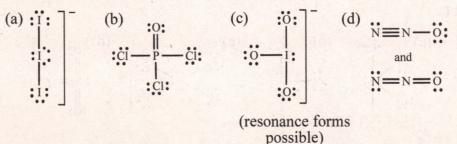

(resonance forms
possible)

(a) The I_3^- molecule is predicted to be linear, so the I—I—I angle should equal 180°. AX_2E_3.

(b) The $POCl_3$ molecule is tetrahedral. All bond angles should be 109.5°. AX_4.

(c) The shape of IO_3^- will be a trigonal pyramid, so the O—I—O bond angles should be less than 109.5°. AX_3E.

(d) The structure of N_2O is linear with a bond angle of 180°. AX_2.

2E.15 The Lewis structures are

(a) :F:
:Cl—C—F:
:F:

(b) :Cl:
:Cl—Te—Cl:
:Cl:

(c) :O:
:F—C—F:

(d) H
H—C—H
H

(a) The shape of CF_3Cl is tetrahedral; all halogen—C—halogen angles should be approximately 109.5°. AX_4;

(b) $TeCl_4$ molecules will be seesaw-shaped with Cl—Te—Cl bond angles of approximately 90° and 120°. AX_4E;

(c) COF_2 molecules will be trigonal planar with F—C—F and O—C—F angles of 120°. AX_3;

(d) CH_3^- ions will be trigonal pyramidal with H—C—H angles of slightly less than 109.5°. AX_3E.

2E.17 (a) slightly less than 120°; (b) 180°; (c) 180°; (d) slightly less then 109.5°

2E.19 (a) tetrahedral, bond angle of 109.5°

(b) tetrahedral about the carbon atoms (109.5°); C—Be—C angle of 180°

(c) angular, H—B—H angle slightly less than 120°

(d) angular, Cl—Sn—Cl angle slightly less than 120°

2E.21

(a) H—C—H and H—C—C angles of 120°.

(b) Linear, 180°

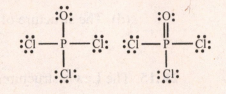

(c) Tetrahedral, 109.5° (Note: Both are acceptable Lewis structures; the presence of the double bond makes the second molecule the more stable Lewis structure.)

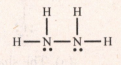

(d) The arrangement of atoms about each N is trigonal pyramidal giving H—N—H and H—N—N bond angles of approximately 107°.

2E.23

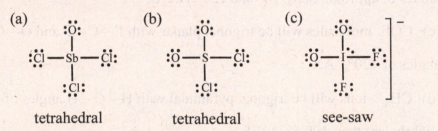

(a) tetrahedral   (b) tetrahedral   (c) see-saw

Note: Other acceptable Lewis structures are possible (those with double bonds between the central atom and oxygen, which are more stable structures due to the lowering of formal charge) but they do not change the shape of the molecule

2E.25 The Lewis structures are

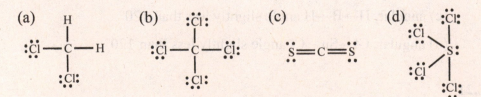

(a)   (b)   (c)   (d)

Molecules (a) and (d) are polar; (b) and (c) are nonpolar.

2E.27 (a) pyridine: polar; (b) ethane: nonpolar; (c) trichloromethane: polar

2E.29 (a) Of the three forms, **1** and **2** are polar; only **3** is nonpolar. This is because the C—Cl bond dipoles arc pointing in exactly opposite directions in **3**. (b) The dipole moment for **1** would be the largest because the C—Cl bond vectors are pointing most nearly in the same direction in **1** (60° apart) whereas in **2** the C—Cl vectors point more away from each other (120°), giving a larger cancellation of dipole.

2F.1 (a) sp^3, orbitals oriented toward corners of a tetrahedron (109.5° apart);

(b) sp, orbitals oriented directly opposite to each other (180° apart);

(c) sp^3d^2, orbitals oriented toward the corners of an octahedron (interorbital angles of 90° and 180°);

(d) sp^2, orbitals oriented toward the corners of an equilateral triangle trigonal planar array (angles $=120°$); trigonal planar

2F.3 (a) 2 σ bonds. 0 π bonds   (b) 2 σ bonds. 1 π bonds, there is also a resonance structure that includes 2 σ bonds. 2 π bonds.

2F.5 (a) sp;   (b) sp^2;   (c) sp^3;   (d) sp^3

2F.7 (a) sp^2; (b) sp^3; (c) sp^3d; (d) sp^3

2F.9 (a) sp^3; (b) sp^3d^2; (c) sp^3d; (d) sp^3

2F.11 The Lewis structure of tetrahedral P_4 is:

(a) Each phosphorous is attached to three other phosphorous atoms and a single lone pair, therefore its hybridization is sp^3.

(b) While each P within this structure is polar (due to the presence of a single lone pair on each phosphorous), P_4 is nonpolar due to the 3D orientations of those lone pairs.

2F.13

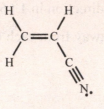

The first two carbons (CH_2 and CH) are sp^2 hybridized with H—C—H and C—C—H angles of 120°. The third carbon (bonded to N) is sp hybridized with a C—C—N angle of 180°.

2F.15 As the s-character of a hybrid orbital increases, the bond angle increases.

2F.17 In formaldehyde, both the C and the O are sp^2 hybridized. The H—C—O bond angle is 120°; the molecule has three sigma bonds (one each connecting the two H and O to the C) and one pi bond (between the C and the O).

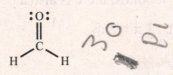

2F.19 Atomic orbitals a and b are mutually orthogonal if

$\int a \cdot b \, d\tau = 0$ (assuming $a \neq b$) where the integration is over all space.

Furthermore, an orbital, a, is normalized if $\int a^2 \, d\tau = 1$.

In this problem, the two hybrid orbitals are:

$h_1 = s + p_x + p_y + p_z$ and $h_2 = s - p_x + p_y - p_z$. Therefore, to show these

two orbitals are orthogonal we must show $\int h_1 h_2 \, d\tau = 0$.

$$\int h_1 h_2 \, d\tau = \int (s + p_x + p_y + p_z)(s - p_x + p_y - p_z) d\tau =$$

$$\int (s^2 - sp_x + sp_y - sp_z + sp_x - p_x^2 + p_x p_y - p_x p_z + sp_y -$$

$$p_x p_y + p_y^2 - p_y p_z + sp_z - p_z p_x + p_z p_y - p_z^2)d\tau$$

Of course, this integral of a sum may be written as a sum of integrals:

$$\int s^2 d\tau - \int sp_x d\tau + \int sp_y d\tau - \int sp_z d\tau + \dots$$

Because the hydrogen wavefunctions are mutually orthogonal, the members of this sum which are integrals of a product of two different wavefunctions are zero. Therefore, this sum of integrals simplifies to:

$$\int s^2 d\tau - \int p_x^2 d\tau + \int p_y^2 d\tau - \int p_z^2 d\tau = 1 - 1 + 1 - 1 = 0$$

(recall that the integral of the square of a normalized wavefunction is one.)

2F.21 We are given: $\lambda = -\dfrac{\cos\theta}{\cos^2(\frac{1}{2}\theta)}$. In the H_2O molecule, the bond angle is

104.5°. Therefore, $\lambda = 0.67$ and the hybridization is $sp^{0.67}$.

2G.1 The molecular orbital diagrams are as follows (only the valence electrons are shown):

(a)   (b)   (c)

(a) Li_2 $BO = \frac{1}{2}(2) = 1$; diamagnetic, no unpaired electrons

(b) Li_2^+ $BO = \frac{1}{2}(1) = \frac{1}{2}$; paramagnetic, one unpaired electron

(c) Li_2^- $BO = \frac{1}{2}(2-1) = \frac{1}{2}$; paramagnetic, one unpaired electron

2G.3 (a) (i) $(\sigma_{2s})^2(\sigma_{2s}{}^*)^2(\sigma_{2p})^2(\pi_{2p})^4(\pi_{2p}{}^*)^4(\sigma_{2p}{}^*)^1$

(ii) $(\sigma_{2s})^2(\sigma_{2s}{}^*)^2(\sigma_{2p})^2(\pi_{2p})^4(\pi_{2p}{}^*)^3$

(iii) $(\sigma_{2s})^2(\sigma_{2s}{}^*)^2(\sigma_{2p})^2\,(\pi_{2p})^4\,(\pi_{2p}{}^*)^4(\sigma_{2p}{}^*)^2$

(b) (i) 0.5;   (i) 1.5;   (iii) 0

(c) (i) and   (ii) are paramagnetic, with one unpaired electron each

(d) σ for (i) and (iii), π for (ii)

2G.5   The charge on C_2^{n-} is -2 and the bond order is 3

$$\left(BO = \frac{1}{2}(2+4+2-2) = 3 \right).$$

2G.7 HeH$^-$ would not be expected to exist because its bond order is 0;

BO = ½ (2 – 2) = 0. HeH$^+$ on the other hand would have a bond order of 1 and

thus be lower in energy.

2G.9   (a) The energy level diagram for N_2 is as follows:

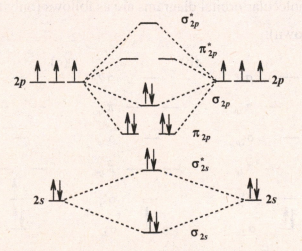

(b) The oxygen atom is more electronegative, which will make its
orbitals lower in energy than those of N. The revised energy-level diagram
is shown below. This will make all of the bonding orbitals closer to O than
to N in energy and will make all the antibonding orbitals closer to N than
to O in energy.

Energy level diagram for NO⁺

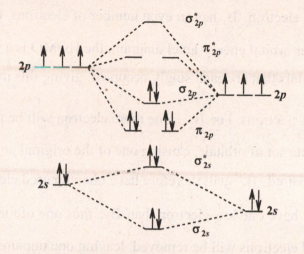

Orbitals on N   Orbitals on O

(c) The electrons in the bonding orbitals will have a higher probability of being at O because O is more electronegative and its orbitals are lower in energy.

2G.11 (a) B_2 (6 valence electrons): $(\sigma_{2s})^2 (\sigma_{2s}^*)^2 (\pi_{2p})^2$, bond order = 1

(b) Be_2 (4 valence electrons): $(\sigma_{2s})^2 (\sigma_{2s}^*)^2$, bond order = 0

(c) F_2 (14 valence electrons): $(\sigma_{2s})^2 (\sigma_{2s}^*)^2 (\sigma_{2p})^2 (\pi_{2p})^4 (\pi_{2p}^*)^4$, bond order = 1

2G.13

$$CO: \left(\sigma_{2s}\right)^2 \left(\sigma_{2s}^*\right)^2 \left(\pi_{2p}\right)^4 \left(\sigma_{2p}\right)^2; \text{ bond order} = 3$$

$$CO^+: \left(\sigma_{2s}\right)^2 \left(\sigma_{2s}^*\right)^2 \left(\pi_{2p}\right)^4 \left(\sigma_{2p}\right)^1; \text{ bond order} = 2.5$$

Due to the higher bond order for CO, it should form a stronger bond and therefore have the higher bond enthalpy

2G.15 (a) – (c) All of these molecules possess unpaired electrons and therefore are paramagnetic. B_2^- and B_2^+ have an odd number of electrons and must, therefore, have at least one unpaired electron; indeed, both have only one unpaired electron. B_2 has an even number of electrons, but in its molecular orbital energy level diagram, the HOMO is a degenerate set of π_{2p} orbitals that are each singly occupied, giving this molecule two unpaired electrons. For B_2^-, one more electron will be placed in this degenerate set of orbitals, causing one of the original unpaired electrons to now be paired. B_2^- will therefore have one unpaired electron. Likewise, B_2^+ will have one less electron than B_2; thus one of the originally unpaired electrons will be removed, leaving one unpaired electron in this molecule as well.

2G.17 (a) F_2 with 14 valence electrons has a valence electron configuration of $(\sigma_{2s})^2(\sigma_{2s}{}^*)^2(\sigma_{2p})^2(\pi_{2p})^4(\pi_{2p}{}^*)^4$ with a bond order of 1. After forming F_2^- from F_2, an electron is added into a $\sigma_{2p}{}^*$ orbital. The addition of an electron to this antibonding orbital will result in a reduction of the bond order to 1/2 (See Problem 2G.3). F_2 will have the stronger bond. (b) B_2 will have an electron configuration of $(\sigma_{2s})^2(\sigma_{2s}{}^*)^2(\pi_{2p})^2$ with a bond order of 1. Removing one electron to form B_2^+ will eliminate one electron in the bonding orbitals, creating a bond order of 1/2. B_2 will have the stronger bond.

2G.19 C_2^+ $(\sigma_{2s})^2(\sigma_{2s}{}^*)^2(\pi_{2p})^3$;

C_2 $(\sigma_{2s})^2(\sigma_{2s}{}^*)^2(\pi_{2p})^4$;

C_2^- $(\sigma_{2s})^2(\sigma_{2s}{}^*)^2(\pi_{2p})^4(\sigma_{2p})^1$;

C_2^- is expected to have the lowest ionization energy because its electron is lost from a higher energy MO (σ_{2p}) than either C_2^+ or C_2 (π_{2p_y}).

2G.21 Given the overlap integral $S = \int \Psi_{A1s} \Psi_{B1s} d\tau$, the bonding orbital

$\Psi = \Psi_{A1s} + \Psi_{B1s}$, and the fact that the individual atomic orbitals are normalized, we are asked to find the normalization constant N which will normalize the bonding orbital Ψ such that:

$$\int N^2 \Psi^2 d\tau = N^2 \int (\Psi_{A1s} + \Psi_{B1s})^2 d\tau = 1$$

$$N^2 \int (\Psi_{A1s} + \Psi_{B1s})^2 d\tau = N^2 \int (\Psi_{A1s}^2 + 2\Psi_{A1s}\Psi_{B1s} + \Psi_{B1s}^2) d\tau$$

$$= N^2 \left(\int \Psi_{A1s}^2 d\tau + 2 \int \Psi_{A1s}\Psi_{B1s} d\tau + \int \Psi_{B1s}^2 d\tau \right)$$

Given the definition of the overlap integral above and the fact that the individual orbitals are normalized, this expression simplifies to:

$$N^2(1 + 2S + 1) = 1$$

Therefore, $N = \sqrt{\dfrac{1}{2 + 2S}}$

2.1 Atoms for which no formal charge is shown have a charge of zero:

(a)

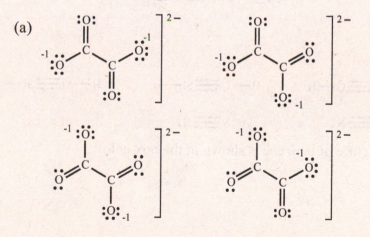

(b) $\left[\overset{+1}{\ddot{Br}} = \ddot{\underset{..}{O}} \right]^+$ (c) $\left[:\overset{-1}{C} \equiv \overset{-1}{C}: \right]^{2-}$

2.3   Iron(III) chloride would be expected to have the greater lattice energy.
The iron(III) ion is higher in charge and smaller in size then the iron(II)
ion, thus the attraction between it and the chloride ion will be greater and
hence the lattice energy will be greater.

2.5

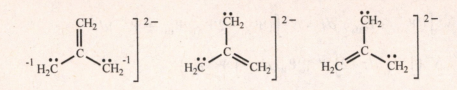

2.7   The Lewis structure for N_5^+ is possible without having multiple formal
charges present. The first of the three resonance structures shown below is
the most important Lewis structure, since in it no two like charges are near
each other.

$$:N\equiv\overset{+1}{N}-\overset{-1}{\underset{\cdot\cdot}{N}}-\overset{+1}{N}\equiv N: \qquad \overset{-1}{\underset{\cdot\cdot}{N}}=\overset{+1}{N}=\overset{}{\underset{\cdot}{N}}-\overset{+1}{N}\equiv N:$$

$$:N\equiv\overset{+1}{N}-\overset{\cdot\cdot}{\underset{}{N}}=\overset{+1}{N}=\overset{-1}{\underset{\cdot\cdot}{N}}$$

2.9

(a) $H-C\equiv C-H$ $H-C\equiv Si-H$ $H-Si\equiv Si-H$

$H-C\equiv N:$ $:N\equiv N:$

(b) The structure of benzene is shown in the box below:

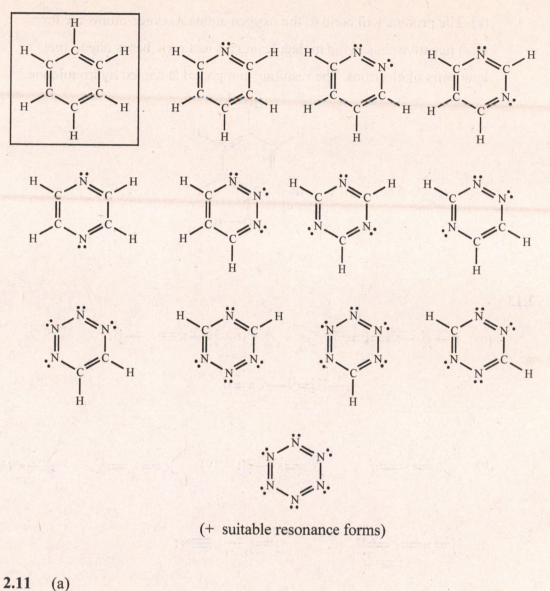

(+ suitable resonance forms)

2.11   (a)

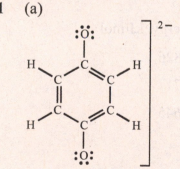

(b) All the atoms have formal charge 0 except the two oxygen atoms, which are -1. The negative charge is most likely to be concentrated at the oxygen atoms.

(c) The protons will bond to the oxygen atoms. Oxygen atoms are the most negative sites in the molecule and act as Lewis bases due to their lone pairs of electrons. The resulting compound is named hydroquinone.

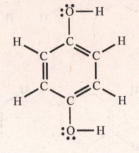

2.13

(a) H—Ö—N≡C≡Ö   H—Ö—N≡C—Ö: ⁺¹ ⁻¹

     H—Ö≡N—C̈≡Ö ⁺¹ ⁻¹

(b) H₂C=S̈=Ö   H₂C=S̈—Ö: ⁺¹ ⁻¹   (c) H₂C=N=N̈ ⁺¹ ⁻¹   H₂C—N≡N: ⁻¹ ⁺¹

(d) Ö=N=N̈ ⁺¹ ⁻¹   Ö=N—C≡N:

2.15

	d(K – X)	Lattice Energy, kJ/mol
Fluoride	271 pm	826
Chloride	319 pm	717
Iodide	358 pm	645

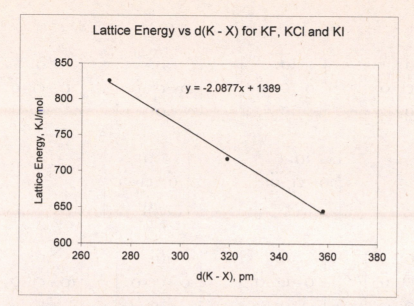

The data fit a straight line with a correlation coefficient of greater than 99%.   (b) From the equation derived for the straight line relationship Lattice Energy $= -1.984\, d_{M-X} + 1356$ and the value of $d_{K-Br} = 338$ pm, we can estimate the lattice energy of KBr to be 693 kJ·mol^{-1}.

(c) The experimental value for the lattice energy for KBr is 689 kJ·mol^{-1}, so the agreement is very good.

2.17

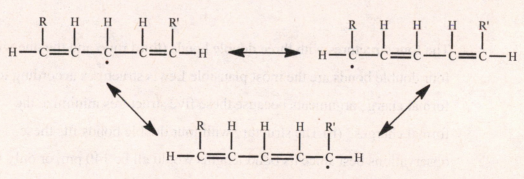

2.19   (a) I: Tl_2O_3; II: Tl_2O; (b) +3; +1; (c) $[Xe]4f^{14}5d^{10}$; $[Xe]\,4f^{14}5d^{10}6s^2$; (d) Because compound II has a lower melting point, it is probably more covalent, which is consistent with the fact that the +3 ion is more polarizing.

2.21 (a)

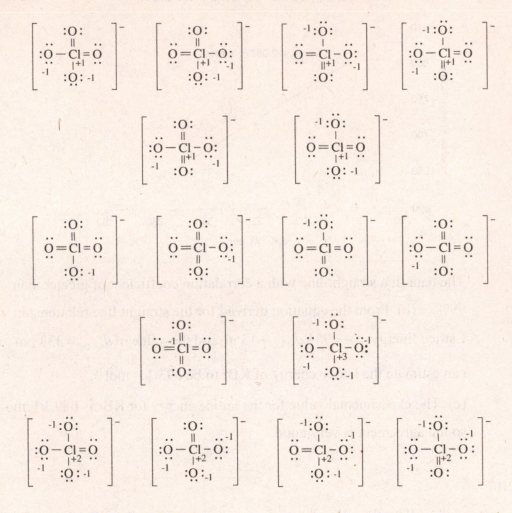

The four structures with three double bonds (third row) and the one with four double bonds are the most plausible Lewis structures according to formal charge arguments because these five structures minimize the formal charges. (b) The structure with four double bonds fits these observations best since its bond lengths would all be 140 pm, or only 4 pm shorter than the observed length. However, the four structures with three double bonds also fit because, if the double bonds are delocalized by resonance, we can estimate the average bond length to be

$\frac{1}{4}(170 \text{ pm}) + \frac{3}{4}(140 \text{ pm}) = 147.5 \text{ pm}$, or just 3.5 pm longer than observed.

(c) +7; The structure with all single bonds fits this criterion best.

(d) Approaches (a) and (b) are consistent but approach (c) is not. This result is reasonable because oxidation numbers are assigned by assuming ionic bonding.

2.23   (a) Compare the length of a S—S bond to that of a Cl—Cl bond. From Table 2D.3, the Cl—Cl bond length is 199 pm; here the S—S bond length in S_2F_2 is reported to be 190 pm. From Table 2D.3 it can be determined that, on average, an X—Y bond is approximately 20 pm longer that a X=Y bond; this suggests that the S—S bond (which is 9 pm shorter than a Cl—Cl single bond) has some double bond character.

(b) and (c) The Lewis structures for the two possible S_2F_2 are:

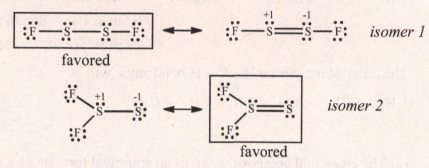

If resonance is occurring, then one would expect that the S—S bond length is indeed between a single and a double bond in length.

2.25   (a) The CN bond in CH_3NH_2 would be expected to be longer. The C—N bond in HCN is a triple bond, in CH_3NH_2 it is a single bond, thus it would be expected to be longer.

(b) The PF bond in PF_3 would be expected to be longer. Bond distance increases with atomic size. Phosphorous is larger then nitrogen, thus it would be expected to have the longer bond.

2.27   (a) The Lewis structures are:

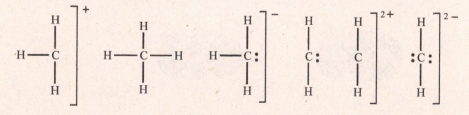

(b) All of these species are expected to be diamagnetic. None are radicals.

(c) The predicted bond angles in each species based upon the Lewis structure and VSEPR theory will be

CH_3^+	AX_3	trigonal planar	120°
CH_4	AX_4	tetrahedral	109.5°
CH_3^-	AX_3E	pyramidal	slightly less than 109.5°
CH_2	AX_2E	angular	slightly less than 120°
CH_2^{2+}	AX_2	linear	180°
CH_2^{2-}	AX_2E_2	angular	less than 109.5°, more so than CH_3^- due to the presence of two lone pairs

The order of increasing H—C—H bond angle will be

$$CH_2^{2-} < CH_3^- < CH_4 < CH_2 < CH_3^+ < CH_2^{2+}$$

2.29 (a) The elemental composition gives an empirical formula of CH_4O, which agrees with the molar mass. There is only one reasonable Lewis structure; this corresponds to the compound methanol. All of the bond angles about carbon should be 109.5°. The bond angles about oxygen should be close to 109.5° but will be somewhat less, due to the repulsions by the lone pairs. (b) Both carbon and oxygen are sp^3 hybridized. (c) The molecule is polar.

2.31   <u>Bonding</u>   <u>Anti bonding</u>

(a)

(b)

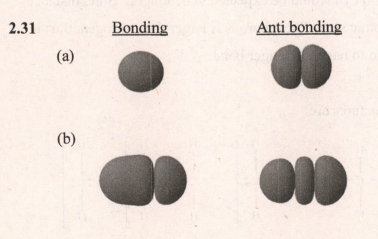

(c) The bonding and antibonding orbitals for HF appear different because a p-orbital from the F atom is used to construct bonding and antibonding orbitals whereas in the H₂ molecule s-orbitals on each atom are used to construct bonding and antibonding orbitals.

2.33   (a) The expected molecular orbital diagram for CF is

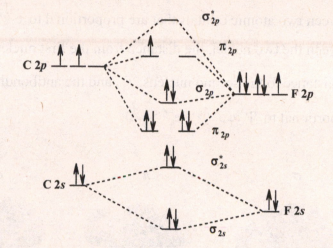

The bond order for the neutral species is 2.5 because one electron occupies a π_{2p}*-orbital. Adding an electron to form CF⁻ will reduce the bond order by 1/2 to 2, while removing an electron from form CF⁺ will increase the bond order to 3. Because bond lengths increase as bond order decreases, the C—F bond length varies in these molecules in the following manner: CF⁺ < CF < CF⁻. (b) The CF⁺ ion will be diamagnetic but both CF and CF⁻ will have unpaired electrons (one in the case of CF and two in the case of CF⁻).

2.35   The Lewis structure of borazine is:

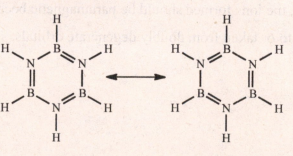

It is nearly identical to that of benzene. It is obtained by replacing alternating C atoms in the benzene structure with B and N (although not shown, each B has a formal charge of -1 and each N has a formal charge of +1). The orbitals at each B and N atom will be sp^2 hybridized.

2.37 The antibonding molecular orbital is obtained by taking the difference between two atomic orbitals that are proportional to e^{-r/a_o}. Halfway between the two nuclei, the distance from the first nucleus, r_1, is equal to the distance to the second nucleus, r_2, and the antibonding orbital is proportional to $\Psi \propto e^{-r/a_o} - e^{-r/a_o} = 0$.

2.39 (a) σ π δ

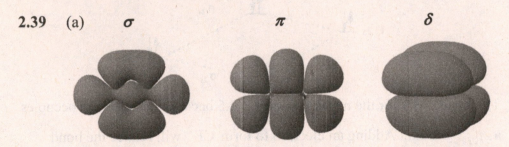

(b) The overlap between the orbitals will decrease as one goes from σ to π to δ, so we expect the bond strengths to decline in the same order.

2.41 The effect of changes (a) and (b) will be similar. The overall bond order will change. In the first case, electrons will be removed from π-orbitals so that the net π-bond order will drop from three to two. The same thing will happen in (b), but because two electrons are added to antibonding orbitals, a net total of one π-bond will be broken. Based on this simple model, the ions formed should be paramagnetic because the electrons are added to or taken from doubly degenerate orbitals.

2.43   (a) The Lewis structure of benzyne is

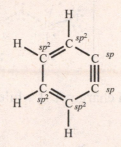

(b) Benzyne would be highly reactive because the two carbon atoms that are *sp* hybridized are constrained to have a very strained structure compared with what their hybridization would like to adopt—namely a linear arrangement. Instead of 180° angles at these carbon atoms, the angles by necessity of being in a six-membered ring are constrained to be close to 120°. A possibility that allows the carbon atoms to adopt more reasonable angles is the formation of a diradical:

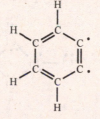

2.45

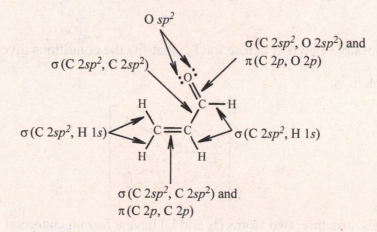

All atoms in this molecule have a formal charge of zero.

2.47   (a) A possible Lewis structure for $[Sb_2F_7]^-$ based on the information provided is as follows:

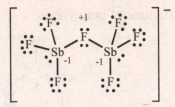

Each Sb has a formal charge of -1 while the bridging F has a formal charge of +1.

(b) Each Sb atom in this structure has a hybridization of sp^3d.

(NOTE: the crystal structure of this ion has been done for both the potassium and the cesium salt; to read more about this controversial ion, see S.H. Mastin and R.R. Ryan (1971) "Crystal Structure of KSb_2F_7. On the Existence of the $[Sb_2F_7]^-$ ion", *Inorganic Chemistry*, **10**, 1757.

2.49 A possible Lewis structure for $[Bi_2Cl_4]^{2-}$ is as follows:

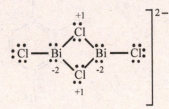

Formal charges for the bridging Cl atoms and the two Bi are shown.

2.51 A possible Lewis structure for I_5^- that fits the conditions given in the problem is:

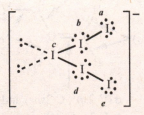

In this structure, two atoms (I_b and I_d) have a formal charge of −1, one atom (I_c) has a formal charge of +1, and two atoms (I_a and I_e) have no formal charge; the ion has an overall charge of −1. To explain the shape, I_b and I_d are sp^3d hybridized (trigonal bipyramid) with the attached atoms occupying axial positions, which results in a bond angle of 180° around each axial atom. I_c has a VSEPR formula of AX_2E_2; its four electron pairs adopt a tetrahedral electron geometry, which is consistent with sp^3 hybridization.

2.53 The molecular orbital diagrams for each of diatomics and their respective cations are shown below:

(a)

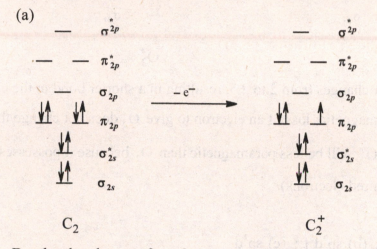

Bond order changes from 2 to 1.5, resulting in a longer bond in the cation. C_2 is diamagnetic; loss of an electron to give C_2^+ makes that species paramagnetic.

(b)

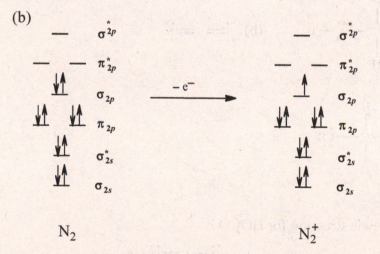

Bond order changes from 3 to 2.5, resulting in a longer bond in the cation. N_2 is diamagnetic; loss of an electron to give N_2^+ makes that species paramagnetic.

(c)

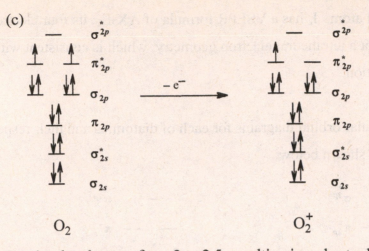

O_2 O_2^+

Bond order changes from 2 to 2.5, resulting in a shorter bond in the cation. O_2 is paramagnetic; loss of an electron to give O_2^+ does not change that (although O_2^+ will be less paramagnetic than O_2 because it possesses fewer unpaired electrons).

2.55   (a) sp^3d^3;   (b) sp^3d^3f;   (c) sp^2d

2.59

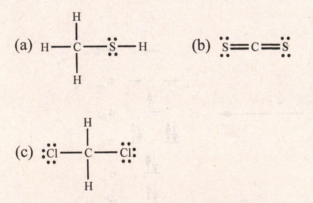

2.61   (a) The Lewis structure for HOCO is:

$$H—\overset{..}{\underset{..}{O}}—C=\overset{..}{\underset{..}{O}}$$

(b) It is a radical since the C has an unpaired electron on it.

2.63   Angles a and c are expected to be approximately 120°. Angle b is expected to be around 109.5°.

2.65   (a) The Lewis structures of NO and NO$_2$ are:

$$\ddot{N}=\ddot{O}\quad \text{(best possible structure)}$$

$$\overset{-1}{:\ddot{O}}-\overset{+1}{N}=\ddot{O} \longleftrightarrow \ddot{O}=\overset{+1}{N}-\overset{-1}{\ddot{O}:}$$

(equivalent resonance structures)

From Table 2D.2, the average bond dissociation energy of a N=O bond is 630 kJ mol^{-1}, which is right in line with the Lewis structure of NO. The bond dissociation energy of each NO bond in NO$_2$ is 469 kJ mol^{-1}, which is about half-way between a N–O double and an N–O single bond, suggesting that the resonance picture of NO$_2$ is a reasonable one.

(b) An N–O single bond should have a bond length of 149 pm while an N–O double bond should have a bond length of 120 pm (values are obtained by summing the respective covalent radii values from Figure 2D.11). From Table 2D.3 the length of a N–O triple bond can be estimated to be between 105 and 110 pm. Since NO itself has a bond length of 115 pm, this suggests that its actual bond order is somewhere between that of a double and a triple bond.

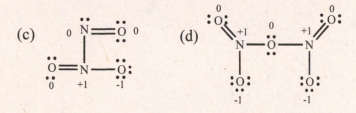

(c)   (d)

(e) N$_2$O$_5$(g) + H$_2$O(l) → 2 HNO$_3$(aq); Nitric acid is produced

(f)

$$4.05 \text{ g N}_2\text{O}_5 \times \frac{1 \text{ mol N}_2\text{O}_5}{108.02 \text{ g N}_2\text{O}_5} \times \frac{2 \text{ mol HNO}_3}{1 \text{ mol N}_2\text{O}_5} = 7.50 \times 10^{-2} \text{ mol HNO}_3$$

$$\text{Molarity} = \frac{7.50 \times 10^{-2} \text{ mol } HNO_3}{1.00 \text{ L}} = 7.50 \times 10^{-2} \text{ mol} \cdot L^{-1} \text{ } HNO_3$$

(g) The oxidation number of nitrogen for the various nitrogen oxides are as follows: NO: +2; NO_2: +4; N_2O_3: +3; and N_2O_5: +5. An oxidizing agent is a species that wants to gain electrons; based on oxidation number, N_2O_5 should be the most potent oxidizing agent of the nitrogen oxides, as it possesses the most positive oxidation number for N of the group.

FOCUS 3
STATES OF MATTER

3A.1 (a) 8×10^9 Pa; (b) 80 kbar; (c) 6×10^7 Torr; (d) 1×10^6 lb·in^{-2}

3A.3 Pressure is defined as the total weight of the air, which is proportional to total air molecules, per unit area. At a higher altitude, there are fewer air molecules per unit area than a same unit area at a lower altitude.

3A.5 (a) The difference in column height will be equal to the difference in pressure between atmospheric pressure and pressure in the gas bulb. If the pressures were equal, the height of the mercury column on the air side and on the apparatus side would be the same. The pressure in the gas bulb is 0.890 atm or 0.890×760 Torr·atm^{-1} = 676 Torr. The difference would be 762 Torr − 676 Torr = 86 Torr = 86 mm Hg. (b) The side attached to the bulb will be higher because the neon pressure is less than the pressure of the atmosphere. (c) If the student had recorded the level in the atmosphere arm to be higher than the level in the bulb arm by 86 mmHg then the pressure in the bulb would have been reported as 762 Torr + 86 Torr = 848 Torr.

3A.7 $d_1 h_1 = d_2 h_2$

$$74.7 \text{ cm} \times \frac{13.6 \text{ g} \cdot \text{cm}^{-3}}{1.10 \text{ g} \cdot \text{cm}^{-3}} = 924 \text{ cm or } 9.24 \text{ m}$$

3A.9 $(20. \text{ in})(10. \text{ in})(14.7 \text{ lb} \cdot \text{in}^{-2}) = 2.9 \times 10^3 \text{ lb}$

3B.1 (a) In Boyle's law, $V \propto \dfrac{1}{P}$; $V = \pi r^2 h$; $h \propto V$

Therefore, $\dfrac{h_1}{h_2} = \dfrac{P_2}{P_1}$, since the atmospheric pressure is 29.85 inHg

$P_1 = (29.85 + 12.0)$ inHg $= 41.85$ inHg

$P_2 = (29.85 + 30.0)$ inHg $= 59.85$ inHg

$h_2 = \dfrac{h_1 P_1}{P_2} = \dfrac{32.0 \text{ in} \times 41.85 \text{ inHg}}{59.85 \text{ inHg}} = 22.4$ in

(b) If the atmospheric pressure is 29.85 inHg, the pressure of gas in tube

(1) is: 29.85 inHg + 12.0 inHg = 41.85 inHg

The pressure of gas in tube (2) is: 29.85 inHg + 30.0 inHg = 59.85 inHg

3.B3 (a) By using the data provided to make a plot (P vs T), the slope = $\dfrac{nR}{V}$;

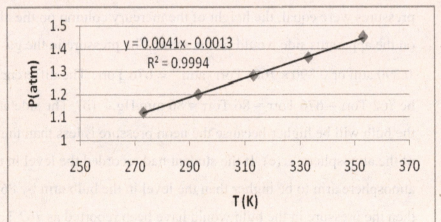

Slope = 0.0041; V = 20 L; or simply use any conditions in the table to

calculate the volume. For example, using the second set of data:

V = nRT/P = (1.0 × 0.082 × 293)/1.20 = 20 L

(b) when additional mole of gases is added into the same volume, the

pressure will be doubled at each T in (a). The plot is:

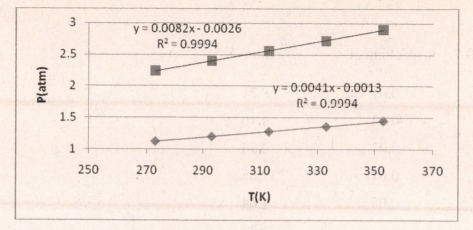

(c) When pressures (Y axis) are the same for both lines, two lines intersect, at which: $0.0082\,T - 0.0026 = 0.0041T - 0.0013$; $T = 0.32$ K

3.B5 (a) Using the ideal gas law with the gas constant R expressed in kPa:

$P(0.3500 \text{ L}) = (0.1500 \text{ mol})(8.314\,51 \text{ L} \cdot \text{kPa} \cdot \text{K}^{-1} \cdot \text{mol}^{-1})(297 \text{ K})$;

$P = 1.06 \times 10^3$ kPa

(b) BrF_3 has a molar mass of 136.91 g·mol^{-1}. We then substitute into the ideal gas equation:

$$\left(\frac{10.0 \text{ Torr}}{760 \text{ Torr} \cdot \text{atm}^{-1}}\right)V$$

$$= \left(\frac{23.9 \times 10^{-3} \text{ g}}{136.91 \text{ g} \cdot \text{mol}^{-1}}\right)(0.082\,06 \text{ L} \cdot \text{atm} \cdot \text{K}^{-1} \cdot \text{mol}^{-1})(373 \text{ K})$$

$V = 0.406 \ L = 4.06 \times 10^2$ mL

(c) $(0.77 \text{ atm})(100.0 \text{ L})$

$$= \left(\frac{m}{64.06 \text{ g} \cdot \text{mol}^{-1}}\right)(0.082\,06 \text{ L} \cdot \text{atm} \cdot \text{K}^{-1} \cdot \text{mol}^{-1})(303 \text{ K})$$

$m = 2.0 \times 10^2$ g

(d) $(129 \text{ kPa})(6.00 \times 10^3 \text{ m}^3)\left(\dfrac{1 \times 10^6 \text{ cm}^3}{\text{m}^3}\right)\left(\dfrac{1 \text{ L}}{1000 \text{ cm}^3}\right)$

$= n(8.314\,51 \text{ L} \cdot \text{kPa} \cdot \text{K}^{-1} \cdot \text{mol}^{-1})(287 \text{ K})$

$n = 3.24 \times 10^5$ mol CH_4

3B.7 (a)

Volume, L	$\dfrac{nR}{V}$, atm·K^{-1}
0.01	8.21
0.02	4.10
0.03	2.74
0.04	2.05
0.05	1.64

(b) The slope is equal to $\dfrac{nR}{V}$

(c) The intercept is equal to 0.00 for all the plots

3B.9 (a) From $P_1V_1 = P_2V_2$, we have

$(2.0 \times 10^5 \text{ kPa}) (7.50 \text{ mL}) = (P_2) (1000 \text{ mL})$; solving for P_2 we get

1.5×10^3 kPa;   (b) similar to (a),

$P_1V_1 = P_2V_2$ or $(643 \text{ Torr}) (54.2 \text{ cm}^3) = (P_2)(7.8 \text{ cm}^3)$, $P_2 = 4.5 \times 10^3$ Torr

3B.11 Because P is a constant, we can use $\dfrac{V_1}{T_1} = \dfrac{V_2}{T_2}$, which gives

$\dfrac{12.4 \, L}{295 \, K} = \dfrac{V_2}{255 \, K}$; $V_2 = 10.7$ L

3B.13 If P and T are constant, then $\dfrac{V_1}{n_1} = \dfrac{V_2}{n_2}$, or $\dfrac{V_1}{0.100 \text{ mol}} = \dfrac{V_2}{0.110 \text{ mol}}$.

Solving for V_2 in terms of V_1, we obtain $V_2 = \dfrac{n_2 V_1}{n_1} = \dfrac{0.110 V_1}{0.10} = 1.10 V_1$.

So the volume must be increased by 10% to keep P and T constant.

3B.15 Slope $= \left(\dfrac{V}{T}\right) = 2.88 \times 10^{-4}$ L·K^{-1}. For a given T, we can find a V value,

vice versa.   Assuming at T = 25°C = 298 K, V = 0.0858 L

$$P = 0.90 \text{ bar} \times \left(\frac{1 \text{ atm}}{1.01325 \text{ bar}} \right) = 0.89 \text{ atm};$$

Based on the equation: $PV = nRT$, $n = \left(\dfrac{\text{mass}}{\text{Mol. mass}} \right)$

$$\text{Mass} = \left(\frac{PV(\text{Mol. mass})}{RT} \right) = \left(\frac{0.89 \text{ atm} \times 0.0858 \text{ L} \times 16.04 \text{ g} \cdot \text{mol}^{-1}}{0.08206 \text{ L} \cdot \text{atm} \cdot \text{mol}^{-1} \cdot \text{K}^{-1} \times 298 \text{K}} \right)$$

$$= 0.050 \text{ g}$$

3B.17 (a) Because P, V, and T all change, we use the relation $\dfrac{P_1 V_1}{T_1} = \dfrac{P_2 V_2}{T_2}$.

Substituting for the appropriate values we get

$$\frac{(0.255 \text{ atm})(35.5 \text{ mL})}{228 \text{ K}} = \frac{(1.00 \text{ atm})(V_2)}{298 \text{ K}}; V_2 = 11.8 \text{ mL}$$

(b) The same relation holds as in (a) but here the final temperature and volume are known:

$$\frac{(0.255 \text{ atm}) (35.5 \text{ mL})}{228 \text{ K}} = \frac{(P_2)(12.0 \text{ mL})}{293 \text{ K}} \cdot P_2 = 0.969 \text{ atm} \quad \text{(c) Similarly,}$$

we can use the same expression, with P and V known and T wanted.

$$\frac{(0.255 \text{ atm})(35.5 \text{ mL})}{228 \text{ K}} = \frac{\left(\dfrac{(500 \text{ Torr})}{760 \text{ Torr} \cdot \text{atm}^{-1}} \right)(12.0 \text{ mL})}{T_2}; T_2 = 199 \text{ K}$$

3B.19 (a) Molar volume =

$$\frac{nRT}{P} = \frac{(1 \text{ mol})(0.082 \, 06 \text{ L} \cdot \text{atm} \cdot \text{K}^{-1} \cdot \text{mol}^{-1})(773 \text{ K})}{1 \text{ atm}} = 63.4 \text{ L/mol}$$

(b) Molar volume =

$$\frac{nRT}{P} = \frac{(1 \text{ mol})(0.082 \, 06 \text{ L} \cdot \text{atm} \cdot \text{K}^{-1} \cdot \text{mol}^{-1})(77 \text{ K})}{1 \text{ atm}} = 6.3 \text{ L/mol}$$

3B.21 Because P, V, and T are state functions, the intermediate conditions are irrelevant to the final states. We can simply use the ideal gas law in the form

$$\frac{P_1 V_1}{T_1} = \frac{P_2 V_2}{T_2}$$

$$\frac{\left(\dfrac{759 \text{ Torr}}{760 \text{ Torr} \cdot \text{atm}^{-1}}\right)(1.00 \text{ L})}{253 \text{ K}} = \frac{\left(\dfrac{252 \text{ Torr}}{760 \text{ Torr} \cdot \text{atm}^{-1}}\right)(V_2)}{1523 \text{ K}}$$

$$V_2 = 18.1 \text{ L}$$

3B.23   Because T is constant, we can use

$$P_1 V_1 = P_2 V_2$$
$$(1.00 \text{ atm})(1.00 \text{ L}) = P_2 (0.239 \text{ L})$$
$$P_2 = 4.18 \text{ atm}$$

3B.25 The volume of each tank is: $3980. \text{ cm}^3 = 3.980 \text{ L}$, $V_{\text{two tanks}} = 7.960 \text{ L}$

The pressure of each tank, $P = 5860. \text{ kPa} \times \left(\dfrac{1 \text{ atm}}{101.325 \text{kPa}}\right) = 57.83 \text{ atm}$

Based on the equation: $PV = nRT$, $n = \left(\dfrac{\text{mass}}{\text{Mol. mass}}\right)$

(Mass of O_2 in two tanks)

$$= \left(\frac{PV(\text{Mol. mass})}{RT}\right) = \left(\frac{57.83 \text{ atm} \times 7.960 \text{ L} \times 32.00 \text{ g} \cdot \text{mol}^{-1}}{0.08206 \text{ L} \cdot \text{atm} \cdot \text{mol}^{-1} \cdot \text{K}^{-1} \times 289 \text{K}}\right) = 621 \text{ g}$$

3B.27 $PV = nRT$

$$\left(\frac{24.5 \text{ kPa}}{101.325 \text{ kPa} \cdot \text{atm}^{-1}}\right)(0.2500 \text{ L})$$

$$= n(0.082\,06 \text{ L} \cdot \text{atm} \cdot \text{K}^{-1} \cdot \text{mol}^{-1})(292.7 \text{ K})$$

$$n = 2.52 \times 10^{-3} \text{ mol}$$

3B.29 (a) $\dfrac{P_1 V_1}{T_1} = \dfrac{P_2 V_2}{T_2}$

$$\dfrac{(104 \text{ kPa})(2.0 \text{ m}^3)}{294.3 \text{ K}} = \dfrac{(52 \text{ kPa}) V_2}{268.2 \text{ K}}$$

$V_2 = 3.6 \text{ m}^3$

(b) $\dfrac{P_1 V_1}{T_1} = \dfrac{P_2 V_2}{T_2}$

$$\dfrac{(104 \text{ kPa})(2.0 \text{ m}^3)}{294.3 \text{ K}} = \dfrac{(0.880 \text{ kPa}) V_2}{221.2 \text{ K}}$$

$V_2 = 1.8 \times 10^2 \text{ m}^3$

3B.31 Density is proportional to the molar mass of the gas as seen from the ideal gas law:

$$PV = nRT$$

$$PV = \dfrac{m}{M} RT$$

$$\text{density} = \text{mass per unit volume} = \dfrac{m}{V} = \dfrac{MP}{RT}$$

The molar masses of the gases in question are

$32.04 \text{ g} \cdot \text{mol}^{-1}$ for N_2H_4; $28.0 \text{ g} \cdot \text{mol}^{-1}$ for N_2;

and $17.03 \text{ g} \cdot \text{mol}^{-1}$ for NH_3;

The densest gas will be the one with the highest molar mass, which in this case is N_2H_4. The order of increasing density will be $NH_3 < N_2 < N_2H_4$.

3B.33 The pressure of the Ar sample will be given by

$$P_{Ar} = \dfrac{nRT}{V} = \dfrac{\left(\dfrac{2.00 \times 10^{-3} \text{g}}{39.95 \text{ g} \cdot \text{mol}^{-1}} \right)(0.082\,06 \text{ L} \cdot \text{atm} \cdot \text{K}^{-1} \cdot \text{mol}^{-1})(293 \text{ K})}{0.050\,0 \text{ L}}$$

$$P_{Kr} = \dfrac{\left(\dfrac{2.00 \times 10^{-3} \text{ g}}{83.80 \text{ g} \cdot \text{mol}^{-1}} \right)(0.082\,06 \text{ L} \cdot \text{atm} \cdot \text{K}^{-1} \cdot \text{mol}^{-1})(T_2)}{0.050\,0 \text{ L}}$$

Because we want the pressure to be the same, we can set these two equal to each other. Because volume, mass of the gases, and the gas constant R are the same on both sides of the equation, they will cancel.

$$\left(\frac{1}{83.80 \text{ g} \cdot \text{mol}^{-1}}\right)(T_2) = \left(\frac{1}{39.95 \text{ g} \cdot \text{mol}^{-1}}\right)(293 \text{ K})$$

Solving for T_2, we obtain temperature = 615 K or 342°C.

3B.35 (a) Density is proportional to the molar mass of the gas as seen from the ideal gas law. See Section 3B.3.

$$d = \frac{(119.37 \text{ g} \cdot \text{mol}^{-1})\left(\dfrac{200 \text{ Torr}}{760 \text{ Torr} \cdot \text{atm}^{-1}}\right)}{(0.082\ 06 \text{ L} \cdot \text{atm} \cdot \text{K}^{-1} \cdot \text{mol}^{-1})(298 \text{ K})} = 1.28 \text{ g} \cdot \text{L}^{-1}$$

(b) $d = \dfrac{(119.37 \text{ g} \cdot \text{mol}^{-1})(1.00 \text{ atm})}{(0.082\ 06 \text{ L} \cdot \text{atm} \cdot \text{K}^{-1} \cdot \text{mol}^{-1})(373 \text{ K})} = 3.90 \text{ g} \cdot \text{L}^{-1}$

3B.37 (a) $M = \dfrac{dRT}{P} = \dfrac{(8.0 \text{ g} \cdot \text{L}^{-1})(0.082\ 06 \text{ L} \cdot \text{atm} \cdot \text{K}^{-1} \cdot \text{mol}^{-1})(300.\text{ K})}{2.81 \text{ atm}}$

$= 70. \text{ g} \cdot \text{mol}^{-1}$

(b) The compound is most likely CHF_3 for which $M = 70 \text{ g} \cdot \text{mol}^{-1}$.

(c) You can use the relationship in (a) to calculate the new density, or you can apply the proportionality changes expected from the change in pressure and temperature to the original density:

$$d_2 = (8.0 \text{ g} \cdot \text{L}^{-1})\left(\frac{1.00 \text{ atm}}{2.81 \text{ atm}}\right)\left(\frac{300.\text{ K}}{298 \text{ K}}\right) = 2.9 \text{ g} \cdot \text{L}^{-1}$$

3B.39 From the analytical data, an empirical formula of CHCl is calculated. The empirical formula mass is $48.47 \text{ g} \cdot \text{mol}^{-1}$. The problem may be solved using the ideal gas law:

$$PV = nRT$$

$$PV = \frac{m}{M}RT$$

$$M = \frac{mRT}{PV}$$

$$M = \frac{(3.557 \text{ g})(0.082\ 06 \text{ L} \cdot \text{atm} \cdot \text{K}^{-1} \cdot \text{mol}^{-1})(273 \text{ K})}{(1.10 \text{ atm})(0.755 \text{ L})} = 95.9 \text{ g} \cdot \text{mol}^{-1}$$

n in the formula $(CHCl)_n$ is equal to

$95.9 \text{g} \cdot \text{mol}^{-1} \div 48.47 \text{g} \cdot \text{mol}^{-1} = 1.98$. The formula

is $C_2H_2Cl_2$.

3B.41 Density is proportional to the molar mass of the gas as seen from the ideal
gas law:

$$PV = nRT$$

$$PV = \frac{m}{M}RT$$

$$\text{density} = \text{mass per unit volume} = \frac{m}{V} = \frac{MP}{RT}$$

$$0.943 \text{ g} \cdot \text{L}^{-1} = \frac{M\left(\dfrac{53.1 \text{ kPa}}{101.325 \text{ kPa} \cdot \text{atm}^{-1}}\right)}{(0.082\ 06 \text{ L} \cdot \text{atm} \cdot \text{K}^{-1} \cdot \text{mol}^{-1})(298 \text{ K})}$$

$$M = 44.0 \text{ g} \cdot \text{mol}^{-1}$$

3C.1 (a) $x_{HCl} = \left(\dfrac{n_{HCl}}{n_{Hcl} + n_{benzene}}\right) = \left(\dfrac{\#\text{of HCl molecule}}{\text{Total \#molecules}}\right) = \left(\dfrac{9}{10}\right) = 0.9$;

$x_{benzene} = 1 - 0.9 = 0.1$

(b) $P_{HCl} = 0.9 \times 0.8 \text{ atm} = 0.72 \text{ atm}$; $P_{benzene} = 0.1 \times 0.8 \text{ atm} = 0.08 \text{ atm}$

3C.3 (a) The molar volume of an ideal gas is 22.4 L at 273.15 K. 1.0 mol of
ideal gas will exert a pressure of 1.0 atm under those conditions. The
partial pressure of $N_2(g)$ will be 1.0 atm. Because there are 2.0 mol of

$H_2(g)$, the partial pressure of $H_2(g)$ will be 2.0 atm.   (b) The total

pressure will be 1.0 atm + 2.0 atm = 3.0 atm.

3C.5 $n_{total} = n_{N_2} + n_{O_2}$; $n_{N_2} = \dfrac{PV}{RT} = \dfrac{(0.50\,\text{bar}) \times \left(\dfrac{1\,\text{atm}}{1.01325\,\text{bar}} \right) \times (1.00\,\text{L})}{(0.08206\,\text{L} \cdot \text{atm} \cdot \text{K}^{-1} \cdot \text{mol}^{-1}) \times (288\,\text{K})}$

$$= 0.021\,\text{mol}$$

$n_{total} = 0.021 + 0.10 = 0.12\,\text{mol}$; $x_{nitrogen} = \left(\dfrac{0.021}{0.12} \right) = 0.18$; $x_{oxygen} = 0.82$

If 0.020 mol were released, then

$n_{left} = 0.10\,\text{mol}$ and $n_{oxygen} = (0.82) \times (0.10\,\text{mol}) = 0.082\,\text{mol}$

$$P_{O_2} = \dfrac{nRT}{V} = \dfrac{(0.082\,\text{mol}) \times (0.08206\,\text{L} \cdot \text{atm} \cdot \text{K}^{-1} \cdot \text{mol}^{-1}) \times (288\,\text{K})}{1.00\,\text{L}}$$

=1.9 atom

3C.7   (a) Of the 756.7 Torr measured, 17.54 Torr will be due to water vapor.

The pressure due to $H_2(g)$ will therefore be

756.7 Torr − 17.54 Torr = 739.2 Torr.   (b) $H_2O(l) \rightarrow H_2(g) + \frac{1}{2}O_2(g)$;

(c) To answer this question, we must determine the number of moles of

H_2 produced in the reaction. Using the partial pressure of H_2 calculated

in part (a) and the ideal gas equation, we can set up the following:

$$\left(\dfrac{739.2\,\text{Torr}}{760\,\text{Torr} \cdot \text{atm}^{-1}} \right)(0.220\,\text{L}) = n(0.08206\,\text{L} \cdot \text{atm} \cdot \text{K}^{-1} \cdot \text{mol}^{-1})(293\,\text{K})$$

Solving for n, we obtain $n = 0.00890$ mol. According to the

stoichiometry of the reaction, half as much oxygen as hydrogen should be

produced, so the number of moles of $O_2 = 0.00445$ mol. The mass of O_2

will be given by $(0.00445\,\text{mol})(32.00\,\text{g} \cdot \text{mol}^{-1}) = 0.142$ g.

3C.9   (a) The number of moles of H_2 needed will be 1.5 times the amount of

NH_3 produced, as seen from the balanced equation:

$\frac{1}{2}N_2(g) + \frac{3}{2}H_2(g) \rightarrow NH_3(g)$

or

$N_2(g) + 3\,H_2(g) \rightarrow 2\,NH_3(g)$

Once the number of moles is known, the volume can be obtained from the ideal gas law.

$$V = \frac{n_{H_2}RT}{P} = \frac{\left(\frac{3}{2}n_{NH_3}\right)RT}{P} = \frac{\left(\left(\dfrac{3\text{ mol }H_2}{2\text{ mol }NH_3}\right)\dfrac{(1.0\times10^3\text{ kg})(10^3\text{ g}\cdot\text{kg}^{-1})}{17.03\text{ g}\cdot\text{mol}^{-1}}\right)RT}{P}$$

$$= \frac{\left(\dfrac{3}{2}\right)\left(\dfrac{10^6\text{ g}}{17.03\text{ g}\cdot\text{mol}^{-1}}\right)(0.082\,06\text{ L}\cdot\text{atm}\cdot\text{K}^{-1}\cdot\text{mol}^{-1})(623\text{ K})}{15.00\text{ atm}}$$

$$= 3.0\times10^5\text{ L}$$

(b) The ideal gas equation

$\dfrac{P_1V_1}{n_1RT_1} = \dfrac{P_2V_2}{n_2RT_2}$ simplifies to $\dfrac{P_1V_1}{T_1} = \dfrac{P_2V_2}{T_2}$

because R and n are constant for this problem.

$$\frac{(15.00\text{ atm})(3.0\times10^5\text{ L})}{623\text{ K}} = \frac{(376\text{ atm})V_2}{(523\text{ K})}$$

$$V_2 = 1.0\times10^4\text{ L}$$

3C.11 We need to find the number of moles of $CH_4(g)$ present in each case.

Because the combustion reaction is the same in both cases, as are the temperature and pressure, the larger number of moles of $CH_4(g)$ should produce the larger volume of $CO_2(g)$. We will use the ideal gas equation to solve for n in the first case:

$$n = \frac{PV}{RT} = \frac{(1.00\text{ atm})(2.00\text{ L})}{(0.082\,06\text{ L}\cdot\text{atm}\cdot\text{K}^{-1}\cdot\text{mol}^{-1})(348\text{ K})} = 0.0700\text{ mol }CH_4$$

2.00 g of CH_4 will be $\dfrac{2.00\text{ g}}{16.04\text{ g}\cdot\text{mol}^{-1}} = 0.124\text{ mol}$

The latter case will have the greater number of moles of CH_4 and should produce the larger amount of $CO_2(g)$.

3C.13 (a) $Cl_2(g) + F_2(g) \rightarrow 2\,ClF(g)$

$$2.00 \text{ mol } F_2 \left(\frac{2 \text{ mol ClF}}{1 \text{ mol } F_2} \right) = 4.00 \text{ mol ClF}$$

$$PV = nRT; \quad V = nRT/P = \frac{4.00 \text{ mol} \times 0.08206 \dfrac{L \cdot atm}{mol \cdot K} \times 298 \text{ K}}{1.00 \text{ atm}} = 97.8 \text{ L}$$

(b) $Cl_2(g) + 3\,F_2(g) \rightarrow 2\,ClF_3(g)$

$$2.00 \text{ mol } F_2 \left(\frac{2 \text{ mol ClF}}{3 \text{ mol } F_2} \right) = 1.33 \text{ mol ClF}_3$$

$$PV = nRT; \quad V = nRT/P = \frac{1.33 \text{ mol} \times 0.08206 \dfrac{L \cdot atm}{mol \cdot K} \times 298 \text{ K}}{1.00 \text{ atm}} = 32.5 \text{ L}$$

(c) $Cl_2(g) + 5\,F_2(g) \rightarrow 2\,ClF_5(g)$

$$2.00 \text{ mol } F_2 \left(\frac{2 \text{ mol ClF}}{5 \text{ mol } F_2} \right) = 0.80 \text{ mol ClF}_3$$

$$PV = nRT; \quad V = nRT/P = \frac{0.80 \text{ mol} \times 0.08206 \dfrac{L \cdot atm}{mol \cdot K} \times 298 \text{ K}}{1.00 \text{ atm}} = 19.6 \text{ L}$$

3C.15 (a) This is a limiting reactant problem. Our first task is to determine the number of moles of NH_3 and HCl that are present to start with. This can be done from the ideal gas equation:

$$PV = nRT$$

$$n_{NH_3} = \frac{PV}{RT} = \frac{\left(\dfrac{100 \text{ Torr}}{760 \text{ Torr} \cdot atm^{-1}} \right)(0.0150 \text{ L})}{(0.082\,06 \text{ L} \cdot atm \cdot K^{-1} \cdot mol^{-1})(303 \text{ K})} = 7.94 \times 10^{-5} \text{ mol}$$

$$n_{HCl} = \frac{PV}{RT} = \frac{\left(\dfrac{150 \text{ Torr}}{760 \text{ Torr} \cdot atm^{-1}} \right)(0.0250 \text{ L})}{(0.082\,06 \text{ L} \cdot atm \cdot K^{-1} \cdot mol^{-1})(298 \text{ K})} = 2.02 \times 10^{-4} \text{ mol}$$

The ammonia is the limiting reactant. The number of moles of $NH_4Cl(s)$ that form will be equal to the number of moles of NH_3 that react. From the molar mass of NH_4Cl ($53.49 \text{ g} \cdot \text{mol}^{-1}$) and the number of moles, we can calculate the mass of NH_4Cl that forms:

$$(7.94 \times 10^{-5} \text{ mol } NH_4Cl(s))(53.49 \text{ g} \cdot \text{mol}^{-1}) = 4.25 \times 10^{-3} \text{ g}$$

(b) There will be $(2.02 \times 10^{-4} \text{ mol} - 7.94 \times 10^{-5} \text{ mol}) = 1.23 \times 10^{-4} \text{ mol HCl}$ left after the reaction. This quantity will exist in a total volume after mixing of 40.0 mL or 0.0400 L. Again, we use the ideal gas law to determine the final pressure:

$$PV = nRT$$

$$P = \frac{(1.23 \times 10^{-4} \text{ mol})(0.082 \, 06 \text{ L} \cdot \text{atm} \cdot \text{K}^{-1} \cdot \text{mol}^{-1})(300 \text{ K})}{0.0400 \text{ L}} = 0.0757 \text{ atm}$$

3D.1 No. The reason is that not all the molecules are traveling at the same velocity. Force $= \left(\dfrac{NmAv_x^2}{V} \right)$, if v_x varies, force will be different.

3D.3 Graham's law of effusion states that the rate of effusion of a gas is inversely proportional to the square root of its molar mass:

$$\text{rate of effusion} = \frac{1}{\sqrt{M}}$$

Diffusion also follows this relationship. If we have two different gases whose rates of diffusion are measured under identical conditions, we can take the ratio

$$\frac{\text{rate}_1}{\text{rate}_2} = \frac{\dfrac{1}{\sqrt{M_1}}}{\dfrac{1}{\sqrt{M_2}}} = \sqrt{\frac{M_2}{M_1}}$$

If a compound takes 1.24 times as long to diffuse as Kr gas, the rate of diffusion of Kr is 1.24 times that of the unknown. We can now use the expression to calculate the molar mass of the unknown, given the mass of Kr:

$$\frac{1.24}{1} = \sqrt{\frac{M_2}{83.80 \text{ g} \cdot \text{mol}^{-1}}}$$

$$M_2 = 129 \text{ g} \cdot \text{mol}^{-1}$$

A mass of $129 \text{ g} \cdot \text{mol}^{-1}$ corresponds to a molecular formula of $C_{10}H_{10}$.

3D.5   The rate of effusion is inversely proportional to the square root of the molar mass. Using a ratio as follows allows us to calculate the time of effusion without knowing the exact conditions of pressure and temperature:

$$\frac{\text{rate}_1}{\text{rate}_2} = \frac{\dfrac{1}{\sqrt{M_1}}}{\dfrac{1}{\sqrt{M_2}}} = \sqrt{\frac{M_2}{M_1}}$$

The rate will be equal to the number of molecules N that effuse in a given time interval. For the conditions given, N will be the same for argon and for the second gas chosen.

$$\frac{\dfrac{N}{\text{time}}}{\dfrac{N}{147 \text{ s}}} = \frac{\dfrac{1}{\text{time}}}{\dfrac{1}{147 \text{ s}}} = \sqrt{\frac{39.95 \text{ g} \cdot \text{mol}^{-1}}{M_1}}$$

In order to calculate the time of effusion, we need to know only the molar mass of the gases.

(a) For CO_2 with a molar mass of

$$44.01 \text{ g} \cdot \text{mol}^{-1} : \frac{\dfrac{1}{\text{time}_{CO_2}}}{\dfrac{1}{147 \text{ s}}} = \sqrt{\frac{39.95 \text{ g} \cdot \text{mol}^{-1}}{44.01 \text{ g} \cdot \text{mol}^{-1}}}$$

time = 154 s

(b) For C_2H_4 with a molar mass of

$$28.05 \text{ g} \cdot \text{mol}^{-1} : \frac{\dfrac{1}{\text{time}_{C_2H_4}}}{\dfrac{1}{147 \text{ s}}} = \sqrt{\frac{39.95 \text{ g} \cdot \text{mol}^{-1}}{28.05 \text{ g} \cdot \text{mol}^{-1}}}$$

time = 123 s

(c) For H_2 with a molar mass of

$$2.01 \text{ g} \cdot \text{mol}^{-1} : \frac{\dfrac{1}{\text{time}_{CO_2}}}{\dfrac{1}{147 \text{ s}}} = \sqrt{\frac{39.95 \text{ g} \cdot \text{mol}^{-1}}{2.01 \text{ g} \cdot \text{mol}^{-1}}}$$

time = 33.0 s

(d) For SO_2 with a molar mass of

$$64.06 \text{ g} \cdot \text{mol}^{-1} : \frac{\dfrac{1}{\text{time}_{CO_2}}}{\dfrac{1}{147 \text{ s}}} = \sqrt{\frac{39.95 \text{ g} \cdot \text{mol}^{-1}}{64.06 \text{ g} \cdot \text{mol}^{-1}}}$$

time = 186 s

3D.7   The formula mass of C_2H_3 is 27.04 g · mol^{-1}. From the effusion data, we can calculate the molar mass of the sample.

$$\frac{\text{rate}_1}{\text{rate}_2} = \frac{\dfrac{1}{\sqrt{M_1}}}{\dfrac{1}{\sqrt{M_2}}} = \sqrt{\frac{M_2}{M_1}}$$

Because time is inversely proportional to rate, we can write alternatively

$$\frac{\dfrac{1}{349 \text{ s}}}{\dfrac{1}{210 \text{ s}}} = \sqrt{\frac{39.95 \text{ g} \cdot \text{mol}^{-1}}{M_1}}$$

$$\frac{210}{349} = \sqrt{\frac{39.95 \text{ g} \cdot \text{mol}^{-1}}{M_1}}$$

$$M_1 = 110 \text{ g} \cdot \text{mol}^{-1}$$

The molar mass is 4.1 times that of the empirical formula mass, so the molecular formula is C_8H_{12}.

3D.9   The root mean square speed is calculated from the following equation:

$$v_{rms} = \sqrt{\frac{3RT}{M}}$$

(a) methane, CH_4, $M = 16.04$ g $\cdot$ mol^{-1}

$$v_{rms} = \sqrt{\frac{3 \times (8.314 \text{ kg} \cdot \text{m}^2 \cdot \text{s}^{-2} \cdot \text{K}^{-1} \cdot \text{mol}^{-1}) \times (253 \text{ K})}{1.604 \times 10^{-2} \text{ kg} \cdot \text{mol}^{-1}}}$$

$$= 627 \text{ m} \cdot \text{s}^{-1}$$

(b) ethane, C_2H_6, $M = 30.07$ g $\cdot$ mol^{-1}

$$v_{rms} = \sqrt{\frac{3 \times (8.314 \text{ kg} \cdot \text{m}^2 \cdot \text{s}^{-2} \cdot \text{K}^{-1} \cdot \text{mol}^{-1}) \times (253 \text{ K})}{3.007 \times 10^{-2} \text{ kg} \cdot \text{mol}^{-1}}}$$

$$= 458 \text{ m} \cdot \text{s}^{-1}$$

(c) propane, C_3H_8, $M = 44.09$ g $\cdot$ mol^{-1}

$$v_{rms} = \sqrt{\frac{3 \times (8.314 \text{ kg} \cdot \text{m}^2 \cdot \text{s}^{-2} \cdot \text{K}^{-1} \cdot \text{mol}^{-1}) \times (253 \text{ K})}{4.409 \times 10^{-2} \text{ kg} \cdot \text{mol}^{-1}}}$$

$$= 378 \text{ m} \cdot \text{s}^{-1}$$

3D.11   Based on equation: $v_{rms} = \left(\frac{3RT}{M}\right)^{1/2}$, If T is constant,

$$(v_{rms})_1 (M_1)^{1/2} = (v_{rms})_2 (M_2)^{1/2}, \quad (550. \text{ m s}^{-1})(16.06)^{1/2} = (v_{rms})_{Kr} \times (83.80)^{1/2}$$

$$(v_{rms})_{Kr} = 241 \text{ m s}^{-1}$$

Note: you do not need to change M into kg/mol in this calculation

3D.13   Use the expression for the root mean square speed to determine the temperature.

$$v_{rms} = \sqrt{\frac{3RT}{M}}$$

$$T = \frac{v_{rms}^2 \cdot M}{3R} = \frac{(375 \text{ m} \cdot \text{s}^{-1})^2 \cdot (4.80 \times 10^{-2} \text{ kg} \cdot \text{mol}^{-1})}{3(8.314 \text{ J} \cdot \text{K}^{-1} \cdot \text{mol}^{-1})}$$

$$= 271 \text{ K}$$

3D.15 Use the expression for the root mean square speed to determine the temperature.

$$v_{rms} = \sqrt{\frac{3RT}{M}}$$

$$T = \frac{v_{rms}^2 \cdot M}{3R} = \frac{(1477 \text{ m} \cdot \text{s}^{-1})^2 \cdot (4.00 \times 10^{-3} \text{ kg} \cdot \text{mol}^{-1})}{3(8.314 \text{ J} \cdot \text{K}^{-1} \cdot \text{mol}^{-1})}$$

$$= 349.9 \text{ K}$$

Use the Maxwell Distribution of Speeds (Equation (8) in 3D.3) appropriately for both gases:

$$\frac{f(v_{He})}{f(v_{Ar})} = \frac{4\pi N \left(\dfrac{M_{He}}{2RT}\right)^{\frac{3}{2}} (v_{He})^2 e^{-M_{He}(v_{He})^2/2RT}}{4\pi N \left(\dfrac{M_{Ar}}{2RT}\right)^{\frac{3}{2}} (v_{Ar})^2 e^{-M_{Ar}(v_{Ar})^2/2RT}}$$

$$= \frac{(M_{He})^{\frac{3}{2}} (v_{He})^2 e^{-M_{He}(v_{He})^2/2RT}}{(M_{Ar})^{\frac{3}{2}} (v_{Ar})^2 e^{-M_{Ar}(v_{Ar})^2/2RT}}$$

$$= \frac{(4.00 \text{ g} \cdot \text{mol}^{-1})^{\frac{3}{2}} (1477 \text{ m} \cdot \text{s}^{-1})^2}{(39.95 \text{ g} \cdot \text{mol}^{-1})^{\frac{3}{2}} (467 \text{ m} \cdot \text{s}^{-1})^2} \times \frac{e^{\frac{-(4.00 \times 10^{-3} \text{ kg} \cdot \text{mol}^{-1}) \times (1477 \text{ m} \cdot \text{s}^{-1})^2}{2 \times (8.314 \text{ J} \cdot \text{K}^{-1} \cdot \text{mol}^{-1}) \times (349.9 \text{ K})}}}{e^{\frac{-(39.95 \times 10^{-3} \text{ kg} \cdot \text{mol}^{-1}) \times (467 \text{ m} \cdot \text{s}^{-1})^2}{2 \times (8.314 \text{ J} \cdot \text{K}^{-1} \cdot \text{mol}^{-1}) \times (349.9 \text{ K})}}}$$

$$= \left(\frac{4.00}{39.95}\right)^{\frac{3}{2}} \left(\frac{1477}{467}\right)^2 (0.9977)$$

$$= 0.316$$

3D.17 (a) The most probable speed is the one that corresponds to the maximum on the distribution curve. (b) The percentage of molecules having the most probable speed decreases as the temperature is raised (the distribution spreads out).

3D.19 Note that Equation 8, the Maxwell distribution of speeds, should read

$$f(v) = 4\pi \left(\frac{M}{2\pi RT} \right)^{3/2} v^2 e^{\frac{-Mv^2}{2RT}}.$$

The first derivative with respect to speed is

$$\frac{d f(v)}{dv} = \left(\frac{M}{2\pi RT} \right)^{3/2} e^{\frac{-Mv^2}{2RT}} \left[8\pi v + 4\pi v^2 \left(\frac{-Mv}{RT} \right) \right] = 0$$

Solving for v gives the most probable speed in terms of T.

$$v_{mp} = \left(\frac{2RT}{M} \right)^{1/2}$$

Substitution into $f(v)$ gives

$$f(v_{mp}) = 4\pi \left(\frac{M}{2\pi RT} \right)^{3/2} \left(\frac{2RT}{M} \right) \cdot e^{-1} = 4\pi \left(\frac{M}{2\pi RT} \right)^{1/2} e^{-1}$$

The problem asks for the temperature that satisfies this relationship:

$$f(v_{mp})_T = \frac{1}{4} f(v_{mp})_{200}$$

$$4\pi \left(\frac{M}{2\pi RT} \right)^{1/2} e^{-1} = \left(\frac{1}{4} \right) 4\pi \left(\frac{M}{2\pi R(200. \text{ K})} \right)^{1/2} e^{-1}$$

$$\left(\frac{1}{T} \right)^{1/2} = \left(\frac{1}{4} \right) \left(\frac{1}{200. \text{ K}} \right)^{1/2}$$

$$\left(\frac{1}{T} \right) = \left(\frac{1}{16} \right) \left(\frac{1}{200. \text{ K}} \right)$$

$$T = 16 \times 200. \text{ K} = 3.20 \times 10^3 \text{ K}$$

3E.1 Hydrogen bonding is important in HF. At low temperatures, this hydrogen bonding causes the molecules of HF to be attracted to each other more strongly, thus lowering the pressure. As the temperature is increased, the hydrogen bonds are broken and the pressure rises more quickly than for an ideal gas. Dimers (two HF molecules bonded to each other) and chains of HF molecules are known to form.

3E.3 (a) Root mean square speed $(v_{rms}) = \sqrt{\dfrac{3RT}{M}}$; the smaller the molecular mass, the larger the v_{rms}. Therefore, H_2 molecules have greater root mean square speed.

(b) NH_3 will have stronger deviation from ideality because of the stronger attraction between NH_3 molecules, which is due to H-bonding among NH_3 molecules.

3E.5 The pressures are calculated very simply from the ideal gas law:

$$P = \frac{nRT}{V} = \frac{(1.00 \text{ mol})(0.082\ 06 \text{ L} \cdot \text{atm} \cdot \text{K}^{-1} \cdot \text{mol}^{-1})(298 \text{ K})}{V}$$

Calculating for the volumes requested, we obtain $P =$ (a) 1.63 atm (ideal gas); (b) 48.9 atm (ideal gas); (c) 489 atm (ideal gas). The calculations can now be repeated using the van der Waals equation:

$$\left(P + \frac{an^2}{V^2}\right)(V - nb) = nRT$$

We can rearrange this to solve for P :

$$P = \left(\frac{nRT}{V - nb}\right) - \left(\frac{an^2}{V^2}\right)$$

$$= \left(\frac{(1.00 \text{ mol})(0.082\ 06 \text{ L} \cdot \text{atm} \cdot \text{K}^{-1} \cdot \text{mol}^{-1})(298 \text{ K})}{V - (1.00 \text{ mol})(0.0429 \text{ L} \cdot \text{mol}^{-1})}\right)$$

$$- \left(\frac{(3.610 \text{ L}^2 \cdot \text{atm} \cdot \text{mol}^{-2})(1.00^2)}{V^2}\right)$$

Using the three values for V, we calculate for $P =$ (a) 1.62 atm (real gas); (b) 39.0 atm (real gas); (c) 2.00×10^3 atm (real gas). Note that at low pressures, the ideal gas law gives essentially the same values as the van der Waals equation, but at high pressures there is a very significant differences.

3E.7

α (bar·L^2·mol^{-2})	Substance
17.813	CH_3CN
3.700	HCl
2.303	CH_4
0.208	Ne

The reason is that α represents the role of molecular attractions and its value will be relatively larger for molecules that attract each other strongly and for large molecules with many electrons.

3E.9 (a) Ideal gas: $PV = nRT$

$$P = \frac{nRT}{V} = \frac{(1.00 \text{ mol})(0.082\ 06 \text{ L} \cdot \text{atm} \cdot \text{K}^{-1} \cdot \text{mol}^{-1})(273.15 \text{ K})}{22.414 \text{ L}}$$

$$= 1.00 \text{ atm}$$

(b) van der Waal's gas:

 (i) at 273.15K in 22.414 L:

The a value for H_2S = 4.544 bar·L^2·mol^{-2} = 4.484 atm·L^2·mol^{-2}

The b value for H_2S = 0.0434 L·mol^{-1}

$$\left(P + a\left(\frac{n}{V}\right)^2\right)(V - nb) = nRT$$

$$P = \frac{nRT}{V - nb} - a\left(\frac{n}{V}\right)^2$$

$$P = \frac{(1.00 \text{ mol})(0.082\ 06 \text{ L} \cdot \text{atm} \cdot \text{K}^{-1} \cdot \text{mol}^{-1})(273.15 \text{ K})}{22.414 \text{ L} - (1.00 \text{ mol})(0.0434 \text{ L} \cdot \text{mol}^{-1})}$$

$$- 4.484 \text{ atm} \cdot \text{L}^2 \cdot \text{mol}^{-2} \left(\frac{1.00 \text{ mol}}{22.414 \text{ L}}\right)^2 = 0.993 \text{ atm}$$

 (ii) at 800. K in 60.0 L:

$$P = \frac{(1.00 \text{ mol})(0.082\ 06 \text{ L} \cdot \text{atm} \cdot \text{K}^{-1} \cdot \text{mol}^{-1})(800 \text{ K})}{60.0 \text{ L} - (1.00 \text{ mol})(0.0434 \text{ L} \cdot \text{mol}^{-1})}$$

$$- 4.484 \text{ atm} \cdot \text{L}^2 \cdot \text{mol}^{-2} \left(\frac{1.00 \text{ mol}}{60.0 \text{ L}} \right)^2 = 1.09 \text{ atm}$$

3E.11 Ammonia: $a = 4.169 \text{ L}^2 \cdot \text{atm} \cdot \text{mol}^{-2}$; $b = 0.0371 \text{ L} \cdot \text{mol}^{-1}$

Oxygen: $a = 1.364 \text{ L}^2 \cdot \text{atm} \cdot \text{mol}^{-2}$; $b = 0.0319 \text{ L} \cdot \text{mol}^{-1}$

Volume (L)	$P_{Ammonia}$ (atm)	P_{Oxygen} (atm)	P_{Ideal} (atm)
0.05	228.05	805.44	489.08
0.2	45.89	111.37	122.27
0.4	41.33	57.91	61.13
0.6	31.86	39.26	40.76
0.8	25.54	29.71	30.57
1	21.23	23.90	24.45

Clearly, the greater deviation from the ideal gas law values occurs at low volumes or higher pressures. Ammonia deviates more strongly and its van der Waals constants are larger than those for oxygen. This may likely arise because ammonia is polar molecule and will have stronger intermolecular interactions and hydrogen bonding.

3E.13 (a) volume of one atom = molar volume ÷ Avogadro's number

$2.38 \times 10^{-2} \text{ L} \cdot \text{mol}^{-1} \div 6.022 \times 10^{23} \text{ atoms} \cdot \text{mol}^{-1} = 3.95 \times 10^{-26} \text{ L} \cdot \text{mol}^{-1}$

$3.95 \times 10^{-26} \text{ L} \cdot \text{mol}^{-1} \left(\frac{1000 \text{ cm}^3}{1 \text{ L}} \right) \left(\frac{1.00 \times 10^{30} \text{ pm}^3}{1 \text{ cm}^3} \right)$

$= 3.95 \times 10^7 \text{ pm}^3 \cdot \text{atom}^{-1}$

$3.95 \times 10^7 \text{ pm}^3 = \frac{4}{3} \pi r^3$; $r = 211 \text{ pm}$

(b) The atomic radius of He is 128 pm (Appendix 2D).

The volume of the He atom, based upon this radius, is

$$V = \frac{4}{3}\pi r^3$$

$$= \frac{4}{3}\pi (128 \text{ pm})^3$$

$$= 8.78 \times 10^6 \text{ pm}^3$$

(c) The difference in these values illustrates that there is no easy definition for the boundaries of an atom. The van der Waals value obtained from the correction for molar volume is considerably larger than the atomic radius, owing perhaps to longer range and weak interactions between atoms. One should also bear in mind that the value for the van der Waals b is a parameter used to obtain a good fit to a curve, and its interpretation is more complicated than a simple molar volume.

3F.1   (a) London forces, dipole-dipole, hydrogen bonding; (b) London forces; (c) London forces, dipole-dipole, hydrogen bonding; (d) London forces, dipole-dipole

3F.3   Only (b) CH_3Cl,   (c) CH_2Cl_2, and   (d) $CHCl_3$ will have dipole-dipole interactions. The molecules CH_4 and CCl_4 do not have dipole moments.

3F.5   (a) NaCl (801°C vs −114.8°C) because it is an ionic compound as opposed to a molecular compound;   (b) butanol (−90°C vs −116°C) due to hydrogen bonding in butanol that is not possible in diethyl ether;   (c) triiodomethane because it will have much stronger London dispersion forces (−82.2°C for trifluoromethane vs. 219°C for triiodomethane);   (d) Methanol (−94 °C vs −169 °C) because of the hydrogen bonding in methanol but not in ethylene.

3F.7   (a) PBr₃ and PF₃ are both trigonal pyramidal and should have similar intermolecular forces, but PBr_3 has the greater number of electrons and should have the higher boiling point. The boiling point of PF_3 is

$-101.5°C$ and that of PBr_3 is $173.2\ °C$;   (b) SO_2 and O_3 are both bent

and have dipole moments. But SO_2 should have the higher boiling point

due its greater number of electrons. SO_2 boils at $-10°C$, whereas O_3 boils

at $-112\ °C$.   (c) BF_3 and BCl_3 are both trigonal planar, so the choice of

higher boiling point depends on the difference in total number of

electrons. BCl_3 should have the higher boiling point ($12.5°C$ vs $-99.9°C$).

3F.9   The interaction energies can be ordered based on the relationship the

energy has to the distance separating the interacting species. Thus ion-ion

interactions are the strongest and are directly proportional to the distance

separating the two interacting species. Ion-dipole energies are inversely

proportional to d^2, whereas dipole-dipole for constrained molecules (i.e.,

solid state) is inversely proportional to d^3. Dipole-dipole interactions

where the molecules are free to rotate become comparable with induced

dipole-induced dipole interactions, which are both inversely related to d^6.

The order thus derived is: (b) dipole-induced dipole $\cong$ (c) dipole-dipole in

the gas phase < (e) dipole-dipole in the solid phase < (a) ion-dipole < (d)

ion-ion.

3F.11   Only molecules with H attached to the electronegative atoms F, N, and O

can hydrogen bond. Additionally, there must be lone pairs available for

the Hs to bond to. This is true only of (d) HNO_2.

3F.13   II because the dipoles are aligned with oppositely charged ends closest to

each other thereby maximizing dipole-dipole attractions.

3F.15   The Lewis structures of AsF_3 and AsF_5 are as follows:

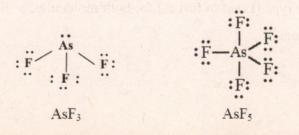

AsF₃   AsF₅

From the Lewis structures of these two molecules, it is clear that AsF_3 is a polar molecule while AsF_5 is not. So the dipole-dipole intermolecular forces in AsF_3 are stronger than the London (dispersion) forces in AsF_5, which results a higher boiling point of AsF_3.

3F.17 The ionic radius of Al^{3+} is 54 pm and that of Be^{2+} is 34 pm. The ratio of energies will be given by

$$E_p \propto \left(\frac{-|Z|\mu}{d^2} \right)$$

$$E_{pAl^{3+}} \propto \left(\frac{-|Z|\mu}{d^2} \right) = \left(\frac{-|3|\mu}{(54+100)^2} \right)$$

$$E_{pBe^{2+}} \propto \left(\frac{-|Z|\mu}{d^2} \right) = \left(\frac{-|2|\mu}{(34+100)^2} \right)$$

The electric dipole moment of the water molecule (μ) will cancel:

$$\text{Ratio} \left(\frac{E_{pAl^{2+}}}{E_{pBe^{2+}}} \right) = \left(\frac{-|3|\mu/(54+100)^2}{-|2|\mu/(34+100)^2} \right) = \left(\frac{3(134)^2}{2(154)^2} \right) = 1.14$$

The attraction of the Al^{3+} ion will be greater than that of the Be^{2+} ion. Even though the Al^{3+} ion has a larger radius, its charge is higher than that of Be^{2+}, making the attraction greater over all.

3F.19 (a) Xenon is larger, with more electrons, giving rise to larger London forces that increase the melting point. (b) Hydrogen bonding in water causes the molecules to be held together more tightly than in diethyl ether. (c) Both molecules have the same molar mass, but pentane is a linear molecule compared with dimethylpropane, which is a compact, spherical molecule. The compactness of the dimethylpropane gives it a lower surface area. That means that the intermolecular attractive forces, which are of the same type (London forces) for both molecules, will have a larger effect for pentane.

3F.21 $F = \dfrac{-dE_P}{dr} = \dfrac{-d}{dr}\left(\dfrac{1}{r^6}\right) = -\left(\dfrac{-6}{r^7}\right) \propto \dfrac{1}{r^7}$

3G.1 (a) As intermolecular forces increase, boiling point will increase because more thermal energy is required to separate molecules apart; (b) viscosity will increase with the increase of intermolecular forces because stronger intermolecular forces will make molecules less mobile (or do not let molecules move past one another easily); (c) surface tension also increases with the increase of intermolecular forces because stronger intermolecular forces tend to pull molecules together and inward.

3G.3 (a) *cis*-Dichloroethene is polar, whereas *trans*-dichloroethene, whose individual bond dipole moments cancel, is nonpolar. Therefore, *cis*-dichloroethene has the greater intermolecular forces and the greater surface tension. (b) Surface tension of liquids decreases with increasing temperature as a result of thermal motion as temperature rises. Increased thermal motion allows the molecules to more easily break away from each other, which manifests itself as decreased surface tension. Therefore, benzene at 20 degrees has a higher surface tension.

3G.5 At 50°C all three compounds are liquids. C_6H_6 (nonpolar) $< C_6H_5SH$ (polar, but no hydrogen bonding) $< C_6H_5OH$ (polar and with hydrogen bonding). The viscosity will show the same ordering as the boiling points, which are 80°C for C_6H_6, 169°C for C_6H_5SH, 182° for C_6H_5OH.

3G.7 CH_4, −162°C; CH_3CH_3, −88.5°C; $(CH_3)_2CHCH_2CH_3$, 28°C; $CH_3(CH_2)_3CH_3$, 36°C; CH_3OH, 64.5°C; CH_3CH_2OH, 78.3 °C; $CH_3CHOHCH_3$, 82.5°C; C_5H_9OH (cyclic, but not aromatic), 140°C; $C_6H_5CH_3OH$ (aromatic ring), 205°C; $OHCH_2CHOHCH_2OH$, 290°C

3G.9 Because of the silanol group (Si-OH) on the glass tube wall, water molecules that are close to the glass tube wall will strongly interact with the silanol group by forming H-bonding and climbing up to certain distance and forming a concave meniscus. Water molecules do not have such interactions with a plastic tube wall because of the hydrophobic character of the plastics, so water molecules will interact among one another through H-bonding and form a convex meniscus.

3G.11 Using $h = \dfrac{2\gamma}{gdr}$ we can calculate the height. For water:

$$r = \frac{1}{2} \text{ diameter} = \frac{1}{2}(0.15 \text{ mm})\left(\frac{1 \text{ m}}{1000 \text{ mm}}\right) = 7.5 \times 10^{-5} \text{ m}$$

$$d = 0.997 \text{ g} \cdot \text{cm}^{-3}\left(\frac{1 \text{ kg}}{1000 \text{ g}}\right)\left(\frac{10^{6} \text{ cm}^{3}}{\text{m}^{3}}\right) = 9.97 \times 10^{2} \text{ kg} \cdot \text{m}^{-3}$$

$$h = \frac{2(72.75 \times 10^{-3} \text{ N} \cdot \text{m}^{-1})}{(9.81 \text{ m} \cdot \text{s}^{-1})(9.97 \times 10^{2} \text{ kg} \cdot \text{m}^{-3})(7.5 \times 10^{-5} \text{ m})} = 0.20 \text{ m or } 200 \text{ mm}$$

Remember that $1 \text{ N} = 1 \text{ kg} \cdot \text{m}^{-1} \cdot \text{s}^{-2}$

For ethanol:

$$d = 0.79 \text{ g} \cdot \text{cm}^{-3}\left(\frac{1 \text{ kg}}{1000 \text{ g}}\right)\left(\frac{10^{6} \text{ cm}^{3}}{\text{m}^{3}}\right) = 7.9 \times 10^{2} \text{ kg} \cdot \text{m}^{-3}$$

$$h = \frac{2(22.8 \times 10^{-3} \text{ N} \cdot \text{m}^{-1})}{(9.81 \text{ m} \cdot \text{s}^{-1})(7.9 \times 10^{2} \text{ kg} \cdot \text{m}^{-3})(7.5 \times 10^{-5} \text{ m})} = 0.078 \text{ m or } 78 \text{ mm}$$

Water will rise to a higher level than ethanol. There are two opposing effects to consider. While the greater density of water than ethanol, acts against it rising as high, it has a much higher surface tension.

3G.13 There are too many ways that these molecules can rotate and twist so that they do not remain rodlike. The molecular backbone when the molecule is stretched out is rodlike, but the molecules tend to curl up on themselves, destroying any possibility of long range order with neighboring molecules.

This is partly due to the fact that the molecules have only single bonds that allow rotation about the bonds, so that each molecule can adopt many configurations. If multiple bonds are present, the bonds are more rigid.

3G.15 Because benzene is isotropic solvent and its viscosity is the same in every direction. However, a liquid crystal solvent is anisotropic and its viscosity is smaller in the direction parallel to the long axis of the molecule than in the "sideway" direction. Since methylbenzene is a small sphere molecule and its diffusion movement does not affected greatly whether solvent is isotropic or anisotropic, it can move and rotate to the same extent in all directions.

3G.17 Substance (a), $C_5H_6N^+Cl^-$, would be a better choice as an ionic liquid solvent. The reason is that it has a larger organic cation with the same anion as that of (b), which has a larger nonpolar region to dissolve nonpolar organic compounds.

3H.1 (a) Glucose will be held in the solid by London forces, dipole-dipole interactions, and hydrogen bonds; benzophenone will be held in the solid by dipole-dipole interactions and London forces; (b) We would expect glucose to have a higher melting point because glucose can experience hydrogen bonding, which is a strong interaction and dominates intermolecular forces, but London forces are the strongest in benzophenone. The melting points for both compounds are: benzophenone (m.p. = 48 °C < glucose (m.p. = 148 – 155 °C)

3H.3 (a) Network; (b) ionic; (c) molecular; (d) molecular; (e) network

3H.5 Substance A: ionic; substance B: metallic; substance C: molecular solid

3H.7   (a) At center: 1 center $\times$ 1 atom $\cdot$ center^{-1} = 1 atom; at 8 corners,

8 corners $\times$ $\frac{1}{8}$ atom $\cdot$ corner^{-1} = 1 atom; total = 2 atoms;   (b) There are

eight nearest neighbors, hence a coordination number of 8;   (c) The

direction along which atoms touch each other is the body diagonal of the

unit cell. This body diagonal will be composed of four times the radius of

the atom. In terms of the unit cell edge length a, the body diagonal will be

$\sqrt{3}\,a$. The unit cell edge length will, therefore, be given by

$$4r = \sqrt{3}\,a \text{ or } a = \frac{4r}{\sqrt{3}} = \frac{4 \cdot (124 \text{ pm})}{\sqrt{3}} = 286 \text{ pm}$$

3H.9   (a) a = length of side for a unit cell; for an fcc unit

cell, $a = \sqrt{8}\,r$ or $2\sqrt{2}\,r$ = 404 pm.

$$V = a^3 = (404 \text{ pm} \times 10^{-12} \text{ m} \cdot \text{pm}^{-1})^3 = 6.59 \times 10^{-29} \text{ m}^3 = 6.59 \times 10^{-23} \text{cm}^3.$$

Because for a fcc unit cell there are four atoms per unit cell, we have

$$\text{mass(g)} = 4 \text{ Al atoms} \times \frac{1 \text{ mol Al atoms}}{6.022 \times 10^{23} \text{ atoms} \cdot \text{mol}^{-1}} \times \frac{26.98 \text{ g}}{\text{mol Al atoms}}$$

$$= 1.79 \times 10^{-22} \text{ g}$$

$$d = \frac{1.79 \times 10^{-22} \text{ g}}{6.59 \times 10^{-23} \text{ cm}^3} = 2.72 \text{ g} \cdot \text{cm}^{-3}$$

(b) $a = \dfrac{4r}{\sqrt{3}} = \dfrac{4 \times 227 \text{ pm}}{\sqrt{3}} = 524 \text{ pm}$

$$V = (524 \times 10^{-12} \text{ m})^3 = 1.44 \times 10^{-28} \text{ m}^3 = 1.44 \times 10^{-22} \text{ cm}^3$$

There are two atoms per bcc unit cell:

$$\text{mass(g)} = 2 \text{ K atoms} \times \frac{1 \text{ mol K atoms}}{6.022 \times 10^{23} \text{ atoms} \cdot \text{mol}^{-1}} \times \frac{39.10 \text{ g}}{\text{mol K atoms}}$$

$$= 1.30 \times 10^{-22} \text{ g}$$

$$d = \frac{1.30 \times 10^{-22} \text{ g}}{1.60 \times 10^{-22} \text{ cm}^3} = 0.813 \text{ g} \cdot \text{cm}^{-3}$$

3H.11 a = length of unit cell edge

$$V = \frac{\text{mass of unit cell}}{d}$$

(a)

$$V = a^3 = \frac{(1 \text{ unit cell})\left(\dfrac{195.09 \text{ g Pt}}{\text{mol Pt}}\right)\left(\dfrac{1 \text{ mol Pt}}{6.022 \times 10^{23} \text{ atoms Pt}}\right)\left(\dfrac{4 \text{ atoms}}{1 \text{ unit cell}}\right)}{21.45 \text{ g} \cdot \text{cm}^3}$$

$$= 6.041 \times 10^{-23} \text{ cm}^3$$

$$a = 3.924 \times 10^{-8} \text{ cm}$$

Because for an fcc cell, $a = \sqrt{8}\, r$, $r = \dfrac{\sqrt{2}\, a}{4} = \dfrac{\sqrt{2}\,(3.92 \times 10^{-8} \text{ cm})}{4}$

$$= 1.387 \times 10^{-8} \text{ cm} = 138.7 \text{ pm}$$

(b) $V = a^3$

$$= \frac{(1 \text{ unit cell})\left(\dfrac{180.95 \text{ g Ta}}{1 \text{ mol Ta}}\right)\left(\dfrac{1 \text{ mol Ta}}{6.022 \times 10^{23} \text{ atoms Ta}}\right)\left(\dfrac{2 \text{ atoms}}{1 \text{ unit cell}}\right)}{16.65 \text{ g} \cdot \text{cm}^3}$$

$$= 3.609 \times 10^{-23} \text{ cm}^3$$

$$a = 3.305 \times 10^{-8} \text{ cm}$$

$$r = \frac{\sqrt{3}\, a}{4} = \frac{\sqrt{3}\left(3.305 \times 10^{-8} \text{ cm}\right)}{4} = 1.431 \times 10^{-8} \text{ cm} = 143.1 \text{ pm}$$

3H.13 Let's assume that rhodium metal structure is *ccp* (*fcc*). In *ccp* structure, there are four atoms in each unit cell. The atomic mass (M) of rhodium is 102.91 g/mol. Therefore, the density of rhodium is:

$d = m/a^3$, m = mass of one unit cell = $4M/N_A$, a = the length of the side of a *fcc* unit cell = $8^{1/2}r$

$$d = \frac{4M/N_A}{(8^{1/2}r)^3} = \frac{4M}{N_A(8^{1/2}r)^3}$$

$$= \frac{4 \text{ atoms} \times (102.91 \text{g} \cdot \text{mol}^{-1})}{(6.022 \times 10^{23} \text{ atom} \cdot \text{mol}^{-1}) \times (8^{1/2} \times 1.34 \times 10^{-8} \text{cm})^3} = 12.55 \text{ g} \cdot \text{cm}^{-3}$$

Then we assume the rhodium metal structure is bcc. In bcc structure, there are two atoms in each unit cell, and a = the length of the side of a

bcc unit cell $= 4r/3^{1/2}$. The density of rhodium is:

$$d = \frac{2M/N_A}{(4r/3^{1/2})^3} = \frac{2M}{N_A(4r/3^{1/2})^3}$$

$$= \frac{2 \text{ atoms} \times (102.91 \text{g} \cdot \text{mol}^{-1})}{(6.022 \times 10^{23} \text{ atom} \cdot \text{mol}^{-1}) \times (4 \times 1.34 \times 10^{-8} \text{cm}/3^{1/2})^3} = 11.53 \text{ g} \cdot \text{cm}^{-3}$$

The density value for the ccp structure is closer to 12. 42 $\text{g} \cdot \text{cm}^{-3}$.

Therefore, rhodium metal is close-packed structure.

3H.15 (a) The volume of the unit cell is

$$V_{\text{unit cell}} = \left(543 \text{ pm} \times \frac{10^{-12} \text{ m}}{\text{pm}} \times \frac{100 \text{ cm}}{\text{m}}\right)^3 = 1.601 \times 10^{-22} \text{ cm}^3$$

mass in unit cell $= (1.601 \times 10^{-22} \text{ cm}^3) \times (2.33 \text{ g} \cdot \text{cm}^{-3}) = 3.73 \times 10^{-22} \text{ g}$

(b) The mass of a Si atom is

$28.09 \text{ g} \cdot \text{mol}^{-1} \div (6.022 \times 10^{23})$ atoms $\cdot \text{mol}^{-1} = 4.665 \times 10^{-23} \text{ g} \cdot \text{atom}^{-1}$.

Therefore, there are $3.73 \times 10^{-22} \text{ g} \div 4.665 \times 10^{-23} \text{ g} \cdot \text{atom}^{-1} = 8$ atoms per

unit cell.

3H.17 Volume of a cylinder $=$ base area $\times$ length $= \pi r^2 l$;

Volume of a triangular prism $=$ base area $\times$ length $=$ (bhl)/2 $= \sqrt{3}r^2 l$ for a

triangle inscribed in the centers of three touching cylinder bases;

Each of the inscribed triangle's 60° angles accounts for 1/6 of the volume

of each of the three touching cylinders, or a total cylinder volume of (3/6)

$\pi r^2 l$. So the percent of the space occupied by the cylinders is (0.5) $\pi r^2 l$

*100/ $\sqrt{3}r^2 l = 0.5\pi/\sqrt{3} = 90.7\%$.

3H.19 In diamond, carbon is sp^3 hybridized and forms a tetrahedral, three-

dimensional network structure, which is extremely rigid. Graphite carbon

is sp^2 hybridized and planar. Its application as a lubricant results from the

fact that the two-dimensional sheets can "slide" across one another, thereby reducing friction. In graphite, the unhybridized p-electrons are free to move from one carbon atom to another, which results in its high electrical conductivity. In diamond, all electrons are localized in sp^3 hybridized C–C σ-bonds, so diamond is a poor conductor of electricity.

3H.21 (a) Carbon atomic radius is 77 pm. The number of carbon atom on the 2.0 cm × 1.0 cm tape is:

$$\frac{2.0 \; cm \times 1.0 \; cm}{\pi (77 \times 10^{-10} \; cm)^2} = 1.1 \times 10^{16} \; C \; atoms$$

(b) $\left(1.1 \times 10^{16} C \quad atoms\right)\left(\dfrac{1 \; mol \quad C \quad atoms}{6.022 \times 10^{23} atoms \cdot mol^{-1}}\right) = 1.8 \times 10^{-8} mol$

$$\left(1.8 \times 10^{-8} mol\right)\left(\frac{1 \; nmol}{10^{-9} mol}\right) = 18 \; nmol$$

3H.23 (a) Indium arsenide has a (4,4)-coordination because each unit cell has four indium and four arsenic; (b) The formula of indium arsenide is InAs.

3H.25 (a) There are eight chloride ions at the eight corners, giving a total of

$$8 \; corners \times \tfrac{1}{8} \; atom \cdot corner^{-1} = 1 \; Cl^- \; ion$$

There is one Cs^+ that lies at the center of the unit cell. All of this ion belongs to the unit cell. The ratio is thus 1 : 1 for an empirical formula of CsCl, with one formula unit per unit cell.

(b) The titanium atoms lie at the corners of the unit cell and at the body center: $8 \; corners \; \times \tfrac{1}{8} \; atom \cdot corner^{-1} + 1 \; at \; body \; center = 2 \; atoms$ per unit cell. Four Oxygen atoms lie on the faces of the unit cell and two lie completely within the unit cell, giving:

$$4 \; atoms \; in \; faces \times \tfrac{1}{2} \; atom \cdot face^{-1} + 2 \; atoms \; wholly \; within \; cell = 4 \; atoms$$

The ratio is thus two Ti per four O, or an empirical formula of TiO_2 with two formula units per unit cell.

(c) The Ti atoms are 6-coordinate and the O atoms are 3-coordinate.

3H.27 (a) Each unit cell has one rhenium $(8 \times \frac{1}{8})$ and three oxide ion $(12 \times \frac{1}{4})$. Therefore, rhenium oxide has a (6,2)-coordination. (b) The formula of rhenium oxide is ReO_3.

3H.29 (a) Ratio $= \dfrac{133 \text{ pm}}{152 \text{ pm}} = 0.875$, predict cesium-chloride structure with (8,8)-coordination; however, rubidium fluoride actually adopts the rock-salt structure

(b) Ratio $= \dfrac{72 \text{ pm}}{140 \text{ pm}} = 0.51$, predict rock-salt structure with (6,6)-coordination

(c) Ratio $= \dfrac{102 \text{ pm}}{196 \text{ pm}} = 0.520$, predict rock-salt structure with (6,6)-coordination

3H.31 (a) In the rock-salt structure, the unit cell edge length is equal to two times the radius of the cation plus two times the radius of the anion. Thus for CaO, $a = 2(100 \text{ pm}) + 2(140 \text{ pm}) = 480 \text{ pm}$. The volume of the unit cell will be given by (converting to cm^3 because density is normally given in terms of $g \cdot cm^{-3}$)

$$V = \left(480 \text{ pm} \times \frac{10^{-12} \text{ m}}{\text{pm}} \times \frac{100 \text{ cm}}{\text{m}} \right)^3 = 1.11 \times 10^{-22} \text{ cm}^3$$

There are four formula units in the unit cell, so the mass in the unit cell will be given by

$$\text{mass in unit cell} = \frac{\left(\dfrac{4 \text{ formula units}}{1 \text{ unit cell}} \right) \times \left(\dfrac{56.08 \text{ g CaO}}{1 \text{ mol CaO}} \right)}{6.022 \times 10^{23} \text{ molecules} \cdot \text{mol}^{-1}} = 3.725 \times 10^{-22} \text{ g}$$

The density will be given by the mass in the unit cell divided by the volume of the unit cell:

$$d = \frac{3.725 \times 10^{-22} \text{ g}}{1.11 \times 10^{-22} \text{ cm}^3} = 3.36 \text{ g} \cdot \text{cm}^{-3}$$

(b) For a cesium chloridelike structure, it is the body diagonal that represents two times the radius of the cation and plus two times the radius of the anion. Thus the body diagonal for CsBr is equal to

$2(167 \text{ pm}) + 2(196) = 726 \text{ pm.}$

For a cubic cell, the body diagonal $= \sqrt{3} \, a = 726 \text{ pm}$

$$a = 419 \text{ pm}$$

$$a^3 = V = \left(419 \text{ pm} \times \frac{10^{-12} \text{ m}}{\text{pm}} \times \frac{100 \text{ cm}}{\text{m}} \right)^3 = 7.36 \times 10^{-23} \text{ cm}^3$$

There is one formula unit of CsBr in the unit cell, so the mass in the unit cell will be given by

$$\text{mass in unit cell} = \frac{\left(\dfrac{1 \text{ formula units}}{1 \text{ unit cell}} \right) \left(\dfrac{212.82 \text{ g CsBr}}{1 \text{ mol CsBr}} \right)}{6.022 \times 10^{23} \text{ molecules} \cdot \text{mol}^{-1}} = 3.534 \times 10^{-22} \text{ g}$$

$$d = \frac{3.534 \times 10^{-22} \text{ g}}{7.36 \times 10^{-23} \text{ cm}^3} = 4.80 \text{ g} \cdot \text{cm}^3$$

3H.33 (a) The smallest possible rectangular unit cell is as follows:

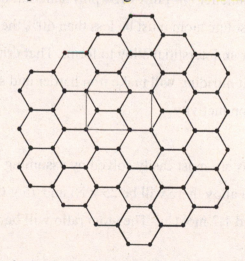

(b) There are four of carbon atoms in each unit cell.

(c) The coordination number is three (C_3).

3H.35 Each unit cell contains four RbI; Total of unit cells = $(6.022 \times 10^{23}$ RbI molecules/mole)/(4 RbI molecules/unit cell) = 1.505×10^{23} unit cells/ mole; The volume of one unit cell = $(732.6 \text{ pm})^3$.

Assume that the edge length of a cubic crystal of RbI that contains one mole RbI is X, then the volume of one mole RbI $= X^3/\text{mole} =$

$$\left(\frac{(732.6 \,\text{pm})^3}{\text{unit cell}} \right) \times \left(\frac{1.505 \times 10^{23} \,\text{unit cell}}{\text{mole}} \right);$$

$$X = (732.6 \text{ pm}) \times \sqrt[3]{1.505 \times 10^{23}} = 3.897 \times 10^{10} \text{ pm} = 3.897 \text{ cm}$$

3I.1 $D_{\text{Al}} = \dfrac{m_1}{V} = \dfrac{26.98n}{V}$;

$$D_{\text{magnalium}} = \frac{m_2}{V} = \frac{(26.89 \times 70\%n + 24.31 \times 30\%n)}{V} = \frac{26.18n}{V}$$

$$D = \frac{D_{\text{magnalium}}}{D_{\text{Al}}} = 0.9703$$

3I.3 Alloys are usually (1) harder and more brittle, and (2) poorer conductors of electricity than the metals from which they are made.

3I.5 (a) The alloy is undoubtedly interstitial because the atomic radius of nitrogen is much smaller (74 pm vs 124 pm) than that of iron. The rule of thumb is that the solute atom must be less than 60% the solvent atom in radius in order for an interstitial alloy to form. That criterion is met here.
(b) We expect that nitriding will make iron harder and stronger, with a lower electrical conductivity.

3I.7 (a) These problems are most easily solved by assuming 100 g of substance. In 100 g of Ni-Cu alloy there will be 25 g Ni and 75 g Cu, corresponding to 0.43 mol Ni and 1.2 mol Cu. The atom ratio will be the same as mole ratio:

$$\frac{1.2 \text{ mol Cu}}{0.43 \text{ mol Ni}} = 2.8 \text{ Cu per Ni}$$

(b) Pewter, which is 7% Sb, 3% Cu, and 90% Sn, will contain 7 g Sb, 3 g Cu, and 90 g Sn per 100 g of alloy. This will correspond to 0.06 mol Sb, 0.05 mol Cu, and 0.76 mol Sn per 100 g alloy. The atom ratio will be 15 Sn: 1.2 Sb : 1 Cu.

3I.9 (a) By viewing the unit cell from different directions, it is clear that it belongs to a hexagonal crystal system. (b) There are eight carbonate ions on edges, giving ¼ × 8 = 2 carbonate ions, plus there are four carbonate ions completely within the unit cell. The total number of carbonate ions in the unit cell is six. Calcium ions lie at the corners of the unit cell ($\frac{1}{8}$ × 8 =1) as well as on the edges (¼ × 4) and within the cell (4). The total number of calcium ions in the cell is six, agreeing with the overall stoichiometry of calcite, $CaCO_3$.

3I.11 $D = \dfrac{m}{V}, \qquad m = DV, \qquad V = \dfrac{m}{D}$

$$V = \frac{m}{D} = \frac{19.28 \text{ g} \cdot \text{cm}^{-3} \times 4.0 \text{ cm}^3}{5.01 \text{ g} \cdot \text{cm}^{-3}} = 15.4 \text{ cm}^3$$

3I.13

Formal charge: Si = 0, O = –1; Oxidation number: Si = +4, O = –2; this is an AX_4 VSEPR structure; therefore, the shape is tetrahedral.

3I.15 The $Si_2O_7^{6-}$ ion is built from two SiO_4^{4-} tetrahedral ions in which the silicate tetrahedral share one O atom. This is the only case in which one O is shared. (b) The pyroxenes, for example, jade, $NaAl(SiO_3)_2$, consist of chains of SiO_4 units in which two O atoms are shared by neighboring units. The repeating unit has the formula SiO_3^{2-}.

3J.1 The conductivity of a semiconductor increases with temperature as increasing numbers of electrons are promoted into the conduction band, whereas the conductivity of a metal will decrease as the motion of atoms will slow down the migration of electrons.

3J.3 The diameter of silver is: 144 pm × 2 = 288 pm; for a 2.4-mm length of one-dimensional line, the total number of Ag atoms:

$$\frac{2.4 \text{ mm} \times \dfrac{1.0 \times 10^9 \text{ pm}}{1 \text{ mm}}}{288 \text{ pm}} = 8.3 \times 10^6 \text{ Ag atoms}$$

Silver has two states since 5s is half filled. So, $n = 4.2 \times 10^6$

3J.5 (a) In and Ga; (b) P and Sb

3J.7 (a) In fluorescence, light absorbed by molecules is immediately emitted, whereas, in phosphorescence, molecules remain in an excited state for a length of time before emitting the absorbed light. So conservatively, emission of light in phosphorescence is much longer than that of fluorescence; (b) mechanistically, fluorescence involves the retention of the relative spin orientation of the excited electron where phosphorescence involves changes of electron spin state so relaxation is a slow process.

3J.9 Paramagnetic elements have at least one unpaired electron.
Sc = $3d^1 4s^2$ → Paramagnetic; Ti = $4s^2 3d^2$ → Paramagnetic
V = $3d^3 4s^2$ → Paramagnetic; Mn= $4s^2 3d^5$ → Paramagnetic

Fe = $3d^6\,4s^2$ → Ferromagnetic; Co= $4s^2\,3d^7$ → Ferromagnetic

Ni= $4s^2\,3d^8$ → Ferromagnetic; Y = $4d^1\,5s^2$ → Paramagnetic

Zr = $4d^2\,5s^2$ → Paramagnetic; Nb = $4d^4\,5s^1$ → Paramagnetic

Mo = $4d^5\,5s^1$ → Paramagnetic; Tc = $4d^5\,5s^2$ → Paramagnetic

Ru= $4d^7\,5s^1$ → Paramagnetic; Rh= $4d^8\,5s^1$ → Paramagnetic

Lu = $6s^2\,4f^{14}\,5d^1$ → Paramagnetic; Hf = $6s^2\,4f^{14}\,5d^2$ → Paramagnetic

Ta = $6s^2\,4f^{14}\,5d^3$ → Paramagnetic; W = $6s^2\,4f^{14}\,5d^4$ → Paramagnetic

Re = $6s^2\,4f^{14}\,5d^5$ → Paramagnetic; Os = $6s^2\,4f^{14}\,5d^6$ → Paramagnetic

Ir = $6s^2\,4f^{14}\,5d^7$ → Paramagnetic; Pt = $4f^{14}\,5d^9\,6s^1$ → Paramagnetic

Au = $4f^{14}\,5d^{10}\,6s^1$ → Paramagnetic; Zn = $3d^{10}\,4s^2$ → Diamagnetic

More information can be found from:
http://www.periodictable.com/Properties/A/MagneticType.html

3J.11 The compound is ferromagnetic below T_c because the magnetization is higher. Above the Curie temperature, the compound is simple paramagnet with randomly oriented spins, but below that temperature, the spins align and the magnetization increases.

3J.13 The reaction is as following (1:1 for CaS):

$CaCl_2(aq)$ + $(NH_4)_2S(aq)$ → $CaS(s)$ + $2\,NH_4Cl(aq)$

Assume 100% yield so the moles of CaS formed

= $(4.0 \times 10^{-3}\,L) \times 0.0015\,mol/L = 6.0 \times 10^{-6}$

Molarity of QDs = $(6.0 \times 10^{-6}\,mole/168)/8.0 \times 10^{-3}\,L = 4.5 \times 10^{-6}\,M$

3J.15 Given the expression of the cubic box:

$$E_{xyz} = \frac{h^2}{8m_e L^2}(n_x^2 + n_y^2 + n_z^2)$$

Where E_{xyz} represents the energy of a given level having n values corresponding to x, y and z we can describe each new level by incrementing each n by one; any levels with the same energy will be degenerate. The three lowest energy levels will therefore be E_{111}, E_{211}, and E_{221}. From this we can arrive at expressions for the energy of each level:

$$E_{111} = \frac{h^2}{8m_eL^2}(1^2+1^2+1^2) = \frac{3h^2}{8m_eL^2}$$

Similarly, the other two energy levels are determined to be:

$$E_{211} = \frac{h^2}{8m_eL^2}(2^2+1^2+1^2) = \frac{3h^2}{4m_eL^2} \quad \text{and}$$

$$E_{221} = \frac{h^2}{8m_eL^2}(2^2+2^2+1^2) = \frac{9h^2}{8m_eL^2}$$

The 211, 121, and 112 levels are degenerate; that is $E_{211} = E_{121} = E_{112}$.

The 221, 122, and 212 levels are degenerate; that is $E_{221} = E_{122} = E_{212}$.

3.1 (a) $x_{Ne} = \left(\frac{6}{10}\right) = 0.6; \quad x_{Ar} = \left(\frac{4}{10}\right) = 0.4$

At constant T and V, $\left(\frac{n_1}{P_1}\right) = \left(\frac{n_2}{P_2}\right) \rightarrow \left(\frac{x_{Ne}}{P_{Ne}}\right) = \left(\frac{x_{Ar}}{P_{Ar}}\right)$

$$P_{Ar} = \left(\frac{(0.4)(420 \text{ Torr})}{0.6}\right) = 280 \text{ Torr}$$

(b) $P_{total} = (280 \text{ Torr}) + (420 \text{ Torr}) = 700 \text{ Torr}$

3.3 The molar mass of glucose is $180.15 \text{ g} \cdot \text{mol}^{-1}$. From this, we can calculate the number of moles of glucose formed and, using the reaction stoichiometry, determine the number of moles of CO_2 needed. With that information and the other information provided in the problem, we can use the ideal gas law to calculate the volume of air that is needed:

$$PV = nRT$$

$$V = \frac{\left[\left(\frac{10.0 \text{ g glucose}}{180.15 \text{ g glucose} \cdot \text{mol}^{-1} \text{ glucose}}\right)\left(\frac{6 \text{ mol } CO_2}{1 \text{ mol glucose}}\right)\right]}{\left(\frac{0.26 \text{ Torr}}{760 \text{ Torr} \cdot \text{atm}^{-1}}\right)}$$

$$\times (0.082\ 06 \text{ L} \cdot \text{atm} \cdot \text{K}^{-1} \cdot \text{mol}^{-1})(298 \text{ K})$$

$$= 2.4 \times 10^4 \text{ L}$$

3.5   Use the ideal gas law to calculate the number of moles of HCl.

$$n = \frac{PV}{RT} = \frac{(690 \text{ Torr})(200 \text{ mL})}{(0.08206 \text{ L} \cdot \text{atm} \cdot \text{K}^{-1} \cdot \text{mL}^{-1})(293 \text{ K})} \cdot \frac{(1 \text{ atm})}{(760 \text{ Torr})} \cdot \frac{(1 \text{ L})}{(1000 \text{ mL})}$$

$$= 7.55 \times 10^{-3} \text{ mol HCl}$$

Since the reaction between HCl and NaOH occurs in a 1:1 mole ratio, this number of moles of NaOH is also present in the volume of NaOH(aq) required to reach the stoichiometric point of the titration. Therefore, the molarity of the NaOH solution is

$$\frac{\text{moles NaOH}}{1 \text{ L of solution}} = \frac{(7.55 \times 10^{-3} \text{ mol NaOH})}{(15.7 \text{ mL})} \cdot \frac{1000 \text{ mL}}{1 \text{ L}}$$

$$= 0.481 \text{ M}$$

3.7   (a) $N_2O_4(g) \rightarrow 2 NO_2(g)$; (b) If all the gas were $N_2O_4(g)$, then the moles can be calculated from the ideal gas equation:

$$P = \frac{nRT}{V}$$

$$= \frac{\left(\dfrac{43.78 \text{ g}}{92.02 \text{ g} \cdot \text{mol}^{-1}}\right)(0.082\,06 \text{ L} \cdot \text{atm} \cdot \text{K}^{-1} \cdot \text{mol}^{-1})(298 \text{ K})}{5.00 \text{ L}}$$

$$= 2.33 \text{ atm}$$

(c) The only difference in the calculation between part (b) and part (c) is that the molar mass of NO_2 is half that of N_2O_4.

$$P = \frac{nRT}{V}$$

$$= \frac{\left(\dfrac{43.78 \text{ g}}{46.01 \text{ g} \cdot \text{mol}^{-1}}\right)(0.082\,06 \text{ L} \cdot \text{atm} \cdot \text{K}^{-1} \cdot \text{mol}^{-1})(298 \text{ K})}{5.00 \text{ L}}$$

$$= 4.65 \text{ atm}$$

(d) Because both N_2O_4 and NO_2 are present, we need to determine some way of calculating the relative amounts of each present. This can be done by taking advantage of the gas law relationships. The total pressure at the end of the reaction will give us the total number of moles present:

$$P_{\text{total}} = 2.96 \text{ atm}$$

$$(2.96 \text{ atm})(5.00 \text{ L}) = n_{\text{total}}(0.082\,06 \text{ L} \cdot \text{atm} \cdot \text{K}^{-1} \cdot \text{mol}^{-1})(298 \text{ K})$$

$$n_{\text{total}} = 0.605 \text{ mol}$$

$$\therefore n_{N_2O4} + n_{NO_2} = 0.605 \text{ mol}$$

This gives us one equation, but we have two unknowns, so another relationship is needed. We can take advantage of knowing the stoichiometry of the reaction. If we assume that all of the gas begins at N_2O_4 and we allow some to react, we can write the following:

Initial amount of N_2O_4 0.476 mol

Amount of N_2O_4 that reacts x (mol)

Amount of NO_2 formed $2x$ (mol)

When the reaction is completed, there will be $0.476 - x$ mole of N_2O_4 and $2x$ mole NO_2. The total number of moles will be given by:

$$(0.476 - x) + 2x = n_{\text{total}}$$
$$0.605 \text{ mol} = 0.476 \text{ mol} + x$$
$$x = 0.129 \text{ mol}$$
$$n_{NO_2} = 2x$$
$$= 2(0.129 \text{ mol})$$
$$= 0.258 \text{ mol}$$
$$n_{N_2O_4} = 0.476 \text{ mol} - x$$
$$= 0.347 \text{ mol}$$
$$x_{NO_2} = \frac{0.258 \text{ mol}}{0.605 \text{ mol}} = 0.426$$
$$x_{N_2O_4} = \frac{0.347 \text{ mol}}{0.605 \text{ mol}} = 0.574$$

3.9   (a) The elemental analyses yield an empirical formula of NH_2. The formula unit has a mass of $16.02 \text{ g} \cdot \text{mol}^{-1}$. The mass, volume, pressure, and temperature data will allow us to calculate the molar mass, using the ideal gas equation:

$$PV = nRT$$

$$PV = \frac{m}{M} RT$$

$$M = \frac{mRT}{PV} = \frac{(0.473 \text{ g})(0.082\ 06 \text{ L} \cdot \text{atm} \cdot \text{K}^{-1} \cdot \text{mol}^{-1})(298 \text{ K})}{(1.81 \text{ atm})(0.200 \text{ L})} = 31.9 \text{ g} \cdot \text{mol}^{-1}$$

The molar mass divided by the mass of the empirical formula mass will give the value of n in the formula

$(NH_2)_n$. $31.9 \text{ g} \cdot \text{mol}^{-1} \div 16.02 \text{ g} \cdot \text{mol}^{-1} = 1.99$, so the molecular formula is N_2H_4, which corresponds to the molecule known as hydrazine.

(b)

$$\text{H} - \overset{\cdot\cdot}{\text{N}} - \overset{\cdot\cdot}{\text{N}} - \text{H}$$
$$\underset{\text{H}}{|} \quad \underset{\text{H}}{|}$$

(c)

$$\frac{rate_A}{rate_B} = \sqrt{\frac{M_B}{M_A}}$$

$$\frac{\dfrac{3.5 \times 10^{-4} \text{ mol}}{15.0 \text{ min}}}{\dfrac{X}{25.0 \text{ min}}} = \sqrt{\frac{32.05 \text{ g} \cdot \text{mol}^{-1}}{17.03 \text{ g} \cdot \text{mol}^{-1}}}$$

$$X = \left(\frac{3.5 \times 10^{-4} \text{ mol}}{15.0 \text{ min}} \right)(25.0 \text{ min}) \sqrt{\frac{17.03 \text{ g} \cdot \text{mol}^{-1}}{32.05 \text{ g} \cdot \text{mol}^{-1}}}$$

$$= 4.2 \times 10^{-4} \text{ mol}$$

3.11   Using v_{rms} = root mean square speed and Equation 5.21a (the Maxwell Distribution of Speeds):

$$\frac{f(10v_{rms})}{f(v_{rms})} = \frac{4\pi N \left(\dfrac{M}{2RT} \right)^{\frac{3}{2}} (10v_{rms})^2 e^{-M(10v_{rms})^2 / 2RT}}{4\pi N \left(\dfrac{M}{2RT} \right)^{\frac{3}{2}} (v_{rms})^2 e^{-M(v_{rms})^2 / 2RT}}$$

$$= \frac{100v_{rms}^2 e^{-M100v_{rms}^2/2RT}}{v_{rms}^2 e^{-Mv_{rms}^2/2RT}} = 100e^{(-M100v_{rms}^2/2RT + Mv_{rms}^2/2RT)}$$

$$= 100e^{-99Mv_{rms}^2/2RT}$$

The ratio is not independent of temperature since the variable T appears in the denominator of a negative exponent on e. It makes sense that the ratio should become bigger at higher temperatures as the distribution spreads out such that the number of molecules with higher speeds increases while the number with lower speeds decreases.

3.13 The molar mass calculation follows from the ideal gas law:

$$PV = nRT$$

$$PV = \frac{m}{M} RT$$

$$M = \frac{mRT}{PV} = \frac{(1.509 \text{ g})(0.08206 \text{ L} \cdot \text{atm} \cdot \text{K}^{-1} \cdot \text{mol}^{-1})(473 \text{ K})}{\left(\dfrac{745 \text{ Torr}}{760 \text{ Torr} \cdot \text{atm}^{-1}}\right)(0.235 \text{ L})}$$

$$= 254 \text{ g} \cdot \text{mol}^{-1}$$

If the molecular formula is OsO_x, then the molar mass will be given by:

$$190.2 \text{ g} \cdot \text{mol}^{-1} + x(16.00 \text{ g} \cdot \text{mol}^{-1}) = 254 \text{ g} \cdot \text{mol}^{-1})$$

$$x = 3.99$$

The formula is OsO_4.

3.15 (a) A, 300K; B, 500K; C, 1000K

$$\text{(b) } v_{rms} = \left(\frac{3RT}{M}\right)^{\frac{1}{2}} = \left(\frac{3 \times 8.314 \text{ J} \cdot \text{K}^{-1} \cdot \text{mol}^{-1} \times 500K}{2.802 \times 10^{-2} \text{ kg} \cdot \text{mol}^{-1}}\right)^{\frac{1}{2}} = 667. \text{ m} \cdot \text{s}^{-1}$$

3.17 (a) $ClNO_2$

(b) $ClNO_2$

(c) Formal charge for N = 5−4 = +1; for Cl = 7−6−1 = 0; for O with double bond with N = 6–4–2 = 0; for O with single bond with

N = 6–6–1= –1;

$$\ddot{O} = N - \ddot{O}: \qquad :\ddot{O} - N = \ddot{O}$$

$$:\overset{..}{\underset{..}{Cl}}: \qquad :\overset{..}{\underset{..}{Cl}}:$$

(d) trigonal planar

3.19 (a) Substitute the van der Waals (vdW) parameters for ammonia as well as the given values of *n, R, P,* and *T* into the vdW equation, then solve for V. Since the equation is cubic, solve graphically for the three roots or use an appropriate program such as Mathcad. Only one of the three roots is physically possible:

$$V^3 + n\left(\frac{RT + bP}{P}\right)V^2 + \left(\frac{n^2 a}{P}\right)V - \frac{n^3 ab}{P} = 0$$

$$V^3 + (0.505 \text{ mol})$$

$$\times \left(\frac{(0.08206 \text{ L} \cdot \text{atm} \cdot \text{K}^{-1} \cdot \text{mol}^{-1})(298 \text{ K}) + (0.0371 \text{L} \cdot \text{mol}^{-1})(95.0 \text{ atm})}{(95.0 \text{ atm})}\right)V^2$$

$$+ \left(\frac{(0.505 \text{ mol})^2(4.225 \text{ bar} \cdot \text{L}^2 \cdot \text{mol}^{-2})}{(95.0 \text{ atm})\left(\dfrac{1.013 \text{ bar}}{1 \text{ atm}}\right)}\right)V$$

$$- \frac{(0.505 \text{ mol})^3(4.169 \text{ L}^2 \cdot \text{atm} \cdot \text{mol}^{-2})(0.0371 \text{ L} \cdot \text{mol}^{-1})}{(95.0 \text{ atm})} = 0$$

$$V^3 + (0.149 \text{ L})V^2 + (0.0112 \text{ L}^2)V - 2.10 \times 10^{-4} \text{ L}^3 = 0$$

$$V = -0.0822, \ -0.0822, \text{ or } 0.0153 \text{ L}$$

but only the positive root is physically possible, so

$$V = 0.0153 \text{ L}$$

(b) Compare the volume calculated in part (a) to that of an ideal gas under the same conditions.

$$PV = nRT$$

$$V = \frac{(0.505 \text{ mol})(0.08206 \text{ L} \cdot \text{atm} \cdot \text{K}^{-1} \cdot \text{mol}^{-1})(298 \text{ K})}{95.0 \text{ atm}} = 0.130 \text{ L}$$

$V_{ideal} = 0.130 \text{ L} > V_{vdW,} = 0.0153 \text{ L}$. Attractive forces dominate because the van der Waals, or "real", gas occupies less volume than the "ideal" gas. If the molecules are attracted to one another they will behave less independently, reducing the effective number of moles of gas.

3.21 (a) $(CH_3)_2 N_2 H_2 (s) + 2 N_2 O_4 (l) \rightarrow 2 CO_2 (g) + 3 N_2 (g) + 4 H_2 O(l)$

$P_{total} = P_{CO_2} + P_{N_2} + P_{H_2O} = 2.50 \text{ atm}$

Since the mole ratio of the gaseous products is 2:3:4 and all three gases are present in the same vessel at the same temperature, the partial pressures will exhibit this ratio as well. Therefore,

$P_{total} = 2x + 3x + 4x = 9x = 2.5 \text{ atm}; \quad x = 0.278 \text{ atm}$

$P_{CO_2} = 2x = 0.556 \text{ atm}$

$P_{N_2} = 3x = 0.834 \text{ atm}$

$P_{H_2O} = 4x = 1.11 \text{ atm}$

(b) Since the volume of the vessel is the same, when CO_2 is removed, only $N_2(g)$ and $H_2O(g)$ remain in the vessel. The total pressure will Decrease: $P_{total} = 2.50 \text{ atm} - 0.556 \text{ atm} = 1.94 \text{ atm}$; but the partial pressure of $N_2(g)$ and $H_2O(g)$ will remain the same as those in part (a).

3.23 (a) The number of moles of nitrogen required to fill the air bag can be found from the ideal gas law.

$$? \text{ mol } N_2 = \frac{pV}{RT} = \frac{(1.37 \text{ atm})(57.0 \text{ L } N_2)}{(0.0821 \text{ L} \cdot \text{atm} \cdot \text{K}^{-1} \cdot \text{mol}^{-1})(298 \text{ K})} = 3.19 \text{ mol } N_2$$

The mass of sodium azide required to produce this amount of nitrogen depends upon the balanced chemical reaction.

$2 NaN_3 (s) \rightarrow 3 N_2 (g) + 2 Na(s)$

$$? \text{ g NaN}_3 = 3.19 \text{ mol } N_2 \left(\frac{2 \text{ mol NaN}_3}{3 \text{ mol } N_2} \right) \left(\frac{65.02 \text{ g NaN}_3}{1 \text{ mol NaN}_3} \right) = 138 \text{ g NaN}_3$$

(b) $v_{rms} = \sqrt{\dfrac{3RT}{M}}$

$$= \left(\dfrac{3(8.314 \text{ J} \cdot \text{K}^{-1} \cdot \text{mol}^{-1})(298 \text{ K})}{(28.0 \text{ g} \cdot \text{mol}^{-1})} \left(\dfrac{\text{kg} \cdot \text{m}^{-2} \cdot \text{s}^{-2}}{\text{J}} \right) \left(\dfrac{1000 \text{ g}}{\text{kg}} \right) \right)^{1/2}$$

$$= 515 \text{ m} \cdot \text{s}^{-1}$$

3.25　(a) The NI_3 Lewis structure is:

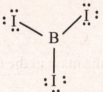

, the molecular shape is trigonal pyramidal, and it is a polar molecule. It can participate in dipole-dipole interactions.

(b) The BI_3 Lewis structure is:

, the molecular shape is trigonal planar, and it is a non-polar molecule. It cannot participate in dipole-dipole interactions.

3.27　(a) Pentane and 2,2-dimethylpropane are isomers; both have the chemical formula C_5H_{12}. We will assume that 2,2-dimethylpropane is roughly spherical and that all the hydrogen atoms lie on this sphere. The surface area of a sphere is given by $A = 4\pi r^2$. For this particular sphere,

$A = 4\pi(254 \text{ pm})^2 = 8.11 \times 10^5 \text{ pm}^2$. For pentane, the surface area of rectangular prism is

$2(295 \text{ pm} \times 766 \text{ pm}) +$

$2(295 \text{ pm} \times 692 \text{ pm}) + 2(692 \text{ pm} \times 766 \text{ pm}) = 1.92 \times 10^6 \text{ pm}^2$.

(b) Pentane has the larger surface area.

(c) The pentane should have the higher boiling point. It has a significantly larger surface area and should have stronger intermolecular forces between the molecules.

3.29 The unit cell for a cubic close-packed lattice is the fcc unit cell. For this cell, the relation between the radius of the atom r and the unit cell edge length a is

$$4r = \sqrt{2}\,a$$

$$a = \frac{4r}{\sqrt{2}}$$

The volume of the unit cell is given by

$$V = a^3 = \left(\frac{4r}{\sqrt{2}}\right)^3$$

If r is given in pm, then a conversion factor to cm is required:

$$V = a^3 = \left(\frac{4r}{\sqrt{2}} \times \frac{10^{-12}\text{ m}}{\text{pm}} \times \frac{100\text{ cm}}{\text{m}}\right)^3$$

Because there are four atoms per fcc unit cell, the mass in the unit cell will be given by

$$\text{mass} = \left(\frac{4\text{ atoms}}{\text{unit cell}}\right)\left(\frac{M}{6.022 \times 10^{23}\text{ atoms}\cdot\text{mol}^{-1}}\right)$$

The density will be given by

$$d = \frac{\text{mass of unit cell}}{\text{volume of unit cell}} = \frac{\left(\dfrac{4\text{ atoms}}{\text{unit cell}}\right) \times \left(\dfrac{M}{6.022 \times 10^{23}\text{ atoms}\cdot\text{mol}^{-1}}\right)}{\left(\dfrac{4\,r}{\sqrt{2}} \times \dfrac{10^{-12}\text{ m}}{\text{pm}} \times \dfrac{100\text{ cm}}{\text{m}}\right)^3}$$

$$= \frac{(2.936 \times 10^5)M}{r^3}$$

or

$$r = \sqrt[3]{\frac{(2.936 \times 10^5)M}{d}}$$

where M is the atomic mass in $\text{g}\cdot\text{mol}^{-1}$ and r is the radius in pm.

For the different gases we calculate the results given in the following table:

Gas	Density (g·cm⁻³)	Molar mass (g·mol⁻¹)	Radius (pm)
Neon	1.20	20.18	170
Argon	1.40	39.95	203
Krypton	2.16	83.80	225
Xenon	2.83	131.30	239
Radon	4.4	222	246

3.31 There are two approaches to this problem. The information given does not specify the radius of the tungsten atom. This value can be looked up in Appendix 2D. We can calculate an answer, however, based simply upon the fact that the density is 19.3 g·cm⁻³ for the bcc cell, by taking the ratio between the expected densities, based upon the assumption that the atomic radius of tungsten will be the same for both. The unit cell for a cubic close-packed lattice is the fcc unit cell. For this cell, the relation between the radius of the atom r and the unit cell edge length a is

$$4r = \sqrt{2}\, a$$

$$a = \frac{4r}{\sqrt{2}}$$

The volume of the unit cell is given by

$$V = a^3 = \left(\frac{4r}{\sqrt{2}}\right)^3$$

If r is given in pm, then a conversion factor to cm is required:

$$V = a^3 = \left(\frac{4r}{\sqrt{2}} \times \frac{10^{-12}\ \text{m}}{\text{pm}} \times \frac{100\ \text{cm}}{\text{m}}\right)^3$$

Because there are four atoms per fcc unit cell, the mass in the unit cell is given by

$$\text{mass} = \left(\frac{4 \text{ atoms}}{\text{unit cell}} \right) \left(\frac{M}{6.022 \times 10^{23} \text{ atoms} \cdot \text{mol}^{-1}} \right)$$

The density is given by

$$d = \frac{\text{mass of unit cell}}{\text{volume of unit cell}} = \frac{\left(\dfrac{4 \text{ atoms}}{\text{unit cell}} \right) \left(\dfrac{M}{6.022 \times 10^{23} \text{ atoms} \cdot \text{mol}^{-1}} \right)}{\left(\dfrac{4r}{\sqrt{2}} \times \dfrac{10^{-12} \text{ m}}{\text{pm}} \times \dfrac{100 \text{ cm}}{\text{m}} \right)^3}$$

$$= \frac{(2.936 \times 10^5)M}{r^3}$$

or

$$r = \sqrt[3]{\frac{(2.936 \times 10^5)M}{d}}$$

where M is the atomic mass in $\text{g} \cdot \text{mol}^{-1}$ and r is the radius in pm.
Likewise, for a body-centered cubic lattice there will be two atoms per
unit cell. For this cell, the relationship between the radius of the atom r
and the unit cell edge length a is derived from the body diagonal of the
cell, which is equal to four times the radius of the atom. The body
diagonal is found from the Pythagorean theorem to be equal to the $\sqrt{3}\, a$.

$$4r = \sqrt{3}\, a$$

$$a = \frac{4r}{\sqrt{3}}$$

The volume of the unit cell is given by

$$V = a^3 = \left(\frac{4r}{\sqrt{3}} \right)^3$$

If r is given in pm, then a conversion factor to cm is required:

$$V = a^3 = \left(\frac{4r}{\sqrt{3}} \times \frac{10^{-12} \text{ m}}{\text{pm}} \times \frac{100 \text{ cm}}{\text{m}} \right)^3$$

Because there are two atoms per bcc unit cell, the mass in the unit cell will
be given by

$$\text{mass} = \left(\frac{2 \text{ atoms}}{\text{unit cell}}\right) \times \left(\frac{M}{6.022 \times 10^{23} \text{ atoms} \cdot \text{mol}^{-1}}\right)$$

The density will be given by

$$d = \frac{\text{mass of unit cell}}{\text{volume of unit cell}} = \frac{\left(\dfrac{2 \text{ atoms}}{\text{unit cell}}\right)\left(\dfrac{M}{6.022 \times 10^{23} \text{ atoms} \cdot \text{mol}^{-1}}\right)}{\left(\dfrac{4\,r}{\sqrt{3}} \times \dfrac{10^{-12} \text{ m}}{\text{pm}} \times \dfrac{100 \text{ cm}}{\text{m}}\right)^3}$$

$$= \frac{\left(\dfrac{2 \text{ atoms}}{\text{unit cell}}\right)\left(\dfrac{M}{6.022 \times 10^{23} \text{ atoms} \cdot \text{mol}^{-1}}\right)}{(2.309 \times 10^{-10}\, r)^3}$$

$$= \frac{(2.698 \times 10^5)M}{r^3}$$

or

$$r = \sqrt[3]{\frac{(2.698 \times 10^5)M}{d}}$$

Setting these two expressions equal and cubing both sides, we obtain

$$\frac{(2.936 \times 10^5)M}{d_{\text{fcc}}} = \frac{(2.698 \times 10^5)M}{d_{\text{bcc}}}$$

The molar mass of tungsten M is the same for both ratios and will cancel from the equation.

$$\frac{(2.936 \times 10^5)}{d_{\text{fcc}}} = \frac{(2.698 \times 10^5)}{d_{\text{bcc}}}$$

Rearranging, we get

$$d_{\text{fcc}} = \frac{(2.936 \times 10^5)}{(2.698 \times 10^5)} d_{\text{bcc}}$$

$$= 1.088\, d_{\text{bcc}}$$

For W, $d_{\text{fcc}} = 1.088 \times 19.3 \text{ g} \cdot \text{cm}^3$

$$= 21.0 \text{ g} \cdot \text{cm}^{-3}$$

3.33 (a) The oxidation state of the titanium atoms must balance the charge on the oxide ions, O^{2-}. The presence of 1.18 O^{2-} ions means that the Ti present must have a charge to compensate the -2.36 charge on the oxide ions. The average oxidation state of Ti is thus $+2.36$.   (b) This is most easily solved by setting up a set of two equations in two unknowns. We know that the total charge on the titanium atoms present must equal 2.36, so if we multiply the charge on each type of titanium by the fraction of titanium present in that oxidation state and sum the values, we should get 2.36:

let x = fraction of Ti^{2+}, y = fraction of Ti^{3+}, then
$$2x + 3y = 2.36$$

Also, because we are assuming all the titanium is either $+2$ or $+3$, the fractions of each present must add up to 1:

$$x + y = 1$$

Solving these two equations simultaneously, we obtain $y = 0.36$, $x = 0.64$.

3.35   (a) True. If this is not the case, the unit cell will not match with other unit cells of the same type when stacked to form the entire lattice.

(b) False. Unit cells do not have to have atoms at the corners.

(c) True. In order for the unit cell to repeat properly, opposite faces must have the same composition.

(d) False. If one face is centered, the opposing face must be centered, but the other faces do not necessarily have to be centered.

3.37   (a) There is only one way to draw unit cells that will repeat to generate the entire lattice. The choice of unit cell is determined by conventions that are beyond the scope of this text (the smallest unit cell that indicates all of the symmetry present in the lattice is typically the one of choice). (b) There are two ways to draw the unit cells as shown below.

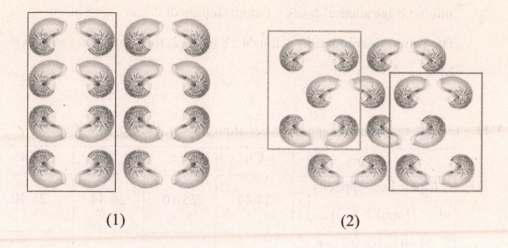

(1)   (2)

3.39   (a) The face diagonal the unit cell equals to 4r (r is the radius of the

buckminsterfullerene). Therefore, $(4r)^2 = 2(142pm)^2$

Solve the equation: $4r = \sqrt{2}\,(142)$ pm; $r = 50.2$ pm

(b) The radius of K^+ is 133 pm. In close-packed cubic unit cell (face

centered cubic unit cell), the tetrahedral holes is smaller than the octahedral

holes. The sizes of tetrahedral holes can be calculated as follows:

One unit cell can be divided into eight subunit cells and each subunit cell

contains one tetrahedral hole. The face diagonal of each subunit cell

$f^2 = (\frac{a}{2})^2 + (\frac{a}{2})^2$ (a is the edge length); the body diagonal $b^2 = f^2 + (\frac{a}{2})^2$

$b^2 = 2(\frac{a}{2})^2 + (\frac{a}{2})^2 = 3(\frac{a}{2})^2;$ $b = \sqrt{3}\,(\frac{a}{2}) = \sqrt{3}\,(\frac{142\,pm}{2}) = 123$ pm

The radius of the tetrahedral hole is: $r_{tetrahedral} + r_{buckminsterfullerene} = \frac{b}{2}$

$r_{tetrahedral} = \frac{123\,pm}{2} - 50.20$ pm $= 11.3$ pm, which is not large enough for a

K^+ ion. Therefore, K^+ must take octahedral holes.

Each unit cell has four octahedral holes and three are occupied by the K^+.

The percentage of holes filled are 75%.

3.41   For M, there is total of one atom in the unit cell from the corners;

cations: 8 corners $\times \frac{1}{8}$ atom $\cdot$ corner^{-1} + 6 faces + $\frac{1}{2}$ atom $\cdot$ face^{-1} = 4 atoms anions: 8 tetrahedral holes $\times$ 1 atom $\cdot$ tetrahedral hole^{-1} = 8 atoms

The cation to anion ratio is thus 4 : 8 or 1 : 2; the empirical formula is MA_2.

3.43   (a) The molar heat capacities per atom for Cu, Fe, Pb, and Zn are:

	Cu	Fe	Pb	Zn
Molar heat capacity $(J \cdot mol^{-1} \cdot K^{-1})$	24.44	25.10	26.44	25.40
Heat capacity per atom $(J \cdot mol^{-1} \cdot K^{-1} \cdot atom^{-1})$	4.06×10^{-23}	4.17×10^{-23}	4.39×10^{-23}	4.22×10^{-23}

Based on the data from above table, for pure metals, the heat capacity per atom for each metal is very similar, which means that the Dulong-Petit Law applies well here (a good approximation).

(b) The heat capacities per mole of atoms of CuO. FeS, PbBr$_2$, and ZnO are:

	CuO	FeS	PbBr$_2$	ZnO
Molar heat capacity $(J \cdot mol^{-1} \cdot K^{-1})$	42.30	50.54	80.12	40.25
Heat capacity per mole of atoms $(J \cdot K^{-1} \cdot mole\ of\ atom^{-1})$	21.15	25.27	26.71	20.12

From the data in the table above, it is clear that the heat capacities per mole of atoms are different for different ionic crystals. Therefore, Dulong and Petit's law only fits elements that have simple crystal structures, with low vibrational frequencies.

(c) Heat capacity per mole of atoms for $CuSO_4$ =

$(100.0\ J \cdot mol^{-1} \cdot K^{-1})(1\ mole\ CuSO_4/6\ mole\ atoms)$

$= 16.67\ \ J \cdot K^{-1} \cdot mole\ of\ atom^{-1}$

Heat capacity per mole of atoms for $PbSO_4$ =

$(103.21 \text{ J·mol}^{-1}\text{·K}^{-1})(1 \text{ mole } PbSO_4/6 \text{ mole atoms})$

= $17.20 \text{ J·K}^{-1}\text{·mole of atom}^{-1}$

From the results in (c) it seems that the atoms in a polyatomic ion act independently because the heat capacities per mole of atoms are almost the same even though they are two different molecules.

3.45 (a) In a simple cubic unit cell there is a total of one atom. The volume of the atom is given by $\frac{4}{3}\pi r^3$. The volume of the unit cell is given by a^3 and $a = 2r$, so the volume of the unit cell is $8r^3$. The fraction of occupied space in the unit cell is given by

$$\frac{\frac{4}{3}\pi r^3}{8r^3} = \frac{\frac{4}{3}\pi}{8} = 0.52$$

Fifty-two percent of the space is occupied, so 48% of this unit cell would be empty.

(b) The percentage of empty space in an fcc unit cell is 26%, so the fcc cell is much more efficient at occupying the space available.

3.47 The Lewis structure of $(HF)_3$ chain is as follows:

$$H\!-\!\ddot{\underset{\cdot\cdot}{F}}\!: \text{ --- } H\!-\!\ddot{\underset{\cdot\cdot}{F}}\!: \text{ --- } H\!-\!\ddot{\underset{\cdot\cdot}{F}}\!:$$

The bond angle is ~ 180 °.

3.49 For a bcc unit cell (see Example 6.3), d = $3^{3/2}$ (55.85 g/mol)/32(6.022×10^{23}/mole)×(1.24×10^{-8}cm)^3 = 7.90 g/cm^3

$(1.5 \text{ cm})^3 \times 7.90 \text{ g/cm}^3 \times (1 \text{ mole}/55.85 \text{ g}) = 0.477 \text{ mole Fe}$

$(1.00 \text{ atm} \times 15.5 \text{ L}) / (0.08206 \text{ L atm /K mol}) (298K) = 0.634 \text{ mol O}_2$

Fe is LR, so $(0.5)(0.477 \text{ mole})(159.70 \text{ g/mole}) = 38.1 \text{ g Fe}_2O_3$ is theoretical yield, or maximum mass that can be produced.

3.51 (a) They are classified at salts based on classical definition (M^{3+} and N^{3-}) (they are called interstitial nitrides). Alloys are a homogeneous (or heterogeneous) mixture or solid solution of two or more metals such as brass. Molecular compounds are compounds that are made of two or more nonmetal elements that are held together by covalent bonds, usually have low melting points. Ceramics are inorganic materials (nonmetallic) that have been harden by heating and are usually used for heat and electricity insulation; (b) the covalent character of MN is shown as the triple bond between nitrogen (N) and d-block metal (M); (c) because Group 11 and 12 metals have their d-orbitals totally filled with electrons and do not have space to hold the electrons from N

3.53 Y: $8 \text{ atoms} \times \frac{1}{8} \text{ atom} \cdot \text{corner}^{-1} = 1 \text{ Y atom}$

Ba: $8 \text{ atoms} \times \frac{1}{4} \text{ atom} \cdot \text{edge}^{-1} = 2 \text{ Ba atoms}$

Cu: 3 Cu atoms completely inside unit cell = 3 Cu atoms

O: $10 \text{ atoms on faces} \times \frac{1}{2} \text{ atom} \cdot \text{face}^{-1} + 2 \text{ atoms}$ completely inside unit cell = 7 O atoms

Formula = $YBa_2Cu_3O_7$

3.55 $SiC_{(s)} + 2NaOH_{(l)} + 2O_{2(g)} \rightarrow Na_2SiO_{3(s)} + H_2O_{(g)} + CO_{2(g)}$

3.57 Plot (a) was obtained from an organic dye and plot (b) was obtained from a quantum dot suspension. In an organic dye solution, the sizes of organic dye molecules are the same and the quantum levels (n) are well fixed, so a distinct absorption spectrum will be obtained for the organic dye molecules. For a quantum dot suspension, however, because of the diameter variation of quantum dots and variable quantum transition levels, the separation of energy levels will vary (as $1/r^2$). Therefore, the wavelength of light that can cause excitation will also vary. In addition, since the sizes of quantum dots vary only slightly, a continuous absorption spectrum is normally observed.

3.59   When semiconductors are at very low temperatures, the valence band is fully occupied and conduction band is empty, so the conductivity is low. As temperature increases in a vacuum, some electrons are promoted from valence band to the conduction band and cause increases in conductivity. However, when semiconductors are heated in oxygen, chemisorptions will happen and adsorbed oxygen on the surface will trap electrons forming oxygen negative ions, which will decrease the electron conductivity.

3.61   (a) The oxidation number of phosphorus in $Li_7P_3S_{11}$ is +5.

(b) The oxidation number of titanium in $BaTiO_3$ is +4.

3.63   (a) Crystalline phase. Because in a substance with glassy phase, their atoms form an intricate network and bonds do not lie in orderly arrays. Therefore, a substance with glassy phase will not cleave along a plane. However, in a substance with crystalline phase, atoms, ions, or molecules lie in an orderly array so a crystalline solid has long-range order, which is allowed to be cleaved along a plane; (b) Crystalline phase. To break apart a crystal of a network solid, covalent bonds must be broken. Therefore, network solids are very hard and rigid. However, a substance with glassy phase has high flexibility due to its noncrystal atomic arrangement; (c) Crystalline phase. The reason is the component atoms, molecules, or ions are in an orderly array and have the same type of neighbors, and the intermolecular forces holding the crystalline solid together are uniform. Therefore, the thermal energy to break every interaction will be the same amount; (d) Glassy phase. Because a substance with glassy phase does not contain tiny crystalline, which can scatter the light away, light can easily pass through.

3.65 When glass is etched by HF, the following reaction happens:

$$SiO_2(s) + 6HF(aq) \rightarrow SiF_6^{2-}(aq) + 2H_3O^+(aq)$$

The oxidation number of silicon in the anion is +4.

3.67 Ionic fluorides react with water to liberate HF, which then reacts with the glass. Glass bottles used to store metal fluorides become brittle and may disintegrate upon standing on the shelf.

3.69 (a) $4\,FeS_2(s) + 11\,O_2(g) \rightarrow 2\,Fe_2O_3(s) + 8\,SO_2(g)$

(b) First we calculate the number of moles of O_2 present at the conditions given, using the ideal gas law:

$$PV = nRT$$

$$(2.33\ \text{atm})(75.0\ \text{L}) = n\,(0.082\,06\ \text{L} \cdot \text{atm} \cdot \text{K}^{-1} \cdot \text{mol}^{-1})(423\ \text{K})$$

$$n = 5.03\ \text{mol}$$

Then, using the stoichiometry of the reaction and the molar mass of Fe_2O_3, we can calculate the mass of Fe_2O_3 produced:

$$\text{mass of } Fe_2O_3 = (159.70\ \text{g } Fe_2O_3 \cdot \text{mol}^{-1}\ Fe_2O_3)(5.03\ \text{mol } O_2)$$

$$\times \left(\frac{2\ \text{mol } Fe_2O_3}{11\ \text{mol } O_2} \right)$$

$$= 146\ \text{g } Fe_2O_3$$

(c) The number of moles of SO_2 produced can be obtained from the stoichiometry:

$$n_{SO_2} = (5.03\ n_{O_2}) \left(\frac{8\ n_{SO_2}}{11\ n_{O_2}} \right) = 3.66\ \text{mol}$$

3.66 mol of SO_2 will dissolve in 5.00 L of water to form a solution that is $(3.66\ \text{mol} \div 5.00\ \text{L}) = 0.732\ \text{M}$ in H_2SO_3, according to the equation

$$SO_2(g) + H_2O(l) \rightarrow H_2SO_3(aq)$$

(d) mass of $SO_2 = 1 \times 10^3$ kg coal $\left(\dfrac{5 \times 10^3 \text{ g FeS}_2}{100 \text{ kg coal}} \right) \left(\dfrac{1 \text{ mol FeS}_2}{119.99 \text{ g FeS}_2} \right)$

$$\times \left(\dfrac{8 \text{ mol SO}_2}{4 \text{ mol FeS}_2} \right) \left(\dfrac{64.07 \text{ g SO}_2}{1 \text{ mol SO}_2} \right)$$

$$= 5 \times 10^4 \text{ g SO}_2 = 50 \text{ kg SO}_2$$

(e) $n_{SO_2} = 5 \times 10^4 \text{ g SO}_2 \left(\dfrac{1 \text{ mol SO}_2}{64.07 \text{ g SO}_2} \right) = 800 \text{ mol SO}_2$

$$V = \frac{nRT}{P} = \frac{(800 \text{ mol}) (0.082\,06 \text{ L} \cdot \text{atm} \cdot \text{K}^{-1} \cdot \text{mol}^{-1})(298 \text{ K})}{(1.00 \text{ atm})}$$

$$= 2 \times 10^4 \text{ L}$$

(f) We find the pressure of $SO_2(g)$ originally present by difference.

The initial data gives us the total number of moles present, whereas the

data for the gas sample after being passed over $CaSO_3(s)$ represents the

number of moles of $N_2(g)$.

$PV = nRT$

$$n_{total} = \frac{PV}{RT} = \frac{(1.09 \text{ atm})(0.500 \text{ L})}{(0.082\,06 \text{ L} \cdot \text{atm} \cdot \text{K}^{-1} \cdot \text{mol}^{-1})(298 \text{ K})} = 0.0223 \text{ mol}$$

$$n_{N_2} = \frac{PV}{RT} = \frac{(1.09 \text{ atm})(0.150 \text{ L})}{(0.082\,06 \text{ L} \cdot \text{atm} \cdot \text{K}^{-1} \cdot \text{mol}^{-1})(323 \text{ K})} = 0.006\,17 \text{ mol}$$

The number of moles of SO_2 gas is

$0.0223 - 0.006\,17$ mol $= 0.0161$ mol.

The partial pressure will be given by the mole fraction multiplied by

the total pressure. The mole fraction of $SO_2(g)$ will be

0.0161 mol $\div 0.0223$ mol $= 0.722$. The pressure due to SO_2 in the

original mixture is $(0.722)(1.09 \text{ atm}) = 0.787$ atm.

The mass of SO_2 will be obtained by multiplying the number of moles of

SO_2 by the molar mass of SO_2 :

$$m_{SO_2} = (0.0161 \text{ mol})(64.06 \text{ g} \cdot \text{mol}^{-1}) = 1.03 \text{ g}$$

(g) By using ideal gas law and van der Waals equation to calculate the pressures at different moles of SO_2, and the results are listed in the following table:

Moles of SO_2	P (ideal) atm	P (real) atm	Percentage deviation
0.100	2.4618	2.407213	2.267647
0.200	4.9236	4.705575	4.633339
0.300	7.3854	6.895579	7.103409
0.400	9.8472	8.97773	9.684746
0.500	12.309	10.95254	12.38485

(h) The percentage deviation is calculated and included in the table in (g)

(i) Intermolecular attraction (a term) has larger effect on the pressure of SO_2.

(j) Based the data from the table in (g), when p = ~5 atm, SO_2 becomes a "real" gas. To be exactly, when observed P =5.04 atm, SO_2 becomes "real" gas. The results can be obtained when moles of SO_2 = 0.215 moles.

FOCUS 4

THERMODYNAMICS

4A.1 (a) isolated; (b) closed; (c) isolated; (d) open; (e) closed; (f) open

4A.3 (a) Work is given by $w = -P_{ext}\Delta V$. The applied external pressure is known, but we must calculate the change in volume given the physical dimensions of the pump and the distance, d, the piston in the pump moves:

$$\Delta V = -\pi r^2 d = \pi (1.5\text{cm})^2 (20\text{cm})(\frac{1\text{L}}{1000\text{cm}^3}) = -0.14\text{L}$$

ΔV is negative because the air in the pump is compressed to a smaller volume; work is then:

$$w = -(2.00 \text{ atm})(-0.14 \text{ L})(\frac{101.325 \text{ J}}{\text{L}\cdot\text{atm}}) = 28 \text{ J}$$

(b) Work on the air is positive by convention as work is done on the air, it is compressed.

(c) The internal energy of a system may be changed by doing work and in the absence of other changes, $\Delta U = w = 28$ J.

4A.5 (a) Irreversible expansion:

$w = -P\Delta V$

$w = -(1.00\text{atm})(1.20\text{L})$

$w = -122\text{J}$

(b) Reversible expansion does more work:

$$w = -nRT \ln \frac{V_2}{V_1}$$

$$25°C = 298.15K$$

$$w = -(0.200 \text{mol})(8.314 \text{J} \cdot \text{K}^{-1} \cdot \text{mol}^{-1})(298.15 \text{K})(\ln \frac{2.40 \text{L}}{1.20 \text{L}})$$

$$w = -344 \text{J}$$

4A.7 (a) The heat change will be made up of two terms: one term to raise the temperature of the copper and the other to raise the temperature of the water:

$$q = (300.0 \text{ g})(4.18 \text{ J} \cdot (°\text{C})^{-1} \cdot \text{g}^{-1})(100.0°\text{C} - 20.0°\text{C})$$
$$+ (400.0 \text{ g})(0.38 \text{ J} \cdot (°\text{C})^{-1} \cdot \text{g}^{-1})(100.0°\text{C} - 20.0°\text{C})$$
$$= 1.00 \times 10^5 \text{ J} + 1.22 \times 10^4 \text{ J} = 1.12 \times 10^5 \text{ J} = 1.12 \times 10^2 \text{ kJ}$$

(b) The percentage of heat attributable to raising the temperature of water

will be: $\left(\dfrac{1.00 \times 10^5 \text{ J}}{1.12 \times 10^5 \text{ J}} \right)(100) = 89\%$

4A.9 heat lost by metal = – heat gained by water

$$(20.0 \text{ g})(T_{\text{final}} - 100.0°\text{C})(0.38 \text{ J} \cdot (°\text{C})^{-1} \cdot \text{g}^{-1})$$
$$= -(50.7 \text{ g})(4.18 \text{ J} \cdot (°\text{C})^{-1} \cdot \text{g}^{-1})(T_{\text{final}} - 22.0°\text{C})$$

$$(T_{\text{final}} - 100.0°\text{C})(7.6 \text{ J} \cdot (°\text{C})^{-1}) = -(212 \text{ J} \cdot (°\text{C})^{-1})(T_{\text{final}} - 22.0°\text{C})$$

$$T_{\text{final}} - 100.0°\text{C} = -28(T_{\text{final}} - 22.0°\text{C})$$

$$T_{\text{final}} + 28\, T_{\text{final}} = 100.0°\text{C} + 616°\text{C}$$

$$29\, T_{\text{final}} = 716°\text{C}$$

$$T_{\text{final}} = 25°\text{C}$$

4A.11 $C_{\text{cal}} = \dfrac{22.5 \text{ kJ}}{23.97°\text{C} - 22.45°\text{C}} = 14.8 \text{ kJ} \cdot (°\text{C})^{-1}$

4A.13

$$C = q / \Delta T$$

$$C = (3.50 \text{ kJ}) / (7.32 \text{ K})$$

$$C = 0.478 \text{ kJ} \cdot \text{K}^{-1}$$

$$q_{\text{calorimeter}} = (C)(\Delta T)$$

$$q_{\text{calorimeter}} = (0.478 \text{ kJ} \cdot \text{K}^{-1})(2.49 \text{ K})$$

$$q_{\text{calorimeter}} = 1.19 \text{ kJ}$$

$$q_{\text{reaction}} + q_{\text{calorimeter}} = 0$$

$$q_{\text{reaction}} = -q_{\text{calorimeter}} = -1.19 \text{ kJ}$$

4B.1 The change in internal energy ΔU is given simply by summing the two energy terms involved in this process. We must be careful, however, that the signs on the energy changes are appropriate. In this case, internal energy will be added to the gas sample by heating and through compression. Therefore, the change in internal energy is:

$$\Delta U = 524 \text{ kJ} + 340 \text{ kJ} = 864 \text{ kJ}$$

4B.3 (a) The internal energy increased by more than the amount of heat added. Therefore, the extra energy must have come from work done on the system.

(b) $w = \Delta U - q = 982 \text{ J} - 492 \text{ J} = +4.90 \times 10^2 \text{ J}.$

4B.5 To get the entire internal energy change, we must sum the changes due to heat and work. In this problem, $q = +5.50 \text{ kJ}$. Work will be given by $w = -P_{ext} \Delta V$ because it is an expansion against a constant opposing pressure:

$$w = -\left(\frac{750 \text{Torr}}{760 \text{Torr} \cdot \text{atm}^{-1}}\right)\left(\frac{1846 \text{mL} - 345 \text{mL}}{1000 \text{mL} \cdot \text{L}^{-1}}\right) = -1.48 \text{L} \cdot \text{atm}$$

To convert to J we use the equivalency of the ideal gas constants:

$$w = -(1.48 \text{L} \cdot \text{atm})\left(\frac{8.314 \text{J} \cdot \text{K} \cdot \text{mol}^{-1}}{0.08206 \text{L} \cdot \text{atm} \cdot \text{K}^{-1} \cdot \text{mol}^{-1}}\right) = -1.50 \times 10^2 \text{J}$$

$$\Delta U = q + w = 5.50\,\text{kJ} - 0.150\,\text{kJ} = 5.35\,\text{kJ}$$

4B.7   Using $\Delta U = q + w$, where $\Delta U = -2573\,\text{kJ}$ and $q = -947\,\text{kJ}$,

$-2573\,\text{kJ} = -947\,\text{kJ} + \text{w}$.

Therefore, $\text{w} = -1626\,\text{kJ}$.

1626 kJ of work can be done by the system on its surroundings.

4B.9   (a) true if no work is done;   (b) always true;   (c) always false;   (d) true only if $w = 0$ (in which case $\Delta U = q = 0$);   (e) always true

4B.11   (a) During melting heat is absorbed and q is positive.

Since the change is occurring at constant temperature $\Delta E = 0$.

Therefore, work is done by the system and w is negative.

(b) During condensation heat is released and q is negative.

Since the change is occurring at constant temperature $\Delta E = 0$.

Therefore, work is done on the system and w is positive.

4B.13   (a) The irreversible work of expansion against a constant opposing pressure is given by

$w = -P_{\text{ex}}\Delta V$

$w = -(1.00\,\text{atm})(6.52\,\text{L} - 4.29\,\text{L})$

$\quad = -2.23\,\text{L}\cdot\text{atm}$

$\quad = -2.23\,\text{L}\cdot\text{atm} \times 101.325\,\text{J}\cdot\text{L}^{-1}\cdot\text{atm}^{-1} = -226\,\text{J}$

(b) An isothermal expansion will be given by

$$w = -nRT\frac{V_2}{V_1}$$

n is calculated from the ideal gas law:

$$n = \frac{PV}{RT} = \frac{(1.79\,\text{atm})(4.29\,\text{L})}{(0.082\,06\,\text{L}\cdot\text{atm}\cdot\text{K}^{-1}\cdot\text{mol}^{-1})(305\,\text{K})} = 0.307\,\text{mol}$$

$$w = -(0.307\,\text{mol})(8.314\,\text{J}\cdot\text{K}^{-1}\cdot\text{mol}^{-1})(305\,\text{K})\ln\frac{6.52}{4.29}$$

$$\quad = -326\,\text{J}$$

Note that the work done is greater when the process is carried out reversibly.

4B.15 The estimate considers only translational and rotational contributions to the molar internal energy which are both $3/2\,RT$ for a nonlinear molecule.

$$3/2\,RT + 3/2\,RT = 3\,RT = 7.44\ \text{kJ.mol}^{-1}$$

4C.1 NO_2. Heat capacity increases with molecular complexity. Therefore, as more atoms are present in the molecule, there are more possible bond vibrations that can absorb added energy.

4C.3 (a) $q_p = nC_p\Delta T = n(5/2\ R)\Delta T$

$765\ \text{J} = (0.820\ \text{mol})(5/2 \times 8.314\ \text{J mol}^{-1}.\text{K}^{-1})(T_f - 298\ \text{K})$

$T_f = 343\ \text{K}$

At constant pressure, $q_p = \Delta H = 765\ \text{J}$

(b) $q_v = nC_v\Delta T = n(3/2\ R)\ \Delta T$

$765\ \text{J} = (0.820\ \text{mol})(3/2 \times 8.314\ \text{J.mol}^{-1}.\text{K}^{-1})(T_f - 298\ \text{K})$

$T_f = 373\ \text{K}$

4C.5 (a) HCN is a linear molecule. Translational contribution is $3/2\ RT$ and rotational contribution is RT.

Total contribution from molecular motions will be $5/2\ R$.

(b) C_2H_6 is a polyatomic, nonlinear molecule. Translational and rotational contributions are both $3/2\ RT$. Total contribution from molecular motions will be $3R$.

(c) Ar is a monatomic ideal gas. The contribution from molecular motions to the heat capacity will be $3/2\ R$.

(d) HBr is a diatomic, linear molecule. Translational contribution is 3/2 RT and rotational contribution is RT. Total contribution from molecular motions will be 5/2 R.

4C.7 (a) $\Delta H_{vap} = \dfrac{4.76 \text{ kJ}}{0.579 \text{ mol}} = 8.22 \text{ kJ} \cdot \text{mol}^{-1}$

(b) $\Delta H_{vap} = \dfrac{21.2 \text{ kJ}}{\left(\dfrac{22.45 \text{ g}}{46.07 \text{ g} \cdot \text{mol}^{-1}}\right)} = 43.5 \text{ kJ} \cdot \text{mol}^{-1}$

4C.9 (a) The heat change will be made up of two terms: one term to raise the temperature of the copper and the other to raise the temperature of the water:

$q = (400.0 \text{ g})(4.18 \text{ J} \cdot (°\text{C})^{-1} \cdot \text{g}^{-1})(100.0°\text{C} - 22.0°\text{C})$
$\quad + (500.0 \text{ g})(0.38 \text{ J} \cdot (°\text{C})^{-1} \cdot \text{g}^{-1})(100.0°\text{C} - 22.0°\text{C})$
$\quad = 1.30 \times 10^5 \text{ J} + 1.48 \times 10^4 \text{ J} = 1.45 \times 10^5 \text{ J} = 1.4 \times 10^2 \text{ kJ}$

(b) The percentage of heat attributable to raising the temperature of water will be

$\left(\dfrac{1.30 \times 10^5 \text{ J}}{1.45 \times 10^5 \text{ J}}\right)(100) = 90\%$

4C.11 This process is composed of two steps: melting the ice at 0°C and then raising the temperature of the liquid water from 0°C to 25°C :

Step 1: $\Delta H = \left(\dfrac{80.0 \text{ g}}{18.02 \text{ g} \cdot \text{mol}^{-1}}\right)(6.01 \text{ kJ} \cdot \text{mol}^{-1}) = 26.7 \text{ kJ}$

Step 2: $\Delta H = (80.0 \text{ g})(4.18 \text{ J} \cdot (°\text{C})^{-1} \cdot \text{g}^{-1})(20.0°\text{C} - 0.0°\text{C}) = 6.69 \text{ kJ}$

Total heat required $= 26.7 \text{ kJ} + 6.69 \text{ kJ} = 33.4 \text{ kJ}$

4C.13 The heat gained by the water in the ice cube will be equal to the heat lost by the initial sample of hot water. The enthalpy change for the water in the ice cube will be composed of two terms: the heat to melt the ice at 0°C and the heat required to raise the ice from 0°C to the final temperature.

$$\text{heat (ice cube)} = \left(\frac{50.0 \text{ g}}{18.02 \text{ g} \cdot \text{mol}^{-1}} \right)(6.01 \times 10^3 \text{ J} \cdot \text{mol}^{-1})$$

$$+ (50.0 \text{ g})(4.184 \text{ J} \cdot (°\text{C})^{-1} \cdot \text{g}^{-1})(T_f - 0°)$$

$$= 1.67 \times 10^4 \text{ J} + (209 \text{ J} \cdot (°\text{C})^{-1})(T_f - 0°)$$

$$\text{heat (water)} = (400 \text{ g})(4.184 \text{ J} \cdot (°\text{C})^{-1} \cdot \text{g}^{-1})(T_f - 45°)$$

$$= (1.67 \times 10^3 \text{ J} \cdot (°\text{C})^{-1})(T_f - 45°)$$

Setting these equal:

$$- (1.67 \times 10^3 \text{ J} \cdot (°\text{C})^{-1})T_f + 7.5 \times 10^4 \text{ J} = 1.67 \times 10^4 \text{ J} + (209 \text{ J} \cdot (°\text{C})^{-1})T_f$$

Solving for T_f:

$$T_f = \frac{5.8 \times 10^4 \text{ J}}{1.88 \times 10^3 \text{ J} \cdot (°\text{C})^{-1}} = 31 °\text{C}$$

4C.15 Based on the $\Delta H_{\text{fus}} = 10.0 \text{ kJ} \cdot \text{mol}^{-1}$ and $\Delta H_{\text{vap}} = 20.0 \text{ kJ} \cdot \text{mol}^{-1}$ and a constant heating rate, melting should occur twice as fast as vaporization. Also required are similar heating slopes for the solid and gas as they both have a heat capacity of 30 J·mol^{-1}. Finally the heating slope for the liquid should be less steep as it has a higher heat capacity of 60 J·mol^{-1}. This leaves (c) as the best match.

4D.1 (a) $\Delta H = (1.25 \text{ mol})(+358.8 \text{ kJ} \cdot \text{mol}^{-1}) = 448 \text{ kJ}$

(b) $\Delta H = \left(\frac{197 \text{ g C}}{12.01 \text{ g} \cdot \text{mol}^{-1} \text{ C}} \right)\left(\frac{358.8 \text{ kJ}}{4 \text{ mol C}} \right) = 1.47 \times 10^3 \text{ kJ}$

(c) $\Delta H = 415 \text{ kJ} = (n_{\text{CS}_2})\left(\frac{358.8 \text{ kJ} \cdot \text{mol}^{-1}}{4 \text{ mol CS}_2} \right)$

$n_{\text{CS}_2} = 4.63 \text{ mol CS}_2$ or $(4.63 \text{ mol})(76.13 \text{ g} \cdot \text{mol}^{-1}) = 352 \text{ g CS}_2$

4D.3   (a) $CO(g) + H_2O(g) \rightarrow CO_2(g) + H_2(g)$

(b)   Bomb calorimeter means volume is fixed. Therefore, $w = 0$.

$\Delta U = q + w$ Need to find q.   Use $q = C\Delta T$

First, work out how much heat was released on burning 1.40 g of carbon monoxide.

$q_{reaction} + q_{calorimeter} = 0$

$q_{reaction} = - q_{calorimeter}$

$= - C\Delta T$

$= - (3.00 \text{ kJ} \cdot (°C)^{-1})(22.799 °C - 22.113 °C)$

$= - 2.058 \text{ kJ}$

$MW(CO) = 12.0107 \text{ g.mol}^{-1} + 15.9994 \text{ g.mol}^{-1} = 28.0101 \text{ g.mol}^{-1}$

$moles(CO) = \dfrac{1.40 \text{ g}}{28.0101 \text{ g.mol}^{-1}} = 4.998 \times 10^{-2} \text{ mol}$

For 1.40 g CO(g), $q = \dfrac{- 2.058 \text{ kJ}}{4.998 \times 10^{-2} \text{ mol}}$

$= - 41.176 \text{ kJ. mol}^{-1} = - 41.2 \text{ kJ. mol}^{-1}$

For the combustion of 1.40 g CO(g), $\Delta U = q = - 41.2 \text{ kJ. mol}^{-1}$

4D.5   From $\Delta H = \Delta U + P\Delta V$ at constant pressure, or $\Delta U = \Delta H - P\Delta V$.

Because $w = -P\Delta V = +22 \text{ kJ}$, we get $-15 \text{ kJ} + 22 \text{ kJ} = \Delta U = +7 \text{ kJ}$.

4D.7   Net production of 1 mole of gas:

$\Delta n = +1 \text{ mol}$

$PV = nRT$

$P\Delta V = \Delta nRT$

$- P\Delta V = -\Delta nRT = -(1 \text{ mol})(8.314 \text{ J.K}^{-1}.\text{mol}^{-1})(298 \text{ K})$

$= -2.48 \text{kJ}$

$\Delta U = q + w = \Delta H - P\Delta V = -318 \text{kJ} - 2.48 \text{kJ} = -320 \text{kJ}$

4D.9 The enthalpy of reaction for the reaction

$$4 \, C_7H_5N_3O_6(s) + 21 \, O_2(g) \rightarrow 28 \, CO_2(g) + 10 \, H_2O(g) + 6 \, N_2(g)$$

may be found using enthalpies of formation:

$$28\left(-393.51 \, \frac{kJ}{mol}\right) + 10\left(-241.82 \, \frac{kJ}{mol}\right) - 4\left(-67 \, \frac{kJ}{mol}\right) = -13168 \, \frac{kJ}{mol}$$

This is the energy released per mole of reaction as written. One fourth of this amount of energy or $3292 \, \dfrac{kJ}{mol}$ will be *released* per mole of TNT consumed. The energy density in kJ per L may be found by dividing this amount of energy with the mass of one mole of TNT and then by multiplying with the density of TNT:

$$\frac{3292 \, \dfrac{kJ}{mol}}{227.14 \, \dfrac{g}{mol}}\left(1.65 \, \frac{g}{cm^3}\right)\left(\frac{10^3 \, cm^3}{1 \, L}\right) = +23.9 \times 10^3 \, \frac{kJ}{L}$$

4D.11 (a) heat absorbed $= (1.55 \, \text{mol NO})\left(\dfrac{180.6 \, kJ}{2 \, \text{mol NO}}\right) = 140 \, kJ$

(b) $n_{N_2} = \dfrac{PV}{RT} = \dfrac{1.00 \, \text{atm} \times 5.45 \, L}{0.082 \, 06 \, L \cdot \text{atm} \cdot K^{-1} \cdot mol^{-1} \times 273 \, K} = 0.243 \, \text{mol}$

$(0.243 \, \text{mol N}_2)\left(\dfrac{180.6 \, kJ}{\text{mol N}_2}\right) = 43.9 \, kJ$

(c) $n_{N_2} = \dfrac{0.492 \, kJ}{180.6 \, kJ \cdot mol^{-1}} = 0.002 \, 72 \, \text{mol}$

$n_{N_2} = (0.002 \, 72 \, \text{mol})(28.02 \, g \cdot mol^{-1}) = 0.0762 \, g$

4D.13

$$3BaO(s) \rightarrow 3Ba(s) + \frac{3}{2}O_2(g) \qquad \Delta H^\circ = 1660.5 \, kJ$$

$$\underline{2Al(s) + \frac{3}{2}O_2(g) \rightarrow Al_2O_3(s) \qquad \Delta H^\circ = -1676 \, kJ}$$

$$3BaO(s) + 2Al(s) \rightarrow Al_2O_3(s) + 3Ba(s) \qquad \Delta H^\circ = -16 \, kJ$$

4D.15 First, write the balanced equations for the reaction given:

$$C_2H_2(g) + \tfrac{5}{2}O_2(g) \longrightarrow 2\,CO_2(g) + H_2O(l) \qquad \Delta H° = -1300\text{ kJ}$$

$$C_2H_6(g) + \tfrac{7}{2}O_2(g) \longrightarrow 2\,CO_2(g) + 3\,H_2O(l) \qquad \Delta H° = -1560\text{ kJ}$$

$$H_2(g) + \tfrac{1}{2}O_2(g) \longrightarrow H_2O(l) \qquad \Delta H° = -286\text{ kJ}$$

The second equation is reversed and added to the first, plus 2× the third:

$$C_2H_2(g) + \tfrac{5}{2}O_2(g) \longrightarrow 2\,CO_2(g) + H_2O(l) \qquad \Delta H° = -1300\text{ kJ}$$

$$2\,CO_2(g) + 3\,H_2O(l) \longrightarrow C_2H_6(g) + \tfrac{7}{2}O_2(g) \qquad \Delta H° = +1560\text{ kJ}$$

$$2[H_2(g) + \tfrac{1}{2}O_2(g) \longrightarrow H_2O(l)] \qquad 2[\Delta H° = -286\text{ kJ}]$$

$$C_2H_2(g) + 2\,H_2(g) \longrightarrow C_2H_6(g)$$

$$\Delta H° = -1300\text{ kJ}\cdot\text{mol}^{-1} + 1560\text{ kJ}\cdot\text{mol}^{-1} + 2(-286\text{ kJ}\cdot\text{mol}^{-1})$$

$$= -312\text{ kJ}\cdot\text{mol}^{-1}$$

4D.17 The reaction enthalpy for this reaction is given by:

$$\Delta H° = 12\,(\Delta H°_f(H_2O, l))$$

$$\quad - [4\,(\Delta H°_f(HNO_3, l)) + 5\,(\Delta H°_f(N_2H_4, l))]$$

$$= 12(-285.83\text{ kJ}\cdot\text{mol}^{-1})$$

$$\quad - [4(-174.10\text{ kJ}\cdot\text{mol}^{-1}) + 5(+50.63\text{ kJ}\cdot\text{mol}^{-1})]$$

$$= -2986.71\text{ kJ}\cdot\text{mol}^{-1}$$

4D.19

$$2\,NH_4Br(s) \rightarrow 2\,NH_3(g) + 2\,HBr(g) \qquad \Delta H° = 2(+188.32\text{ kJ})$$

$$2\,NH_3(g) \rightarrow N_2(g) + 3\,H_2(g) \qquad \Delta H° = +92.22\text{ kJ}$$

$$\underline{N_2(g) + 4\,H_2(g) + Br_2(l) \rightarrow 2\,NH_4Br(s) \qquad \Delta H° = -541.66\text{ kJ}}$$

$$H_2(g) + Br_2(l) \rightarrow 2\,HBr(g) \qquad \Delta H° = -72.80\text{ kJ}$$

4D.21 (a) $3 NO_2(g) + H_2O(l) \longrightarrow 2 HNO_3(aq) + NO(g)$

$\Delta H°_r = 2 \Delta H°_f(HNO_3, aq)$
$\qquad + \Delta H°_f(NO, g - [3 \Delta H°_f(NO_2, g) + \Delta H°_f(H_2O, l)]$
$\qquad = 2(-207.36 \text{ kJ} \cdot \text{mol}^{-1}) + 90.25 \text{ kJ} \cdot \text{mol}^{-1} - [3(+33.18 \text{ kJ} \cdot \text{mol}^{-1})$
$\qquad\qquad + (-285.83 \text{ kJ} \cdot \text{mol}^{-1})]$
$\qquad = -138.18 \text{ kJ}$

(b) $B_2O_3(s) + 3 CaF_2(s) \longrightarrow 2 BF_3(g) + 3 CaO(s)$

$\Delta H°_r = 2 \Delta H°_f(BF_3, g) + 3 \Delta H°_f(CaO, s)$
$\qquad - [\Delta H°_f(B_2O_3, s) + 3\Delta H°_f(CaF_2, s)]$
$\qquad = 2(-1137.0 \text{ kJ} \cdot \text{mol}^{-1}) + 3(-635.09 \text{ kJ} \cdot \text{mol}^{-1})$
$\qquad - [-1272.8 \text{ kJ} \cdot \text{mol}^{-1} + 3(-1219.6 \text{ J} \cdot \text{mol}^{-1})]$
$\qquad = +752.3 \text{ kJ}$

(c) $H_2S(aq) + 2 KOH(aq) \longrightarrow K_2S(aq) + 2 H_2O(l)$

$\Delta H°_r = \Delta H°_f(K_2S, aq) + 2 \Delta H°_f(H_2O, l)$
$\qquad - [\Delta H°_f(H_2S, aq) + 2 \Delta H°_f(KOH, aq)]$
$\qquad = -417.5 \text{ kJ} \cdot \text{mol}^{-1} + 2(-285.83 \text{ kJ} \cdot \text{mol}^{-1})$
$\qquad - [-39.7 \text{ kJ} \cdot \text{mol}^{-1} + 2(-482.37 \text{ kJ} \cdot \text{mol}^{-1})]$
$\qquad = +15.28 \text{ kJ}$

4D.23 From Appendix 2A, $\Delta H°_f(NO) = +90.25 \text{ kJ}$

The reaction we want is

$N_2(g) + \frac{5}{2} O_2(g) \longrightarrow N_2O_5(g)$

Adding the first reaction to half of the second gives

$2 NO(g) + O_2(g) \longrightarrow 2 NO_2(g) \qquad\qquad \Delta H° = -114.1 \text{ kJ}$

$2 NO_2(g) + \frac{1}{2} O_2(g) \longrightarrow N_2O_5(g) \qquad\qquad \Delta H° = -55.1 \text{ kJ}$

$2 NO(g) + \frac{3}{2} O_2(g) \longrightarrow N_2O_5(g) \qquad\qquad -169.2 \text{ kJ}$

The enthalpy of this reaction equals the enthalpy of formation of $N_2O_5(g)$

minus twice the enthalpy of formation of NO, so we can write

$-169.2 \text{ kJ} = \Delta H°_f(N_2O_5) - 2(+90.25 \text{ kJ})$
$\Delta H°_f(N_2O_5) = +11.3 \text{ kJ}$

4D.25 Combustion of CH_4: $CH_4(g) + 2O_2(g) \rightarrow CO_2(g) + 2H_2O(g)$

First, calculate the differences in heat capacities:

$\Delta C_p = [(1 \text{ mol}) C_{P,m} (CO_2, g) + (2 \text{ mol}) C_{P,m} (H_2O, g)]$

$\qquad - [(1 \text{ mol}) C_{P,m} (CH_4, g) + (2 \text{ mol}) C_{P,m} (O_2, g)]$

$\Delta C_p = [(1 \text{ mol})(37.11 \text{ J.K}^{-1}.\text{mol}^{-1}) + (2 \text{ mol})(33.58 \text{ J.K}^{-1}.\text{mol}^{-1})]$

$\qquad - [(1 \text{ mol})(35.69 \text{ J.K}^{-1}.\text{mol}^{-1}) + (2 \text{ mol})(29.36 \text{ J.K}^{-1}.\text{mol}^{-1})]$

$\Delta C_p = 9.86 \text{ J.K}^{-1}$

Evaluate: $T_2 - T_1$

$T_2 - T_1 = (1200 \text{ K} - 298 \text{ K}) = 902 \text{ K}$

Calculate the enthalpy change at the final temperature from

$\Delta H°(T_2) = \Delta H°(T_1) + (T_2 - T_1) \Delta C_p$

$\Delta H°(1200 \text{ K}) = -802 \text{ kJ} + (902 \text{ K})(9.86 \text{ J.K}^{-1})$

$\qquad = -802 \text{ kJ} + 8.893 \text{ kJ}$

$\qquad = -793.11 \text{ kJ}$

4D.27 (a) The enthalpy of vaporization is the enthalpy change associated with the conversion $C_6H_6(l) \longrightarrow C_6H_6(g)$ at constant pressure. The value at 298.2 K will be given by

$\Delta H°_{\text{vaporization at 298 K}} = \Delta H°_f (C_6H_6, g) - \Delta H°_f (C_6H_6, l)$

$\qquad\qquad = 82.93 \text{ kJ} \cdot \text{mol}^{-1} - (49.0 \text{ kJ} \cdot \text{mol}^{-1})$

$\qquad\qquad = 33.93 \text{ kJ} \cdot \text{mol}^{-1}$

(b) In order to take into account the difference in temperature, we need to use the heat capacities of the reactants and products in order to raise the temperature of the system to 353.2 K. We can rewrite the reactions as follows, to emphasize temperature, and then combine them according to Hess's law:

$C_6H_6(l)_{\text{at 298 K}} \longrightarrow C_6H_6(g)_{\text{at 298 K}} \qquad \Delta H° = 33.93 \text{ kJ}$

$C_6H_6(l)_{\text{at 298 K}} \longrightarrow C_6H_6(l)_{\text{at 353.2 K}} \qquad \Delta H° = (1 \text{ mol})(353.2 \text{ K} - 298.2 \text{ K})$

$\qquad\qquad\qquad\qquad\qquad\qquad\qquad\qquad (136.1 \text{ J} \cdot \text{mol}^{-1} \cdot \text{K}^{-1})$

$\qquad\qquad\qquad\qquad\qquad\qquad\qquad\qquad = 7.48 \text{ kJ}$

$$C_6H_6(g)_{\text{at } 298 \text{ K}} \longrightarrow C_6H_6(g)_{\text{at } 353.2 \text{ K}} \quad \Delta H° = (1 \text{ mol})(353.2 \text{ K} - 298.2 \text{ K})$$
$$(81.67 \text{ J} \cdot \text{mol}^{-1} \cdot \text{K}^{-1})$$
$$= 4.49 \text{ kJ}$$

To add these together to get the overall equation at 353.2 K, we must reverse the second equation:

$$C_6H_6(l)_{\text{at } 298 \text{ K}} \longrightarrow C_6H_6(g)_{\text{at } 298 \text{ K}} \quad \Delta H° = 33.93 \text{ kJ}$$

$$C_6H_6(l)_{\text{at } 353.2 \text{ K}} \longrightarrow C_6H_6(l)_{\text{at } 298 \text{ K}} \quad \Delta H° = -7.48 \text{ kJ}$$

$$C_6H_6(g)_{\text{at } 298 \text{ K}} \longrightarrow C_6H_6(g)_{\text{at } 353.2 \text{ K}} \quad \Delta H° = 4.49 \text{ kJ}$$

$$C_6H_6(l)_{\text{at } 353.2 \text{ K}} \longrightarrow C_6H_6(g)_{\text{at } 353.2 \text{ K}}$$
$$\Delta H° = 33.93 \text{ kJ} \cdot \text{mol}^{-1} - 7.48 \text{ kJ} \cdot \text{mol}^{-1} + 4.49 \text{ kJ} \cdot \text{mol}^{-1}$$
$$= 30.94 \text{ kJ} \cdot \text{mol}^{-1}$$

(c) The value in the table is $30.8 \text{ kJ} \cdot \text{mol}^{-1}$ for the enthalpy of vaporization of benzene. The value is close to that calculated as corrected by heat capacities. At least part of the error can be attributed to the fact that heat capacities are not strictly constant with temperature.

4D.29

For the reaction: $A + 2B \rightarrow 3C + D$ the molar enthalpy of reaction at temperature 2 is given by:

$$\Delta H_{r,2}° = H_{m,2}° (\text{products}) - H_{m,2}° (\text{reactants})$$

$$= 3H_{m,2}°(C) + H_{m,2}°(D) - H_{m,2}°(A) - 2H_{m,2}°(B)$$

$$= 3[H_{m,1}°(C) + C_{p,m}(C)(T_2 - T_1)] + [H_{m,1}°(D) + C_{p,m}(D)(T_2 - T_1)]$$
$$- [H_{m,1}°(A) + C_{p,m}(A)(T_2 - T_1)] - 2[H_{m,1}°(B) + C_{p,m}(B)(T_2 - T_1)]$$

$$= 3H_{m,1}°(C) + H_{m,1}°(D) - H_{m,1}°(A) - 2H_{m,1}°(B)$$

$$+ [3C_{p,m}(C) + C_{p,m}(D) - C_{p,m}(A) - 2C_{p,m}(B)](T_2 - T_1)$$

$$= \Delta H_{r,1}° + [3C_{p,m}(C) + C_{p,m}(D) - C_{p,m}(A) - 2C_{p,m}(B)](T_2 - T_1)$$

Finally, $\Delta H_{r,2}° = \Delta H_{r,1}° + \Delta C_p(T_2 - T_1)$, which is Kirchhoff's law.

4E.1   For the reaction $Na_2O(s) \longrightarrow 2\,Na^+(g) + O^{2-}(g)$

$$\Delta H_L = 2\,\Delta H°_f(Na, g) + \Delta H°_f(O, g) + 2\,I_1(Na)$$
$$-E_{ea1}(O) - E_{ea2}(O) - \Delta H_f(Na_2O(s)$$

$$\Delta H_L = 2(107.32\ kJ\cdot mol^{-1}) + 249\ kJ\cdot mol^{-1} + 2(494\ kJ\cdot mol^{-1})$$
$$-141\ kJ\cdot mol^{-1} + 844\ kJ\cdot mol^{-1} + 409\ kJ\cdot mol^{-1}$$

$$\Delta H_L = 2564\ kJ\cdot mol^{-1}$$

4E.3   (a) $\Delta H_L = \Delta H°_f(Na, g) + \Delta H°_f(Cl, g) + I_1(Na)$
$$- E_{ea}\ of\ Cl - \Delta H_f(NaCl(s))$$

$$787\ kJ\cdot mol^{-1} = 108\ kJ\cdot mol^{-1} + 122\ kJ\cdot mol^{-1} + 494\ kJ\cdot mol^{-1}$$
$$- 349\ kJ\cdot mol^{-1} - \Delta H_f(NaCl(s))$$

$$\Delta H_f(NaCl(s)) = -412\ kJ\cdot mol^{-1}$$

(b) $\Delta H_L = \Delta H°_f(K, g) + \Delta H°_f(Br, g) + I_1(K)$
$$- E_{ea}(Br) - \Delta H_f(KBr(s))$$

$$\Delta H_L = 89\ kJ\cdot mol^{-1} + 97\ kJ\cdot mol^{-1} + 418\ kJ\cdot mol^{-1}$$
$$- 325\ kJ\cdot mol^{-1} + 394\ kJ\cdot mol^{-1}$$

$$= 673\ kJ\cdot mol^{-1}$$

(c) $\Delta H_L = \Delta H°_f(Rb, g) + \Delta H°_f(F, g) + I_1(Rb) - E_{ea}(F) - \Delta H_f(RbF(s))$

$$774\ kJ\cdot mol^{-1} = \Delta H°_f(Rb, g) + 79\ kJ\cdot mol^{-1}$$
$$+ 402\ kJ\cdot mol^{-1} - 328\ kJ\cdot mol^{-1} + 558\ kJ\cdot mol^{-1}$$

$$\Delta H°_f(Rb, g) = 63\ kJ\cdot mol^{-1}$$

4E.5   (a) break: $3\ mol\ C \equiv C$ bonds $3(837)\ kJ\cdot mol^{-1}$

   form: $6\ mol\ C\text{⋯}C$ bonds $-6(518)\ kJ\cdot mol^{-1}$

   Total $-597\ kJ\cdot mol^{-1}$

(b) break: $4\ mol\ C—H$ bonds $4(412)\ kJ\cdot mol^{-1}$

           $4\ mol\ Cl—Cl$ bonds $4(242)\ kJ\cdot mol^{-1}$

   form: $4\ mol\ C—Cl$ bonds $-4(338)\ kJ\cdot mol^{-1}$

           $4\ mol\ H—Cl$ bonds $-4(431)\ kJ\cdot mol^{-1}$

   Total $-460\ kJ\cdot mol^{-1}$

(c) The number and types of bonds on both sides of the equations are equal, so we expect the enthalpy of the reaction to be essentially 0.

4E.7 (a) break:

break:	1 mol N—N triple bonds	$(1\ mol)(944\ kJ \cdot mol^{-1})$
	3 mol F—F bonds	$(3\ mol)(158\ kJ \cdot mol^{-1})$
form:	6 mol N—F bonds	$(6\ mol)(-270\ kJ.mol^{-1})$
	Total	$-202\ kJ.mol^{-1}$

(b) break:	1 mol C=C bonds	$(1\ mol)(612\ kJ \cdot mol^{-1})$
	1 mol O—H bonds	$(1\ mol)(463\ kJ \cdot mol^{-1})$
form:	1 mol C—C bonds	$-(1\ mol)(348\ kJ \cdot mol^{-1})$
	1 mol C—O bonds	$-(1\ mol)(360\ kJ \cdot mol^{-1})$
	1 mol C—H bonds	$-(1\ mol)(412\ kJ \cdot mol^{-1})$
	Total	$-45\ kJ \cdot mol^{-1}$

(c) break:	1 mol C—H bonds	$(1\ mol)(412\ kJ \cdot mol^{-1})$
	1 mol Cl—Cl bonds	$(1\ mol)(242\ kJ \cdot mol^{-1})$
form:	1 mol C—Cl bonds	$-(1\ mol)(338\ kJ \cdot mol^{-1})$
	1 mol H—Cl bonds	$-(1\ mol)(431\ kJ \cdot mol^{-1})$
	Total	$-115\ kJ \cdot mol^{-1}$

4E.9 The value that we want is given simply by the difference between three isolated C=C bonds and three isolated C—C single bonds, versus six resonance-stabilized bonds:

3 C=C bonds + 3 C—C bonds = 3(348 kJ) + 3(612 kJ) = 2880 kJ

6 resonance-stabilized bonds = 6(518 kJ) = 3108 kJ

As can be seen, the six resonance-stabilized bonds are more stable by ca. 228 kJ.

4F.1   Assume the transfer of energy as heat is reversible.

(a) rate of entropy generation $= \dfrac{\Delta S_{surr}}{time} = -\dfrac{q_{rev}}{time \cdot T}$

$$= -\dfrac{\text{rate of heat generation}}{T}$$

$$= \dfrac{-(100. \text{ J} \cdot \text{s}^{-1})}{293 \text{ K}} = 0.341 \text{ J} \cdot \text{K}^{-1} \cdot \text{s}^{-1}$$

(b) $\Delta S_{day} = (0.341 \text{ J} \cdot \text{K}^{-1} \cdot \text{s}^{-1})(60 \text{ sec} \cdot \text{min}^{-1})(60 \text{ min} \cdot \text{hr}^{-1})(24 \text{ hr} \cdot \text{day}^{-1})$

$$= 29.5 \text{ kJ} \cdot \text{K}^{-1} \cdot \text{day}^{-1}$$

(c) Less, because in the equation $\Delta S = \dfrac{-\Delta H}{T}$, if T is larger, ΔS is smaller.

4F.3   (a) $\Delta S = \dfrac{q_{rev}}{T} = \dfrac{65 \text{ J}}{298 \text{ K}} = 0.22 \text{ J} \cdot \text{K}^{-1}$

(b) $\Delta S = \dfrac{65 \text{ J}}{373 \text{ K}} = 0.17 \text{ J} \cdot \text{K}^{-1}$

(c) Since T is in the denominator the entropy change is smaller at higher temperatures. The same amount of heat has a greater effect on entropy changes when transferred at lower temperatures.

4F.5   Assume the process is isothermal and reversible and use the relationship

$dS = \dfrac{dq}{T}$. Because the process is isothermal, $\Delta U = 0$, hence

$q = -w$, where $w = -PdV$. Making this substitution, we obtain

$$dS = \dfrac{P \, dV}{T} = \dfrac{nRT}{TV} dV = \dfrac{nR}{V} dV$$

$$\therefore \Delta S = nR \ln \dfrac{V_2}{V_1}$$

Substituting the known quantities, we obtain

$$\Delta S = (5.25 \text{ mol})(8.314 \text{ J} \cdot \text{K}^{-1} \cdot \text{mol}^{-1}) \ln \dfrac{34.058 \text{ L}}{24.252 \text{ L}}$$

$$= 14.8 \text{ J} \cdot \text{K}^{-1}$$

4F.7 (a) The relationship to use is $dS = \dfrac{dq}{T}$. At constant pressure, we can

$$dq = n\,C_P\,dT :$$

substitute

$$dS = \dfrac{n\,C_p\,dT}{T}$$

Upon integration, this gives $\Delta S = n\,Cp \ln \dfrac{T_2}{T_1}$. The answer is calculated by

simply plugging in the known quantities. Remember that for an ideal

monatomic gas

$$C_P = \tfrac{5}{2}R :$$

$$\Delta S = (1.00 \text{ mol})(\tfrac{5}{2} \times 8.314 \text{ J} \cdot \text{K}^{-1} \cdot \text{mol}^{-1}) \ln \dfrac{431.0 \text{ K}}{310.8 \text{ K}} = 6.80 \text{ J} \cdot \text{K}^{-1}$$

(b) A similar analysis using C_V gives $\Delta S = n\,C_V \ln \dfrac{T_2}{T_1}$, where C_V for a

monatomic ideal gas is $\tfrac{3}{2}R :$

$$\Delta S = (1.00 \text{ mol})(\tfrac{3}{2} \times 8.314 \text{ J} \cdot \text{K}^{-1} \cdot \text{mol}^{-1}) \ln \dfrac{431.0 \text{ K}}{310.8 \text{ K}} = 4.08 \text{ J} \cdot \text{K}^{-1}$$

4F.9

$$\Delta S = nR \ln \dfrac{V_2}{V_1}$$

Since P is the inverse of V for ideal gases we can write:

$$\Delta S = nR \ln \dfrac{P_1}{P_2}$$

$$= (1.50 \text{ mol})(8.314 \text{ J.K}^{-1}.\text{mol}^{-1}) \ln \dfrac{15.0 \text{ atm}}{0.500 \text{ atm}}$$

$$= 42.4 \text{ J.K}^{-1}$$

4F.11 First, calculate the decrease in entropy resulting from the decrease in

volume. Then, calculate the increase in entropy resulting from the increase in temperature. Then, add these to get the net entropy change. Assume ideal behavior and 1 mol N_2 gas.

$$\Delta S = nR \ln \frac{V_2}{V_1}$$

$$= (1.00 \text{ mol})(8.314 \text{ J.K}^{-1}.\text{mol}^{-1}) \ln \frac{0.500 \text{ L}}{3.00 \text{ L}}$$

$$= -14.897 \text{ J.K}^{-1} \quad (-14.9 \text{ J.K}^{-1} \text{ using 3 sig. fig.})$$

$$\Delta S = nR \ln \frac{T_2}{T_1}$$

$$= (1.00 \text{ mol})(8.314 \text{ J.K}^{-1}.\text{mol}^{-1}) \ln \frac{301.25 \text{ K}}{291.65 \text{ K}}$$

$$= 0.269 \text{ J.K}^{-1}$$

Net change in entropy, $\Delta S_{net} = (-14.897 + 0.269) \text{ J.K}^{-1} = -14.6 \text{ J.K}^{-1}$

(Rounding off at the end using 3 sig. fig.)

4F.13 (a) $\Delta S° = \dfrac{q}{T} = \dfrac{\Delta H°}{T} = \dfrac{1.00 \text{ mol} \times (-6.01 \text{ kJ} \cdot \text{mol}^{-1})}{273.2 \text{ K}} = -22.0 \text{ J} \cdot \text{K}^{-1}$

(b) $\Delta S = \dfrac{q}{T} = \dfrac{\Delta H}{T} = \dfrac{\dfrac{50.0 \text{ g}}{46.07 \text{ g} \cdot \text{mol}^{-1}} \times 43.5 \text{ kJ} \cdot \text{mol}^{-1}}{351.5 \text{ K}} = +134 \text{ J} \cdot \text{K}^{-1}$

4F.15 (a) Trouton's rule indicates that the entropy of vaporization for a number of organic liquids is approximately $85 \text{ J} \cdot \text{K}^{-1} \cdot \text{mol}^{-1}$. Using this information and the relationship

$$T_B = \frac{\Delta H°_{vap}}{\Delta S°_{vap}} = \frac{21.51 \times 10^3 \text{ J} \cdot \text{mol}^{-1}}{85 \text{ J} \cdot \text{K}^{-1} \cdot \text{mol}^{-1}} = 253 \text{ K}.$$

(b) The experimental boiling point of dimethyl ether is 248 K, which is in reasonably close agreement. The value $85 \text{ J.K}^{-1}.\text{mol}^{-1}$ is an average value for the entropy of vaporization of organic liquids and therefore deviations are expected when using this average value for individual organic liquids.

4F.17 The entropy of vaporization of water at 85 °C may be carried out through a series of three reversible steps. Namely, reversibly heating the reactants to 100 °C, carrying out the phase change at this temperature, and finally cooling the products back to 85 °C. The sum of the ΔS's for these three steps will be equivalent to vaporizing water at 85 °C in one irreversible step.

Step 1, heating the reactants to 100 °C:

$$\Delta S_1 = C_{P,m} \ln\left(\frac{T_2}{T_1}\right) = \left(75.3 \text{ J} \cdot \text{K}^{-1} \cdot \text{mol}^{-1}\right)\ln\left(\frac{373 \text{ K}}{358 \text{ K}}\right) = 3.09 \text{ J} \cdot \text{K}^{-1} \cdot \text{mol}^{-1}$$

Step 2, the entropy of vaporization of H_2O at 100 °C is 109.0 J $\cdot$ K^{-1} $\cdot$ mol^{-1}

Step 3, cooling the products to 85 °C:

$$\Delta S_3 = C_{P,m} \ln\left(\frac{T_2}{T_1}\right) = \left(33.6 \text{ J} \cdot \text{K}^{-1} \cdot \text{mol}^{-1}\right)\ln\left(\frac{358 \text{ K}}{373 \text{ K}}\right) = -1.38 \text{ J} \cdot \text{K}^{-1} \cdot \text{mol}^{-1}$$

Therefore, the molar entropy of vaporization is H_2O at 85 °C is:

$$\Delta S_{v,m} = \Delta S_1 + \Delta S_2 + \Delta S_3 = 111 \text{ J} \cdot \text{K}^{-1} \cdot \text{mol}^{-1}.$$

4G.1 (a)

$$W = 1^{64} = 1$$
$$S = k_B \ln W = k_B \ln 1 = 0$$

(b)

$$S = k_B \ln W$$
$$W = 4^{64} = 3.403 \times 10^{38}$$
$$S = (1.3806 \times 10^{-23} \text{ J} \cdot \text{K}^{-1})(\ln(3.403 \times 10^{38}))$$
$$S = 1.22 \times 10^{-21} \text{ J} \cdot \text{K}^{-1}$$

4G.3 COF_2. COF_2 and BF_3 are both trigonal planar molecules, but it would be possible for the molecule to be disordered with the fluorine and oxygen atoms occupying the same locations. Because all the groups attached to boron are identical, such disorder is not possible.

4G.5   For the cis compound there will be 12 different orientations:

For the trans compound there will only be 3 different orientations.

Comparing the Boltzmann entropy calculations for the cis and trans forms:

cis:

$$S = k \ln 12^{6.02 \times 10^{23}} = (1.38 \times 10^{-23} \text{ J} \cdot \text{K}^{-1}) \ln 12^{6.02 \times 10^{23}}$$

$$S = 20.6 \text{ J} \cdot \text{K}^{-1}$$

trans:

$$S = k \ln 3^{6.02 \times 10^{23}} = (1.38 \times 10^{-23} \text{ J} \cdot \text{K}^{-1}) \ln 3^{6.02 \times 10^{23}}$$

$$S = 9.13 \text{ J} \cdot \text{K}^{-1}$$

The cis form should have the higher residual entropy.

4G.7   There are six orientations of an SO_2F_2 molecule as shown below:

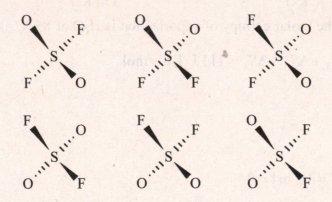

The Boltzmann expression for one mole of SO_2F_2 molecules having six

possible orientations is:

$$S = k \ln 6^{6.02 \times 10^{23}} = (1.38 \times 10^{-23} \text{ J} \cdot \text{K}^{-1}) \ln 6^{6.02 \times 10^{23}}$$

$$S = 14.9 \text{ J} \cdot \text{K}^{-1}$$

4G.9   (a) $W = 3$

(b) $W = 12$

(c)   Initially one of the three atom systems had two atoms in higher energy states. In part (b) the system will be at equilibrium when each three atom system has one quantum of energy. Therefore, energy will flow from the system with two quanta to the system with none.

4H.1 (a) HBr(g), because Br is more massive and contains more elementary particles than F in HF.

(b) NH$_3$(g), because it has greater complexity, being a molecule rather than a single atom.

(c) I$_2$ (l), because molecules in liquids are more randomly oriented than molecules in solids.

(d) 1.0 mol Ar(g) at 1.00 atm, because it will occupy a larger volume than 1.0 mol of Ar(g) at 2.00 atm.

4H.3 It is easy to order H$_2$O in its various phases because entropy will increase when going from a solid to a liquid to a gas. The main question concerns where to place C(s, diamond) in this order, and that will essentially become a question of whether C(s, diamond) should have more or less entropy than H$_2$O(s), because we would automatically expect C(s, diamond) to have less entropy than any liquid. Because water is a molecular substance held together in the solid phase by weak hydrogen bonds, and in C(s, diamond) the carbon is more rigidly held in place and will have less entropy.

In summary, C(s, diamond) < H$_2$O(s) < H$_2$O(l) < H$_2$O(g).

4H.5 (a) Iodine is expected to have higher entropy due to its larger mass and consequently a larger number of fundamental particles.

(b) When we consider the two structures, it is clear that 1-pentene will have more flexibility in its framework than cyclopentane, which will be comparatively rigid. Therefore, we predict 1-pentene to have a higher entropy.

(c) Ethene (or ethylene) is a gas and polyethylene is a solid, so we automatically expect ethene to have a higher entropy. Also, for the same mass, a sample of ethene will be composed of many small molecules, whereas polyethylene will be made up of fewer but larger molecules.

4H.7   (a) Entropy should decrease because the number of moles of gas is less on the product side of the reaction.

(b) Entropy should increase because the dissolution of the solid copper phosphate will increase the randomness of the copper and phosphate ions.

(c) Entropy should decrease as the total number of moles decreases.

4H.9 $\Delta S_B < \Delta S_C < \Delta S_A$. The change in entropy for container A is greater than that for container B or C due to the greater number of particles. The change in entropy in container C is greater than that of container B because of the disorder due to the vibrational motion of the molecules in container C.

4H.11   (a) $H_2(g) + \frac{1}{2}O_2(g) \longrightarrow H_2O(l)$

$$\Delta S^\circ_f = \sum S^\circ_{products} - \sum S^\circ_{reactants}$$
$$= 69.91 \text{ J} \cdot \text{K}^{-1} \cdot \text{mol}^{-1} - [130.68 \text{ J} \cdot \text{K}^{-1} \cdot \text{mol}^{-1} + \frac{1}{2}(205.14 \text{ J} \cdot \text{K}^{-1} \cdot \text{mol}^{-1})]$$
$$= -163.34 \text{ J} \cdot \text{K}^{-1} \cdot \text{mol}^{-1}$$

The entropy change is negative because the number of moles of gas has decreased by 1.5. Note that the absolute entropies of the elements are not 0, and that the entropy change for the reaction in which a compound is formed from the elements is also not 0.

(b) $CO(g) + \frac{1}{2}O_2(g) \longrightarrow CO_2(g)$

$$\Delta S^\circ_r = 213.74 \text{ J} \cdot \text{K}^{-1} \cdot \text{mol}^{-1}$$
$$- [197.67 \text{ J} \cdot \text{K}^{-1} \cdot \text{mol}^{-1} + \frac{1}{2}(205.14 \text{ J} \cdot \text{K}^{-1} \cdot \text{mol}^{-1})]$$
$$= -86.5 \text{ J} \cdot \text{K}^{-1} \cdot \text{mol}^{-1}$$

The entropy change is negative because the number of moles of gas has decreased by 0.5.

(c) $CaCO_3(s) \longrightarrow CaO(s) + CO_2(g)$

$$\Delta S°_r = 39.75 \text{ J} \cdot \text{K}^{-1} \cdot \text{mol}^{-1} + 213.74 \text{ J} \cdot \text{K}^{-1} \cdot \text{mol}^{-1} - [+92.9 \text{ J} \cdot \text{K}^{-1} \cdot \text{mol}^{-1}]$$
$$= +160.6 \text{ J} \cdot \text{K}^{-1} \cdot \text{mol}^{-1}$$

The entropy change is positive because the number of moles of gas has increased by 1.

(d) $4 \text{ KClO}_3(s) \longrightarrow 3 \text{ KClO}_4(s) + \text{KCl}(s)$

$$\Delta S°_r = 3(151.0 \text{ J} \cdot \text{K}^{-1} \cdot \text{mol}^{-1}) + 82.59 \text{ J} \cdot \text{K}^{-1} \cdot \text{mol}^{-1}$$
$$- [4(143.1 \text{ J} \cdot \text{K}^{-1} \cdot \text{mol}^{-1})]$$
$$= -36.8 \text{ J} \cdot \text{K}^{-1} \cdot \text{mol}^{-1}$$

It is not immediately obvious, but the four moles of solid products are more ordered than the four moles of solid reactants.

4H.13 $dS = \dfrac{dq_{rev}}{T} = \dfrac{C_{P,m} dT}{T}$ and, therefore, $\Delta S = \displaystyle\int_{T_1}^{T_2} \dfrac{C_{P,m}}{T} dT$.

If $C_{P,m} = a + bT + c/T^2$, then

$$\Delta S = \int_{T_1}^{T_2} \frac{a + bT + c/T^2}{T} dT$$

$$= \int_{T_1}^{T_2}\left(\frac{a}{T} + b + \frac{c}{T^3}\right) dT = \left(a\ln(T) + bT - \frac{c}{2T^2}\right)\Bigg|_{T_1}^{T_2}$$

$$= a\ln\left(\frac{T_2}{T_1}\right) + b(T_2 - T_1) - \frac{c}{2}\left(\frac{1}{T_2^2} - \frac{1}{T_1^2}\right)$$

ΔS(true) for heating graphite from 298 K to 400 K is

$$= (16.86 \text{ J} \cdot \text{K}^{-1} \cdot \text{mol}^{-1})\ln\left(\frac{400 \text{ K}}{298 \text{ K}}\right)$$

$$+ (0.00477 \text{ J} \cdot \text{K}^{-2} \cdot \text{mol}^{-1})(400 \text{ K} - 298 \text{ K})$$

$$- \frac{(-8.54 \times 10^5 \text{ J} \cdot \text{K} \cdot \text{mol}^{-1})}{2}\left(\frac{1}{(400 \text{ K})^2} - \frac{1}{(298 \text{ K})^2}\right)$$

$$= 3.31 \text{ J} \cdot \text{K}^{-1} \cdot \text{mol}^{-1}$$

If we assume a constant heat capacity at the mean temperature of 350 K:

$$C_{P,m} = (16.86 \text{ J} \cdot \text{K}^{-1} \cdot \text{mol}^{-1}) + (0.00477 \text{ J} \cdot \text{K}^{-2} \cdot \text{mol}^{-1})(350 \text{ K})$$

$$+ \frac{(-8.54 \times 10^5 \text{ J} \cdot \text{K} \cdot \text{mol}^{-1})}{(350 \text{ K})^2}$$

$$= 11.6 \text{ J} \cdot \text{K}^{-1} \cdot \text{mol}^{-1}$$

and

$$\Delta S(\text{mean}) = C_{P,m} \ln \frac{T_2}{T_1} = (11.6 \text{ J} \cdot \text{K}^{-1} \cdot \text{mol}^{-1}) \ln\left(\frac{400 \text{ K}}{298 \text{ K}}\right) = 3.41 \text{ J} \cdot \text{K}^{-1} \cdot \text{mol}^{-1}$$

Not integrating the heat capacity leads to roughly a 3.0% error in ΔS.

4I.1

$$\Delta S = \frac{q}{T}$$

$$\Delta S = \frac{-40000 \text{ J}}{800 \text{ K}}$$

$$\Delta S = -50.0 \text{ J} \cdot \text{K}^{-1}$$

$$\Delta S = \frac{+40000 \text{ J}}{200 \text{ K}}$$

$$\Delta S = +200 \text{ J} \cdot \text{K}^{-1}$$

$$\Delta S_{\text{tot}} = -50.0 \text{ J} \cdot \text{K}^{-1} + 200 \text{ J} \cdot \text{K}^{-1} = +150 \text{ J} \cdot \text{K}^{-1}$$

4I.3 (a) The value can be estimated from

$$\Delta H^\circ{}_{\text{vap}} = T\Delta S^\circ{}_{\text{vap}}$$

$$\Delta H^\circ{}_{\text{vap}} = (353 \text{ K})(85 \text{ J} \cdot \text{mol}^{-1} \cdot \text{K}^{-1})$$

$$= +30. \text{ kJ} \cdot \text{mol}^{-1}$$

(b) $\Delta S^\circ{}_{\text{surr}} = -\dfrac{\Delta H^\circ{}_{\text{system}}}{T}$

$$\Delta S^\circ{}_{\text{surr}} = -\left(\frac{10 \text{ g}}{78.11 \text{ g} \cdot \text{mol}^{-1}}\right)\left(\frac{30 \text{ kJ} \cdot \text{mol}^{-1}}{353 \text{ K}}\right) = -11 \text{ J} \cdot \text{K}^{-1}$$

4I.5 Assume no change in volume. To calculate the change in entropy for the hot and cold water, the amount of energy which flows from one to the other must first be calculated:

$q_c = -q_h$

$= n_c \cdot C_{p,m}(H_2O) \cdot (T_f - T_{i,c}) = -n_h \cdot C_{p,m}(H_2O) \cdot (T_f - T_{i,h})$

where n_c and n_h are the moles of cold and hot water, respectively, and

$T_{i,c}$ and $T_{i,h}$ are the initial temperatures of the cold and hot water,

respectively. Dividing both sides by $C_{p,m}(H_2O)$ we obtain:

$n_c \cdot (T_f - T_{i,c}) = -n_h \cdot (T_f - T_{i,h})$.

The moles of hot and cold water are:

$$n_c = \frac{50\ g}{18.015\ g \cdot mol^{-1}} = 2.78\ mol \text{ and } n_h = \frac{65\ g}{18.015\ g \cdot mol^{-1}} = 3.61\ mol$$

and the final temperature T_f is therefore:

$$T_f = \frac{(2.78\ mol \cdot 293.15\ K) + (3.61\ mol \cdot 323.15\ K)}{(2.78\ mol + 3.61\ mol)} = 310.1\ K.$$

With T_f, we can calculate ΔS for the hot and cold water, and the total ΔS

for the entire system:

$$\Delta S_c = n_c \cdot C_{p,m} \ln\left(\frac{T_f}{T_{i,c}}\right) = (2.78\ mol)(75.3\ J \cdot K^{-1} \cdot mol^{-1})\ln\left(\frac{310.1\ K}{293\ K}\right)$$

$$= +11.9\ J \cdot K^{-1}$$

$$\Delta S_h = n_h \cdot C_{p,m} \ln\left(\frac{T_f}{T_{i,h}}\right) = (3.61\ mol)(75.3\ J \cdot K^{-1} \cdot mol^{-1})\ln\left(\frac{310.1\ K}{323\ K}\right)$$

$$= -11.1\ J \cdot K^{-1}$$

Therefore, ΔS_{tot} is :

$$\Delta S_{tot} = \Delta S_c + \Delta S_h = +0.8\ J \cdot K^{-1}.$$

4I.7 (a) The change in entropy will be given by

$$\Delta S_{surr} = \frac{-\Delta H_{system}}{T} = \frac{-1.00\ mol \times 8.2 \times 10^3\ J \cdot mol^{-1}}{111.7\ K} = -73\ J \cdot K^{-1}$$

$$\Delta S_{system} = \frac{\Delta H_{system}}{T} = \frac{1.00\ mol \times 8.2 \times 10^3\ J \cdot mol^{-1}}{111.7\ K} = +73\ J \cdot K^{-1}$$

(b) $\Delta S_{surr} = \dfrac{-\Delta H_{system}}{T} = \dfrac{-1.00\ mol \times 4.60 \times 10^3\ J \cdot mol^{-1}}{158.7\ K} = -29.0\ J \cdot K^{-1}$

$$\Delta S_{system} = \frac{\Delta H_{system}}{T} = \frac{1.00 \text{ mol} \times 4.60 \times 10^3 \text{ J} \cdot \text{mol}^{-1}}{158.7 \text{ K}} = +29.0 \text{ J} \cdot \text{K}^{-1}$$

(c) $$\Delta S_{surr} = \frac{-\Delta H_{system}}{T} = \frac{-(1.00 \text{ mol} \times -4.60 \times 10^3 \text{ J} \cdot \text{mol}^{-1})}{158.7 \text{ K}}$$
$$= +29.0 \text{ J} \cdot \text{K}^{-1}$$

$$\Delta S_{system} = \frac{\Delta H_{system}}{T} = \frac{1.00 \text{ mol} \times -4.60 \times 10^3 \text{ J} \cdot \text{mol}^{-1}}{158.7 \text{ K}} = -29.0 \text{ J} \cdot \text{K}^{-1}$$

4I.9 (a) The total entropy change is given by $\Delta S_{tot} = \Delta S_{surr} + \Delta S$. ΔS for an isothermal, reversible process is calculated from

$$\Delta S = \frac{q_{rev}}{T} = \frac{-w_{rev}}{T} = nR \ln \frac{V_2}{V_1}.$$ To do the calculation we need the value of

n, which is obtained by use of the ideal gas law:

$(4.95 \text{ atm})(1.67 \text{ L}) = n(0.082\ 06 \text{ L} \cdot \text{atm} \cdot \text{K}^{-1} \cdot \text{mol}^{-1})(323 \text{ K});\ n = 0.312 \text{ mol}.$

$\Delta S = (0.312 \text{ mol})(8.314 \text{ J} \cdot \text{K}^{-1} \cdot \text{mol}^{-1}) \ln \frac{7.33 \text{ L}}{1.67 \text{ L}} = +3.84 \text{ J} \cdot \text{K}^{-1}.$ Because the

process is reversible, $\Delta S_{tot} = 0$, so $\Delta S_{surr} = -\Delta S = -3.84 \text{ J} \cdot \text{K}^{-1}.$

(b) For the irreversible process, ΔS is the same, $+3.84 \text{ J} \cdot \text{K}^{-1}.$ No work is

done in free expansion (see Section 6.6) so $w = 0$. Because $\Delta U = 0$, it

follows that $q = 0$. Therefore, no heat is transferred into the surroundings,

and their entropy is unchanged: $\Delta S_{surr} = 0$. The total change in entropy is

therefore $\Delta S_{tot} = +3.84 \text{ J} \cdot \text{K}^{-1}.$

4I.11 Spontaneous change means $\Delta S_{tot} > 0$.

A vapor to liquid change is condensation and it releases heat. However, as

shown in the diagram, the temperature of the system does not change.

Therefore, the released heat must have left system to the surroundings.

Since heat flows from hot to cold, the temperature of the surroundings

(T_{surr}) must be lower than the temperature of the system (T_{sys}), and since

$$\Delta S_{sys} = \frac{q_{rev}}{T_{sys}} \text{ and } \Delta S_{surr} = \frac{q_{rev}}{T_{surr}}, \text{ then } \Delta S_{tot} > 0.$$

4J.1 Exothermic reactions tend to be spontaneous because the result is an increase in the entropy of the surroundings. Using the mathematical relationship $\Delta G_r = \Delta H_r - T\Delta S_r$, it is clear that if ΔH_r is large and negative compared with ΔS_r, then the reaction will generally be spontaneous.

4J.3 (a)

$\Delta G = \Delta H_{vap} - T\Delta S_{vap}$

$-15.0 + 273.15 = 258.15 \text{ K}$

$\Delta G = 23.4 \text{ kJ} \cdot \text{mol}^{-1} - [(258.15 \text{ K})(97.6 \text{ J} \cdot \text{K}^{-1} \cdot \text{mol}^{-1})]$

$= 23.4 \text{ kJ} \cdot \text{mol}^{-1} - 25.195 \text{ kJ} \cdot \text{mol}^{-1}$

$= -1.8 \text{ kJ} \cdot \text{mol}^{-1}$

The reaction has a negative change in Gibbs free energy and is therefore spontaneous at −15.0 °C and 1 atm.

(b)

$\Delta G = \Delta H_{vap} - T\Delta S_{vap}$

$-45.0 + 273.15 = 228.15 \text{ K}$

$\Delta G = 23.4 \text{ kJ} \cdot \text{mol}^{-1} - [(228.15 \text{ K})(97.6 \text{ J} \cdot \text{K}^{-1} \cdot \text{mol}^{-1})]$

$= 23.4 \text{ kJ} \cdot \text{mol}^{-1} - 22.267 \text{ J} \cdot \text{mol}^{-1}$

$= +1.1 \text{ kJ} \cdot \text{mol}^{-1}$

The reaction has a positive change in Gibbs free energy and is therefore not spontaneous at − 45.0 °C and 1 atm.

4J.5 (a) $\frac{1}{2}N_2(g) + \frac{3}{2}H_2(g) \longrightarrow NH_3(g)$

$$\Delta H°_r = \Delta H°_f(NH_3) = -46.11 \text{ kJ} \cdot \text{mol}^{-1}$$

$$\Delta S°_r = S°_m(NH_3, g) - [\frac{1}{2}S°_m(N_2, g) + \frac{3}{2}S°_m(H_2, g)]$$

$$= 192.45 \text{ J} \cdot \text{K}^{-1} \cdot \text{mol}^{-1} - [\frac{1}{2}(191.61 \text{ J} \cdot \text{K}^{-1} \cdot \text{mol}^{-1})$$

$$+ \frac{3}{2}(130.68 \text{ J} \cdot \text{K}^{-1} \cdot \text{mol}^{-1})]$$

$$= -99.38 \text{ J} \cdot \text{K}^{-1} \cdot \text{mol}^{-1}$$

$$\Delta G°_r = -46.11 \text{ kJ} \cdot \text{mol}^{-1} - (298 \text{ K})(-99.38 \text{ J} \cdot \text{K}^{-1} \cdot \text{mol}^{-1})/(1000 \text{ J} \cdot \text{kJ}^{-1})$$

$$= -16.49 \text{ kJ} \cdot \text{mol}^{-1}$$

$$S°_m(NH_3) = 192.45 \text{ J} \cdot \text{K}^{-1} \cdot \text{mol}^{-1}$$

$\Delta S°_f(NH_3)$ is negative because several gas molecules combine to form 1

NH_3 molecule.

(b) $H_2(g) + \frac{1}{2}O_2(g) \longrightarrow H_2O(g)$

$$\Delta H°_r = \Delta H°_f(H_2O, g) = -241.82 \text{ kJ} \cdot \text{mol}^{-1}$$

$$\Delta S°_r = S°_m(H_2O, g) - [S°_m(H_2, g) + \frac{1}{2}S°_m(O_2, g)]$$

$$= 188.83 \text{ J} \cdot \text{K}^{-1} \cdot \text{mol}^{-1} - [130.68 \text{ J} \cdot \text{K}^{-1} \cdot \text{mol}^{-1} + \frac{1}{2}(205.14 \text{ J} \cdot \text{K}^{-1} \cdot \text{mol}^{-1})]$$

$$= -44.42 \text{ J} \cdot \text{K}^{-1} \cdot \text{mol}^{-1}$$

$$\Delta G°_r = -241.82 \text{ kJ} \cdot \text{mol}^{-1} - (298 \text{ K})(-44.42 \text{ J} \cdot \text{K}^{-1} \cdot \text{mol}^{-1})/(1000 \text{ J} \cdot \text{kJ}^{-1})$$

$$= -228.58 \text{ kJ} \cdot \text{mol}^{-1}$$

$$S°_m(H_2O, g) = 188.83 \text{ J} \cdot \text{K}^{-1} \cdot \text{mol}^{-1}$$

$\Delta S°_f(H_2O, g)$ is a negative number because there is a reduction in the

number of gas molecules in the reaction when $S°_m$ is positive.

(c) $C(s, \text{graphite}) + \frac{1}{2}O_2(g) \longrightarrow CO(g)$

$$\Delta H°_r = \Delta H°_f(CO, g) = -110.53 \text{ kJ} \cdot \text{mol}^{-1}$$

$$\Delta S°_r = S°_m(CO, g) - [S°_m(C, s) + \frac{1}{2}S°_m(O_2, g)]$$

$$= 197.67 \text{ J} \cdot \text{K}^{-1} \cdot \text{mol}^{-1} - [5.740 \text{ J} \cdot \text{K}^{-1} \cdot \text{mol}^{-1} + \frac{1}{2}(205.14 \text{ J} \cdot \text{K}^{-1} \cdot \text{mol}^{-1})]$$

$$= +89.36 \text{ J} \cdot \text{K}^{-1} \cdot \text{mol}^{-1}$$

$$\Delta G°_r = -110.53 \text{ kJ} \cdot \text{mol}^{-1} - (298 \text{ K})(89.36 \text{ J} \cdot \text{K}^{-1} \cdot \text{mol}^{-1})/(1000 \text{ J} \cdot \text{kJ}^{-1})$$

$$= -137.2 \text{ kJ} \cdot \text{mol}^{-1}$$

$$S°_m(CO, g) = 197.67 \text{ J} \cdot \text{K}^{-1} \cdot \text{mol}^{-1}$$

The $S°_m(CO, g)$ is larger than $\Delta S°_f(CO, g)$ because in the formation reaction the number of moles of gas is reduced.

(d) $\frac{1}{2} N_2(g) + O_2(g) \longrightarrow NO_2(g)$

$\Delta H°_r = \Delta H°_f(NO_2) = +33.18 \text{ kJ} \cdot \text{mol}^{-1}$

$\Delta S°_r = S°_m(NO_2, g) - [\frac{1}{2} S°_m(N_2, g) + S°_m(O_2, g)]$

$\quad = 240.06 \text{ J} \cdot \text{K}^{-1} \cdot \text{mol}^{-1} - [\frac{1}{2}(191.61 \text{ J} \cdot \text{K}^{-1} \cdot \text{mol}^{-1}) + 205.14 \text{ J} \cdot \text{K}^{-1} \cdot \text{mol}^{-1}]$

$\quad = -60.89 \text{ J} \cdot \text{K}^{-1} \cdot \text{mol}^{-1}$

$\Delta G°_r = 33.18 \text{ kJ} \cdot \text{mol}^{-1} - (298 \text{ K})(-60.89 \text{ J} \cdot \text{K}^{-1} \cdot \text{mol}^{-1})/(1000 \text{ J} \cdot \text{kJ}^{-1})$

$\quad = +51.33 \text{ kJ} \cdot \text{mol}^{-1}$

$S°_m(NO_2, g) = 240.06 \text{ J} \cdot \text{K}^{-1} \cdot \text{mol}^{-1}$

The $\Delta S°_f(NO_2, g)$ is somewhat negative due to the reduction in the number of gas molecules during the reaction. For all of these, the important point is that the $S°_m$ value of a compound is not the same as the $\Delta S°_f$ for the formation of that compound. $\Delta S°_f$ is often negative because one is bringing together a number of elements to form that compound.

4J.7 (a) $2 H_2O_2(l) \longrightarrow 2 H_2O(l) + O_2(g)$

$\Delta S°_r = 2S°_m(H_2O, l) + S°_m(O_2, g) - 2S°_m(H_2O_2, l)$

$\quad = 2(69.91 \text{ J} \cdot \text{K}^{-1} \cdot \text{mol}^{-1}) + 205.14 \text{ J} \cdot \text{K}^{-1} \cdot \text{mol}^{-1}$

$\quad\quad - 2(109.6 \text{ J} \cdot \text{K}^{-1} \cdot \text{mol}^{-1})$

$\quad = +125.8 \text{ J} \cdot \text{K}^{-1} \cdot \text{mol}^{-1}$

$\Delta H°_r = 2\Delta H°_f(H_2O, aq) - 2\Delta H°_f(H_2O_2, l)$

$\quad = 2(-285.83 \text{ kJ} \cdot \text{mol}^{-1}) - 2(-187.78 \text{ kJ} \cdot \text{mol}^{-1})$

$\quad = -196.10 \text{ kJ} \cdot \text{mol}^{-1}$

$\Delta G°_r = 2\Delta G°_f(H_2O, aq) - 2\Delta G°_f(H_2O_2, l)$

$\quad = 2(-237.13 \text{ kJ} \cdot \text{mol}^{-1}) - 2(-120.35 \text{ kJ} \cdot \text{mol}^{-1})$

$\quad = -233.56 \text{ kJ} \cdot \text{mol}^{-1}$

$\Delta G°_r$ can also be calculated from $\Delta S°_r$ and $\Delta H°_r$ using the relationship:

$$\Delta G^{\circ}_{r} = \Delta H^{\circ}_{r} - T\Delta S^{\circ}_{r}$$

$$= -196.1 \, kJ \cdot mol^{-1} - (298 \, K)(+125.8 \, J \cdot K^{-1} \cdot mol^{-1})/(1000 \, J \cdot kJ^{-1})$$

$$= -233.6 \, kJ \cdot mol^{-1}$$

(b) $2 \, F_2(g) + 2 \, H_2O(l) \longrightarrow 4 \, HF(aq) + O_2(g)$

$$\Delta S^{\circ}_{r} = 4S^{\circ}_{m}(HF, aq) + S^{\circ}_{m}(O_2, g) - [2S^{\circ}_{m}(F_2, g) + 2S^{\circ}_{m}(H_2O, l)]$$

$$= 4(88.7 \, J \cdot K^{-1} \cdot mol^{-1}) + 205.14 \, J \cdot K^{-1} \cdot mol^{-1}$$

$$- [2(202.78 \, J \cdot K^{-1} \cdot mol^{-1}) + 2(69.91 \, J \cdot K^{-1} \cdot mol^{-1})]$$

$$= +14.6 \, J \cdot K^{-1} \cdot mol^{-1}$$

$$\Delta H^{\circ}_{r} = 4\Delta H^{\circ}_{f}(HF, aq) - 2\Delta H^{\circ}_{f}(H_2O, l)$$

$$= 4(-330.08 \, kJ \cdot mol^{-1}) - 2(-285.83 \, kJ \cdot mol^{-1})$$

$$= -748.66 \, kJ \cdot mol^{-1}$$

$$\Delta G^{\circ}_{r} = 4\Delta G^{\circ}_{f}(HF, aq) - 2\Delta G^{\circ}_{f}(H_2O, l)$$

$$= 4(-296.82 \, kJ \cdot mol^{-1}) - 2(-237.13 \, kJ \cdot mol^{-1})$$

$$= -713.02 \, kJ \cdot mol^{-1}$$

ΔG°_{r} can also be calculated from ΔS°_{r} *and* ΔH°_{r} using the relationship:

$$\Delta G^{\circ}_{r} = \Delta H^{\circ}_{r} - T\Delta S^{\circ}_{r}$$

$$= -748.66 \, kJ \cdot mol^{-1} - (298 \, K)(14.6 \, J \cdot K^{-1} \cdot mol^{-1})/(1000 \, J \cdot kJ^{-1})$$

$$= -753.01 \, kJ \cdot mol^{-1}$$

4J.9   (a) Balanced equation: $2 \, Fe_3O_4(s) + \frac{1}{2}O_2(g) \rightarrow 3 \, Fe_2O_3(s)$

$$\Delta H^{\circ}_{r} = 3(-824.2 \, kJ \cdot mol^{-1}) - [2(-1118.4 \, kJ \cdot mol^{-1})]$$

$$= -235.8 \, kJ \cdot mol^{-1}$$

$$\Delta S^{\circ}_{r} = 3(87.40 \, J \cdot K^{-1} \cdot mol^{-1}) - [2(146.4 \, J \cdot K^{-1} \cdot mol^{-1})$$

$$+ \frac{1}{2}(205.14 \, J \cdot K^{-1} \cdot mol^{-1})]$$

$$= -133.17 \, J \cdot K^{-1} \cdot mol^{-1}$$

$$\Delta G^{\circ}_{r} = 3(-742.2 \, kJ \cdot mol^{-1}) - [2(-1015.4 \, kJ \cdot mol^{-1})]$$

$$= -195.8 \, kJ \cdot mol^{-1}$$

ΔG_{r} may also be calculated from ΔH°_{r} and ΔS°_{r} (the numbers calculated differ slightly from the two methods due to rounding differences):

$$\Delta G^{\circ}_{r} = \Delta H^{\circ}_{r} - T\Delta S^{\circ}_{r}$$

$$= -235.8 \text{ kJ.mol}^{-1} - (298 \text{ K})(-133.17 \text{ J.K}^{-1}.\text{mol}^{-1})/(1000 \text{ J.kJ}^{-1})$$

$$= -196.1 \text{ kJ.mol}^{-1}$$

(b) $\Delta H°_r = -1208.09 \text{ kJ} \cdot \text{mol}^{-1} - [-1219.6 \text{ kJ} \cdot \text{mol}^{-1}]$

$\quad = 11.5 \text{ kJ} \cdot \text{mol}^{-1}$

$\Delta S°_r = -80.8 \text{ J} \cdot \text{K}^{-1} \cdot \text{mol}^{-1} - [68.87 \text{ J} \cdot \text{K}^{-1} \cdot \text{mol}^{-1}]$

$\quad = -149.7 \text{ J} \cdot \text{K}^{-1} \cdot \text{mol}^{-1}$

$\Delta G°_r = -1111.15 \text{ kJ} \cdot \text{mol}^{-1} - [-1167.3 \text{ kJ} \cdot \text{mol}^{-1}]$

$\quad = +56.2 \text{ kJ} \cdot \text{mol}^{-1}$

or

$\Delta G°_r = \Delta H°_r - T\Delta S°_r$

$\quad = 11.5 \text{ kJ} \cdot \text{mol}^{-1} - (298 \text{ K})(-149.7 \text{ J} \cdot \text{K}^{-1} \cdot \text{mol}^{-1})/(1000 \text{ J} \cdot \text{kJ}^{-1})$

$\quad = 56.1 \text{kJ} \cdot \text{mol}^{-1}$

(c) $\Delta H°_r = 9.16 \text{ kJ} \cdot \text{mol}^{-1} - [2(33.18 \text{ kJ} \cdot \text{mol}^{-1})] = -57.20 \text{ kJ} \cdot \text{mol}^{-1}$

$\Delta S°_r = 304.29 \text{ J} \cdot \text{K}^{-1} \cdot \text{mol}^{-1} - [2(240.06 \text{ J} \cdot \text{K}^{-1} \cdot \text{mol}^{-1})]$

$\quad = -175.83 \text{ J} \cdot \text{K}^{-1} \cdot \text{mol}^{-1}$

$\Delta G°_r = 97.89 \text{ kJ} \cdot \text{mol}^{-1} - [2(51.31 \text{ kJ} \cdot \text{mol}^{-1})] = -4.73 \text{ kJ} \cdot \text{mol}^{-1}$

or

$\Delta G°_r = \Delta H°_r - T\Delta S°_r$

$\quad = -57.2 \text{ kJ} \cdot \text{mol}^{-1} - (298 \text{ K})(-175.83 \text{ J} \cdot \text{K}^{-1} \cdot \text{mol}^{-1})/(1000 \text{ J} \cdot \text{kJ}^{-1})$

$\quad = -4.80 \text{ kJ} \cdot \text{mol}^{-1}$

4J.11 Use the relationship $\Delta G°_r = \Sigma \Delta G°_f \text{(products)} - \Sigma \Delta G°_r \text{(reactants)}$:

(a) $\Delta G°_r = 2\Delta G°_f (SO_3, g) - [2\Delta G°_f (SO_2, g)]$

$\quad = 2(-371.06 \text{ kJ} \cdot \text{mol}^{-1}) - [2(-300.19 \text{ kJ} \cdot \text{mol}^{-1})]$

$\quad = -141.74 \text{ kJ} \cdot \text{mol}^{-1}$

The reaction is spontaneous.

(b) $\Delta G°_r = \Delta G°_f (CaO, s) + \Delta G°_f (CO_2, g) - \Delta G°_f (CaCO_3, s)$

$\quad = (-604.03 \text{ kJ} \cdot \text{mol}^{-1}) + (-394.36 \text{ kJ} \cdot \text{mol}^{-1})$

$\quad \quad - (-1128.8 \text{ kJ} \cdot \text{mol}^{-1})$

$\quad = +130.4 \text{ kJ} \cdot \text{mol}^{-1}$

The reaction is not spontaneous.

(c) $\Delta G°_r = 16\Delta G°_f (CO_2, g) + 18\Delta G°_f (H_2O, l) - [2\Delta G°_f (C_8H_{18}, l)]$

$= 16(-394.36 \text{ kJ} \cdot \text{mol}^{-1}) + 18(-237.13 \text{ kJ} \cdot \text{mol}^{-1})$

$- [2(6.4 \text{ kJ} \cdot \text{mol}^{-1})]$

$= -10\ 590.9 \text{ kJ} \cdot \text{mol}^{-1}$

The reaction is spontaneous.

4J.13 The standard free energies of formation of the compounds are:

(a) $PCl_5(g)$, $-305.0 \text{ kJ} \cdot \text{mol}^{-1}$; (b) $HCN(g)$, $+124.7 \text{ kJ} \cdot \text{mol}^{-1}$;

(c) $NO(g)$, $+86.55 \text{ kJ} \cdot \text{mol}^{-1}$; (d) $SO_2(g)$, $-300.19 \text{ kJ} \cdot \text{mol}^{-1}$.

Those compounds with a positive free energy of formation are unstable with respect to the elements. Thus (a) and (d) are thermodynamically stable.

4J.15 To understand what happens to $\Delta G°_r$ as temperature is raised, we use the relationship $\Delta G°_r = \Delta H°_r - T\Delta S°_r$. From this it is clear that the free energy of the reaction becomes less favorable (more positive) as temperature increases, only if $\Delta S°_r$ is a negative number. Therefore, we need only to find out whether the standard entropy of formation of the compound is a negative number. Assuming constant standard pressure, this is calculated for each compound as follows:

(a) $P(s) + \frac{5}{2}Cl_2(g) \longrightarrow PCl_5(g)$

$\Delta S°_r = S°_m (PCl_5, g) - [S°_m (P, s) + \frac{5}{2} S°_m (Cl_2, g)]$

$= 364.6 \text{ J} \cdot \text{K}^{-1} \cdot \text{mol}^{-1} - [41.09 \text{ J} \cdot \text{K}^{-1} \cdot \text{mol}^{-1} + \frac{5}{2}(223.07 \text{ J} \cdot \text{K}^{-1} \cdot \text{mol}^{-1})]$

$= -234.2 \text{ J} \cdot \text{K}^{-1} \cdot \text{mol}^{-1}$

The compound is less stable at higher temperatures.

(b) $C(s), \text{graphite} + \frac{1}{2}N_2(g) + \frac{1}{2}H_2(g) \longrightarrow HCN(g)$

$$\Delta S^{\circ}_r = S^{\circ}_m(HCN, g) - [S^{\circ}_m(C, s) + \tfrac{1}{2} S^{\circ}_m(N_2, g) + \tfrac{1}{2} S^{\circ}_m(H_2, g)]$$

$$= 201.78 \text{ J} \cdot \text{K}^{-1} \cdot \text{mol}^{-1} - [5.740 \text{ J} \cdot \text{K}^{-1} \cdot \text{mol}^{-1}$$

$$+ \tfrac{1}{2}(191.61 \text{ J} \cdot \text{K}^{-1} \cdot \text{mol}^{-1}) + \tfrac{1}{2}(130.68 \text{ J} \cdot \text{K}^{-1} \cdot \text{mol}^{-1})]$$

$$= +34.90 \text{ J} \cdot \text{K}^{-1} \cdot \text{mol}^{-1}$$

HCN(g) is more stable at higher T.

(c) $\tfrac{1}{2} N_2(g) + \tfrac{1}{2} O_2(g) \longrightarrow NO(g)$

$$\Delta S^{\circ}_r = S^{\circ}_m(NO, g) - [\tfrac{1}{2} S^{\circ}_m(N_2, g) + \tfrac{1}{2} S^{\circ}_m(O_2, g)]$$

$$= 210.76 \text{ J} \cdot \text{K}^{-1} \cdot \text{mol}^{-1} - [\tfrac{1}{2}(191.61 \text{ J} \cdot \text{K}^{-1} \cdot \text{mol}^{-1})$$

$$+ \tfrac{1}{2}(205.14 \text{ J} \cdot \text{K}^{-1} \cdot \text{mol}^{-1})]$$

$$= +12.38 \text{ J} \cdot \text{K}^{-1} \cdot \text{mol}^{-1}$$

NO(g) is more stable as T increases.

(d) $S(s) + O_2(g) \longrightarrow SO_2(g)$

$$\Delta S^{\circ}_r = S^{\circ}(SO_2, g) - [S^{\circ}(S, s) + S^{\circ}(O_2, g)]$$

$$= 248.22 \text{ J} \cdot \text{K}^{-1} \cdot \text{mol}^{-1} - [31.80 \text{ J} \cdot \text{K}^{-1} \cdot \text{mol}^{-1}$$

$$+ 205.14 \text{ J} \cdot \text{K}^{-1} \cdot \text{mol}^{-1}]$$

$$= +11.28 \text{ J} \cdot \text{K}^{-1} \cdot \text{mol}^{-1}$$

$SO_2(g)$ is more stable as T increases.

4J.17 In order to find ΔG°_r at a temperature other than 298 K, we must first calculate ΔH°_r and ΔS°_r and then use the relationship

$\Delta G^{\circ}_r = \Delta H^{\circ}_r + T\Delta S^{\circ}_r$ to calculate ΔG°_r.

(a) In order to determine the range over which the reaction will be spontaneous, we consider the relative signs of ΔH°_r and ΔS°_r and their effect on ΔG°_r.

$$\Delta H^{\circ}{}_{r} = 2\Delta H^{\circ}{}_{f}(BF_3, g) + 3\Delta H^{\circ}{}_{f}(H_2O, l) - [\Delta H^{\circ}{}_{f}(B_2O_3, s)$$
$$+ 6\Delta H^{\circ}{}_{f}(HF, g)]$$
$$= 2(-1137.0 \text{ kJ} \cdot \text{mol}^{-1}) + 3(-285.83 \text{ kJ} \cdot \text{mol}^{-1})$$
$$- [(-1272.8 \text{ kJ} \cdot \text{mol}^{-1}) + 6(-271.1 \text{ kJ} \cdot \text{mol}^{-1})]$$
$$= -232.1 \text{ kJ} \cdot \text{mol}^{-1}$$
$$\Delta S^{\circ}{}_{r} = 2S^{\circ}{}_{m}(BF_3, g) + 3S^{\circ}{}_{m}(H_2O, l)$$
$$- [S^{\circ}{}_{m}(B_2O_3, S) + 6S^{\circ}{}_{m}(HF, g)]$$
$$= 2(254.12 \text{ J} \cdot \text{K}^{-1} \cdot \text{mol}^{-1}) + 3(69.91 \text{ J} \cdot \text{K}^{-1} \cdot \text{mol}^{-1})$$
$$- [53.97 \text{ J} \cdot \text{K}^{-1} \cdot \text{mol}^{-1} + 6(173.78 \text{ J} \cdot \text{K}^{-1} \cdot \text{mol}^{-1})]$$
$$= -378.68 \text{ J} \cdot \text{K}^{-1} \cdot \text{mol}^{-1}$$
$$\Delta G^{\circ}{}_{r} = -232.1 \text{ J} \cdot \text{K}^{-1} \cdot \text{mol}^{-1} - (353 \text{ K})(-378.68 \text{ J} \cdot \text{K}^{-1} \cdot \text{mol}^{-1})/(1000 \text{ J} \cdot \text{kJ}^{-1})$$
$$= -98.42 \text{ kJ} \cdot \text{mol}^{-1}$$

Because $\Delta H^{\circ}{}_{r}$ is negative and $\Delta S^{\circ}{}_{r}$ is also negative, we expect the reaction to be spontaneous at low temperatures, where the term $T\Delta S^{\circ}{}_{r}$ will be less than $\Delta H^{\circ}{}_{r}$. To find the temperature of the cutoff, we calculate the temperature at which $\Delta G^{\circ}{}_{r} = 0$. For this reaction, that temperature is

$$\Delta G^{\circ}{}_{r} = 0 = -232.1 \text{ kJ} \cdot \text{mol}^{-1} - (T)(-378.68 \text{ J} \cdot \text{K}^{-1} \cdot \text{mol}^{-1})/(1000 \text{ J} \cdot \text{kJ}^{-1})$$
$$T = 612.9 \text{ K}$$

The reaction should be spontaneous below 612.9 K.

(b) $\Delta H^{\circ}{}_{r} = \Delta H^{\circ}{}_{f}(CaCl_2, aq) + \Delta H^{\circ}{}_{f}(C_2H_2, g) - [\Delta H^{\circ}{}_{f}(CaC_2, s)$
$$+ 2\Delta H^{\circ}{}_{f}(HCl, aq)]$$
$$= (-877.1 \text{ kJ} \cdot \text{mol}^{-1}) + 226.73 \text{ kJ} \cdot \text{mol}^{-1}$$
$$- [(-59.8 \text{ kJ} \cdot \text{mol}^{-1}) + 2(-167.16 \text{ kJ} \cdot \text{mol}^{-1})]$$
$$= -256.3 \text{ kJ} \cdot \text{mol}^{-1}$$
$$\Delta S^{\circ}{}_{r} = S^{\circ}{}_{m}(CaCl_2, aq) + S^{\circ}{}_{m}(C_2H_2, g)$$
$$- [S^{\circ}{}_{m}(CaC_2, s) + 2S^{\circ}{}_{m}(HCl, aq)]$$
$$= 59.8 \text{ J} \cdot \text{K}^{-1} \cdot \text{mol}^{-1} + 200.94 \text{ J} \cdot \text{K}^{-1} \cdot \text{mol}^{-1}$$
$$- [69.96 \text{ J} \cdot \text{K}^{-1} \cdot \text{mol}^{-1} + 2(56.5 \text{ J} \cdot \text{K}^{-1} \cdot \text{mol}^{-1})]$$
$$= +77.8 \text{ J} \cdot \text{K}^{-1} \cdot \text{mol}^{-1}$$
$$\Delta G^{\circ}{}_{r} = -256.2 \text{ kJ} \cdot \text{mol}^{-1} - (353 \text{ K})(+77.8 \text{ J} \cdot \text{K}^{-1} \cdot \text{mol}^{-1})/(1000 \text{ J} \cdot \text{kJ}^{-1})$$
$$= -283.7 \text{ kJ} \cdot \text{mol}^{-1}$$

Because $\Delta H°_r$ is negative and $\Delta S°_r$ is positive, the reaction will be spontaneous at all temperatures.

(c)

$\Delta H°_r = \Delta H°_f (C (s), \text{diamond}) = +1.895 \text{ kJ} \cdot \text{mol}^{-1}$

$\Delta S°_r = S°_m (C (s), \text{diamond}) - S°_m (C (s), \text{graphite})$

$\qquad = +2.377 \text{ J} \cdot \text{K}^{-1} \cdot \text{mol}^{-1} - 5.740 \text{ J} \cdot \text{K}^{-1} \cdot \text{mol}^{-1}$

$\qquad = -3.363 \text{ J} \cdot \text{K}^{-1}$

$\Delta G°_r = +1.895 \text{ kJ} \cdot \text{mol}^{-1} - (353 \text{ K})(-3.363 \text{ J} \cdot \text{K}^{-1} \cdot \text{mol}^{-1})/(1000 \text{ J} \cdot \text{kJ}^{-1})$

$\qquad = +3.082 \text{ kJ} \cdot \text{mol}^{-1}$

Because $\Delta H°_r$ is positive and $\Delta S°_r$ is negative, the reaction will be nonspontaneous at all temperatures. Note: This calculation is for atmospheric pressure. Diamond can be produced from graphite at elevated pressures and high temperatures.

4.1 This process involves five separate steps: (1) raising the temperature of the ice from $-5.042 °C$ to $0.00 °C$, (2) melting the ice at $0.00°C$, (3) raising the temperature of the liquid water from $0.00°C$ to $100.00°C$, (4) vaporizing the water at $100.00°C$, and (5) raising the temperature of the water vapor from $100.00°C$ to $150.35°C$.

Step 1:

$\Delta H = (42.30 \text{ g})(2.03 \text{ J} \cdot (°C)^{-1} \cdot g^{-1})(0.00 °C - (-5.042°C)) = 0.433 \text{ kJ}$

Step 2: $\Delta H = \left(\dfrac{42.30 \text{ g}}{18.02 \text{ g} \cdot \text{mol}^{-1}} \right)(6.01 \text{ kJ} \cdot \text{mol}^{-1}) = 14.1 \text{ kJ}$

Step 3:

$\Delta H = (42.30 \text{ g})(4.18 \text{ J} \cdot (°C)^{-1} \cdot g^{-1})(100.00 °C - 0.00°C) = 17.7 \text{ kJ}$

Step 4: $\Delta H = \left(\dfrac{42.30 \text{ g}}{18.02 \text{ g} \cdot \text{mol}^{-1}} \right)(40.7 \text{ kJ} \cdot \text{mol}^{-1}) = 95.5 \text{ kJ}$

Step 5:

$$\Delta H = (42.30\ \text{g})(2.01\ \text{J} \cdot (°\text{C})^{-1} \cdot \text{g}^{-1})(150.35\ °\text{C} - 100.00°\text{C}) = 4.3\ \text{kJ}$$

The total heat required

$$= 0.4\ \text{kJ} + 14.1\ \text{kJ} + 17.7\ \text{kJ} + 95.5\ \text{kJ} + 4.3\ \text{kJ}$$

$$= 132.0\ \text{kJ}$$

4.3 Appendix 2A provides us with the heat of formation of $I_2(g)$ at 298K ($+62.44\ \text{kJ} \cdot \text{mol}^{-1}$) and the heat capacities of $I_2(g)$ ($36.90\ \text{J} \cdot \text{K}^{-1} \cdot \text{mol}^{-1}$) and $I_2(s)$ ($54.44\ \text{J} \cdot \text{K}^{-1} \cdot \text{mol}^{-1}$). We can calculate the ΔH_{sub}^{0} at 298K:

$$I_2(s) \longrightarrow I_2(g) \qquad\qquad \Delta H_{sub}^{0} = +62.44\ \text{kJ} \cdot \text{mol}^{-1}$$

We can calculate the enthalpy of fusion from the relationship

$$\Delta H_{sub}^{0} = \Delta H_{fus}^{0} + \Delta H_{vap}^{0}$$

but these values need to be at the same temperature. To correct the value for the fact that we want all the numbers for 298K, we need to alter the heat of vaporization, using the heat capacities for liquid and gaseous iodine.

$$I_2(l) \text{ at } 184.3°\text{C} \longrightarrow I_2(g) \text{ at } 184.3°\text{C} \qquad \Delta H_{vap}^{0} = +41.96\ \text{kJ} \cdot \text{mol}^{-1}$$

From Section 6.22, we find the following relationship

$$\Delta H_{r,2}^{0} = \Delta H_{r,1}^{0} + \Delta C_{P,m}^{0}(T_2 - T_1)$$

$$\Delta H_{vap,298K}^{0} = \Delta H_{vap,475.5K}^{0} + (C_{P,m}^{0}(I_2,g) - C_{P,m}^{0}(I_2,l))(T_2 - T_1)$$

$$\Delta H_{vap,298K}^{0} = +41.96\ \text{kJ} \cdot \text{mol}^{-1}$$

$$+ (36.90\ \text{J} \cdot \text{K}^{-1} \cdot \text{mol}^{-1} - 80.7\ \text{J} \cdot \text{K}^{-1} \cdot \text{mol}^{-1})(298\text{K} - 475.5\text{K})$$

$$= +49.73\ \text{kJ} \cdot \text{mol}^{-1}$$

So, at 298K:

$$+62.44\ \text{kJ} \cdot \text{mol}^{-1} = \Delta H_{fus}^{0} + 49.73\ \text{kJ} \cdot \text{mol}^{-1}$$

$$\Delta H_{fus}^{0} = +12.71\ \text{kJ} \cdot \text{mol}^{-1}$$

4.5 First, we'll work out how much heat was required to raise the temperature

of the water sample: $q = gC\Delta T = (150\,g)\,(4.18\,J.°C^{-1}.g^{-1})(5.00°C) = 3135\,J$

Since both the water and ice samples were at $0.00°$ C and the water sample

took 0.5 h to get to $5.00°$ C, we can estimate that the ice took 10.0 h to

melt. If 3135 J of heat were transferred in 0.5 h, then the amount of heat

transferred in 10.0 h is,

$$\frac{10.0\,h}{0.5\,h} \times 3135\,J = 62.7\,kJ$$
$$\text{(for 150 g sample).}$$

$$\frac{150\,g}{18.0\,g.\,mol^{-1}} = 8.33\,mol\,H_2O \qquad \frac{62.7\,kJ}{8.33\,mol} = 7.53\,kJ.\,mol^{-1}$$

Which, given the assumptions, is fairly close to $6.01\,kJ.mol^{-1}$.

4.7 (a) First, we must balance the chemical reaction:

$$C_6H_6\,(l) + \tfrac{15}{2}\,O_2\,(g) \longrightarrow 6\,CO_2\,(g) + 3\,H_2O(l)$$

For 1 mol $C_6H_6\,(l)$ burned, the change in the number of moles of gas is

$(6.00 - 7.50)\,mol = -1.50\,mol = \Delta n$

$$w = -P\Delta V = -P\left(\frac{\Delta nRT}{P}\right) = -\Delta nRT$$

$$w = -(-1.50\,mol)\,(8.314\,J\cdot K^{-1}\cdot mol^{-1})(298\,K) = +3.716 \times 10^3\,J = +3.72\,kJ$$

(b) $\Delta H_c = 6(-393.51\,kJ\cdot mol^{-1}) + 3(-285.83\,kJ\cdot mol^{-1}) - (+49.0\,kJ\cdot mol^{-1})$
$\qquad = -3267.5\,kJ$

(c) $\Delta U° = \Delta H° + w = (-3267.5 + 3.72)\,kJ = -3263.8\,kJ$

4.9 (a) Using the estimation that $3R = C$:

$$3R = C = (0.392\,J\cdot K^{-1}\cdot g^{-1})(M)$$

$$M = \frac{3R}{0.392\,J\cdot K^{-1}\cdot g^{-1}} = 63.6\,g\,mol^{-1}$$

This molar mass indicates that the atomic solid is Cu(s)

(b) From Example 5.3 we find that the density of a substance which forms

a face-centered cubic unit cell is given by: $d = \dfrac{4M}{8^{3/2} N_A r^3}$.

Therefore, we expect the density of copper to be:

$$d = \frac{4\,(63.55\ \text{g} \cdot \text{mol}^{-1})}{8^{3/2}\,(6.022 \times 10^{23}\ \text{mol}^{-1})(1.28 \times 10^{-8}\ \text{cm})^3} = 8.90\ \text{g} \cdot \text{cm}^{-3}$$

4.11

(a)   (1)   (2)   (3)

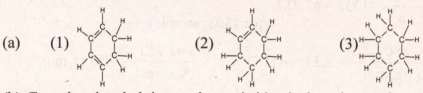

(b) From bond enthalpies, each step is identical, as the number and types

of bonds broken and formed are the same:

break:	1 mol C=C bonds	612 kJ
	1 mol H—H bonds	436 kJ
form:	1 mol C—C bonds	−348 kJ
	2 mol C—H bonds	2(−412 kJ)
Total:		−124 kJ

The total energy change should be equal to the sum of the three steps or

$3\,(-124\ \text{kJ}) = -372\ \text{kJ}.$

(c) The Hess's law calculation using standard enthalpies of formation is

easily performed on the composite reaction:

$\text{C}_6\text{H}_6\,(\text{l}) + 3\ \text{H}_2\,(\text{g}) \longrightarrow \text{C}_6\text{H}_{12}\,(\text{l})$

$\Delta H^\circ_{\text{r}} = \Sigma\,\Delta H^\circ_{\text{f}}\,(\text{products}) - \Sigma\,\Delta H^\circ_{\text{f}}\,(\text{reactants})$

$= \Delta H^\circ_{\text{f}}\,(\text{cyclohexane}) - \Delta H^\circ_{\text{f}}\,(\text{benzene})$

$= -156.4\ \text{kJ} \cdot \text{mol}^{-1} - (+49.0\ \text{kJ} \cdot \text{mol}^{-1})$

$= -205.4\ \text{kJ} \cdot \text{mol}^{-1}$

(d) The hydrogenation of benzene is much less exothermic than predicted

by bond enthalpy estimations. Part of this difference can be due to the

inherent inaccuracy of using average values, but the difference is so large

that this cannot be the complete explanation. As may be expected, the

resonance energy of benzene makes it more stable than would be expected by treating it as a set of three isolated double and three isolated single bonds. The difference in these two values [−205 kJ − (−372 kJ) = 167 kJ] is a measure of how much more stable benzene is than the Kekulé structure would predict.

4.13   (a) The combustion reaction is

$$C_{60}(s) + 60\,O_2(g) \longrightarrow 60\,CO_2(g)$$

The enthalpy of formation of $C_{60}(s)$ will be given by

$$\Delta H^\circ_c = 60\,\Delta H^\circ_f(CO_2,g) - \Delta H^\circ_f(C_{60},s)$$
$$-25\,937\ \text{kJ} = 60\ \text{mol} \times (-393.51\ \text{kJ} \cdot \text{mol}^{-1}) - \Delta H^\circ_f(C_{60},s)$$
$$\Delta H^\circ_f(C_{60},s) = +2326\ \text{kJ} \cdot \text{mol}^{-1}$$

(b) The bond enthalpy calculation is

$60\,C(gr) \longrightarrow 60\,C(g)$	$(60)(+717\ \text{kJ} \cdot \text{mol}^{-1})$
Form 60 mol C—C bonds	$-60(348\ \text{kJ} \cdot \text{mol}^{-1})$
Form 30 mol C=C bonds	$-30(612\ \text{kJ} \cdot \text{mol}^{-1})$
$C_{60}(g) \longrightarrow C_{60}(s)$	$-233\ \text{kJ}$
$60\,C(gr) \longrightarrow C_{60}(s)$	$+3547\ \text{kJ}$

(c) From the experimental data, the enthalpy of formation of C_{60} shows that it is *more* stable by (3547 kJ − 2326 kJ) = 1221 kJ than predicted by the isolated bond model.

(d) 1221 kJ ÷ 60 = 20 kJ per carbon atom

(e) 150 kJ ÷ 6 = 25 kJ per carbon atom

(f) Although the comparison of the stabilization of benzene with that of C_{60} should be treated with caution, it does appear that there is slightly less stabilization per carbon atom in C_{60} than in benzene. This fits with expectations, as the C_{60} molecule is forced by its geometry to be curved. This means that the overlap of the *p*-orbitals, which gives rise to the delocalization that results in resonance, will not be as favorable as in the planar benzene molecule. Another perspective on this is obtained by noting that the C atoms in C_{60} are forced to be partially sp^3 hybridized because they cannot be rigorously planar as required by sp^2 hybridization.

4.15   To determine the enthalpy of the reaction we must start with a balanced chemical reaction and determine the limiting reagent:

$2HCl(aq) + Zn(s) \rightarrow H_2(g) + ZnCl_2(aq)$.

$0.800 \text{ L} \cdot 0.500 \text{ M HCl} = 0.400 \text{ mol HCl}$

$\dfrac{8.5 \text{ g}}{65.37 \text{g} \cdot \text{mol}^{-1}} = 0.130 \text{ mol Zn}$

Examining the reaction stoichiometry and the initial quantities of HCl and Zn, we note that Zn is the limiting reagent (0.260 mol of HCl is needed to completely react with 0.130 moles of Zn). The enthalpy of reaction may be obtained using tabulated enthalpies of formation:

$\Delta H_r = -153.89 \dfrac{\text{kJ}}{\text{mol}} + 2\left(-167.16 \dfrac{\text{kJ}}{\text{mol}}\right) - 2\left(-167.16 \dfrac{\text{kJ}}{\text{mol}}\right) - 0$

$= -153.89 \dfrac{\text{kJ}}{\text{mol}}$

This is the enthalpy per mole of zinc consumed. Therefore, the energy released by the reaction of 8.5 g of zinc is:

$\left(-153.89 \dfrac{\text{kJ}}{\text{mol}}\right)(0.130 \text{ mol}) = -20.0 \text{ kJ}$

The change in the temperature of the water is then:

$-20000 \text{ J} = \left(-4.184 \dfrac{\text{J}}{^\circ\text{C g}}\right)(800 \text{ g})\Delta T$

$\Delta T = 5.98\,^\circ\text{C}$ and $T_f = 25\,^\circ\text{C} + 5.98\,^\circ\text{C} = 31\,^\circ\text{C}$

4.17   (a) $V_{\text{init}} = \dfrac{nRT}{P} = \dfrac{(0.060 \text{ mol})(0.0820578 \text{ L} \cdot \text{atm} \cdot \text{K}^{-1} \cdot \text{mol}^{-1})(298.15 \text{ K})}{1.00 \text{ atm}}$

$= 1.5 \text{ L}$

(b) The combustion reaction is: $2 SO_2(g) + O_2(g) \rightarrow 2 SO_3(g)$. If equal molar amounts of SO_2 and O_2 are mixed, as in this case, SO_2 is the limiting reagent.

(c) The total number of moles remaining in the container will be: 0.030 mol $SO_3(g)$ + 0.015 mol $O_2(g)$ = 0.045 mol of gas at the end of the reaction. The final volume will, therefore, be:

$$V_f = \frac{nRT}{P} = \frac{(0.045 \text{ mol})(0.0820578 \text{ L} \cdot \text{atm} \cdot \text{K}^{-1} \cdot \text{mol}^{-1})(298.15 \text{ K})}{1.00 \text{ atm}}$$

$$= 1.1 \text{ L}$$

(d) $\Delta V = 1.1 \text{ L} - 1.5 \text{ L} = -0.4 \text{ L}$

$w = -P\Delta V = (1.00 \text{ atm})(-0.4 \text{ L})(101.325 \text{ J} \cdot \text{L}^{-1} \cdot \text{atm}^{-1})$

$= 40 \text{ J}$ of work done on the system (work is positive)

(e) The enthalpy of reaction may be found using standard enthalpies of formation and the balanced equation given above:

$\Delta H_r = 2(-395.72 \text{ kJ} \cdot \text{mol}^{-1}) - 2(-296.83 \text{ kJ} \cdot \text{mol}^{-1}) = -197.78 \text{ kJ} \cdot \text{mol}^{-1}$.

If 0.030 mol of SO_2 are consumed, then enthalpy change is:

$(0.030 \text{ mol SO}_2)\left(\frac{-197.78 \text{ kJ}}{2 \text{ mol SO}_2}\right) = -2966.7 \text{ kJ} = -3.0 \text{ kJ} (2 \text{ sf}) = -3000 \text{ J}$.

(f) $\Delta U_r = q + w = -3000 \text{ J} + 40 \text{ J} = -2960 \text{ J}$

4.19 (a) The average kinetic energy is obtained from the expression: average kinetic energy $= \frac{3}{2}RT$.

(a) $4103.2 \text{ J} \cdot \text{mol}^{-1}$;

(b) $4090.7 \text{ J} \cdot \text{mol}^{-1}$;

(c) $4103.2 \text{ J} \cdot \text{mol}^{-1} - 4090.7 \text{ J} \cdot \text{mol}^{-1} = 12.5 \text{ J} \cdot \text{mol}^{-1}$

4.21 According to the second law of thermodynamics, the formation of complex molecules from simpler precursors would not be spontaneous, because such processes create order from disorder. If there is an external input of energy, however, a more ordered system could be created. One of the challenges remaining to evolutionary biology is to explain how the exceedingly complex and highly organized biological structures were created from randomly occurring chemical reactions.

4.23 The strategy here is to determine the amount of energy per photon and the amount of energy needed to heat the water. Dividing the latter by the former will give the number of photons needed. Energy per photon is given by:

$$E = \frac{hc}{\lambda} = \frac{(6.626 \times 10^{-34} \text{ J} \cdot \text{s})(2.9979 \times 10^8 \text{ m} \cdot \text{s}^{-1})}{4.50 \times 10^{-3} \text{ m}} = 4.41 \times 10^{-23} \text{ J} \cdot \text{photon}^{-1}$$

The energy needed to heat the water is:

$$350 \text{ g } (4.184 \text{ J} \cdot \text{g}^{-1} \cdot °\text{C}^{-1}) (100.0°\text{C} - 25.0°\text{C}) = 1.10 \times 10^5 \text{ J}$$

The number of photons needed is therefore:

$$\frac{1.10 \times 10^5 \text{ J}}{4.41 \times 10^{-23} \text{ J} \cdot \text{photon}^{-1}} = 2.49 \times 10^{27} \text{ photons}$$

4.25

$$1 \text{ ounce} = 28.3496 \text{ grams}$$

$$(17.0 \text{ kJ} \cdot \text{g}^{-1})(28.3495 \text{ g}/1 \text{ oz}) = 481.9415 \text{ kJ} \cdot \text{oz}^{-1}$$

$$(481.9415 \text{ kJ} \cdot \text{oz}^{-1})(2 \text{ oz}) = 963.883 \text{ kJ}$$

Energy = (Rate of combustion)(time)

Time = Energy/Rate of combustion

Time = 963.883 kJ/ 15 kJ.min^{-1} = 64 minutes

4.27 (a) $\text{heat} = \dfrac{(7.0\text{ cm} \times 6.0\text{ cm} \times 5.0\text{ cm})(1.5\text{ g}\cdot\text{cm}^{-3}\text{ C})}{12.01\text{ g}\cdot\text{mol}^{-1}\text{ C}}(-394\text{ kJ}\cdot\text{mol}^{-1}\text{ C})$

$= -1.0 \times 10^4\text{ kJ}$

(b) $\text{mass of water} = \dfrac{(1.0 \times 10^7\text{ J})}{(4.18\text{ J}\cdot(°C)^{-1}\cdot g^{-1})(100°C - 25°C)} = 3.2 \times 10^4\text{ g}$

4.29 (a)

$$2H_2S(g) + 3O_2(g) \rightarrow 2SO_2(g) + 2H_2O(l)$$
$$4H_2S(g) + 2SO_2(g) \rightarrow 6S(s) + 4H_2O(l)$$
$$\overline{6H_2S(g) + 3O_2(g) \rightarrow 6S(s) + 6H_2O(l)}$$

or

$$2H_2S(g) + O_2(g) \rightarrow 2S(s) + 2H_2O(l)$$

(b)

$$\dfrac{60.0 \times 10^3\text{ g}}{32.065\text{ g.mol}^{-1}} = 1871.2\text{ mol} = 1.87 \times 10^3\text{ mol}$$

$$\Delta H°(\text{reaction}) = \Sigma\Delta H_f°(\text{products}) - \Sigma\Delta H_f°(\text{reactants})$$

$$= 2(-2.85.83\text{ kJ.mol}^{-1}) - 2(-20.63\text{ kJ.mol}^{-1})$$

$$= -571.66\text{ kJ.mol}^{-1} + 41.26\text{ kJ.mol}^{-1}$$

$$= -530.40\text{ kJ.mol}^{-1}$$

Since -530.40 kJ.mol^{-1} reaction produces 2 moles of S(s), then the enthalpy change associated with the production of 60.0 kg of S(s) would be -4.96×10^5 kJ.

(c) highly exothermic reaction and the reactor would need to be cooled

4.31 (a) The reaction enthalpy is obtained by Hess's law:

$\Delta H°_r = \Delta H°_f (\text{CO, g}) - \Delta H°_f (\text{H}_2\text{O, g})$

$\Delta H°_r = (1)(-110.53\text{ kJ}\cdot\text{mol}^{-1}) - (1)(-241.82\text{ kJ}\cdot\text{mol}^{-1})$

$\Delta H°_r = +131.29\text{ kJ}\cdot\text{mol}^{-1}$

endothermic

(b) The number of moles of H_2 produced is obtained from the ideal gas

law: $n = \dfrac{PV}{RT} = \dfrac{\left(\dfrac{500 \text{ Torr}}{760 \text{ Torr} \cdot \text{atm}^{-1}}\right)(200 \text{ L})}{(0.082\,06 \text{ L} \cdot \text{atm} \cdot \text{K}^{-1} \cdot \text{mol}^{-1})(338 \text{ K})} = 4.74 \text{ mol}$

The enthalpy change accompanying the production of this amount of

hydrogen will be given by: $\Delta H = (4.74 \text{ mol})(131.29 \text{ kJ} \cdot \text{mol}^{-1}) = 623 \text{ kJ}$.

4.33   (a) The number of moles burned may be obtained by taking the difference
in the number of moles of gas present in the tank before and after the drive
using the ideal gas equation:

$$n_1 - n_2 = \frac{P_1 V}{RT} - \frac{P_2 V}{RT} = (P_1 - P_2)\left(\frac{V}{RT}\right)$$

$$= (16.0 \text{ atm} - 4.0 \text{ atm})\left(\frac{30.0 \text{ L}}{(0.0820574 \text{ L} \cdot \text{atm} \cdot \text{K}^{-1} \cdot \text{mol}^{-1})(298 \text{ K})}\right)$$

$$= 14.7 \text{ mol}$$

(b) From a table of enthalpies of combustion, the enthalpy of combustion
of H_2 is found to be -286 kJ.mol^{-1}. The energy change is, therefore,

$(14.7 \text{ mol})(-286 \text{ kJ} \cdot \text{mol}^{-1}) = -4.20 \times 10^3 \text{ kJ}$.

4.35 $\Delta U = q + w = \Delta H - P\Delta V$

Net production of 2/3 mole of gas; $\Delta n = +\dfrac{2}{3}\text{mol}$

$PV = nRT$

$P\Delta V = \Delta nRT$

$-P\Delta V = -\Delta nRT = -(\dfrac{2}{3}\text{mol})(8.314 \text{ J.K}^{-1}.\text{mol}^{-1})(298 \text{ K})$

$= -1.65 \text{ kJ}$

$\therefore \Delta U = \dfrac{1}{3}(-318 \text{ kJ}) - 1.65 \text{ kJ} = -108 \text{ kJ}$

4.37 (a) $\Delta G° < 0$

(b) $\Delta H°$ unable to tell

(c) $\Delta S°$ unable to tell

(d) $\Delta S_{total} > 0$

4.39 According to Trouton's rule, the entropy of vaporization of an organic liquid is a constant of approximately $85 \ \text{J} \cdot \text{mol}^{-1} \cdot \text{K}^{-1}$. The relationship between entropy of fusion, enthalpy of fusion, and melting point is given by $\Delta S°_{fus} = \dfrac{\Delta H°_{fus}}{T_{fus}}$.

For Pb : $\Delta S°_{fus} = \dfrac{5100 \ \text{J}}{600 \ \text{K}} = 8.50 \ \text{J} \cdot \text{K}^{-1}$

For Hg : $\Delta S°_{fus} = \dfrac{2290 \ \text{J}}{234 \ \text{K}} = 9.79 \ \text{J} \cdot \text{K}^{-1}$

For Na : $\Delta S°_{fus} = \dfrac{2640 \ \text{J}}{371 \ \text{K}} = 7.12 \ \text{J} \cdot \text{K}^{-1}$

These numbers are reasonably close but clearly much smaller than the value associated with Trouton's rule. One would need to compare calculated values for more metallic elements to determine if an average could be used.

4.41 This is best answered by considering the reaction that interconverts the two compounds

$$4 \ \text{Fe}_3\text{O}_4(s) + \text{O}_2(g) \longrightarrow 6 \ \text{Fe}_2\text{O}_3(s)$$

We calculate $\Delta G°_r$ using data from Appendix 2A:

$\Delta G°_r = 6\Delta G°_f(\text{Fe}_2\text{O}_3, s) - [4\Delta G°_f(\text{Fe}_3\text{O}_4, s)]$

$\Delta G°_r = 6(-742.2 \ \text{kJ} \cdot \text{mol}^{-1}) - [4(-1015.4 \ \text{kJ} \cdot \text{mol}^{-1})]$

$= -391.6 \ \text{kJ} \cdot \text{mol}^{-1}$

Because $\Delta G°_r$ is negative, the process is spontaneous at 25°C.

Therefore, Fe_2O_3 is thermodynamically more stable.

4.43   (a) $H_2(g) + Br_2(g) \longrightarrow 2\,HBr(g)$

Use thermodynamic data at 25 °C from the appendix, and the relationship

$\Delta G°_r = \Sigma\,\Delta G°_f\,(\text{products}) - \Sigma\,\Delta G°_r\,(\text{reactants})$:

$\Delta G°_r = 2\,\Delta G°_f\,(\text{HBr, g}) - [\Delta G°_f\,(\text{H}_2,\text{ g}) + \Delta G°_f\,(\text{Br}_2,\text{ g})]$

$\qquad\quad = 2\,(-53.45\text{ kJ.mol}^{-1}) - [0 + 3.11\text{ kJ.mol}^{-1}]$

$\qquad\quad = -110.0\text{ kJ.mol}^{-1}$

(b)   STP: 0°C or 273.15K and 1 atm

moles of H_2 gas:

$$n = \frac{PV}{RT} = \frac{(1\text{ atm})(0.120\text{ L})}{(0.0821\text{ L.atm.K}^{-1}.\text{mol}^{-1})(273.15\text{ K})} = 5.35 \times 10^{-3}\text{ mol}$$

A stoichiometric amount of bromine gas is 5.35×10^{-3} mol.

Moles of HBr(g) formed $= 2(5.35 \times 10^{-3}\text{ mol}) = 1.07 \times 10^{-2}$ mol

Molar concentration of the resulting hydrobromic acid,

$$M = \frac{n}{V} = \frac{0.0107\text{ mol}}{(0.150\text{L})} = 7.14 \times 10^{-2}\text{ mol.L}^{-1}$$

4.45   (a) Because the enthalpy change for dissolution is positive, the entropy change of the surroundings must be a negative number

$$\left(\Delta S°_{surr} = -\frac{\Delta H°_{system}}{T}\right).$$ Because spontaneous processes are accompanied

by an increase in entropy, the change in enthalpy does not favor the dissolution process.

(b) In order for the process to be spontaneous (because it occurs readily, we know it is spontaneous), the entropy change of the system must be positive.

(c) Positional disorder is dominant.

(d) Because the surroundings participate in the solution process only as a source of heat, the entropy change of the surroundings is primarily a result of the dispersal of thermal motion.

(e) The driving force for the dissolution is the dispersal of matter, resulting in an overall positive ΔS.

4.47 (a)(i) $\Delta G_r^o = \sum \Delta G_f^o (products) - \sum \Delta G_f^o (reactants) = -120.35 \text{ kJ.mol}^{-1}$

(ii) $\Delta G_r^o = 2(-120.35) - 2(-237.13) = +233.56 \text{ kJ.mol}^{-1}$

Method (i) releases more energy per mole of O_2.

(b) Method (i) has the more negative standard Gibbs free energy.

(c) No.

4.49 The entries all correspond to aqueous ions. The fact that they are negative is due to the reference point that has been established. Because ions cannot actually be separated and measured independently, a reference point that defines $S^o_m (H^+, aq) = 0$ has been established. This definition is then used to calculate the standard entropies for the other ions. The fact that they are negative will arise in part because the solvated ion $M(H_2O)_x^{n+}$ will be more ordered than the isolated ion and solvent molecules ($M^{n+} + x\, H_2O$).

4.51 (a) In order to calculate the free energy at different temperatures, we need to know ΔH° and ΔS° for the process: $H_2O(l) \longrightarrow H_2O(g)$

$\Delta H^\circ_r = \Delta H^\circ_f (H_2O, g) - \Delta H^\circ_f (H_2O, l)$

$\quad = (-241.82 \text{ kJ} \cdot \text{mol}^{-1}) - [-285.83 \text{ kJ} \cdot \text{mol}^{-1}]$

$\quad = 44.01 \text{ kJ} \cdot \text{mol}^{-1}$

$\Delta S^\circ_r = S^\circ_m (H_2O, g) - S^\circ_m (H_2O, l)$

$\quad = 188.83 \text{ J} \cdot \text{K}^{-1} \cdot \text{mol}^{-1} - [69.91 \text{ J} \cdot \text{K}^{-1} \cdot \text{mol}^{-1}]$

$\quad = 118.92 \text{ J} \cdot \text{K}^{-1} \cdot \text{mol}^{-1}$

$\Delta G^\circ_r = \Delta H^\circ_r - T\Delta S^\circ_r$

$\quad = 44.01 \text{ kJ} \cdot \text{mol}^{-1} - T(118.92 \text{ J} \cdot \text{K}^{-1} \cdot \text{mol}^{-1})/(1000 \text{ J} \cdot \text{kJ}^{-1})$

$T(K) \Delta G^\circ_r (kJ)$

298   8.57 kJ

373   − 0.35 kJ

423   – 6.29 kJ

The reaction goes from being nonspontaneous near room temperature to being spontaneous above 100°C.

(b) The value at 100°C should be exactly 0, because this is the normal boiling point of water.

(c) The discrepancy arises because the enthalpy and entropy values calculated from the tables are not rigorously constant with temperature. Better values would be obtained using the actual enthalpy and entropy of vaporization measured at the boiling point.

4.53

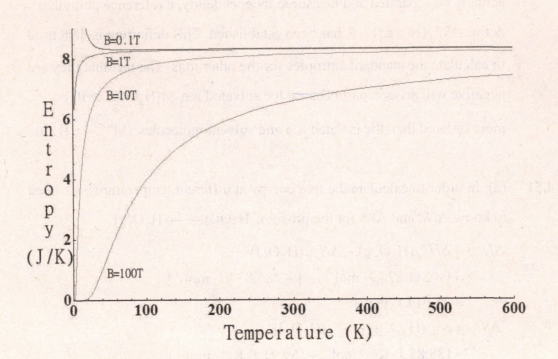

4.55 $p_{down} = p_{up}$

By inspection of the equations in exercise 9.94,

$e^{-\Delta E/kT} = 1$ for $p_{down} = p_{up}$. $p_{down} = p_{up}$

$$\frac{1}{1+1} = \frac{1}{1+1}$$

Since ΔE is the energy difference between the low energy and high energy spin states it is a positive value. Since ΔE is a positive value, $e^{-\Delta E/kT}$ will tend 1 as T $\rightarrow \infty$.

4.57 (a) Since standard molar entropies increase with temperature (more translational, vibrational and rotational motion), the $-T\Delta S_m^o$ term becomes more negative at higher temperatures.

(b) For gases with translational motion, their increase in standard molar entropy, on heating, is much larger than for solids and liquids.

Therefore, their $-T\Delta S_m^o$ term is more negative.

4.59 Dehydrogenation of cyclohexane to benzene:

$$C_6H_{12}(l) \rightarrow C_6H_6(l) + 3H_2(g)$$

$$\Delta G^\circ_r = \Delta G^\circ_f (C_6H_6, l) - \Delta G^\circ_f (C_6H_{12}, l)$$

$$= 124.3 \text{ kJ} \cdot \text{mol}^{-1} - 26.7 \text{ kJ} \cdot \text{mol}^{-1}$$

$$= +97.6 \text{ kJ} \cdot \text{mol}^{-1}$$

The reaction of ethene with hydrogen:

$$C_2H_4(g) + H_2(g) \rightarrow C_2H_6(g)$$

$$\Delta G^\circ_r = \Delta G^\circ_f (C_2H_6, g) - \Delta G^\circ_f (C_2H_4, g)$$

$$= (-32.82 \text{ kJ} \cdot \text{mol}^{-1}) - 68.15 \text{ kJ} \cdot \text{mol}^{-1}$$

$$= -100.97 \text{ kJ} \cdot \text{mol}^{-1}$$

Combine these two reactions so that $C_2H_4(g)$ accepts the hydrogen that is formed in the dehydrogenation reaction:

$C_6H_{12}(l) \rightarrow C_6H_6(l) + 3H_2(g)$	$\Delta G^\circ_r = +97.6 \text{ kJ} \cdot \text{mol}^{-1}$
$+ 3[C_2H_4(g) + H_2(g) \rightarrow C_2H_6(g)]$	$\Delta G^\circ_r = 3(-100.97 \text{ kJ} \cdot \text{mol}^{-1})$

$$C_6H_{12}(l) + 3C_2H_4(g) \rightarrow C_6H_6(l) + 3C_2H_6(g) \quad \Delta G^\circ_r = -205.31 \text{ kJ} \cdot \text{mol}^{-1}$$

By combining the above two reactions the overall process becomes spontaneous.

4.61   Assuming standard state conditions, ΔG_r^o for $C_2H_4(g) + H_2O(g) \rightarrow$ $CH_3CH_2OH(l)$ is:

$$\Delta G^o_r = \left(-174.8 \text{ kJ} \cdot \text{mol}^{-1}\right) - \left(68.15 \text{ kJ} \cdot \text{mol}^{-1}\right) - \left(- 228.57 \text{ kJ} \cdot \text{mol}^{-1}\right)$$

$$= -14.4 \text{ kJ} \cdot \text{mol}^{-1}$$ Negative ΔG^o_r indicates a spontaneous reaction.

ΔG_r^o for $C_2H_6(g) + H_2O(g) \rightarrow CH_3CH_2OH(l) + H_2(g)$ is:

$$\Delta G^o_r = \left(-174.8 \text{ kJ} \cdot \text{mol}^{-1}\right) - \left(-32.82 \text{ kJ} \cdot \text{mol}^{-1}\right) - \left(- 228.57 \text{ kJ} \cdot \text{mol}^{-1}\right)$$

$$= +86.6 \text{ kJ} \cdot \text{mol}^{-1}$$ Positive ΔG^o_r indicates a nonspontaneous reaction.

Reaction A is spontaneous, but reaction B is not spontaneous at any temperature.

4.63   (a)

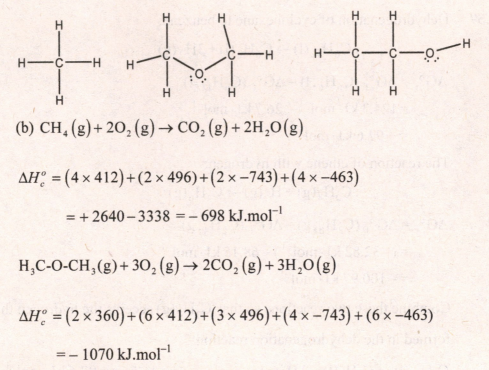

(b) $CH_4(g) + 2O_2(g) \rightarrow CO_2(g) + 2H_2O(g)$

$$\Delta H^o_c = (4 \times 412) + (2 \times 496) + (2 \times -743) + (4 \times -463)$$

$$= + 2640 - 3338 = - 698 \text{ kJ.mol}^{-1}$$

$H_3C\text{-}O\text{-}CH_3(g) + 3O_2(g) \rightarrow 2CO_2(g) + 3H_2O(g)$

$$\Delta H^o_c = (2 \times 360) + (6 \times 412) + (3 \times 496) + (4 \times -743) + (6 \times -463)$$

$$= - 1070 \text{ kJ.mol}^{-1}$$

Given: $CH_3CH_2OH(l) \rightarrow CH_3CH_2OH(g)$ $\Delta H^o_{vap} = 43.5 \text{ kJ mol}^{-1}$

$CH_3CH_2OH(g) + 3O_2(g) \rightarrow 2CO_2(g) + 3H_2O(g)$

$$\Delta H^o_c = (360) + (463) + (348) + (5 \times 412) + (3 \times 496) + (4 \times -743) + (6 \times -463)$$

$$= - 1031 \text{ kJ.mol}^{-1}$$

For the burning of 1 mol

$CH_3CH_2OH(l)$, $\Delta H_c^{\circ} = 43.5 - 1031 = -988$ kJ.mol^{-1}.

The burning of 1 mole of dimethyl ether releases the most heat.

(c) $\dfrac{-890 \text{ kJ.mol}^{-1}}{16.01 \text{ g.mol}^{-1}} = -55.6$ kJ.g^{-1}

$\dfrac{-1368 \text{ kJ.mol}^{-1}}{46.02 \text{ g.mol}^{-1}} = -29.73$ kJ.g^{-1}

$\dfrac{-5471 \text{ kJ.mol}^{-1}}{114.22 \text{ g.mol}^{-1}} = -47.90$ kJ.g^{-1}

Methane as it releases the most heat per gram.

(b) mass of 10.00 L octane = $(0.70 \text{ g.mL}^{-1})(10{,}000 \text{ mL}) = 7.0 \times 10^3$ g

$\dfrac{-5471 \text{ kJ.mol}^{-1}}{114.22 \text{ g.mol}^{-1}} = -47.90$ kJ.g^{-1}

heat released = $(7.0 \times 10^3 \text{ g})(-47.90 \text{ kJ.g}^{-1}) = -3.35 \times 10^5$ kJ

moles of methane gas = $\dfrac{3.35 \times 10^5 \text{ kJ}}{890 \text{ kJ.mol}^{-1}} = 376.7$ mol $= 377$ mol

$V = \dfrac{nRT}{P} = \dfrac{(376.7 \text{ mol})(8.20574 \times 10^{-2} \text{ L.atm.K}^{-1}.\text{mol}^{-1})(298 \text{ K})}{10.00 \text{ atm}}$

   $= 921$ L

(e) methane gas, -890 kJ.mol^{-1} CO_2 (less CO_2)

ethanol liquid, -684 kJ.mol^{-1} CO_2 (more CO_2)

octane liquid, -684 kJ.mol^{-1} CO_2 (more CO_2)

FOCUS 5

EQUILIBRIUM

5A.1   Ammonia can hydrogen bond with itself; this increase in intermolecular forces will cause the vapor pressure to be lower than expected for NH_3. Phosphine cannot hydrogen bond with itself therefore PH_3 will have the higher vapor pressure.

5A.3   (a) 87°C   (b) 113°C

5A.5   Table 8.3 contains the enthalpy of vaporization and the boiling point of methanol (at which the vapor pressure = 1 atm). Using this data and the equation

$$\ln \frac{P_2}{P_1} = -\frac{\Delta H°_{vap}}{R} \left[\frac{1}{T_2} - \frac{1}{T_1} \right]$$

$$\ln \left(\frac{P_{25.0°C}}{1} \right) = -\frac{35\ 300\ \text{J} \cdot \text{mol}^{-1}}{8.314\ \text{J} \cdot \text{K}^{-1} \cdot \text{mol}^{-1}} \left[\frac{1}{298.2\ \text{K}} - \frac{1}{337.2\ \text{K}} \right]$$

$P_{25.0°C} = 0.19$ atm or 1.5×10^2 Torr

5A.7   At the normal boiling of a material its vapor pressure will be 1.00 atm (101.325 kPa). Since we know the vapor pressure at −28°C (245.15 K) is 17.0 kPa and the $\Delta H°_{vap} = 23.77$ kJ · mol^{-1} we can use the relationship

$\ln \dfrac{P_2}{P_1} = \dfrac{\Delta H°_{vap}}{R} \left[\dfrac{1}{T_1} - \dfrac{1}{T_2} \right]$ to calculate the temperature at which the vapor

pressure will be 101.325 kPa (be careful with your units):

$$\ln \frac{101.325\ \text{kPa}}{17.0\ \text{kPa}} = \frac{23.77\ \text{kJ} \cdot \text{mol}^{-1}}{8.314\ 47 \times 10^{-3}\ \text{kJ} \cdot \text{K}^{-1} \cdot \text{mol}^{-1}} \left(\frac{1}{245.15\ \text{K}} - \frac{1}{T_2} \right);$$

solving for T_2, we get the normal boiling point of BCl_3 to be 289 K, or 16°C.

5A.9   (a) The quantities ΔH°_{vap} and ΔS°_{vap} can be calculated using the

relationship $\ln P = -\dfrac{\Delta H^\circ_{vap}}{R} \cdot \dfrac{1}{T} + \dfrac{\Delta S^\circ_{vap}}{R}$

Because we have two temperatures with corresponding vapor pressures, we can set up two equations with two unknowns and solve for ΔH°_{vap} and ΔS°_{vap}. If the equation is used as is, P must be expressed in atm, which is the standard reference state. Remember that the value used for P is really activity, which for pressure is P divided by the reference state of 1 atm, so that the quantity inside the ln term is dimensionless.

$$(8.314 \text{ J} \cdot \text{K}^{-1} \cdot \text{mol}^{-1}) \ln\left(\frac{58 \text{ Torr}}{760 \text{ Torr}}\right) = -\frac{\Delta H^\circ_{vap}}{250.4 \text{ K}} + \Delta S^\circ_{vap}$$

$$(8.314 \text{ J} \cdot \text{K}^{-1} \cdot \text{mol}^{-1}) \ln\left(\frac{512 \text{ Torr}}{760 \text{ Torr}}\right) = -\frac{\Delta H^\circ_{vap}}{298.2 \text{ K}} + \Delta S^\circ_{vap}$$

which give, upon combining terms,

$$-21.39 \text{ J} \cdot \text{K}^{-1} \cdot \text{mol}^{-1} = (-0.003\,994 \text{ K}^{-1})\,\Delta H^\circ_{vap} + \Delta S^\circ_{vap}$$

$$-3.284 \text{ J} \cdot \text{K}^{-1} \cdot \text{mol}^{-1} = (-0.003\,353 \text{ K}^{-1})\Delta H^\circ_{vap} + \Delta S^\circ_{vap}$$

Subtracting one equation from the other will eliminate the ΔS°_{vap} term and allow us to solve for ΔH°_{vap} :

$$-18.11 \text{ J} \cdot \text{K}^{-1} \cdot \text{mol}^{-1} = (-0.000\,641 \text{ K}^{-1})\,\Delta H^\circ_{vap}$$

$$\Delta H^\circ_{vap} = +28.3 \text{ kJ} \cdot \text{mol}^{-1}$$

(b) We can then use ΔH°_{vap} to calculate ΔS°_{vap} using either of the two equations:

$$-21.39 \text{ J} \cdot \text{K}^{-1} \cdot \text{mol}^{-1} = -0.003\,994 \text{ K}^{-1}\,(28\,300 \text{ J} \cdot \text{mol}^{-1}) + \Delta S^\circ_{vap}$$

$$\Delta S^\circ_{vap} = 91.6 \text{ J} \cdot \text{K}^{-1} \cdot \text{mol}^{-1}$$

$$-3.284 \text{ J} \cdot \text{K}^{-1} \cdot \text{mol}^{-1} = -0.003\,353 \text{ K}^{-1}\,(28\,300 \text{ J} \cdot \text{mol}^{-1}) + \Delta S^\circ_{vap}$$

$$\Delta S^\circ_{vap} = 91.6 \text{ J} \cdot \text{K}^{-1} \cdot \text{mol}^{-1}$$

(c) The ΔG°_{vap} is calculated using $\Delta G^\circ_r = \Delta H^\circ_r - T\Delta S^\circ_r$

$$\Delta G^\circ_r = +28.3 \text{ kJ} \cdot \text{mol}^{-1} - (298 \text{ K})(91.6 \text{ J} \cdot \text{K}^{-1} \cdot \text{mol}^{-1})/(1000 \text{ J} \cdot \text{kJ}^{-1})$$

$$\Delta G^\circ_r = +1.0 \text{ kJ} \cdot \text{mol}^{-1}$$

(d) The boiling point can be calculated using one of several methods. The easiest to use is the one developed in the last chapter:

$$\Delta G°_{vap} = \Delta H°_{vap} - T_B \, \Delta S°_{vap} = 0$$

$$\Delta H°_{vap} = T_B \, \Delta S°_{vap} \text{ or } T_B = \frac{\Delta H°_{vap}}{\Delta S°_{vap}}$$

$$T_B = \frac{(28.3 \text{ kJ} \cdot \text{mol}^{-1})(1000 \text{ J} \cdot \text{kJ}^{-1})}{91.6 \text{ J} \cdot \text{K}^{-1} \cdot \text{mol}^{-1}} = 309 \text{ K or } 36°C$$

Alternatively, we could use the relationship $\ln \dfrac{P_2}{P_1} = -\dfrac{\Delta H°_{vap}}{R}\left[\dfrac{1}{T_2} - \dfrac{1}{T_1}\right]$.

Here we would substitute, in one of the known vapor pressure points, the value of the enthalpy of vaporization and the condition that $P = 1$ atm at the normal boiling point.

5A.11 (a) The quantities $\Delta H°_{vap}$ and $\Delta S°_{vap}$ can be calculated, using the relationship $\ln P = -\left(\dfrac{\Delta H°_{vap}}{R}\right)\dfrac{1}{T} + \dfrac{\Delta S°_{vap}}{R}$. Because we have two temperatures with corresponding vapor pressures (we know that the vapor pressure = 1 atm at the boiling point), we can set up two equations with two unknowns and solve for $\Delta H°_{vap}$ and $\Delta S°_{vap}$. If the equation is used as is, P must be expressed in atm which is the standard reference state. Remember that the value used for P is really activity which, for pressure, is P divided by the reference state of 1 atm, so that the quantity inside the ln term is dimensionless.

$$(8.314 \text{ J} \cdot \text{K}^{-1} \cdot \text{mol}^{-1}) \ln (1) = -\frac{\Delta H°_{vap}}{315.58 \text{ K}} + \Delta S°_{vap}$$

$$(8.314 \text{ J} \cdot \text{K}^{-1} \cdot \text{mol}^{-1}) \ln \left(\frac{140 \text{ Torr}}{760 \text{ Torr}}\right) = -\frac{\Delta H°_{vap}}{273.2 \text{ K}} + \Delta S°_{vap}$$

which give, upon combining terms,

$$0 \text{ J} \cdot \text{K}^{-1} \cdot \text{mol}^{-1} = (-0.003\,169 \text{ K}^{-1})\Delta H°_{vap} + \Delta S°_{vap}$$

$$-14.06 \text{ J} \cdot \text{K}^{-1} \cdot \text{mol}^{-1} = (-0.003\,660 \text{ K}^{-1})\Delta H°_{vap} + \Delta S°_{vap}$$

Subtracting one equation from the other will eliminate the $\Delta S°_{vap}$ term and allow us to solve for $\Delta H°_{vap}$ of iodomethane.

$$-14.06 \text{ J} \cdot \text{K}^{-1} \cdot \text{mol}^{-1} = (-0.000\,491 \text{ K}^{-1})\,\Delta H°_{vap}$$

$$\Delta H°_{vap} = +28.6 \text{ kJ} \cdot \text{mol}^{-1}$$

(b) We can then use $\Delta H°_{vap}$ to calculate $\Delta S°_{vap}$ using either of the two equations:

$$0 = (-0.003\,169 \text{ K}^{-1})(28\,600 \text{ J} \cdot \text{mol}^{-1}) + \Delta S°_{vap}$$

$$\Delta S°_{vap} = 90.6 \text{ J} \cdot \text{K}^{-1} \cdot \text{mol}^{-1}$$

$$-14.06 \text{ J} \cdot \text{K}^{-1} \cdot \text{mol}^{-1} = (-0.003\,660 \text{ K}^{-1})(28\,600 \text{ J} \cdot \text{mol}^{-1}) + \Delta S°_{vap}$$

$$\Delta S°_{vap} = 90.6 \text{ J} \cdot \text{K}^{-1} \cdot \text{mol}^{-1}$$

(c) The vapor pressure of CH_3I at another temperature is calculated using

$$\ln\frac{P_2}{P_1} = -\frac{\Delta H°_{vap}}{R}\left[\frac{1}{T_2} - \frac{1}{T_1}\right]$$

We need to insert the calculated value of the enthalpy of vaporization and one of the known vapor pressure points:

$$\ln\left(\frac{P_{\text{at } 25.0°C}}{1 \text{ atm}}\right) = -\frac{28\,600 \text{ J} \cdot \text{mol}^{-1}}{8.314 \text{ J} \cdot \text{K}^{-1} \cdot \text{mol}^{-1}}\left[\frac{1}{298.2 \text{ K}} - \frac{1}{315.58 \text{ K}}\right]$$

$$P_{\text{at } 25.0°C} = 0.53 \text{ atm or } 4.0 \times 10^2 \text{ Torr}$$

5B.1 (a) vapor;   (b) liquid;   (c) vapor

5B.3 (a) 2.4 K;   (b) about 10 atm;   (c) 5.5 K;   (d) no

5B.5 (a) At the lower pressure triple point, liquid helium-I and -II are in equilibrium with helium gas; at the higher pressure triple point, liquid helium-I and -II are in equilibrium with solid helium.   (b) helium-I

5B.7 The pressure increase would bring CO_2 into the solid region.

5C.1   The free energy of the solvent in the NaCl solution will always be lower than the free energy of pure water; if given enough time, all of the water from the "pure water" beaker will become part of the NaCl solution, leaving an empty beaker.

5C.3   (a) $P = x_{solvent} \times P_{pure\ solvent}$. At 100°C, the normal boiling point of water, the vapor pressure of water is 1.00 atm. If the mole fraction of sucrose is 0.100, then the mole fraction of water is 0.900:

$P = 0.900 \times 1.000$ atm $= 0.900$ atm or 684 Torr

(b)   The molarity of the sucrose solution must be converted to mole fraction. For dilute aqueous solutions we can assume that the molarity will approximately equal the molality. The molality can be converted to mole fraction. If the molality is $0.100\ mol \cdot kg^{-1}$, then there will be 0.100 mol sucrose per 1000 g of water.

$$x_{H_2O} = \frac{n_{H_2O}}{n_{H_2O} + n_{sucrose}} = \frac{\dfrac{1000\ g}{18.02\ g \cdot mol^{-1}}}{\dfrac{1000\ g}{18.02\ g \cdot mol^{-1}} + 0.100\ mol} = 0.998$$

$P = 0.998 \times 1.000$ atm $= 0.998$ atm or 758 Torr

5C.5   The mole fraction of benzene in the mixture can be calculated using Raoult's law :

$$P_{benzene\ soln} = x_{benzene} \times P_{pure\ benzene};\ \text{so}\ x_{benzene} = \frac{P_{benzene\ soln}}{P_{pure\ benzene}}$$

$$x_{benzene} = \frac{75.0\ Torr}{94.6\ Torr} = 0.793$$

Since the sum of the mole fractions of solute and solvent must equal 1 and since we know the moles of benzene that are present calculation of the moles of solute added is straightforward:

$$x_{unknown} = 1 - x_{benzene} = 1 - 0.793 = 0.207$$

$$x_{unknown} = \frac{n_{unknown}}{n_{unknown} + n_{benzene}} = \frac{n_{unknown}}{n_{unknown} + 0.300} = 0.207$$

Solving this gives $n_{unknown} = 7.80 \times 10^{-2}$ mol.

5C.7 (a) From the relationship $P = x_{solvent} \times P_{pure\ solvent}$ we can calculate the mole fraction of the solvent: 94.8 Torr = $x_{solvent}$ (100.0 Torr); $x_{solvent} = 0.948$. The mole fraction of the unknown compound will be $1.000 - 0.948 = 0.052$.

(b) The molar mass can be calculated by using the definition of mole fraction for either the solvent or solute. In this case, the math is slightly easier if the definition of mole fraction of the solvent is used:

$$x_{solvent} = \frac{n_{solvent}}{n_{unknown} + n_{solvent}}$$

$$0.948 = \frac{\dfrac{100\ g}{78.11\ g \cdot mol^{-1}}}{\dfrac{100\ g}{78.11\ g \cdot mol^{-1}} + \dfrac{8.05\ g}{M_{unknown}}}$$

$$M_{unknown} = \frac{8.05\ g}{\left[\left(\dfrac{100\ g}{78.11\ g \cdot mol^{-1}}\right) \Big/ 0.948\right] - \left(\dfrac{100\ g}{78.11\ g \cdot mol^{-1}}\right)} = 115\ g \cdot mol^{-1}$$

5C.9 In order to solve for the partial pressures of each component and the total pressure we need to first find the mole fraction of each component:

$$x_{benzene} = \frac{1.00\ mol}{1.00\ mol + 0.400\ mol} = 0.714$$

$$x_{toluene} = 1 - x_{benzene} = 0.286$$

$$P_{benzene\ soln} = x_{benzene} \times P_{pure\ benzene} = (0.714 \times 94.6\ Torr) = 67.5\ Torr$$

$$P_{toluene\ soln} = x_{toluene} \times P_{pure\ toluene} = (0.286 \times 29.1\ Torr) = 8.32\ Torr$$

$$P_{total} = P_{benzene} + P_{toluene} = 67.5\ Torr + 8.32\ Torr = 75.8\ Torr$$

5C.11 (a) To determine the vapor pressure of the solution, we need to know the mole fraction of each component.

$$x_{benzene} = \frac{1.50\ mol}{1.50\ mol + 0.50\ mol} = 0.75$$

$$x_{toluene} = 1 - x_{benzene} = 0.25$$

$$P_{total} = 0.75\ (94.6\ Torr) + 0.25\ (29.1\ Torr) = 78.2\ Torr$$

The vapor phase composition will be given by

$$x_{benzene,\ vapor} = \frac{P_{benzene}}{P_{total}} = \frac{0.75(94.6\ Torr)}{78.2\ Torr} = 0.91$$

$$x_{toluene,\ vapor} = 1 - 0.91 = 0.09$$

The vapor is richer in the more volatile benzene, as expected.

(b) The procedure is the same as in (a) but the number of moles of each component must be calculated first:

$$n_{benzene} = \frac{15.0\ g}{78.11\ g \cdot mol^{-1}} = 0.192$$

$$n_{toluene} = \frac{64.3\ g}{92.14\ g \cdot mol^{-1}} = 0.698$$

$$x_{benzene} = \frac{0.192\ mol}{0.192\ mol + 0.698\ mol} = 0.216$$

$$x_{toluene} = 1 - x_{benzene} = 0.784$$

$$P_{total} = 0.216\ (94.6\ Torr) + 0.784(29.1\ Torr) = 43.2\ Torr$$

The vapor phase composition will be given by

$$x_{benzene,\ vapor} = \frac{P_{benzene}}{P_{total}} = \frac{0.216\ (94.6\ Torr)}{43.2\ Torr} = 0.473$$

$$x_{toluene,\ vapor} = 1 - 0.473 = 0.527$$

5C.13 To calculate this quantity, we must first find the mole fraction of each that will be present in the mixture. This value is obtained from the relationship

$$P_{total} = \left[x_{1,1\text{-dichloroethane}} \left(P_{pure\ 1,1\text{-dichloroethane}} \right) \right]$$

$$+ \left[x_{1,1\text{-dichlorotetrafluoroethane}} \left(P_{pure\ 1,1\text{-dichlorotetrafluoroethane}} \right) \right]$$

$$157\ Torr = \left[x_{1,1\text{-dichloroethane}} \left(228\ Torr \right) \right] + \left[x_{1,1\text{-dichlorotetrafluoroethane}} \left(79\ Torr \right) \right]$$

$$157\ Torr = \left[x_{1,1\text{-dichloroethane}} \left(228\ Torr \right) \right] + \left[(1 - x_{1,1\text{-dichloroethane}})(79\ Torr) \right]$$

$$157\ Torr = 79\ Torr + \left[x_{1,1\text{-dichloroethane}} \left(228\ Torr \right) - x_{1,1\text{-dichloroethane}} \left(79\ Torr \right) \right]$$

$$78\ Torr = x_{1,1\text{-dichloroethane}} \left(149\ Torr \right)$$

$$x_{1,1\text{-dichloroethane}} = 0.52$$

$$x_{1,1\text{-dichlorotetrafluoroethane}} = 1 - 0.52 = 0.48$$

To calculate the number of grams of 1,1-dichloroethane (m), we use the definition of mole fraction. Mathematically, it is simpler to use the

$$x_{1,1\text{-dichlorotetrafluoroethane}} \quad \text{definition:}$$

$$n_{1,1\text{-dichlorotetrafluoroethane}} = \frac{100.0 \text{ g}}{170.92 \text{ g} \cdot \text{mol}^{-1}} = 0.5851 \text{ mol}$$

$$x_{1,1\text{-dichlorotetrafluoroethane}} = \frac{0.5851 \text{ mol}}{0.5851 \text{ mol} + \dfrac{m}{98.95 \text{ g} \cdot \text{mol}^{-1}}} = 0.48$$

$$0.5851 \text{ mol} = 0.48 \left[0.5851 \text{ mol} + \frac{m}{98.95 \text{ g} \cdot \text{mol}^{-1}} \right]$$

$$0.5851 \text{ mol} - \left[0.48 \, (0.5851 \text{ mol}) \right] = 0.48 \left[\frac{m}{98.95 \text{ g} \cdot \text{mol}^{-1}} \right]$$

$$0.3042 \text{ mol} = (0.004\,851 \text{ mol} \cdot \text{g}^{-1}) \, m$$

$$m = 63 \text{ g}$$

5C.15 Raoult's law applies to the vapor pressure of the mixture, so positive deviation means that the vapor pressure is higher than expected for an ideal solution. Negative deviation means that the vapor pressure is lower than expected for an ideal solution. Negative deviation will occur when the interactions between the different molecules are somewhat stronger than the interactions between molecules of the same kind   (a) For methanol and ethanol, we expect the types of intermolecular attractions in the mixture to be similar to those in the component liquids, so that an ideal solution is predicted.   (b) For HF and H_2O, the possibility of intermolecular hydrogen bonding between water and HF would suggest that negative deviation would be observed, which is the case. HF and H_2O form an azeotrope that boils at 111°C, a temperature higher than the boiling point of either HF (19.4°C) or water.   (c) Because hexane is nonpolar and water is polar with hydrogen bonding, we would expect a mixture of these two to exhibit positive deviation (the interactions between the different molecules would be weaker than the intermolecular forces between like molecules). Hexane and water do form an azeotrope that boils at 61.6°C, a temperature below the boiling point of either hexane or water.

5D.1   (a) KCl is an ionic solid, so water would be the best choice;   (b) CCl_4 is nonpolar, so the best choice is benzene;   (c) CH_3COOH is polar, so water is the better choice.

5D.3   (a) hydrophilic, because NH_2 is polar, and has a lone pair and H atoms that can participate in hydrogen bonding to water molecules;

(b) hydrophobic, because the CH_3 group is not very polar;

(c) hydrophobic, because the Br group is not very polar;

(d) hydrophilic, because the carboxylic acid group has lone pairs on oxygen and an acidic proton that can participate in hydrogen bonding to water molecules.

5D.5   (a) The solubility of $O_2(g)$ in water is $1.3 \times 10^{-3} \ mol \cdot L^{-1} \cdot atm^{-1}$.

$$solubility \ at \ 50 \ kPa = \left(\frac{50 \ kPa}{101.325 \ kPa \cdot atm^{-1}} \right) (1.3 \times 10^{-3} \ mol \cdot L^{-1} \cdot atm^{-1})$$

$$= 6.4 \times 10^{-4} \ mol \cdot L^{-1}$$

(b) The solubility of $CO_2(g)$ in water is $2.3 \times 10^{-2} \ mol \cdot L^{-1} \cdot atm^{-1}$.

$$solubility \ at \ 500 \ Torr = \left(\frac{500 \ Torr}{760 \ Torr \cdot atm^{-1}} \right) (2.3 \times 10^{-2} \ mol \cdot L^{-1} \cdot atm^{-1})$$

$$= 1.5 \times 10^{-2} \ mol \cdot L^{-1}$$

(c) solubility =

$$(0.10 \ atm)(2.3 \times 10^{-2} \ mol \cdot L^{-1} \cdot atm^{-1}) = 2.3 \times 10^{-3} \ mol \cdot L^{-1}.$$

5D.7   (a) $(4 \ mg \cdot L^{-1})(1000 \ mL \cdot L^{-1})(1.00 \ g \cdot mL^{-1})(1 \ kg/1000 \ g)$

$= 4 \ mg \cdot kg^{-1}$ or 4 ppm

(b) The solubility of $O_2(g)$ in water is $1.3 \times 10^{-3} \ mol \cdot L^{-1} \cdot atm^{-1}$ which can be converted to parts per million as follows: In 1.00 L (corresponding to 1 kg of solution) there will be $1.3 \times 10^{-3} \ mol \ O_2$; thus

$$(1.3 \times 10^{-3}\,\text{mol} \cdot \text{kg}^{-1} \cdot \text{atm}^{-1}\,O_2)(32.00\,\text{g} \cdot \text{mol}^{-1}\,O_2)(10^3\,\text{mg} \cdot \text{g}^{-1}) =$$

$$42\,\text{mg} \cdot \text{kg}^{-1} \cdot \text{atm}^{-1}\ \text{or}\ 42\,\text{ppm} \cdot \text{atm}^{-1}$$

$$4\,\text{mg} \div 41\,\text{mg} \cdot \text{kg}^{-1} \cdot \text{atm}^{-1} = 0.1\,\text{atm}$$

(c) $P = \dfrac{0.1\,\text{atm}}{0.21} = 0.5\,\text{atm}$

5D.9 (a) By Henry's law, the concentration of CO_2 in solution will double.

(b) No change in the equilibrium will occur; the partial pressure of CO_2 is unchanged and the concentration is unchanged.

5D.11 The solubility of CO_2 can be determined using Henry's law:

$$S_{CO_2} = k_H \times P = (2.3 \times 10^{-2}\,\text{mol} \cdot \text{L}^{-1} \cdot \text{atm}^{-1})(3.60\,\text{atm})$$

$$n_{CO_2} = S_{CO_2} \times 0.420\,\text{L} = 3.48 \times 10^{-2}\,\text{mol}\ CO_2$$

This is equal to 1.5 g of CO_2.

5D.13 (a) Because it is exothermic, the enthalpy change must be negative;

(b) $Li_2SO_4(s) \rightarrow 2\,Li^+(aq) + SO_4^{2-}(aq) + \text{heat}$; (c) Given that

$$\Delta H_L + \Delta H_{\text{hydration}} = \Delta H\ \text{of solution, the enthalpy of hydration should be}$$

larger. If the lattice energy were greater, the overall process would be endothermic.

5D.15 To answer this question we must first determine the molar enthalpies of solution and multiply this by the number of moles of solid dissolved to get the actual amount of heat released.

(a) $\Delta H^\circ_{sol} = 3.9\,\text{kJ} \cdot \text{mol}^{-1}$

$$\Delta H = \left(\frac{10.0\,\text{g NaCl}}{58.44\,\text{g} \cdot \text{mol}^{-1}}\right)(3.9\,\text{kJ} \cdot \text{mol}^{-1}) = 0.67\,\text{kJ or}\ 6.7 \times 10^2\,\text{J}$$

(b) $\Delta H^\circ_{sol} = -7.5\,\text{kJ} \cdot \text{mol}^{-1}$

$$\Delta H = \left(\frac{10.0\,\text{g NaI}}{149.89\,\text{g} \cdot \text{mol}^{-1}}\right)(-7.5\,\text{kJ} \cdot \text{mol}^{-1}) = -0.50\,\text{kJ} = -5.0 \times 10^2\,\text{J}$$

(c) $\Delta H^\circ_{sol} = -329\,\text{kJ} \cdot \text{mol}^{-1}$

$$\Delta H = \left(\frac{10.0 \text{ g AlCl}_3}{133.33 \text{ g} \cdot \text{mol}^{-1}} \right) (-329 \text{ kJ} \cdot \text{mol}^{-1}) = -24.7 \text{ kJ}$$

(d) $\Delta H_{sol}^{\circ} = 6.6 \text{ kJ} \cdot \text{mol}^{-1}$

$$\Delta H = \left(\frac{10.0 \text{ g NH}_4\text{NO}_3}{80.05 \text{ g} \cdot \text{mol}^{-1}} \right) (6.6 \text{ kJ} \cdot \text{mol}^{-1}) = +0.82 \text{ kJ} = +8.2 \times 10^2 \text{ J}$$

5D.17 A *foam* is a colloid formed by suspending a gas in a liquid or a solid matrix, while a *sol* is a suspension of a solid in a liquid. Some examples of foams are Styrofoam^{TM} and soapsuds; examples of sols are muddy water and mayonnaise.

5D.19 Colloids will reflect or scatter light while true solutions do not (this is known as the Tyndall effect).

5E.1 (a) $m_{NaCl} = \dfrac{\left(\dfrac{25.0 \text{ g NaCl}}{58.44 \text{ g} \cdot \text{mol}^{-1}} \right)}{0.5000 \text{ kg}} = 0.856 \ m$

(b) $m_{NaOH} = \dfrac{\left(\dfrac{\text{mass NaOH}}{40.00 \text{ g} \cdot \text{mol}^{-1}} \right)}{0.345 \text{ kg}} = 0.18 \ m$

mass NaOH = 2.5 g

(c) $m_{NaCl} = \dfrac{\left(\dfrac{0.978 \text{ g urea}}{60.06 \text{ g} \cdot \text{mol}^{-1}} \right)}{(285 \text{ mL})(1 \text{ g} \cdot \text{mL}^{-1})(10^{-3} \text{ kg} \cdot \text{g}^{-1})} = 0.0571 \ m$

5E.3 If the mole fraction of ethylene glycol is 0.250, then the mole fraction of water must be 0.750; thus 1 mol of solution contains 0.250 mol of ethylene glycol and 0.750 mol of water. Converting water to kg:

$0.750 \text{ mol} \times 18.02 \text{ g} \cdot \text{mol}^{-1} \times 10^{-3} \text{ kg} \cdot \text{g}^{-1} = 0.0135 \text{ kg water}.$

The molality of the solution is $\dfrac{0.250 \text{ mol ethylene glycol}}{0.0135 \text{ kg water}} = 18.5 \ m.$

5E.5 (a) 1 kg of 5.00% K_3PO_4 will contain 50.0 g K_3PO_4 and 950.0 g H_2O.

$$\frac{\left(\dfrac{50.0 \text{ g } K_3PO_4}{212.27 \text{ g} \cdot \text{mol}^{-1} \text{ } K_3PO_4}\right)}{0.950 \text{ kg}} = 0.248 \, m$$

(b) The mass of 1.00 L of solution will be 1043 g, which will contain

1043 g × 0.0500 = 52.2 g K_3PO_4.

$$\frac{\left(\dfrac{52.2 \text{ g } K_3PO_4}{212.27 \text{ g} \cdot \text{mol}^{-1} \text{ } K_3PO_4}\right)}{1.00 \text{ L}} = 0.246 \text{ M}$$

5E.7 In 1.000 L of 14.8 M NH_3 (aq) there will be 1000 mL × 0.901 g · mL^{-1} or

901 g of solution. The mass of NH_3 present in this solution will be

14.8 mol NH_3 × 17.03 g NH_3 · mol^{-1} or 252 g of NH_3, meaning there are

649 g of water in this 1 L solution of 14.8 M NH_3. Thus the molality of

this solution is $\dfrac{14.8 \text{ mol } NH_3}{0.649 \text{ kg } H_2O} = 22.8 \, m$.

5E.9 (a) If x_{MgCl_2} is 0.0120, then there are 0.0120 mol $MgCl_2$ for every 0.9880

mol H_2O. The mass of water will be

18.02 g · mol^{-1} × 0.9880 mol = 17.80 g or 0.01780 kg.

$$m_{Cl^-} = \frac{(0.0120 \text{ mol } MgCl_2)(2 \text{ mol } Cl^- / 1 \text{ mol } MgCl_2)}{0.01780 \text{ kg solvent}} = 1.35 \, m.$$

(b) $m_{NaOH} = \dfrac{\left(\dfrac{6.75 \text{ g NaOH}}{40.00 \text{ g} \cdot \text{mol}^{-1}}\right)}{0.325 \text{ kg solvent}} = 0.519 \, m$

(c) 1.000 L of 15.00 M HCl (aq) will contain 15.00 mol with a mass of

15.00 mol × 36.46 g · mol^{-1} = 546.9 g of HCl. The density of the 1.000 L of

solution is 1.0745 g · cm^{-3} so the total mass in the solution is 1074.5 g.

This leaves 1074.5 g − 546.9 g = 527.6 g as water. So,

$$\frac{15.00 \text{ mol HCl}}{0.5276 \text{ kg solvent}} = 28.43 \, m$$

5E.11 (a) Molar mass of $CaCl_2 \cdot 6\,H_2O = 219.08\ g \cdot mol^{-1}$, which consists of 110.98 g $CaCl_2$ and 108.10 g of water. Assuming the density of water to be $1.00\ g \cdot mL^{-1}$,

$$m_{CaCl_2} = \frac{x\ mol\ CaCl_2 \cdot 6\,H_2O}{0.500\ kg + x(6 \times 0.018\,02\ kg\ H_2O)}$$

Note: $18.02\ g\ H_2O = 0.018\,02\ kg\ H_2O = 1.000\ mol\ H_2O$

x = number of moles of $CaCl_2 \cdot 6\,H_2O$ needed to prepare a solution of molality m_{CaCl_2}, in which each mole of $CaCl_2 \cdot 6\,H_2O$ produces 6 x (0.018 02 kg) of water as solvent (assuming we begin with 0.500 kg of water). For a 0.125 m solution of $CaCl_2 \cdot 6\,H_2O$,

$$0.125\ m = \frac{x}{0.500\ kg + x(6)(0.01802\ kg\ H_2O)}$$

$x = 0.0625\ mol + x(0.0135\ mol)$

$x - 0.0135x = 0.0625\ mol$

$0.986\ x = 0.0625\ mol$

$x = 0.0634\ mol\ CaCl_2 \cdot 6\,H_2O$

$$\therefore (0.0634\ mol\ CaCl_2 \cdot 6\,H_2O)\left(\frac{219.08\ g\ CaCl_2 \cdot 6\,H_2O}{1\ mol\ CaCl_2 \cdot 6\,H_2O}\right) = 13.9\ g\ CaCl_2 \cdot 6\,H_2O$$

(b) Molar mass of $NiSO_4 \cdot 6\,H_2O = 262.86\ g \cdot mol^{-1}$, which consists of 154.77 g $NiSO_4$ and 108.09 g H_2O.

$$m_{NiSO_4} = \frac{x\ mol\ NiSO_4 \cdot 6\,H_2O}{0.500\ kg + x(6 \times 0.018\,02\ kg\ H_2O)}$$

where x = number of moles of $NiSO_4 \cdot 6H_2O$ needed to prepare a solution of molality m_{NiSO_4}, in which each mole of $NiSO_4 \cdot 6H_2O$ produces $6(0.018\ 02\ kg)$ of water as solvent. Assuming we begin with 0.500 kg H_2O, for a 0.22 m solution of $NiSO_4 \cdot 6H_2O$,

$$0.22 \, m = \frac{x}{0.500 \, \text{kg} + x(6)(0.01802 \, \text{kg H}_2\text{O})}$$

$$x = 0.11 \, \text{mol} + x(0.0238 \, \text{mol})$$

$$x - 0.0238 \, x = 0.11 \, \text{mol}$$

$$0.976 \, x = 0.11 \, \text{mol}$$

$$x = 0.11 \, \text{mol NiSO}_4 \cdot 6 \, \text{H}_2\text{O}$$

$$\therefore (0.11 \, \text{mol NiSO}_4 \cdot 6 \, \text{H}_2\text{O}) \left(\frac{262.86 \, \text{g NiSO}_4 \cdot 6 \, \text{H}_2\text{O}}{1 \, \text{mol NiSO}_4 \cdot 6 \, \text{H}_2\text{O}} \right) = 29 \, \text{g NiSO}_4 \cdot 6 \, \text{H}_2\text{O}$$

5F.1 $\Delta T_f = 179.8°\text{C} - 176.9°\text{C} = 2.9°\text{C} \text{ or } 2.9 \, \text{K}$

$$\Delta T_f = k_f m$$

$$2.9 \, \text{K} = (39.7 \, \text{K} \cdot \text{kg} \cdot \text{mol}^{-1}) m$$

$$2.9 \, \text{K} = (39.7 \, \text{K} \cdot \text{kg} \cdot \text{mol}^{-1}) \left(\frac{\left(\dfrac{1.14 \, \text{g}}{M_{\text{unknown}}} \right)}{0.100 \, \text{kg}} \right)$$

$$\frac{(0.100 \, \text{kg})(2.9 \, \text{K})}{39.7 \, \text{K} \cdot \text{kg} \cdot \text{mol}^{-1}} = \frac{1.14 \, \text{g}}{M_{\text{unknown}}}$$

$$M_{\text{unknown}} = \frac{(1.14 \, \text{g})(39.7 \, \text{K} \cdot \text{kg} \cdot \text{mol}^{-1})}{(0.100 \, \text{kg})(2.9 \, \text{K})} = 1.6 \times 10^2 \, \text{g} \cdot \text{mol}^{-1}$$

5F.3 (a) A 1.00% aqueous solution of NaCl will contain 1.00 g of NaCl for 99.0 g of water. To use the freezing point depression equation, we need the molality of the solution:

$$\text{molality} = \frac{\left(\dfrac{1.00 \, \text{g}}{58.44 \, \text{g} \cdot \text{mol}^{-1}} \right)}{0.0990 \, \text{kg}} = 0.173 \, \text{mol} \cdot \text{kg}^{-1}$$

$$\Delta T_f = i k_f m$$

$$\Delta T_f = i \, (1.86 \, \text{K} \cdot \text{kg} \cdot \text{mol}^{-1})(0.173 \, \text{mol} \cdot \text{kg}^{-1}) = 0.593 \, \text{K}$$

$$i = 1.84$$

(b) molality of all solute species (undissociated NaCl(aq) plus Na^+

(aq) + Cl^- (aq)) $= 1.84 \times 0.173 \, \text{mol} \cdot \text{kg}^{-1} = 0.318 \, \text{mol} \cdot \text{kg}^{-1}$.

(c) If all the NaCl had dissociated, the total molality in solution would have been 0.346 mol·kg^{-1}, giving an i value equal to 2. If no dissociation had taken place, the molality in solution would have equaled 0.173 mol·kg^{-1}.

$$\text{NaCl (aq)} \rightleftharpoons \text{Na}^+ \text{(aq)} + \text{Cl}^- \text{(aq)}$$
$$(0.173 \text{ mol·kg}^{-1} - x) \qquad x \qquad\quad x$$

0.173 mol·kg^{-1} $- x + x + x = 0.318$ mol·kg^{-1}

0.173 mol·kg^{-1} $+ x = 0.318$ mol·kg^{-1}

$x = 0.145$ mol·kg^{-1}

$$\% \text{ dissociation} = \left(\frac{0.145 \text{ mol·kg}^{-1}}{0.173 \text{ mol·kg}^{-1}} \right)(100) = 83.8\%$$

5F.5 The compound that freezes at the lower temperature will have the largest $\left| \Delta T_f \right|$ (assuming that $i = 1$, since we are told they are both molecular compounds). Since the same mass of each compound is present in the same amount of solvent for each solution, the compound with the lower molar mass will have the larger number of moles and therefore the larger $\left| \Delta T_f \right|$ (since the compound with the largest molar amount present will have the largest amount of solute particles present). Since compound A caused the water to freeze at a lower T than compound B, compound A must have the lower molar mass. Compound B must therefore have the greater molar mass.

5F.7 For an electrolyte that dissociates into two ions, the van 't Hoff i factor will be 1 plus the degree of dissociation, in this case 0.075. This can be readily seen for the general case MX. Let A = initial concentration of MX (if none is dissociated) and let Y = the concentration of MX that subsequently dissociates:

$$\text{MX (aq)} \rightleftharpoons \text{M}^{n+} \text{(aq)} + \text{X}^{n-} \text{(aq)}$$
$$A - Y \qquad\qquad Y \qquad\quad Y$$

The total concentration of solute species is $(A - Y) + Y + Y = A + Y$

The value of i will then be equal to A + Y or 1.075.

The freezing point change is then simple to calculate:

$$\Delta T_f = ik_f m$$
$$= 1.075 \ (1.86 \ \text{K} \cdot \text{kg} \cdot \text{mol}^{-1})(0.10 \ \text{mol} \cdot \text{kg}^{-1})$$
$$= 0.20$$

Freezing point of the solution will be $0.00°C - 0.20°C = -0.20°C$.

5F.9 (a) $\Pi = iRT \, (\text{molarity})$

$$= 1 \left(0.082 \ 06 \ \text{L} \cdot \text{atm} \cdot \text{K}^{-1} \cdot \text{mol}^{-1}\right) \left(293 \ \text{K}\right) \left(0.010 \ \text{mol} \cdot \text{L}^{-1}\right)$$

$$= 0.24 \ \text{atm or} \ 1.8 \times 10^2 \ \text{Torr}$$

(b) Because HCl is a strong acid, it should dissociate into two ions, H^+ and Cl^-, so $i = 2$.

$$\Pi = 2(0.082 \ 06 \ \text{L} \cdot \text{atm} \cdot \text{K}^{-1} \cdot \text{mol}^{-1})(293 \ \text{K})(1.0 \ \text{mol} \cdot \text{L}^{-1})$$
$$= 48 \ \text{atm}$$

(c) $CaCl_2$ should dissociate into 3 ions in solution, therefore $i = 3$.

$$\Pi = 3(0.082 \ 06 \ \text{L} \cdot \text{atm} \cdot \text{K}^{-1} \cdot \text{mol}^{-1})(293 \ \text{K})(0.010 \ \text{mol} \cdot \text{L}^{-1})$$
$$= 0.72 \ \text{atm or} \ 5.5 \times 10^2 \ \text{Torr}$$

5F.11 The polypeptide is a nonelectrolye, so $i = 1$.

$$\Pi = iRT \, (\text{molarity})$$

$$\Pi = \frac{3.74 \ \text{Torr}}{760 \ \text{Torr} \cdot \text{atm}^{-1}} = 4.92 \times 10^{-3} \ \text{atm}$$

$$4.92 \times 10^{-3} \ \text{atm} = 1(0.082 \ 06 \ \text{L} \cdot \text{atm} \cdot \text{K}^{-1} \cdot \text{mol}^{-1})(300 \ \text{K}) \left(\frac{(0.40 \ \text{g}/M_{unknown})}{1.0 \ \text{L}} \right)$$

$$M_{unknown} = \frac{(0.082 \ 06 \ \text{L} \cdot \text{atm} \cdot \text{K}^{-1} \cdot \text{mol}^{-1})(300 \ \text{K})(0.40 \ \text{g})}{(4.92 \times 10^{-3} \ \text{atm})(1.0 \ \text{L})}$$

$$= 2.0 \times 10^3 \ \text{g} \cdot \text{mol}^{-1}$$

5F.13 We assume the polymer to be a nonelectrolyte, so $i = 1$.

$$\Pi = iRT \text{(molarity)}$$

$$\Pi = \frac{6.3 \text{ Torr}}{760 \text{ Torr} \cdot \text{atm}^{-1}} = 1(0.08206 \text{ L} \cdot \text{atm} \cdot \text{K}^{-1} \cdot \text{mol}^{-1})(293 \text{ K})\left(\frac{(0.20 \text{ g}/M_{unknown})}{0.100 \text{ L}}\right)$$

$$M_{unknown} = \frac{(0.08206 \text{ L} \cdot \text{atm} \cdot \text{K}^{-1} \cdot \text{mol}^{-1})(293 \text{ K})(0.20 \text{ g})(760 \text{ Torr} \cdot \text{atm}^{-1})}{(6.3 \text{ Torr})(0.100 \text{ L})}$$

$$= 5.8 \times 10^3 \text{ g} \cdot \text{mol}^{-1}$$

5F.15 (a) $C_{12}H_{22}O_{11}$ should be a nonelectrolyte, so $i = 1$.

$$\Pi = iRT \text{(molarity)}$$

$$= 1(0.082\ 06 \text{ L} \cdot \text{atm} \cdot \text{K}^{-1} \cdot \text{mol}^{-1})(293 \text{ K})(0.050 \text{ mol} \cdot \text{L}^{-1})$$

$$= 1.2 \text{ atm}$$

(b) NaCl dissociates to give 2 ions in solution, so $i = 2$.

$$\Pi = iRT \text{(molarity)}$$

$$= 2(0.082\ 06 \text{ L} \cdot \text{atm} \cdot \text{K}^{-1} \cdot \text{mol}^{-1})(293 \text{ K})(0.0010 \text{ mol} \cdot \text{L}^{-1})$$

$$= 0.048 \text{ atm or } 36 \text{ Torr}$$

(c) AgCN dissociates in solution to give two ions (Ag^+ and CN^-), so $i = 2$.

$$\Pi = iRT \text{(molarity)}$$

We must assume that the AgCN does not significantly affect either the volume or density of the solution, which is reasonable given the very small amount of it that dissolves.

$$\Pi = 2(0.082\ 06 \text{ L} \cdot \text{atm} \cdot \text{K}^{-1} \cdot \text{mol}^{-1})(293 \text{ K})\left(\frac{\left(\dfrac{2.3 \times 10^{-5} \text{ g}}{133.89 \text{ g} \cdot \text{mol}^{-1}}\right)}{\left(\dfrac{100 \text{ g H}_2\text{O}}{1 \text{ g} \cdot \text{cm}^{-3} \text{ H}_2\text{O} \times 1000 \text{ cm}^3 \cdot \text{L}^{-1}}\right)}\right)$$

$$= 8.3 \times 10^{-5} \text{ atm}$$

5G.1 (a) False. Equilibrium is dynamic. At equilibrium, the concentrations of reactants and products will not change, but the reaction will continue to proceed in both directions.

(b) False. Equilibrium reactions are affected by the presence of both products and reactants.

(c) False. The value of the equilibrium constant is not affected by the amounts of reactants or products added as long as the temperature is constant.

(d) True.

5G.3 (a) $K = \dfrac{(P_{C_2H_4Cl_2})^2 (P_{H_2O})^2}{(P_{C_2H_4})^2 (P_{O_2})(P_{HCl})^4}$

(b) $K = \dfrac{(P_{N_2})^7 (P_{H_2O})^6}{(P_{NH_3})^4 (P_{NO})^6}$

5G.5 (a) Flask 3 represents the point of reaction equilibrium.

(b) percent decomposition $= (6/11) \times 100 = 54.5\%$

(c) Decomposition: $X_2 \rightleftharpoons 2X$

If 54.5% of X_2 has decomposed by equilibrium then the amount of X_2 at equilibrium will be $(0.10 \text{ bar})(1 - 0.545) = 0.0455$ bar. Since the decomposition is a 1:2 ratio, for every 1 mole of X_2 that decomposes there will be 2 moles of X produced. The amount of X at equilibrium will be $2[(0.10)(0.545)] = 0.109$ bar.

$$K = \frac{(P_X)^2}{P_{X_2}} = \frac{(0.109)^2}{(0.0455)} = 0.26$$

5G.7 (a) $CH_4(g) + 2\,O_2(g) \leftrightarrow CO_2(g) + 2\,H_2O(g)$

$$K = \frac{(P_{CO_2})(P_{H_2O})^2}{(P_{CH_4})(P_{O_2})^2}$$

(b) $I_2(g) + 5\,F_2(g) \leftrightarrow 2\,IF_5(g)$

$$K = \frac{(P_{IF_5})^2}{(P_{I_2})(P_{F_2})^5}$$

(c) $2\,NO_2(g) + F_2(g) \leftrightarrow 2\,FNO_2(g)$

$$K = \frac{(P_{FNO_2})^2}{(P_{F_2})(P_{NO_2})^2}$$

5G.9 (a) Because the volume is the same, the number of moles of O_2 is larger in the second experiment.   (b) Because K is a constant and the denominator is larger in the second case, the numerator must also be larger; so the partial pressure of O_2 is larger in the second case.

(c) Although $(P_{O_2})^3/(P_{O_3})^2$ is the same, $(P_{O_2})/(P_{O_2})$ will be different, a result seen by solving for K in each case.   (d) Because K is a constant, $(P_{O_2})^3/(P_{O_3})^2$ is the same.   (e) Because $(P_{O_2})^3/(P_{O_3})^2$ is the same in (d), its reciprocal, $(P_{O_3})^2/(P_{O_2})^3$, must be the same.

5G.11 (a) $\dfrac{1}{P_{BCl_3}^{\ 2}}$

(b) $[H_3PO_4]^4 [H_2S]^{10}$

(c) $\dfrac{P_{BrF_3}^{\ 2}}{P_{Br_2} \, P_{F_2}^{\ 3}}$

5G.13 The free energy at a specific set of conditions is given by

$$\Delta G_r = \Delta G°_r + RT \ln Q$$
$$\Delta G_r = -RT \ln K + RT \ln Q$$

$$\Delta G_r = -RT \ln K + RT \ln \frac{[I]^2}{[I_2]}$$

$$= -(8.314 \text{ J} \cdot \text{K}^{-1} \cdot \text{mol}^{-1})(1200 \text{ K}) \ln (6.8)$$

$$+ (8.314 \text{ J} \cdot \text{K}^{-1} \cdot \text{mol}^{-1})(1200 \text{ K}) \ln \frac{(0.98)^2}{(0.13)}$$

$$= 8.3 \times 10^{1} \text{ kJ} \cdot \text{mol}^{-1}$$

Because ΔG_r is positive, the reaction will be spontaneous to produce I_2.

5G.15   The free energy at a specific set of conditions is given by

$$\Delta G_r = \Delta G°_r + RT \ln Q$$

$$\Delta G_r = -RT \ln K + RT \ln Q$$

$$\Delta G_r = -RT \ln K + RT \ln \frac{[NH_3]^2}{[N_2][H_2]^3}$$

$$= -(8.314 \text{ J} \cdot \text{K}^{-1} \cdot \text{mol}^{-1})(400 \text{ K}) \ln (41)$$

$$+ (8.314 \text{ J} \cdot \text{K}^{-1} \cdot \text{mol}^{-1})(400 \text{ K}) \ln \frac{(21)^2}{(4.2)(1.8)^3}$$

$$= -27 \text{ kJ} \cdot \text{mol}^{-1}.$$

Because ΔG_r is negative, the reaction will proceed to form products.

5G.17

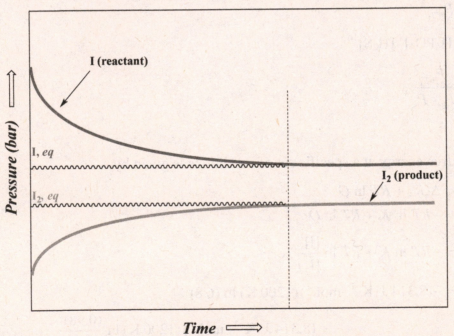

5G.19   (a) $\Delta G°_r = -RT \ln K$

$$= -(8.314 \text{ J} \cdot \text{K}^{-1} \cdot \text{mol}^{-1})(1200 \text{ K}) \ln 6.8 = -19 \text{ kJ} \cdot \text{mol}^{-1}$$

(b) $\Delta G°_r = -RT \ln K = -(8.314 \text{ J} \cdot \text{K}^{-1} \cdot \text{mol}^{-1})(298 \text{ K}) \ln 1.1 \times 10^{-12}$

$$= +68 \text{ kJ} \cdot \text{mol}^{-1}$$

5G.21 To answer these questions, we will first calculate $\Delta G°$ for each reaction and then use that value in the expression $\Delta G° = -RT \ln K$.

(a) $2 H_2(g) + O_2(g) \longrightarrow 2 H_2O(g)$

$\Delta G°_r = 2 \times \Delta G°_f(H_2O, g) = 2(-228.57 \text{ kJ} \cdot \text{mol}^{-1}) = -457.14 \text{ kJ}$

$\Delta G°_r = -RT \ln K$ or

$\ln K = -\dfrac{\Delta G°_r}{RT}$

$\ln K = -\dfrac{-457\,140 \text{ J}}{(8.314 \text{ J} \cdot \text{K}^{-1} \cdot \text{mol}^{-1})(298.15 \text{ K})} = +184.42$

$K - 1 \times 10^{80}$

(b) $2 CO(g) + O_2(g) \longrightarrow 2 CO_2(g)$

$\Delta G°_r = 2 \times \Delta G°_f(CO_2, g) - [2 \times \Delta G°_f(CO, g)]$

$\qquad = 2(-394.36 \text{ kJ} \cdot \text{mol}^{-1}) - [2(-137.17 \text{ kJ} \cdot \text{mol}^{-1})]$

$\qquad = -514.38 \text{ kJ}$

$\ln K = -\dfrac{-514\,380 \text{ J}}{(8.314 \text{ J} \cdot \text{K}^{-1} \cdot \text{mol}^{-1})(298.15 \text{ K})} = +207.5$

$K = 1 \times 10^{90}$

(c) $CaCO_3(s) \longrightarrow CaO(s) + CO_2(g)$

$\Delta G°_r = \Delta G°_f(CaO, s) + \Delta G°_f(CO_2, g) - [\Delta G°_f(CaCO_3(s)]$

$\qquad = (-604.03 \text{ kJ} \cdot \text{mol}^{-1}) + (-394.36 \text{ kJ} \cdot \text{mol}^{-1}) - [-1128.8 \text{ kJ} \cdot \text{mol}^{-1}]$

$\qquad = +130.4 \text{ kJ}$

$\ln K = -\dfrac{+130\,400 \text{ J}}{(8.314 \text{ J} \cdot \text{K}^{-1} \cdot \text{mol}^{-1})(298.15 \text{ K})} = -52.6$

$K = 1 \times 10^{-23}$

5H.1 For the reaction written as $N_2(g) + 3 H_2(g) \rightarrow 2 NH_3(g)$ Eq. 1,

$K = \dfrac{P_{NH_3}{}^2}{P_{N_2} P_{H_2}{}^3} = 41.$

(a) For the reaction written as $2 NH_3(g) \rightarrow N_2(g) + 3 H_2(g)$ Eq. 2,

$K = \dfrac{P_{N_2} P_{H_2}{}^3}{P_{NH_3}{}^2}.$

This is $\dfrac{1}{K_{Eq.\,1}} = \dfrac{1}{41} = 0.024$.

(b) For the reaction written as $\frac{1}{2}\,N_2(g) + \frac{3}{2}\,H_2(g) \rightarrow NH_3(g)$ Eq. 3,

$$K_{Eq.\,3} = \dfrac{P_{NH_3}}{P_{N_2}^{1/2}\,P_{H_2}^{3/2}} = \sqrt{K_{Eq.\,1}} = \sqrt{41} = 6.4.$$

Note that Eq. 3 = $\frac{1}{2}$ Eq. 1 and thus $K_{Eq.\,3} = K_{Eq.\,1}^{1/2}$.

(c) For the reaction written as $2\,N_2(g) + 6\,H_2(g) \rightarrow 4\,NH_3(g)$ Eq. 4,

$$K_{Eq.\,4} = \dfrac{P_{NH_3}^{4}}{P_{N_2}^{2}\,P_{H_2}^{6}} = K_{Eq.\,1}^{2} = 41^2 = 1.7 \times 10^3.$$

Note that Eq. 4 = 2 Eq. 1 and thus $K_{Eq.\,3} = K_{Eq.\,1}^{2}$.

5H.3 $\quad 2\,BrCl(g) \rightleftharpoons Br_2(g) + Cl_2(g) \qquad\qquad K = 377 \qquad\qquad (1)$

$\qquad\quad H_2(g) \;+\; Cl_2(g) \rightleftharpoons 2\,HCl(g) \qquad\quad K = 4.0 \times 10^{31} \qquad (2)$

$\qquad(1) + (2): \quad 2\,BrCl(g) + H_2(g) \rightleftharpoons Br_2(g) + 2\,HCl(g)$

K for this reaction $= K_1 \times K_2 \;=\; 377 \times 4.0 \times 10^{31} \;=\; 1.5 \times 10^{34}$.

5H.5 (a) $K = \dfrac{P_{NO}^{2}\,P_{Cl_2}}{P_{NOCl}^{2}} = 1.8 \times 10^{-2}$

$$K = \left(\dfrac{T}{12.027\ K}\right)^{\Delta n} K_C$$

$$K_C = \left(\dfrac{12.027\ K}{T}\right)^{\Delta n} K = \left(\dfrac{12.027\ K}{500\ K}\right)^{(3-2)} 1.8 \times 10^{-2} = 4.3 \times 10^{-4}$$

(b) $K = P_{CO_2} = 167$

$$K_C = \left(\dfrac{12.027\ K}{T}\right)^{\Delta n} K = \left(\dfrac{12.027\ K}{1073\ K}\right)^{(1)} 167 = 1.87$$

5I.1 $K_C = \dfrac{[BrCl]^2}{[Cl_2][Br_2]}$

$0.031 = \dfrac{(0.145)^2}{(0.495)[Br_2]}$

$[Br_2] = \dfrac{(0.145)^2}{(0.495)(0.031)} = 1.4 \text{ mol} \cdot L^{-1}$

5I.3 $H_2(g) + I_2(g) \rightarrow 2\,HI(g) \qquad K_C = 160$

$K_C = \dfrac{[HI]^2}{[H_2][I_2]}$

$160 = \dfrac{(2.21 \times 10^{-3})^2}{[H_2](1.46 \times 10^{-3})}$

$[H_2] = \dfrac{(2.21 \times 10^{-3})^2}{(160)(1.46 \times 10^{-3})}$

$[H_2] = 2.1 \times 10^{-5} \text{ mol} \cdot L^{-1}$

5I.5 $K = \dfrac{P_{PCl_3} P_{Cl_2}}{P_{PCl_5}}$

$25 = \dfrac{P_{PCl_3}(5.43)}{1.18}$

$P_{PCl_3} = \dfrac{(25)(1.18)}{5.43} = 5.4$

$P_{PCl_3} = 5.4 \text{ bar}$

5I.7   First we must calculate K for the reaction, which can be done using data from Appendix 2A:

$\Delta G^\circ_r = 2 \times \Delta G^\circ_f(NO, g) = 2 \times 86.55 \text{ kJ} \cdot \text{mol}^{-1} = 173.1 \text{ kJ} \cdot \text{mol}^{-1}$

$\Delta G^\circ = -RT \ln K$

$\ln K = -\dfrac{\Delta G^\circ}{RT}$

$\ln K = -\dfrac{+173\,100 \text{ J} \cdot \text{mol}^{-1}}{(8.314 \text{ J} \cdot \text{K}^{-1} \cdot \text{mol}^{-1})(298 \text{ K})} = -69.9$

$K = 4 \times 10^{-31}$

Because $Q > K$, the reaction will tend to proceed to produce reactants.

5I.9 (a) $K = \dfrac{p_{HI}^2}{p_{H_2} \cdot p_{I_2}} = 160$

$Q = \dfrac{(0.10)^2}{(0.20)(0.10)} = 0.50$

(b) $Q \neq K, \therefore$ the system is not at equilibrium.

(c) Because $Q < K$, more products will be formed.

5I.11 (a) $K_C = \dfrac{[SO_3]^2}{[SO_2]^2[O_2]} = 1.7 \times 10^6$

$Q_C = \dfrac{\left[\dfrac{1.0 \times 10^{-4}}{0.500}\right]^2}{\left[\dfrac{1.20 \times 10^{-3}}{0.500}\right]^2 \left[\dfrac{5.0 \times 10^{-4}}{0.500}\right]} = 6.9$

(b) Because $Q_C < K_C$, more products will tend to form, which will result

in the formation of more SO_3.

5I.13 (a) The balanced equation is $Cl_2(g) \rightarrow 2\,Cl(g)$.

The initial concentration of $Cl_2(g)$ is $\dfrac{0.0020 \text{ mol } Cl_2}{2.0 \text{ L}} = 0.0010 \text{ mol} \cdot L^{-1}$.

Concentration $(mol \cdot L^{-1})$	$Cl_2(g) \rightarrow$	$2\,Cl(g)$
initial	0.0010	0
change	$-x$	$+2x$
equilibrium	$0.0010 - x$	$+2x$

$$K_C = \frac{[Cl]^2}{[Cl_2]} = \frac{(2x)^2}{(0.0010 - x)} = 1.2 \times 10^{-7}$$

$$4x^2 = (1.2 \times 10^{-7})(0.0010 - x)$$

$$4x^2 + (1.2 \times 10^{-7})x - (1.2 \times 10^{-10}) = 0$$

$$x = \frac{-(1.2 \times 10^{-7}) \pm \sqrt{(1.2 \times 10^{-7})^2 - 4(4)(-1.2 \times 10^{-10})}}{2 \times 4}$$

$$x = \frac{-(1.2 \times 10^{-7}) \pm 4.4 \times 10^{-5}}{8}$$

$$x = -5.5 \times 10^{-6} \text{ or } +5.5 \times 10^{-6}$$

The negative answer is not meaningful, so we choose

$x = +5.5 \times 10^{-6}$ mol·L^{-1}. The concentration of Cl_2 is essentially

unchanged because $0.0010 - 5.5 \times 10^{-6} \cong 0.0010$. The concentration of Cl

atoms is $2 \times (5.5 \times 10^{-6}) = 1.1 \times 10^{-5}$ mol·L^{-1}.

(b) The balanced equation is: $F_2(g) \rightarrow 2\,F(g)$.

The problem is worked in an identical fashion to (a) but the equilibrium

constant is now 1.2×10^{-4}.

The initial concentration of $F_2(g)$ is $\dfrac{0.0020 \text{ mol } F_2}{2.0 \text{ L}} = 0.0010$ mol·L^{-1}.

Concentration (mol·L^{-1})	$F_2(g)$	$\rightarrow$	$2\,F(g)$
initial	0.0010		0
change	$-x$		$+2x$
equilibrium	$0.0010 - x$		$+2x$

$$K_C = \frac{[F]^2}{[F_2]} = \frac{(2x)^2}{(0.0010 - x)} = 1.2 \times 10^{-4}$$

$$4x^2 = (1.2 \times 10^{-4})(0.0010 - x)$$

$$4x^2 + (1.2 \times 10^{-4})x - (1.2 \times 10^{-7}) = 0$$

$$x = \frac{-(1.2 \times 10^{-4}) \pm \sqrt{(1.2 \times 10^{-4})^2 - 4(4)(-1.2 \times 10^{-7})}}{2 \times 4}$$

$$x = \frac{-(1.2 \times 10^{-4}) \pm 1.4 \times 10^{-3}}{8}$$

$$x = -1.9 \times 10^{-4} \text{ or } +1.6 \times 10^{-4}$$

The negative answer is not meaningful, so we choose

$x = +1.6 \times 10^{-4}$ mol $\cdot$ L^{-1}. The concentration of F_2 is

$0.0010 - 1.6 \times 10^{-4} = 8 \times 10^{-4}$ mol $\cdot$ L^{-1}. The concentration of F atoms is

$2 \times (1.6 \times 10^{-4}) = 3.2 \times 10^{-4}$ mol $\cdot$ L^{-1}.

(c) Cl_2 is more stable. This can be seen even without the aid of the

calculation from the larger equilibrium constant for the dissociation for F_2

compared to Cl_2.

5I.15 Starting concentration of $NH_3 = \dfrac{0.400 \text{ mol}}{2.00 \text{ L}} = 0.200$ mol $\cdot$ L^{-1}.

Concentration (mol $\cdot$ L^{-1})	$NH_4HS(s) \rightarrow$	$NH_3(g)$	$+$	$H_2S(g)$
initial	—	0.200		0
change	—	$+x$		$+x$
final	—	$0.200 + x$		$+x$

$K_C = [NH_3][H_2S] = (0.200 + x)(x)$

$1.6 \times 10^{-4} = (0.200 + x)(x)$

$x^2 + 0.200x - 1.6 \times 10^{-4} = 0$

$x = \dfrac{-(+0.200) \pm \sqrt{(+0.200)^2 - (4)(1)(-1.6 \times 10^{-4})}}{2}$

$x = \dfrac{-0.200 \pm 0.2016}{2} = +0.0008 \text{ or } -0.2008$

The negative root is not meaningful, so we choose $x = 8 \times 10^{-4}$ mol $\cdot$ L^{-1}

(note that in order to get this number we have had to ignore our normal

significant figure conventions).

$[NH_3] = +0.200$ mol $\cdot$ L^{-1} $+ 8 \times 10^{-4}$ mol $\cdot$ L^{-1} $= 0.200$ mol $\cdot$ L^{-1}

$[H_2S] = 8 \times 10^{-4}$ mol $\cdot$ L^{-1}

Alternatively, we could have assumed that

$x \ll 0.2$, the $0.200x = 1.6 \times 10^{-4}$, $x = 8.0 \times 10^{-4}$.

5I.17 The initial concentrations of N_2 and O_2 are equal at 0.114 mol $\cdot$ L^{-1}

because the vessel has a volume of 1.00 L.

Concentrations (mol·L⁻¹) $N_2(g)$ + $O_2(g)$ → $2NO(g)$

initial	0.114	0.114	0
change	$-x$	$-x$	$+2x$
final	$0.114 - x$	$0.114 - x$	$+2x$

$$K_C = \frac{[NO]^2}{[N_2][O_2]} = \frac{(2x)^2}{(0.114-x)(0.114-x)} = \frac{(2x)^2}{(0.114-x)^2}$$

$$1.00 \times 10^{-5} = \frac{(2x)^2}{(0.114-x)^2}$$

$$\sqrt{1.00 \times 10^{-5}} = \sqrt{\frac{(2x)^2}{(0.114-x)^2}}$$

$$3.16 \times 10^{-3} = \frac{(2x)}{(0.114-x)}$$

$$2x = (3.16 \times 10^{-3})(0.114-x)$$

$$2.00316\,x = 3.60 \times 10^{-4}$$

$$x = 1.8 \times 10^{-4}$$

$[NO] = 2x = 2 \times 1.8 \times 10^{-4} = 3.6 \times 10^{-4}$ mol·L⁻¹; the concentrations of N_2

and O_2 remain essentially unchanged at 0.114 mol·L⁻¹.

5I.19 The initial concentrations of H_2 and I_2 are

$$[H_2] = \frac{0.400 \text{ mol}}{3.00 \text{ L}} = 0.133 \text{ mol·L}^{-1}; [I_2] = \frac{1.60 \text{ mol}}{3.00 \text{ L}} = 0.533 \text{ mol·L}^{-1}$$

Concentrations (mol·L⁻¹) $H_2(g)$ + $I_2(g)$ → $2HI(g)$

initial	0.133	0.533	0
change	$-x$	$-x$	$+2x$
final	$0.133 - x$	$0.533 - x$	$+2x$

At equilibrium, 60.0% of the H_2 had reacted, so 40.0% of the H_2

remains:

$$(0.400)(0.133 \text{ mol·L}^{-1}) = 0.133 \text{ mol·L}^{-1} - x$$

$$x = 0.133 \text{ mol·L}^{-1} - (0.400)(0.133 \text{ mol·L}^{-1})$$

$$x = 0.080 \text{ mol·L}^{-1}$$

At equilibrium: $[H_2] = 0.133 \text{ mol·L}^{-1} - 0.080 \text{ mol·L}^{-1} = 0.053 \text{ mol·L}^{-1}$

$[I_2] = 0.533 \text{ mol} \cdot L^{-1} - 0.080 \text{ mol} \cdot L^{-1} = 0.453 \text{ mol} \cdot L^{-1}$

$[HI] = 2 \times 0.080 \text{ mol} \cdot L^{-1} = 0.16 \text{ mol} \cdot L^{-1}$

$$K_C = \frac{[HI]^2}{[H_2][I_2]} = \frac{0.16^2}{(0.053)(0.453)} = 1.1$$

5I.21 Initial concentrations of CO and O_2 are given by

$$[CO] = \frac{\left(\dfrac{0.28 \text{ g CO}}{28.01 \text{ g} \cdot \text{mol}^{-1} \text{CO}}\right)}{2.0 \text{ L}} = 5.0 \times 10^{-3} \text{ mol} \cdot L^{-1}$$

$$[O_2] = \frac{\left(\dfrac{0.032 \text{ g O}_2}{32.00 \text{ g} \cdot \text{mol}^{-1} \text{ O}_2}\right)}{2.0 \text{ L}} = 5.0 \times 10^{-4} \text{ mol} \cdot L^{-1}.$$

Concentration (mol·L⁻¹)	2 CO(g)	+	O_2(g)	→	2 CO_2(g)
initial	5.0×10^{-3}		5.0×10^{-4}		0
change	$-2x$		$-x$		$+2x$
final	$5.0 \times 10^{-3} - 2x$		$5.0 \times 10^{-4} - x$		$+2x$

$$K_C = \frac{[CO_2]^2}{[CO]^2[O_2]}$$

$$= \frac{(2x)^2}{(5.0 \times 10^{-3} - 2x)^2 (5.0 \times 10^{-4} - x)}$$

$$= \frac{4x^2}{(4x^2 - 0.020x + 2.5 \times 10^{-5})(5.0 \times 10^{-4} - x)}$$

$$0.66 = \frac{4x^2}{-4x^3 + 0.022x^2 - 3.5 \times 10^{-5} x + 1.25 \times 10^{-8}}$$

$$4x^2 = (0.66)(-4x^3 + 0.022x^2 - 3.5 \times 10^{-5} x + 1.25 \times 10^{-8})$$

$$6.06x^2 = -4x^3 + 0.022x^2 - 3.5 \times 10^{-5} x + 1.25 \times 10^{-8}$$

$$0 = -4x^3 - 6.04x^2 - 3.5 \times 10^{-5} x + 1.25 \times 10^{-8}$$

$$x = 4.3 \times 10^{-5}$$

$[CO_2] = 8.6 \times 10^{-5} \text{ mol} \cdot L^{-1}; [CO] = 4.9 \times 10^{-3} \text{ mol} \cdot L^{-1}$

$[O_2] = 4.6 \times 10^{-4} \text{ mol} \cdot L^{-1}$

5I.23 We can calculate changes according to the reaction stoichiometry:

Amount (mol)	$CO(g)$ +	$3 H_2(g)$ $\rightarrow$	$CH_4(g)$ +	$H_2O(g)$
initial	2.00	3.00	0	0
change	$-x$	$-3x$	$+x$	$+x$
final	$2.00 - x$	$3.00 - 3x$	0.478	$+x$

According to the stoichiometry, 0.478 mol = x; therefore, at equilibrium, there are 2.00 mol − 0.478 mol = 1.52 mol CO, 3.00 − 3(0.478 mol) = 1.57 mol H_2, and 0.478 mol H_2O. To employ the equilibrium expression, we need either concentrations or pressures; because K_C is given, we will choose to express these as concentrations. This calculation is easy because $V = 10.0$ L:

$[CO] = 0.152$ mol·L^{-1}; $[H_2] = 0.157$ mol·L^{-1}; $[CH_4] = 0.0478$ mol·L^{-1};

$[H_2O] = 0.0478$ mol·L^{-1}

$$K_C = \frac{[CH_4][H_2O]}{[CO][H_2]^3} = \frac{(0.0478)(0.0478)}{(0.152)(0.157)^3} = 3.88$$

5I.25 First, we calculate the initial concentrations of each species:

$$[SO_2] = [NO] = \frac{0.100 \text{ mol}}{5.00 \text{ L}} = 0.0200 \text{ mol·L}^{-1};$$

$$[NO_2] = \frac{0.200 \text{ mol}}{5.00 \text{ L}} = 0.0400 \text{ mol·L}^{-1}; [SO_3] = \frac{0.150 \text{ mol}}{5.00 \text{ L}} = 0.0300 \text{ mol·L}^{-1}$$

We can use these values to calculate Q in order to see which direction the reactions will go:

$$Q = \frac{(0.0200)(0.0300)}{(0.0200)(0.0400)} = 0.75$$

Because $Q < K_C$, the reaction will proceed to produce more products.

Concentration (mol·L^{-1})

	$SO_2(g)$ +	$NO_2(g)$ $\rightarrow$	$NO(g)$ +	$SO_3(g)$
initial	0.0200	0.0400	0.0200	0.0300
change	$-x$	$-x$	$+x$	$+x$
final	$0.0200 - x$	$0.0400 - x$	$0.0200 + x$	$0.0300 + x$

$$K_C = \frac{[NO][SO_3]}{[SO_2][NO_2]}$$

$$85.0 = \frac{(0.0200 + x)(0.0300 + x)}{(0.0200 - x)(0.0400 - x)}$$

$$= \frac{x^2 + 0.0500\,x + 0.000\,600}{x^2 - 0.0600\,x + 0.000\,800}$$

$$85.0\,(x^2 - 0.0600\,x + 0.000\,800) = x^2 + 0.0500\,x + 0.000\,600$$

$$85.0\,x^2 - 5.10\,x + 0.0680 = x^2 + 0.0500\,x + 0.000\,600$$

$$84.0\,x^2 - 5.15\,x + 0.0674 = 0$$

$$x = \frac{-(-5.15) \pm \sqrt{(-5.15)^2 - (4)(84.0)(0.0674)}}{(2)(84.0)} = \frac{+5.15 \pm 1.97}{168}$$

$$x = +0.0424 \text{ or } +0.0189$$

The root 0.0424 is not meaningful because it is larger than the concentration of NO_2. The root of choice is therefore 0.0189.

At equilibrium:

$$[SO_2] = 0.0200 \text{ mol} \cdot L^{-1} - 0.0189 \text{ mol} \cdot L^{-1} = 0.0011 \text{ mol} \cdot L^{-1}$$

$$[NO_2] = 0.0400 \text{ mol} \cdot L^{-1} - 0.0189 \text{ mol} \cdot L^{-1} = 0.0211 \text{ mol} \cdot L^{-1}$$

$$[NO] = 0.0200 \text{ mol} \cdot L^{-1} + 0.0189 \text{ mol} \cdot L^{-1} = 0.0389 \text{ mol} \cdot L^{-1}$$

$$[SO_3] = 0.0300 \text{ mol} \cdot L^{-1} + 0.0189 \text{ mol} \cdot L^{-1} = 0.0489 \text{ mol} \cdot L^{-1}$$

To check, we can put these numbers back into the equilibrium constant expression:

$$K_C = \frac{[NO][SO_3]}{[SO_2][NO_2]}$$

$$\frac{(0.0389)(0.0489)}{(0.0011)(0.0211)} = 82.0$$

Compared with $K_C = 85.0$, this is reasonably good agreement given the nature of the calculation. We can check to see, by trial and error, if a better answer could be obtained. Because the K_C value is low for the concentrations we calculated, we can choose to alter x slightly so that this ratio becomes larger. If we let $x = 0.0190$, the concentrations of NO and SO_3 are increased to 0.0390 and 0.0490, and the concentrations of SO_2 and NO_2 are decreased to 0.0010 and 0.0200 (the stoichiometry of the reaction is maintained by calculating the concentrations in this fashion).

Then the quotient becomes 91.0, which is further from the value for K_C than the original answer. So, although the agreement is not the best with the numbers we obtained, it is the best possible, given the limitation on the number of significant figures we are allowed to use in the calculation.

5I.27   (a)   The initial concentrations are

$$[PCl_5] = \frac{1.50 \text{ mol}}{0.500 \text{ L}} = 3.00 \text{ mol} \cdot L^{-1}; [PCl_3] = \frac{3.00 \text{ mol}}{0.500 \text{ L}} = 6.00 \text{ mol} \cdot L^{-1};$$

$$[Cl_2] = \frac{0.500 \text{ mol}}{0.500 \text{ L}} = 1.00 \text{ mol} \cdot L^{-1}.$$

First calculate Q:

$$Q = \frac{[PCl_5]}{[PCl_3][Cl_2]} = \frac{3.00}{(6.00)(1.00)} = 0.500$$

Because $Q \neq K$, the reaction is not at equilibrium.

(b)   Because $Q < K_C$, the reaction will proceed to form products.

(c)

Concentrations ($mol \cdot L^{-1}$)	$PCl_3(g)$	+	$Cl_2(g)$	$\rightarrow$	$PCl_5(g)$
initial	6.00		1.00		3.00
change	$-x$		$-x$		$+x$
final	$6.00 - x$		$1.00 - x$		$3.00 + x$

$$K_C = \frac{[PCl_5]}{[PCl_3][Cl_2]}$$

$$0.56 = \frac{3.00 + x}{(6.00 - x)(1.00 - x)} = \frac{3.00 + x}{x^2 - 7x + 6.00}$$

$$(0.56)(x^2 - 7x + 6.00) = 3.00 + x$$

$$0.56x^2 - 3.92x + 3.36 = 3.00 + x$$

$$0.56x^2 - 4.92x + 0.36 = 0$$

$$x = \frac{-(-4.92) \pm \sqrt{(-4.92)^2 - (4)(0.56)(0.36)}}{(2)(0.56)} = \frac{+4.92 \pm 4.48}{1.12}$$

$$x = 9.2 \text{ or } 0.071$$

Because the root 9.2 is larger than the amount of PCl_3 or Cl_2 available, it is physically meaningless and can be discarded. Thus, $x = 0.071 \text{ mol} \cdot L^{-1}$, giving

$$[PCl_5] = 3.00 \text{ mol} \cdot L^{-1} + 0.07 \text{ mol} \cdot L^{-1} = 3.07 \text{ mol} \cdot L^{-1}$$

$$[PCl_3] = 6.00 \text{ mol} \cdot L^{-1} - 0.07 \text{ mol} \cdot L^{-1} = 5.93 \text{ mol} \cdot L^{-1}$$

$$[Cl_2] = 1.00 \text{ mol} \cdot L^{-1} - 0.07 \text{ mol} \cdot L^{-1} = 0.93 \text{ mol} \cdot L^{-1}$$

The number can be checked by substituting them back into the equilibrium constant expression:

$$K_C = \frac{[PCl_5]}{[PCl_3][Cl_2]}$$

$$\frac{(3.07)}{(5.93)(0.93)} \overset{?}{=} 0.56$$

$$0.56 \overset{?}{=} 0.56$$

5I.29 Pressures (bar) $2 \text{ HCl(g)} \quad \rightarrow \quad H_2(g) \quad + \quad Cl_2(g)$

initial	0.22	0	0
change	$-2x$	$+x$	$+x$
final	$0.22 - 2x$	$+x$	$+x$

$$K = \frac{P_{H_2} P_{Cl_2}}{P_{HCl}{}^2}$$

$$3.2 \times 10^{-34} = \frac{(x)(x)}{(0.22 - 2x)^2}$$

Because the equilibrium constant is small, assume that $x \ll 0.22$:

$$3.2 \times 10^{-34} = \frac{x^2}{(0.22)^2}$$

$$x^2 = (3.2 \times 10^{-34})(0.22)^2$$

$$x = \sqrt{(3.2 \times 10^{-34})(0.22)^2}$$

$$x = \pm 3.9 \times 10^{-18}$$

The negative root is not physically meaningful and can be discarded. x is small compared with 0.22, so the initial assumption was valid. The pressures at equilibrium are

$$P_{HCl} = 0.22 \text{ bar}; P_{H_2} = P_{Cl_2} = 3.9 \times 10^{-18} \text{ bar}$$

The values can be checked by substituting them into the equilibrium expression:

$$\frac{(3.9 \times 10^{-18})(3.9 \times 10^{-18})}{(0.22)^2} \stackrel{?}{=} 3.2 \times 10^{-34}$$

$$3.1 \times 10^{-34} \stackrel{\checkmark}{=} 3.2 \times 10^{-34}$$

$$P_{HCl} = 0.22 \text{ bar}; P_{H_2} = P_{Cl_2} = 3.9 \times 10^{-18} \text{ bar}$$

The numbers agree very well for a calculation of this type.

5I.31 (a) To determine on which side of the equilibrium position the conditions lie, we will calculate Q:

$$[CO] = \frac{0.342 \text{ mol}}{3.00 \text{ L}} = 0.114 \text{ mol} \cdot L^{-1}; [H_2] = \frac{0.215 \text{ mol}}{3.00 \text{ L}} = 0.0717 \text{ mol} \cdot L^{-1};$$

$$[CH_3OH] = \frac{0.125 \text{ mol}}{3.00 \text{ L}} = 0.0417 \text{ mol} \cdot L^{-1}$$

$$Q = \frac{[CH_3OH]}{[CO][H_2]^2} = \frac{0.0417}{(0.114)(0.0717)^2} = 71.1$$

Because $Q > K_C$, the reaction will proceed to produce more of the reactants, which means that the concentration of methanol will decrease.

(b)

Concentrations $(mol \cdot L^{-1})$	CO(g) +	2 H$_2$(g)	$\rightarrow$	CH$_3$OH(g)
initial	0.114	0.0717		0.0417
change	+x	+2x		−x
final	0.0114 + x	0.0717 + 2x		0.0417 − x

$$K_C = \frac{0.0417 - x}{(0.0114 + x)(0.0717 + 2x)^2} = 1.1 \times 10^{-2}$$

$$0.0417 - x = (1.1 \times 10^{-2})(0.114 + x)(0.0717 + 2x)^2$$

$$= (1.2 \times 10^{-3} + 1.1 \times 10^{-2} x)(4x^2 + 0.287x + 5.14 \times 10^{-3})$$

$$= 4.4 \times 10^{-2} x^3 + 8.0 \times 10^{-3} x^2 + 4.0 \times 10^{-4} x + 6.2 \times 10^{-6}$$

$$0 = 4.4 \times 10^{-2} x^3 + 8.0 \times 10^{-3} x^2 + 1.00x - 0.0417$$

This equation can be solved approximately, simply by inspection: It is clear that the x term will be very much larger than the x^3 and the x^2 terms because their coefficients are very small compared with 1.00. This leads to a prediction that $x = 0.0417 \text{ mol} \cdot L^{-1}$ to within the accuracy of the data. Essentially all of the CH_3OH will react, so that

$[CO] = 0.114 \ mol \cdot L^{-1} + 0.0417 \ mol \cdot L^{-1} = 0.156 \ mol \cdot L^{-1};$

$[H_2] = 0.0717 \ mol \cdot L^{-1} + 2(0.0417 \ mol \cdot L^{-1}) = 0.155 \ mol \cdot L^{-1}.$ The mathematical situation is odd in that clearly a $[CH_3OH] = 0$ will not satisfy the equilibrium constant. Knowing that the methanol concentration is very small compared with the CO and H_2 concentrations, we can now back-calculate to get a concentration value that will satisfy the equilibrium expression:

$$K_C = \frac{y}{(0.156)(0.155)^2} = 1.1 \times 10^{-2}$$

$$y = (1.1 \times 10^{-2})(0.156)(0.155)^2 = 4.1 \times 10^{-5}$$

Alternatively, the cubic equation can be solved with the aid of a graphing calculator.

5I.33 $\quad \dfrac{25.0 \ g \ NH_4(NH_2CO_2)}{78.07 \ g \cdot mol^{-1} \ NH_4(NH_2CO_2)} = 0.320 \ mol \ NH_4(NH_2CO_2)$

$\dfrac{0.0174 \ g \ CO_2}{44.01 \ g \cdot mol^{-1} CO_2} = 3.95 \times 10^{-4} \ mol \ CO_2$

2 mol NH_3 are formed per mol of CO_2, so mol $NH_3 = 2 \times 3.95 \times 10^{-4}$
$\quad = 7.90 \times 10^{-4}.$

$$K_C = [NH_3]^2[CO_2] = \left(\frac{7.90 \times 10^{-4}}{0.250}\right)^2 \left(\frac{3.95 \times 10^{-4}}{0.250}\right) = 1.58 \times 10^{-8}$$

5I.35 (c) $K = p^2/(1.0 - 2p)^2$

5J.1 (a) According to Le Chatelier's principle, an increase in the partial pressure of CO_2 will result in creation of reactants, which will decrease the H_2 partial pressure.

(b) According to Le Chatelier's principle, if the CO pressure is reduced, the reaction will shift to form more CO, which will decrease the pressure of CO_2.

(c) According to Le Chatelier's principle, if the concentration of CO is increased, the reaction will proceed to form more products, which will result in a higher pressure of H_2.

(d) The equilibrium constant for the reaction is unchanged because it is unaffected by any change in concentration.

5J.3 (a) According to Le Chatelier's principle, increasing the concentration of NO will cause the reaction to form reactants in order to reduce the concentration of NO; the amount of water will decrease.

(b) For the same reason as in (a), the amount of O_2 will increase.

(c) According to Le Chatelier's principle, removing water will cause the reaction to shift toward products, resulting in the formation of more NO.

(d) According to Le Chatelier's principle, removing a reactant will cause the reaction to shift in the direction to replace the removed substance; the amount of NH_3 should increase.

(e) According to Le Chatelier's principle, adding ammonia will shift the reaction to the right, but the equilibrium constant, which is a constant, will not be affected.

(f) According to Le Chatelier's principle, removing NO will cause the formation of more products; the amount of NH_3 will decrease.

(g) According to Le Chatelier's principle, adding reactants will promote the formation of products; the amount of oxygen will decrease.

5J.5 As per Le Chatelier's principle, whether increasing the pressure on a reaction will affect the distribution of species within an equilibrium mixture of gases depends largely upon the difference in the number of moles of gases between the reactant and product sides of the equation. If there is a net increase in the amount of gas, then applying pressure will shift the reaction toward reactants in order to remove the stress applied by increasing the pressure. Similarly, if there is a net decrease in the amount of gas, applying pressure will cause the formation of products. If the number of moles of gas is the same on the product and reactant side, then changing the pressure will have little or no effect on the equilibrium distribution of species present. Using this information, we can apply it to the specific reactions given. The answers are: (a) reactants; (b) reactants; (c) reactants; (d) no change (there is the same number of moles of gas on both sides of the equation); (e) reactants.

5J.7 First, we calculate the equilibrium constant for the conditions given.

$$K = \frac{(23.72)^2}{(3.11)(1.64)^3} = 41.0, \text{ which corresponds to the reaction written as}$$

$$N_2(g) + 3 H_2(g) \rightarrow 2 NH_3(g)$$

We then set up the table of anticipated changes upon introduction of the nitrogen:

	$N_2(g)$	+	$3 H_2(g)$	$\rightarrow$	$2 NH_3(g)$
initial	4.68 bar		1.64 bar		23.72 bar
change	$-x$		$-3x$		$+2x$
total	$4.68 - x$		$1.64 - 3x$		$23.72 + 2x$

$$41.0 = \frac{(23.72 + 2x)^2}{(4.68 - x)(1.64 - 3x)^3}$$

The equation can be solved using a graphing calculator, other computer software, or by trial and error. The solution is $x = 0.0656$ and the pressures of gases are $P_{N_2} = 4.61$ bar, $P_{H_2} = 1.44$ bar, $P_{NH_3} = 23.85$ bar.

5J.9 (a) If the pressure of NO (a product) is increased, the reaction will shift to form more reactants; the pressure of NH_3 should increase.

(b) If the pressure of NH_3 (a reactant) is decreased, then the reaction will shift to form more reactants; the pressure of O_2 should increase.

5J.11 If a reaction is exothermic, raising the temperature will tend to shift the reaction toward reactants, whereas if the reaction is endothermic, a shift toward products will be observed. For the specific examples given, (a) and (b) are endothermic (the values for (b) can be calculated, but we know that it requires energy to break an X—X bond, so those processes will all be endothermic) and raising the temperature should favor the formation of products; (c) and (d) are exothermic and raising the temperature should favor the formation of reactants.

5J.13 Even though numbers are given, we do not need to do a calculation to answer this qualitative question. Because the equilibrium constant for the formation of ammonia is smaller at the higher temperature, raising the temperature will favor the formation of reactants. Less ammonia will be present at higher temperature, assuming no other changes occur to the system (i.e., the volume does not change, no reactants or products are added or removed from the container, etc.).

5J.15 Because we want the equilibrium constant at two temperatures, we will need to calculate ΔH°_r and ΔS°_r for each reaction:

(a) $NH_4Cl \rightarrow NH_3(g) + HCl(g)$

$\Delta H^\circ_r = \Delta H^\circ_f(NH_3, g) + \Delta H^\circ_f(HCl, g) - \Delta H^\circ_f(NH_4Cl, s)$

$\Delta H^\circ_r = (-46.11 \text{ kJ} \cdot \text{mol}^{-1}) + (-92.31 \text{ kJ} \cdot \text{mol}^{-1}) - (-314.43 \text{ kJ} \cdot \text{mol}^{-1})$

$\Delta H^\circ_r = 176.01 \text{ kJ} \cdot \text{mol}^{-1}$

$\Delta S^\circ_r = S^\circ(NH_3, g) + S^\circ(HCl, g) - S^\circ(NH_4Cl, s)$

$\Delta S^\circ_r = 192.45 \text{ J} \cdot \text{K}^{-1} \cdot \text{mol}^{-1} + 186.91 \text{ J} \cdot \text{K}^{-1} \cdot \text{mol}^{-1} - 94.6 \text{ J} \cdot \text{K}^{-1} \cdot \text{mol}^{-1}$

$\Delta S^\circ_r = 284.8 \text{ J} \cdot \text{K}^{-1} \cdot \text{mol}^{-1}$

$\Delta G^\circ_r = \Delta H^\circ_r - T\Delta S^\circ_r$

At 298 K:

$\Delta G^\circ_{r(298 \text{ K})} = 176.01 \text{ kJ} - (298 \text{ K})(284.8 \text{ J} \cdot \text{K}^{-1})/(1000 \text{ J} \cdot \text{kJ}^{-1})$

$= 91.14 \text{ kJ} \cdot \text{mol}^{-1}$

$\Delta G^\circ_{r(298 \text{ K})} = -RT \ln K$

$$\ln K = -\frac{\Delta G^\circ_{r(298\,K)}}{RT}$$

$$= -\frac{91140\,J}{(8.314\,J\cdot K^{-1})(298\,K)} = -36.8$$

$K = 1 \times 10^{-16}$

At 423 K:

$$\Delta G^\circ_{r(423\,K)} = 176.01\,kJ - (423\,K)(284.8\,J\cdot K^{-1})/(1000\,J\cdot kJ^{-1})$$

$$= 55.54\,kJ\cdot mol^{-1}$$

$$\Delta G^\circ_{r(423\,K)} = -RT\ln K$$

$$\ln K = -\frac{\Delta G^\circ_{r(423\,K)}}{RT}$$

$$= -\frac{55\,540\,J}{(8.314\,J\cdot K^{-1})(423\,K)} = -15.8$$

$K = 1 \times 10^{-7}$

(b) $H_2(g) + D_2O(l) \rightarrow D_2(g) + H_2O(l)$

$\Delta H^\circ_r = \Delta H^\circ_f(H_2O, l) - [\Delta H^\circ_f(D_2O, l)]$

$\Delta H^\circ_r = (-285.83\,kJ\cdot mol^{-1}) - [-294.60\,kJ\cdot mol^{-1}]$

$\Delta H^\circ_r = 8.77\,kJ\cdot mol^{-1}$

$\Delta S^\circ_r = S^\circ(D_2, g) + S^\circ(H_2O, l) - [S^\circ(H_2, g) + S^\circ(D_2O, l)]$

$\Delta S^\circ_r = 144.96\,J\cdot K^{-1}\cdot mol^{-1} + 69.91\,J\cdot K^{-1}\cdot mol^{-1}$

$\qquad - [130.68\,J\cdot K^{-1}\cdot mol^{-1} + 75.94\,J\cdot K^{-1}\cdot mol^{-1}]$

$\Delta S^\circ_r = 8.25\,J\cdot K^{-1}\cdot mol^{-1}$

At 298 K:

$$\Delta G^\circ_{r(298\,K)} = 8.77\,kJ\cdot mol^{-1} - (298\,K)(8.25\,J\cdot K^{-1}\cdot mol^{-1})/(1000\,J\cdot kJ^{-1})$$

$$= 6.31\,kJ\cdot mol^{-1}$$

$$\Delta G^\circ_{r(298\,K)} = -RT\ln K$$

$$\ln K = -\frac{\Delta G^\circ_{r(298\,K)}}{RT}$$

$$= -\frac{6310\,J}{(8.314\,J\cdot K^{-1})(298\,K)} = -2.55$$

$K = 7.8 \times 10^{-2}$

At 423 K:

$$\Delta G^{\circ}_{r(423\,K)} = 8.77\ \text{kJ}\cdot\text{mol}^{-1} - (423\ \text{K})(8.25\ \text{J}\cdot\text{K}^{-1}\cdot\text{mol}^{-1})/(1000\ \text{J}\cdot\text{kJ}^{-1})$$
$$= 5.28\ \text{kJ}\cdot\text{mol}^{-1}$$

$$\ln K = -\frac{5280\ \text{J}}{(8.314\ \text{J}\cdot\text{K}^{-1})(423\ \text{K})} = -1.50$$

$$K = 0.22$$

5J.17 $K = (RT)^{\Delta n} K_c$

$$\ln\frac{K_2}{K_1} = \frac{\Delta H_r^{\circ}}{R}\left(\frac{1}{T_1} - \frac{1}{T_2}\right)$$

$$\ln\left(\frac{(RT_2)^{\Delta n} K_{c2}}{(RT_1)^{\Delta n} K_{c1}}\right) = \frac{\Delta H_r^{\circ}}{R}\left(\frac{1}{T_1} - \frac{1}{T_2}\right)$$

$$\Delta n \ln\left(\frac{T_2}{T_1}\right) + \ln\left(\frac{K_{c2}}{K_{c1}}\right) = \frac{\Delta H_r^{\circ}}{R}\left(\frac{1}{T_1} - \frac{1}{T_2}\right)$$

$$\ln\left(\frac{K_{c2}}{K_{c1}}\right) = \frac{\Delta H_r^{\circ}}{R}\left(\frac{1}{T_1} - \frac{1}{T_2}\right) - \Delta n \ln\left(\frac{T_2}{T_1}\right)$$

5.1   Water is a polar molecule and as a result will orient itself differently around cations and anions, aligning its dipole in such a way as to present the most favorable interaction possible:

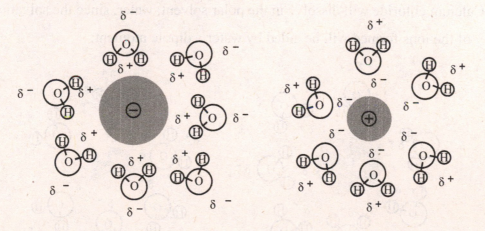

5.3   (a) stronger;   (b) low;   (c) high;   (d) weaker;   (e) weak, low;   (f) low;   (g) strong, high

5.5   (a, b) Viscosity and surface tension decrease with increasing temperature; at high temperatures the molecules readily move away from their neighbors because of increased kinetic energy. (c, d) Evaporation rate and vapor pressure increase with increasing temperature because the kinetic energy of the molecules increase with temperature, and the molecules are more likely to escape into the gas phase.

5.7 (a) Butane (a nonpolar molecule) will dissolve in the nonpolar solvent (tetrachloromethane).

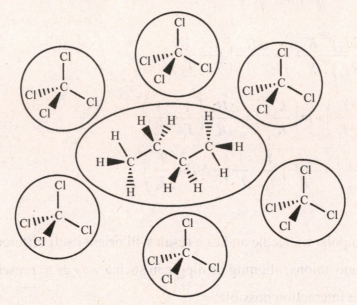

(b) Calcium chloride will dissolve in the polar solvent, water, since the solvation of the ions formed will be aided by water's dipole moment:

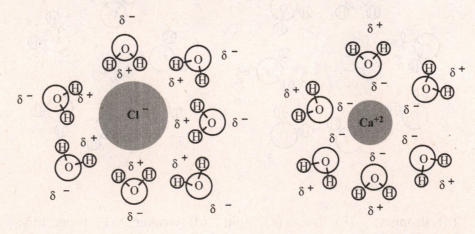

5.9 $\Delta T = k_f m;$ also $m = \dfrac{n_{solute}}{mass_{solvent(kg)}}$ and $n_{solute} = \dfrac{mass_{solute}}{M_{solute}}$

so

$$\Delta T = k_f \frac{mass_{solute}}{M_{solute}(mass_{solvent(kg)})}$$

Solving for M_{solute},

$$M_{solute} = \frac{k_f (mass_{solute})}{\Delta T_f (mass_{solvent(kg)})}$$

(a) If $mass_{solute}$ appears greater, M_{solute} appears greater than actual molar mass, as $mass_{solute}$ occurs in the numerator above. Also, the ΔT measured will be smaller because less solute will actually be dissolved. This has the same effect as increasing the apparent M_{solute}.

(b) Because the true $mass_{solvent} = d \times V$, if $d_{solvent}$ is less than $1.00 \text{ g} \cdot \text{cm}^{-3}$, then the true $mass_{solvent}$ will be less than the assumed mass. M_{solute} is inversely proportional to $mass_{solvent}$, so an artificially high $mass_{solvent}$ will lead to an artificially low M_{solute}.

(c) If true freezing point is higher than the recorded freezing point, true $\Delta T <$ assumed ΔT, or assumed $\Delta T >$ true ΔT, and M_{solute} appears less than actual M_{solute}, as ΔT occurs in the denominator.

(d) If not all solute dissolved, the true $mass_{solute} <$ assumed $mass_{solute}$ or assumed $mass_{solute} >$ true $mass_{solute}$, and M_{solute} appears greater than the actual M_{solute}, as $mass_{solute}$ occurs in the numerator.

5.11 (a) The 5.22 cm or 52.2 mm rise for an aqueous solution must be converted to Torr or mmHg in order to be expressed into consistent units.

$$52.2 \text{ mm} \left(\frac{0.998 \text{ g} \cdot \text{cm}^{-3}}{13.6 \text{ g} \cdot \text{cm}^{-3}} \right) = 3.83 \text{ mmHg or } 3.83 \text{ Torr}$$

The molar mass can be calculated using the osmotic pressure equation:

$$\Pi = iRT \times molarity$$

Assume that the protein is a nonelectrolyte with $i = 1$ and that the amount of protein added does not significantly affect the volume of the solution.

$$\Pi = 1(0.082\ 06\ \text{L} \cdot \text{atm} \cdot \text{K}^{-1} \cdot \text{mol}^{-1})(293\ \text{K})\left(\frac{\left(\dfrac{0.010\ \text{g}}{M_{\text{protein}}}\right)}{0.010\ \text{L}}\right) = \frac{3.83\ \text{Torr}}{760\ \text{Torr} \cdot \text{atm}^{-1}}$$

$$M_{\text{protein}} = \frac{1(0.082\ 06\ \text{L} \cdot \text{atm} \cdot \text{K}^{-1} \cdot \text{mol}^{-1})(293\ \text{K})(0.010\ \text{g})(760\ \text{Torr} \cdot \text{atm}^{-1})}{(0.010\ \text{L})(3.83\ \text{Torr})}$$

$$M_{\text{protein}} = 4.8 \times 10^3\ \text{g} \cdot \text{mol}^{-1}$$

(b) The freezing point can be calculated using the relationship $\Delta T_f = ik_f m$:

$$\Delta T_f = 1(1.86\ \text{K} \cdot \text{kg} \cdot \text{mol}^{-1})\left(\frac{\left(\dfrac{0.010\ \text{g}}{4.8 \times 10^3\ \text{g} \cdot \text{mol}^{-1}}\right)}{\left(\dfrac{10\ \text{mL} \times 1.00\ \text{g} \cdot \text{mL}^{-1}}{1000\ \text{g} \cdot \text{kg}^{-1}}\right)}\right)$$

$$= 3.9 \times 10^{-4}\ \text{K or } 3.9 \times 10^{-4}\ ^\circ\text{C}$$

The freezing point will be $0.00^\circ\text{C} - 3.9 \times 10^{-4}\ ^\circ\text{C} = -3.9 \times 10^{-4}\ ^\circ\text{C}$.

(c) The freezing point change is so small that it cannot be measured accurately, so osmotic pressure would be the preferred method for measuring the molecular weight.

5.13 (a) The data in Appendix 2A can be used to calculate the change in enthalpy and entropy for the vaporization of methanol:

$$CH_3OH(l) \rightleftharpoons CH_3OH(g)$$

$$\Delta H^\circ_{\text{vap}} = \Delta H^\circ_f(CH_3OH, g) - \Delta H^\circ_f(CH_3OH, l)$$

$$= (-200.66\ \text{kJ} \cdot \text{mol}^{-1}) - (-238.86\ \text{kJ} \cdot \text{mol}^{-1})$$

$$= 38.20\ \text{kJ} \cdot \text{mol}^{-1}$$

$$\Delta S^\circ_{\text{vap}} = S^\circ_m(CH_3OH(g)) - S^\circ_m(CH_3OH(l))$$

$$= 239.81\ \text{J} \cdot \text{K}^{-1} \cdot \text{mol}^{-1} - 126.8\ \text{J} \cdot \text{K}^{-1} \cdot \text{mol}^{-1}$$

$$= 113.0\ \text{J} \cdot \text{K}^{-1} \cdot \text{mol}^{-1}$$

To derive the general equation, we start with the expression that $\Delta G^\circ_{\text{vap}} = -RT \ln P$, where P is the vapor pressure of the solvent. Because

$\Delta G°_{vap} = \Delta H°_{vap} - T\Delta S°_{vap}$, this is the relationship to use to determine the temperature dependence of $\ln P$:

$$\Delta H°_{vap} - T\Delta S°_{vap} = -RT \ln P$$

This equation can be rearranged to give

$$\ln P = -\frac{\Delta H°_{vap}}{R} \cdot \frac{1}{T} + \frac{\Delta S°_{vap}}{R}$$

To create an equation specific to methanol, we can plug in the actual values of R, $\Delta H°_{vap}$, and $\Delta S°_{vap}$:

$$\ln P = -\frac{38\,200 \text{ J} \cdot \text{mol}^{-1}}{8.314 \text{ J} \cdot \text{K}^{-1} \cdot \text{mol}^{-1}} \cdot \frac{1}{T} + \frac{113.0 \text{ J} \cdot \text{K}^{-1} \cdot \text{mol}^{-1}}{8.314 \text{ J} \cdot \text{K}^{-1} \cdot \text{mol}^{-1}}$$

$$= -\frac{4595 \text{ K}}{T} + 13.59$$

(b) The relationship to plot is $\ln P$ versus $\frac{1}{T}$. This should result in a straight line whose slope is $-\frac{\Delta H°_{vap}}{R}$ and whose intercept is $\frac{\Delta S°_{vap}}{R}$. The pressure must be given in atm for this relationship, because atm is the standard state condition.

(c) Because we have already determined the equation, it is easiest to calculate the vapor by inserting the value of 0.0°C or 273.2 K :

$$\ln P = -\frac{4595 \text{ K}}{T} + 13.59 = -\frac{4595 \text{ K}}{273.2 \text{ K}} + 13.59 = -16.82 + 13.59 = -3.23$$

$P = 0.040$ atm or 30. Torr

(d) As in (c) we can use the equation to find the point where the vapor pressure of methanol = 1 atm.

$$\ln P = \ln 1 = 0 = -\frac{4595 \text{ K}}{T} + 13.59$$

$T = 338.1$ K

5.15 (a) $\Delta T_f = 1.72 \text{ K} = ik_f m = (1.86 \text{ K} \cdot \text{kg} \cdot \text{mol}^{-1})\left(\dfrac{n_{aa}}{0.95 \text{ kg}}\right)$

where n_{aa} are the moles of acetic acid in solution.

Assuming $i = 1$:

$$n_{aa} = \frac{(1.72\ K)(0.95\ kg)}{(1.86\ K \cdot kg \cdot mol^{-1})} = 0.878\ mol$$

Experimentally, the molar mass of acetic acid is, therefore:

$$\frac{50\ g}{0.878\ mol} = 56.9\ g \cdot mol^{-1}$$

This experimental molar mass of acetic acid is less than the known molecular mass of the acid ($60.0\ g \cdot mol^{-1}$) indicating that the acid is dissociating in solution giving an $i > 1$.

(b) As in part (a) the experimental molar mass is first found:

$$\Delta T_f = 2.32\ K = ik_f m = (5.12\ K \cdot kg \cdot mol^{-1})\left(\frac{n_{aa}}{0.95\ kg}\right)$$

where n_{aa} are the moles of acetic acid in solution.

Assuming $i = 1$: $n_{aa} = \dfrac{(2.32\ K)(0.95\ kg)}{(5.12\ K \cdot kg \cdot mol^{-1})} = 0.430\ mol.$

Experimentally, the molar mass of acetic acid is, therefore:

$$\frac{50\ g}{0.430\ mol} = 116\ g \cdot mol^{-1}.$$

This experimental molar mass of acetic acid is significantly higher than the known molar mass of the acid ($60.0\ g \cdot mol^{-1}$) indicating that the acid is dissociating in solution giving an $i < 1$. A van 't Hoff factor less than 1 indicates that acetic acid is _dimerizing_ in the benzene, or acetic acid molecules are aggregating together in solution.

5.17 (a) The plot of the data is shown below. On this plot, the slope $= -\dfrac{\Delta H^\circ_{vap}}{R}$

and the intercept $= \dfrac{\Delta S^\circ_{vap}}{R}$.

Temp. (K)	T^{-1} (K^{-1})	Vapor pressure (Torr)	V.P. (atm)	$\ln P$
190	0.005 26	3.2	0.0042	−5.47
228	0.004 38	68	0.089	−2.41
250	0.004 00	240	0.316	−1.15
273	0.003 66	672	0.884	−0.123

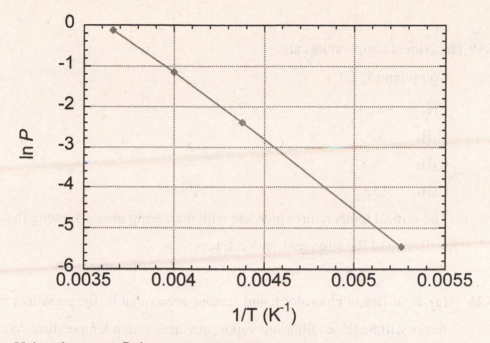

Using the curve fitting program:

$$y = -3358.714x + 12.247$$

(b) $-\dfrac{\Delta H^\circ_{vap}}{R} = -3359$

$$\Delta H^\circ_{vap} = (3359)(8.314\ \text{J} \cdot \text{K}^{-1} \cdot \text{mol}^{-1}) = 28\ \text{kJ} \cdot \text{mol}^{-1}$$

(c) $\dfrac{\Delta S^\circ_{vap}}{R} = 12.25$

$$\Delta S^\circ_{vap} = (12.25)(8.314\ \text{J} \cdot \text{K}^{-1} \cdot \text{mol}^{-1}) = 1.0 \times 10^2\ \text{J} \cdot \text{K}^{-1} \cdot \text{mol}^{-1}$$

(d) The normal boiling point will be the temperature at which the vapor pressure = 1 atm or at which the ln 1 = 0. This will occur when

$$T^{-1} = 0.0036 \text{ or } T = 2.7 \times 10^2\ \text{K}.$$

(e) This is most easily done by using an equation derived from ΔH°_{vap} and

ΔS°_{vap} :

$$\ln P = -\frac{\Delta H^\circ_{vap}}{R} \cdot \frac{1}{T} + \frac{\Delta S^\circ_{vap}}{R}$$

$$\ln \frac{15\ \text{Torr}}{760\ \text{Torr}} = -\frac{28\,000\ \text{J} \cdot \text{mol}^{-1}}{8.314\ \text{J} \cdot \text{K}^{-1} \cdot \text{mol}^{-1}} \cdot \frac{1}{T} + \frac{100\ \text{J} \cdot \text{K}^{-1} \cdot \text{mol}^{-1}}{8.314\ \text{J} \cdot \text{K}^{-1} \cdot \text{mol}^{-1}}$$

$$T = 2.1 \times 10^2\ \text{K}$$

5.19 The critical temperatures are

Compound T_C (°C)

CH_4 −82.1

C_2H_6 32.2

C_3H_8 96.8

C_4H_{10} 152

The critical temperatures increase with increasing mass, showing the influence of the stronger London forces.

5.21 (a) If sufficient chloroform and acetone are available, the pressures in the flasks will be the equilibrium vapor pressures at that temperature. We can calculate these amounts using the ideal gas equation:

$$\left(\frac{195 \text{ Torr}}{760 \text{ Torr} \cdot \text{atm}^{-1}}\right)(1.00 \text{ L}) =$$

$$\left(\frac{m_{chloroform}}{119.37 \text{ g} \cdot \text{mol}^{-1} \text{ chloroform}}\right)(0.08206 \text{ L} \cdot \text{atm} \cdot \text{K}^{-1} \cdot \text{mol}^{-1})(298 \text{ K})$$

$$m_{chloroform} = 1.25 \text{ g}$$

$$\left(\frac{225 \text{ Torr}}{760 \text{ Torr} \cdot \text{atm}^{-1}}\right)(1.00 \text{ L}) =$$

$$\left(\frac{m_{acetone}}{58.08 \text{ g} \cdot \text{mol}^{-1} \text{ acetone}}\right)(0.08206 \text{ L} \cdot \text{atm} \cdot \text{K}^{-1} \cdot \text{mol}^{-1})(298 \text{ K})$$

$$m_{acetone} = 0.703 \text{ g}$$

In both cases, sufficient compound is available to achieve the vapor pressure; flask A will have a pressure of 195 Torr and flask B will have a pressure of 222 Torr.

(b) When the stopcock is opened, some chloroform will move into flask B and acetone will move into flask A to restore the equilibrium vapor pressure. Additionally, however, some acetone vapor will dissolve in the liquid chloroform and vice versa. Ultimately the system will reach an equilibrium state in which the compositions of the liquid phases in both flasks are the same and the gas phase composition is uniform. The gas phase and liquid phase compositions will be established by Raoult's law.

It is conceptually most convenient for this calculation to start by putting all the material into one liquid phase. Such a solution would have the following composition:

$$x_{acetone} = \frac{\dfrac{35.0 \text{ g}}{58.08 \text{ g} \cdot \text{mol}^{-1} \text{ acetone}}}{\left(\dfrac{35.0 \text{ g}}{58.08 \text{ g} \cdot \text{mol}^{-1} \text{ acetone}}\right) + \left(\dfrac{35.0 \text{ g}}{119.37 \text{ g} \cdot \text{mol}^{-1} \text{ chloroform}}\right)}$$

$$x_{acetone} = 0.67$$

Similarly:

$$x_{chloroform} = \frac{\dfrac{35.0 \text{ g}}{119.37 \text{ g} \cdot \text{mol}^{-1} \text{ chloroform}}}{\left(\dfrac{35.0 \text{ g}}{58.08 \text{ g} \cdot \text{mol}^{-1} \text{ acetone}}\right) + \left(\dfrac{35.0 \text{ g}}{119.37 \text{ g} \cdot \text{mol}^{-1} \text{ chloroform}}\right)}$$

$$x_{chloroform} = 0.33$$

($x_{chloroform}$ can also be determined by remembering $x_{chloroform} = 1 - x_{acetone}$).

This gives the composition of the liquid phase. The composition of the gas phase will be determined from the pressures of the gases:

$$P_{acetone, gas} = x_{acetone, liquid} \cdot P_{vp \text{ pure acetone}} = (0.67)(222 \text{ Torr}) = 149 \text{ Torr}$$

$$P_{chloroform, gas} = x_{chloroform, liquid} \cdot P_{vp \text{ pure chloroform}} = (0.33)(195 \text{ Torr}) = 64 \text{ Torr}$$

$$x_{acetone, gas} = \frac{P_{acetone}}{P_{acetone} + P_{chloroform}}$$

$$x_{acetone, gas} = \frac{149 \text{ Torr}}{149 \text{ Torr} + 64 \text{ Torr}} = 0.70$$

$$x_{chloroform, gas} = 1 - x_{acetone, gas} = 0.30$$

The gas phase composition will, therefore, be slightly richer in acetone than in chloroform. The total pressure in the flask will be 213 Torr.

(c) The solution shows negative deviation from Raoult's law. This means that the molecules of acetone and chloroform attract each other slightly more than molecules of the same kind. Under such circumstances, the vapor pressure is lower than expected from the ideal calculation. This will give rise to a high-boiling azeotrope. The gas phase composition will also be slightly different from that calculated from the ideal state, but whether acetone or chloroform would be richer in the gas phase depends on which side of the azeotrope the composition lies. Because we are not given the composition of the azeotrope, we cannot state which way the values will vary.

5.23 The vapor pressure is more sensitive if ΔH_{vap} is small. The fact that ΔH_{vap} is small indicates that it takes little energy to volatilize the sample, which means that the intermolecular forces are weaker. Hence we expect the ratio P_2/P_1 in the Clausius-Clapeyron equation to be larger.

5.25 $5\% \text{ glucose} = \dfrac{5.0 \text{ g glucose}}{1.0 \times 10^2 \text{ g solution}}$

assume the solution density $\approx 1 \text{ g} \cdot \text{mL}^{-1}$

Then $\left(\dfrac{5.0 \text{ g glucose}}{1.0 \times 10^2 \text{ mL}}\right)\left(\dfrac{1 \text{ mL}}{10^{-3} \text{ L}}\right)\left(\dfrac{1 \text{ mol glucose}}{180.16 \text{ g} \cdot \text{mol}^{-1}}\right) = 0.28 \text{ mol} \cdot \text{L}^{-1}$

and

$\Pi = iRTM = (1)(0.082\ 06 \text{ L} \cdot \text{atm} \cdot \text{K}^{-1} \cdot \text{mol}^{-1})(310 \text{ K})(0.28 \text{ mol} \cdot \text{L}^{-1})$
      $= 7.1 \text{ atm}$

5.27   (a) $\left(\dfrac{25.0 \text{ Torr}}{31.83 \text{ Torr}}\right)(100) = 78.5\%$;   (b) At 25°C the vapor pressure of water is only 23.76 Torr, so some of the water vapor in the air would condense as dew or fog.

5.29   (a) Dextrose has a molar mass of 180.15 $\text{g} \cdot \text{mol}^{-1}$; NaCl has a molar mass of 58.44 $\text{g} \cdot \text{mol}^{-1}$. Using these values, we get the following:

$\dfrac{40.0 \text{ g dextrose}}{1 \text{ L}} \times \dfrac{1 \text{ mol dextrose}}{180.15 \text{ g dextrose}} = 0.222 \text{ mol} \cdot \text{L}^{-1} = 0.222 \text{ M}$ and

$\dfrac{1.75 \text{ g NaCl}}{1 \text{ L}} \times \dfrac{1 \text{ mol NaCl}}{58.44 \text{ g NaCl}} = 2.99 \times 10^{-2} \text{ mol} \cdot \text{L}^{-1} = 2.99 \times 10^{-2} \text{ M}$

(b) Assuming that the total osmotic pressure is the sum of the osmotic pressures due to each component,

$$\Pi_{tot} = i_{dextrose} RT (\text{molarity})_{dextrose} + i_{NaCl} RT (\text{molarity})_{NaCl}$$

$$= \left[i_{dextrose} (\text{molarity})_{dextrose} + i_{NaCl} (\text{molarity})_{NaCl} \right] RT$$

$$= \left[1 (0.222 \text{ mol} \cdot \text{L}^{-1}) + 2 (2.99 \times 10^{-2} \text{ mol} \cdot \text{L}^{-1}) \right] (0.082\ 06 \text{ L} \cdot \text{atm} \cdot \text{K}^{-1} \cdot \text{mol}^{-1})(298 \text{ K})$$

$$= 6.89 \text{ atm}$$

5.31 (a) Mannitol has a molar mass of $180.15 \text{ g} \cdot \text{mol}^{-1} = 180.15 \text{ mg} \cdot \text{mmol}^{-1}$.

Remembering that $\text{molarity} = \text{mol} \cdot \text{L}^{-1} = \text{mmol} \cdot \text{mL}^{-1}$ we get

$$\frac{180 \text{ mg mannitol}}{1 \text{ mL}} \times \frac{1 \text{ mmol mannitol}}{180.15 \text{ mg mannitol}} = 1.0 \text{ mmol} \cdot \text{mL}^{-1} = 1.0 \text{ M}$$

(b) Assuming an ambient temperature of 25 °C (298 K),

$$\Pi = iRT (\text{molarity})$$

$$= 1 (0.082\ 06 \text{ L} \cdot \text{atm} \cdot \text{K}^{-1} \cdot \text{mol}^{-1})(298 \text{ K})(1.0 \text{ mol} \cdot \text{L}^{-1})$$

$$= 24 \text{ atm}$$

5.33 The change increased the concentration of $X(g)$ and decreased the concentration of $X_2(g)$ in Flask 2. Increasing the temperature (a) would increase the formation of $X(g)$ because the reaction is endothermic. Adding $X(g)$ atoms or decreasing the volume (increasing pressure) would increase the formation of $X_2(g)$, which did not occur. Adding a catalyst would not favor the formation of either $X(g)$ or $X_2(g)$. Also, the value of the equilibrium constant is larger in Flask 2 than in Flask 1, which is consistent with an increase in temperature of an endothermic reaction.

5.35 (a) $2A(g) \rightarrow B(g) + 2C(g)$

(b) $K = \dfrac{P_B (P_C)^2}{(P_A)^2} = \dfrac{(5/100)(10/100)^2}{(18/100)^2} = 1.54 \times 10^{-2}$

5.37 To answer this question we must calculate Q:

$$Q = \frac{[NH_3]^2}{[N_2][H_2]^3} = \frac{(0.500)^2}{(3.00)(2.00)^3} = 0.0104$$

Because $Q \neq K$, the system is not at equilibrium, and because $Q < K$, the reaction will proceed to produce more products.

5.39 (a) The initial concentration of $NO_2(g)$ is 0.020 mol·L^{-1}

Concentration (mol·L^{-1})	2 $NO_2(g)$	→	$N_2O_4(g)$
initial	0.020		0
change	-2x		+x
equilibrium	0.020 – 2x		+x

$$K_C = \frac{[N_2O_4]}{[NO_2]^2} = \frac{x}{(0.020 - 2x)^2} = \frac{1}{6.1 \times 10^{-3}} = 1.6 \times 10^2$$

Note: $(0.020 - 2x)^2 = 4.0 \times 10^{-4} - 0.08x + 4x^2$

$x = 0.064 - 13x + 6.4 \times 10^2 x^2$

$6.4 \times 10^2 x^2 - 14 x + 0.064 = 0$

Solve the quadratic equation: $x = +0.0065$ or $+0.015$

It is obvious that $x = +0.015$ is not meaningful, because a negative NO_2 concentration will be obtained at equilibrium:

$0.020 - 2x = 0.020 - 2(0.015) = -0.01$ mol·L^{-1}

so we choose $x = 0.0065$ mol·L^{-1}

$[N_2O_4] = 0.0065$ mol·L^{-1}

$[NO_2] = 0.020 - 2x = 0.020 - 2(0.0065) = 7.0 \times 10^{-3}$ mol·L^{-1}

(b) If the volume is decreased by half ($V = 500.$ mL). The initial concentration of NO_2 would be 0.040 mol·L^{-1}. By following the same procedure in (a):

Concentration (mol·L^{-1})	2 $NO_2(g)$	→	$N_2O_4(g)$
initial	0.040		0
change	-2x		+x
equilibrium	0.040 – 2x		+x

$$K_C = \frac{[N_2O_4]}{[NO_2]^2} = \frac{x}{(0.040 - 2x)^2} = \frac{1}{6.1 \times 10^{-3}} = 1.6 \times 10^2$$

Note: $(0.040 - 2x)^2 = 1.6 \times 10^{-3} - 0.16x + 4x^2$

$x = 0.26 - 26x + 6.4 \times 10^2 x^2$

$6.4 \times 10^2 x^2 - 27 x + 0.26 = 0$

Solve the quadratic equation: $x = +0.015$ or $+0.027$

It is obvious that $x = +0.027$ is not meaningful, because a negative NO_2 concentration will be obtained at equilibrium:

$$0.040 - 2x = 0.040 - 2(0.027) = -0.014 \text{ mol·L}^{-1}$$

so we choose $x = 0.015 \text{ mol·L}^{-1}$

$$[N_2O_4] = 0.015 \text{ mol·L}^{-1}$$

$$[NO_2] = 0.040 - 2x = 0.040 - 2(0.015) = 0.010 \text{ mol·L}^{-1}$$

5.41

Pressure	$PCl_5(g)$	$\rightarrow$	$PCl_3(g)$	$+$	$Cl_2(g)$
initial	n				
change	$-n\alpha$		$+n\alpha$		$+n\alpha$
final	$n(1-\alpha)$		$+n\alpha$		$+n\alpha$

$$K = \frac{(n\alpha)(n\alpha)}{n(1-\alpha)} = \frac{n^2\alpha^2}{n(1-\alpha)} = \frac{n\alpha^2}{(1-\alpha)}$$

$$P = n(1-\alpha) + n\alpha + n\alpha = n(1+\alpha)$$

$$n = \frac{P}{1+\alpha}$$

$$K = \frac{n\alpha^2}{(1-\alpha)} = \left(\frac{P}{1+\alpha}\right)\left(\frac{\alpha^2}{1-\alpha}\right) = \frac{P\alpha^2}{1-\alpha^2}$$

$$(1-\alpha^2)K = P\alpha^2$$

$$K - K\alpha^2 = P\alpha^2$$

$$K = P\alpha^2 + K\alpha^2$$

$$\alpha^2 = \frac{K}{P+K}$$

$$\alpha = \sqrt{\frac{K}{P+K}}$$

(a) For the specific conditions $K = 4.96$ and $P = 0.50$ bar,

$$\alpha = \sqrt{\frac{4.96}{0.50 + 4.96}} = 0.953.$$

(b) For the specific conditions $K = 4.96$ and $P = 1.00$ bar,

$$\alpha = \sqrt{\frac{4.96}{1.00 + 4.96}} = 0.912.$$

5.43   (a) If $K = 1.00$, then $\Delta G°$ must be equal to 0 ($\Delta G° = -RT \ln K$).

(b) This can be calculated by determining the values for $\Delta H°$ and $\Delta S°$ at 25°C.

$$\Delta H° = -393.51 \, kJ \cdot mol^{-1} - [(-110.53 \, kJ \cdot mol^{-1}) + (-241.82 \, kJ \cdot mol^{-1})]$$
$$= -41.16 \, kJ \cdot mol^{-1}$$

$$\Delta S° = 130.68 \, J \cdot K^{-1} \cdot mol^{-1} + 213.74 \, J \cdot K^{-1} \cdot mol^{-1}$$
$$- [197.67 \, J \cdot K^{-1} \cdot mol^{-1} + 188.83 \, J \cdot K^{-1} \cdot mol^{-1}]$$
$$= -42.08 \, J \cdot K^{-1} \cdot mol^{-1}$$

$$\Delta G = (-41.16 \, kJ \cdot mol^{-1})(1000 \, J \cdot kJ^{-1}) - T(-42.08 \, J \cdot K^{-1} \cdot mol^{-1}) = 0$$
$$T = 978 \, K \text{ (or 705°C)}$$

(c)

	$CO(g)$ +	$H_2O(g)$	$\rightarrow CO_2(g)$ +	$H_2(g)$
	10.00 bar	10.00 bar	5.00 bar	5.00 bar
change	$-x$	$-x$	$+x$	$+x$
net	$10.00 - x$	$10.00 - x$	$5.00 + x$	$5.00 + x$

$$K = \frac{P_{H_2} P_{CO_2}}{P_{CO} P_{H_2O}} = \frac{(5.00 + x)^2}{(10.00 - x)^2} = 1.00$$

$$x = 2.50 \, bar$$

All pressures are equal to 7.50 bar.

(d) First, check Q to determine the direction of the reaction:

$$Q = \frac{(10.00)(5.00)}{(6.00)(4.00)} = 2.08$$

Because Q is greater than 1, the reaction will shift to produce reactants.

	$CO(g)$ +	$H_2O(g)$	$\rightarrow$	$CO_2(g)$ +	$H_2(g)$
	6.00 bar	4.00 bar		10.00 bar	5.00 bar
change	$+x$	$+x$		$-x$	$-x$
net	$6.00 + x$	$4.00 + x$		$10.00 - x$	$5.00 - x$

$$\frac{(10.00-x)(5.00-x)}{(6.00+x)(4.00+x)}=1$$

$$(10.00-x)(5.00-x)=(6.00+x)(4.00+x)$$

$$x^2-15.00x+50.0=x^2+10.00x+24.0$$

$$25.00x=26.0$$

$$x=1.04 \text{ bar}$$

$$P_{CO(g)}=7.04 \text{ bar}; P_{H_2O(g)}=5.04 \text{ bar}; P_{CO_2(g)}=8.96 \text{ bar}; P_{H_2(g)}=3.96 \text{ bar}$$

5.45 (a) These values are easily calculated from the relationship $\Delta G° = -RT$ $\ln K$. For the atomic species, the free energy of the reaction will be $\frac{1}{2}$ of this value because the equilibrium reactions are for the formation of two moles of halogen atoms.

The results:

Halogen	Bond Dissociation Energy (kJ·mol^{-1})	$\Delta G°$ (kJ·mol^{-1})
fluorine	146	19.1
chlorine	230	47.9
bromine	181	42.8
iodine	139	5.6

(b)

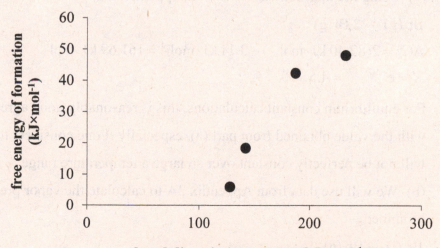

There is a correlation between the bond dissociation energy and the free energy of formation of the atomic species, but the relationship is clearly not linear.

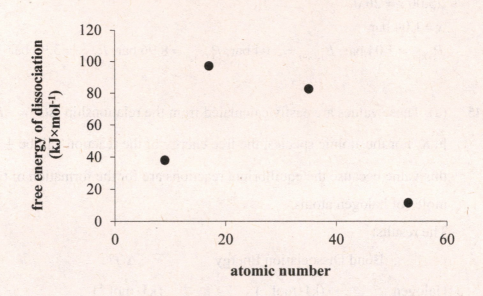

For the heavier three halogens, there is a trend to decreasing free energy of formation of the atoms as the element becomes heavier, but fluorine is anomalous. The F—F bond energy is lower than expected, owing to repulsions of the lone pairs of electrons on the adjacent F atoms because the F—F bond distance is so short.

5.47 (a) Using the thermodynamic data in Appendix 2A:

$Br_2(g) \rightarrow 2 Br(g)$

$\Delta G° = 2(82.40 \, kJ \cdot mol^{-1}) - 3.11 \, kJ \cdot mol^{-1} = 161.69 \, kJ \cdot mol^{-1}$

$K = e^{-\Delta G°/RT} = 4.5 \times 10^{-29}$

For equilibrium constant calculations, this is reasonably good agreement with the value obtained from part (a), especially if one considers that $\Delta H°$ will not be perfectly constant over so large a temperature range.

(b) We will use data from Appendix 2A to calculate the vapor pressure of bromine:

$Br_2(l) \rightarrow Br_2(g)$

$\Delta G° = 3.11 \, kJ \cdot mol^{-1}$

$K = e^{-\Delta G°/RT} = 0.285$

The vapor pressure of bromine will, therefore, be 0.285 bar or 0.289 atm. Remember that because the standard state for the thermodynamic quantities is 1 bar, the values in K will be derived in bar as well.

(c) $\quad 4.5 \times 10^{-29} = \dfrac{P_{Br(g)}^2}{P_{Br_2\,(g)}} = \dfrac{P_{Br(g)}^2}{0.285 \text{ bar}}$

$P_{Br(g)}^2 = 3.6 \times 10^{-15}$ bar or 3.6×10^{-15} atm

(d) Use the ideal gas law:

$PV = nRT$

$(0.289 \text{ atm})V = (0.0100 \text{ mol})(0.082\,06 \text{ L}\cdot\text{atm}\cdot\text{K}^{-1}\cdot\text{mol}^{-1})(298 \text{ K})$

$V = 0.846$ L or 846 mL

5.49 (a) $\quad K_C = \dfrac{[NO_2]^2}{[N_2O_4]} = \dfrac{(2.13)^2}{0.405} = 11.2$

$K = (RT)^{\Delta n} K_c$

$K = [(8.314 \text{ J}\cdot\text{mol}^{-1}\cdot\text{K}^{-1})(298\text{K})]\,(11.2) = 2.77 \times 10^4$

(b) If NO_2 is added, the equilibrium will shift to produce more N_2O_4. The amount of NO_2 will be greater than initially present, but less than the $3.13 \text{ mol}\cdot\text{L}^{-1}$ present immediately upon making the addition. K_C will not be affected.

(c)

Concentrations ($\text{mol}\cdot\text{L}^{-1}$)	N_2O_4	$\rightarrow$	$2\,NO_2$
initial	0.405		3.13
change	$+x$		$-2x$
final	$0.405 + x$		$3.13 - 2x$

$$11.2 = \frac{(3.13 - 2x)^2}{0.405 + x}$$

$$(11.2)(0.405 + x) = (3.13 - 2x)^2$$

$$11.2x + 4.536 = 4x^2 - 12.52x + 9.797$$

$$4x^2 - 23.7x + 5.26 = 0$$

$$x = \frac{-(-23.7) \pm \sqrt{(-23.7)^2 - (4)(4)(5.26)}}{(2)(4)} = \frac{23.7 \pm 21.9}{8}$$

$$x = 5.70 \text{ or } 0.23$$

At equilibrium, $[N_2O_4] = 0.405 \text{ mol} \cdot L^{-1} + 0.23 \text{ mol} \cdot L^{-1} = 0.64 \text{ mol} \cdot L^{-1}$

$[NO_2] = 3.13 \text{ mol} \cdot L^{-1} - 2(0.23 \text{ mol} \cdot L^{-1}) = 2.67 \text{ mol} \cdot L^{-1}$.

$$\frac{[NO_2]^2}{[N_2O_4]} = \frac{(2.67)^2}{0.64} = 11.1 \approx K_C$$

These concentrations are consistent with the predictions in (b).

5.51 To find the vapor pressure, we first calculate $\Delta G°$ for the conversion of the liquid to the gas at 298 K, using the free energies of formation found in the appendix:

$$\Delta G°_{H_2O(l) \to H_2O(g)} = \Delta G°_{f(H_2O(g))} - \Delta G°_{f(H_2O(l))}$$

$$= (-228.57 \text{ kJ} \cdot \text{mol}^{-1}) - [-237.13 \text{ kJ} \cdot \text{mol}^{-1}]$$

$$= 8.56 \text{ kJ} \cdot \text{mol}^{-1}$$

$$\Delta G°_{D_2O(l) \to D_2O(g)} = \Delta G°_{f(D_2O(g))} - \Delta G°_{f(D_2O(l))}$$

$$= (-234.54 \text{ kJ} \cdot \text{mol}^{-1}) - [-243.44 \text{ kJ} \cdot \text{mol}^{-1}]$$

$$= 8.90 \text{ kJ} \cdot \text{mol}^{-1}$$

The equilibrium constant for these processes is the vapor pressure of the liquid:

$$K = P_{H_2O} \text{ or } K = P_{D_2O}$$

Using $\Delta G° = -RT \ln K$, we can calculate the desired values.

For H_2O:

$$\ln K = -\frac{\Delta G°}{RT} = -\frac{8560 \text{ J} \cdot \text{mol}^{-1}}{(8.314 \text{ J} \cdot \text{K}^{-1} \cdot \text{mol}^{-1})(298 \text{ K})} = -3.45$$

$$K = 0.032 \text{ bar}$$

$$0.032 \text{ bar} \times \frac{1 \text{ atm}}{1.013\,25 \text{ bar}} \times \frac{760 \text{ Torr}}{1 \text{ atm}} = 24 \text{ Torr}$$

For D_2O:

$$\ln K = -\frac{\Delta G°}{RT} = -\frac{8900 \text{ J} \cdot \text{mol}^{-1}}{(8.314 \text{ J} \cdot \text{K}^{-1} \cdot \text{mol}^{-1})(298 \text{ K})} = -3.59$$

$$K = 0.028 \text{ bar}$$

$$0.028 \text{ bar} \times \frac{1 \text{ atm}}{1.013\,25 \text{ bar}} \times \frac{760 \text{ Torr}}{1 \text{ atm}} = 21 \text{ Torr}$$

The answer is that D has a lower zero point energy than H. This makes the $D_2O—D_2O$ "hydrogen bond" stronger than the $H_2O—H_2O$ hydrogen bond. Because the hydrogen bond is stronger, the intermolecular forces are stronger, the vapor pressure is lower, and the boiling point is higher. Potential energy curves for the O—H and O—D bonds as a function of distance:

ΔE = energy required to break O—H or O—D bond

5.53

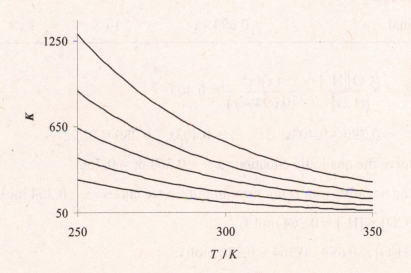

From the plot, it is clear that the larger the equilibrium constant, the more sensitive to the change of temperature.

5.55 (a) $\Delta G_r^o = -RT \ln K$; $\ln K = -\dfrac{\Delta G_r^o}{RT}$

$\Delta G_r^o = (-191.28 \text{ kJ} \cdot \text{mol}^{-1}) - (-198.08 \text{ kJ} \cdot \text{mol}^{-1}) = 6.8 \text{ kJ} \cdot \text{mol}^{-1}$

$\ln K = -\dfrac{6800 \text{ J} \times \text{mol}^{-1}}{(8.314 \text{ J} \times \text{K}^{-1} \times \text{mol}^{-1})(900 \text{ K})} = -0.909;$

$K_C = 0.403$

Since K_C is quite small, there will not be large amount of hydrogen gas produced when the reaction reaches to equilibrium. To increase the amount of hydrogen gas production, temperature needs to be increased.

(b) $5.20 \times 10^3 \text{ g C} \times \dfrac{1 \text{ mol C}}{12.011 \text{ g C}} = 433 \text{ mol C}$

$125 \text{ g H}_2\text{O} \times \dfrac{1 \text{ mol H}_2\text{O}}{18.016 \text{ g H}_2\text{O}} = 6.94 \text{ mol H}_2\text{O}$

H_2O is limiting. Conc. of $H_2O = 6.94 \text{ mol}/10.0 \text{ L} = 0.694 \text{ mol} \cdot \text{L}^{-1}$

Concentration $(\text{mol} \cdot \text{L}^{-1})$	$H_2O\,(g)$	$CO\,(g)$	$H_2\,(g)$
initial	0.694	0	0
change	$-x$	$+x$	$+x$
final	$0.694 - x$	$+x$	$+x$

$K_C = \dfrac{[CO][H_2]}{[H_2O]} = \dfrac{(x)(x)}{(0.694 - x)} = 0.403$

$x^2 = 0.280 - 0.403x;$ $x^2 + 0.403x - 0.280 = 0$

Solve the quadratic equation: $x = +0.364$ or -0.766

The negative root is not meaningful, so we choose $x = 0.364 \text{ mol} \cdot \text{L}^{-1}$

$[CO] = [H_2] = 0.364 \text{ mol} \cdot \text{L}^{-1}$

$[H_2O] = 0.694 - 0.364 = 0.330 \text{ mol} \cdot \text{L}^{-1}$

5.57 (a) (i) Increasing the amount of a reactant will push the equilibrium toward the products. More NO_2 will form. (ii) Removing a product will pull the equilibrium toward products. More NO_2 will form. (iii) Increasing total pressure by adding an inert gas not change the relative partial pressures. There will be no change in the amount of NO_2. (b) Since there are two moles of gas on both sides of the reaction, the volume cancels out of the equilibrium constant expression and it is possible to use moles directly for each component.

At equilibrium, $SO_3 = 0.245$ moles $- 0.240$ moles $= 0.005$ moles, $SO_2 = 0.240$ moles and $NO_2 = 0.240$ moles since the reaction is 1:1:1:1. The original number of moles of NO is set to x.

$K = 6.0 \times 10^3 = (0.240)(0.240)/(0.005)(x - 0.240) = (11.52)/(x - 0.240)$

$x = (11.52/6.0 \times 10^3) + 0.240$

$x = 0.242$ moles of NO

5.59 (a) Using the thermodynamic data in Appendix 2A:

$2\ H_2S(g) + SO_2(g) \leftrightarrow 3\ S(s) + 2\ H_2O(g)$

$\Delta G^o = 2(-228.57\ kJ \cdot mol^{-1}) - [2(-33.56\ kJ \cdot mol^{-1}) + (-300.19\ kJ \cdot mol^{-1})]$

$= -89.83\ kJ \cdot mol^{-1}$

$\ln K = -\dfrac{\Delta G^0}{RT} = -\dfrac{-89.83 \times 10^3\ J \cdot mol^{-1}}{(8.314 J \cdot K^{-1} \cdot mol^{-1})(298 K)} = 36.26$

$K = 5.6 \times 10^{15}$

Part (b) The effect of each change to the equilibrium is:

(a) S is solid and already present in excess in the original equilibrium. Adding more solid will not affect the equilibrium so the amount of H_2O will not change.

(b) According to Le Chatelier's principle, increasing the concentration of H_2S will cause the reaction to form more products in order to reduce the concentration of H_2S; some of the SO_2 will be consumed. Therefore, the amount of SO_2 will decrease.

(c) Removing a product will shift the reaction toward the formation of more products; the amount of SO_2 should decrease.

(d) Removing the reactant SO_2 will cause the reaction to shift toward the production of more reactants; the amount of S should decrease since it will be used in the production of more reactants.

(e) The equilibrium constant will be unaffected by changes in the concentrations of any of the species present.

(f) Decreasing volume (or increase pressure) will cause the reaction to shift to the side which has less number of gas molecules. In this specific case, the reaction will shift toward the product side. The amount of SO_2 should decrease.

(g) Based on the expression in (a), increasing temperature will decrease equilibrium constant, shifting reaction to the reactant side. The amount of SO_2 should increase.

5.61   (a) According to Le Chatelier's principle, adding a product should cause a shift in the equilibrium toward the reactants side of the equation.

(b) Because there are equal numbers of moles of gas on both sides of the equation, there will be little or no effect on compressing the system.

(c) If the amount of CO_2 is increased, this will cause the reaction to shift toward the formation of products.

(d) Because the reaction is endothermic, raising the temperature will favor the formation of products.

(e) If the amount of $C_6H_{12}O_6$ is removed, this will cause the reaction to shift toward the formation of products.

(f) Because water is a liquid, it is by definition present at unit concentration, so changing the amount of water will not affect the reaction. As long as the glucose solution is dilute, its concentration can be considered unchanged.

(g) Decreasing the partial pressure of a reactant (CO_2) will favor the production of more reactants.

5.63 (a) The hybridization of the carbon atoms in citric acid are:

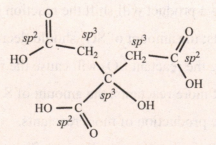

(b) yes; all four −OH groups present can participate in hydrogen bonding; the remaining three oxygens can accept H−bonds.

(c) since citric acid can hydrogen bond with itself it should have large intermolecular forces and should be a solid. The ability to both donate and accept hydrogen bonds should also make it soluble in water.

(d) 100.0 g of a 0.9% NaCl solution will contain 0.9 g of NaCl (MM = 58.44 g/mol), which is 1.54×10^{-2} mol of NaCl. If NaCl completely dissociates, the moles of solute present is 3.08×10^{-2} mol of solute in 100.0 mL, making the concentration of normal saline 0.3 M.

(e) To make a 500.0 mL sports drink isotonic, you need it to have a total molarity of 0.3 M (from part d). This means that the amount of solute needed is: 0.3 $M \times$ 0.5000 L = 0.15 mol total solute. Addition of 1.0 g of NaCl to the 500 mL sports drink will account for 0.034 mol of solute (twice the moles of NaCl added), leaving 0.116 mol of glucose being needed to achieve isotonicity. Since glucose has a molar mass of 180 g/mol, this means that in addition to the 1.0 g of NaCl, 21 g of glucose needs to be added to the 500.0 mL sports drink.

(f) 300.0 mL of a 1.00% boric acid = 300.0 g of solution (assuming the density of the solution is 1.00 g cm^{-3}), meaning there are 3.00 g of $B(OH)_3$ or 4.85×10^{-2} mol of $B(OH)_3$ in the solution (since $M_{boric\ acid}$ = 61.81 g/mol). To be isotonic, 300.0 mL should have 0.3 $M \times$ (0.3000L) = 0.09 mol total solute. Subtracting out the amount of boric acid present means that 0.04 mol of solute needed to be added to achieve isotonicity; since 1 mol of NaCl provides 2 mol of solute, 0.02 mol, or 1.17 g, of NaCl must be added to the boric acid solution.

FOCUS 6
REACTIONS

6A.1 (a) $CH_3NH_3^+$; (b) $NH_2NH_3^+$; (c) H_2CO_3; (d) CO_3^{2-};

(e) $C_6H_5O^-$; (f) $CH_3CO_2^-$

6A.3 For all parts (a) to (e), H_2O and H_3O^+ form a conjugate acid-base pair in which H_2O is the base and H_3O is the acid.

(a) $H_2SO_4(aq) + H_2O(l) \rightleftharpoons H_3O^+(aq) + HSO_4^-(aq)$

H_2SO_4 and $HSO_4^-(aq)$ form a conjugate acid-base pair in which

H_2SO_4 is the acid and $HSO_4^-(aq)$ is the base.

(b) $C_6H_5NH_3^+(aq) + H_2O(l) \rightleftharpoons H_3O^+(aq) + C_6H_5NH_2(aq)$

$C_6H_5NH_3^+$ and $C_6H_5NH_2(aq)$ form a conjugate acid-base pair in which

$C_6H_5NH_3^+$ is the acid and $C_6H_5NH_2(aq)$ is the base.

(c) $H_2PO_4^-(aq) + H_2O(l) \rightleftharpoons H_3O^+(aq) + HPO_4^{2-}(aq)$

$H_2PO_4^-(aq)$ and $HPO_4^{2-}(aq)$ form a conjugate acid-base pair in which

$H_2PO_4^-(aq)$ is the acid and $HPO_4^{2-}(aq)$ is the base.

(d) $HCOOH(aq) + H_2O(l) \rightleftharpoons H_3O^+(aq) + HCO_2^-(aq)$

$HCOOH(aq)$ and $HCO_2^-(aq)$ form a conjugate acid-base pair in which

$HCOOH(aq)$ is the acid and $HCO_2^-(aq)$ is the base.

(e) $NH_2NH_3^+(aq) + H_2O(l) \rightleftharpoons H_3O^+(aq) + NH_2NH_2(aq)$

$NH_2NH_3^+(aq)$ and $NH_2NH_2(aq)$ form a conjugate acid-base pair in

which $NH_2NH_3^+(aq)$ is the acid and $NH_2NH_2(aq)$ is the base.

6A.5 (a) Brønsted acid: HNO_3; Brønsted base: HPO_4^{2-}

(b) conjugate base to HNO_3: NO_3^-

conjugate acid to HPO_4^{2-}: $H_2PO_4^-$

6A.7 (a) $HClO_3$ (chloric acid); conjugate base: ClO_3^-

(b) HNO_2 (nitrous acid); conjugate base: NO_2^-

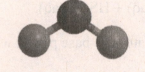

6A.9 (a) proton transferred from NH_4^+ to H_2O, NH_4^+ (acid), H_2O (base);

(b) proton transferred from NH_4^+ to I^-, NH_4^+ (acid), I^- (base);

(c) no proton transferred;

(d) proton transferred from NH_4^+ to NH_2^-, NH_4^+ (acid), NH_2^- (base)

6A.11 (a) HCO_3^- as an acid: $HCO_3^-(aq) + H_2O(l) \rightleftharpoons H_3O^+(aq) + CO_3^{2-}(aq)$,

HCO_3^- (acid) and CO_3^{2-} (base), H_2O (base) and H_3O^+ (acid);

HCO_3^- as a base: $H_2O(l) + HCO_3^-(aq) \rightleftharpoons H_2CO_3(aq) + OH^-(aq)$,

HCO_3^- (base) and H_2CO_3 (acid), H_2O (acid) and OH^- (base)

(b) HPO_4^{2-} as an acid: $HPO_4^{2-}(aq) + H_2O(l) \rightleftharpoons H_3O^+(aq) + PO_4^{3-}(aq)$,

HPO_4^{2-} (acid) and PO_4^{3-} (base), H_2O (base) and H_3O^+ (acid);

HPO_4^{2-} as a base: $HPO_4^{2-}(aq) + H_2O(l) \rightleftharpoons H_2PO_4^-(aq) + OH^-(aq)$,

HPO_4^{2-} (base) and $H_2PO_4^-$ (acid), H_2O (acid) and OH^- (base)

6A.13 The Lewis structures of (a) to (e) are as follows:

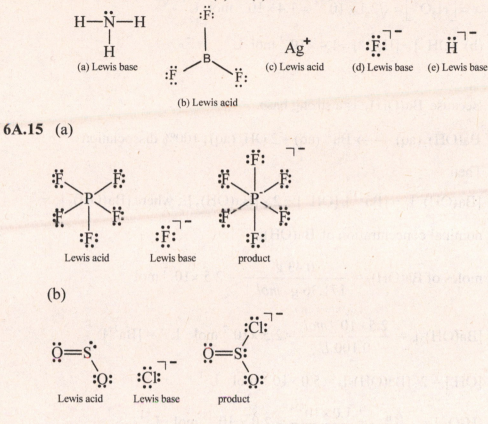

6A.15 (a)

Lewis acid   Lewis base   product

(b)

Lewis acid   Lewis base   product

6A.17 (a) basic; (b) acidic; (c) amphoteric; (d) amphoteric

6A.19 In each case, use $K_w = [H_3O^+][OH^-] = 1.0 \times 10^{-14}$, then

$$[OH^-] = \frac{K_w}{[H_3O^+]} = \frac{1.0 \times 10^{-14}}{[H_3O^+]}.$$

(a) $[OH^-] = \dfrac{1.0 \times 10^{-14}}{0.02} = 5.0 \times 10^{-13} \text{ mol} \cdot \text{L}^{-1}$

(b) $[OH^-] = \dfrac{1.0 \times 10^{-14}}{1.0 \times 10^{-5}} = 1.0 \times 10^{-9} \text{ mol} \cdot \text{L}^{-1}$

(c) $[OH^-] = \dfrac{1.0 \times 10^{-14}}{3.1 \times 10^{-3}} = 3.2 \times 10^{-12} \text{ mol} \cdot \text{L}^{-1}$

6A.21 (a) $K_w = 2.1 \times 10^{-14} = [H_3O^+][OH^-] = x^2$, where $x = [H_3O^+] = [OH^-]$

$$x = \left[H_3O^+ \right] = \sqrt{2.1 \times 10^{-14}} = 1.4 \times 10^{-7} \text{ mol} \cdot L^{-1}$$

(b) $[OH^-] = [H_3O^+] = 1.4 \times 10^{-7} \text{ mol} \cdot L^{-1}$

6A.23 Because $Ba(OH)_2$ is a strong base,

$$Ba(OH)_2(aq) \longrightarrow Ba^{2+}(aq) + 2\,OH^-(aq), \text{ 100\% dissociation}$$

Then

$[Ba(OH)_2]_0 = [Ba^{2+}]$, $[OH^-] = 2 \times [Ba(OH)_2]_0$, where $[Ba(OH)_2]_0 =$

nominal concentration of $Ba(OH)_2$.

$$\text{moles of } Ba(OH)_2 = \frac{0.43 \text{ g}}{171.36 \text{ g} \cdot mol^{-1}} = 2.5 \times 10^{-3} \text{ mol}$$

$$[Ba(OH)_2]_0 = \frac{2.5 \times 10^{-3} \, mol}{0.100 \, L} = 2.5 \times 10^{-2} \text{ mol} \cdot L^{-1} = [Ba^{2+}]$$

$$[OH^-] = 2 \times [Ba(OH)_2]_0 = 5.0 \times 10^{-2} \text{ mol} \cdot L^{-1}$$

$$[H_3O^+] = \frac{K_W}{[OH^-]} = \frac{1.0 \times 10^{-14}}{5.0 \times 10^{-2}} = 2.0 \times 10^{-13} \text{ mol} \cdot L^{-1}$$

6B.1 HCl is a strong acid. If concentration decreases to 12% of its initial value

(i.e., $0.12 \times [HCl]_0$), the pH change $(\Delta pH) = -\log(\frac{0.12 \times [HCl]_0}{[HCl]_0}) = 0.92$

(i.e., pH will increase by 0.92).

6B.3 (a) The pH of the desired solution $= -\log(0.025) = 1.6$

(b) The actual pH $= -\log(\frac{200.0 \text{ mL} \times 0.025 \text{ M}}{250.0}) = 1.7$

6B.5 (a) $[HNO_3] = [H_3O^+] = 0.0146 \text{ mol} \cdot L^{-1}$

$pH = -\log(0.0146) = 1.84$, $pOH = 14.00 - (-1.84) = 12.16$

(b) $[HCl] = [H_3O^+] = 0.11 \text{ mol} \cdot L^{-1}$

$pH = -\log(0.11) = 0.96,\quad pOH = 14.00 - 0.96 = 13.04$

(c) $[OH^-] = 2 \times [Ba(OH)_2] = 2 \times 0.0092\ \text{M} = 0.018\ \text{mol} \cdot L^{-1}$

$pOH = -\log(0.018) = 1.74,\quad pH = 14.00 - 1.74 = 12.26$

(d) $[KOH]_0 = [OH^-]$

$[OH^-] = \left(\dfrac{2.00\ \text{mL}}{500\ \text{mL}}\right) \times (0.175\ \text{mol} \cdot L^{-1}) = 7.0 \times 10^{-4}\ \text{mol} \cdot L^{-1}$

$pOH = -\log(7.0 \times 10^{-4}) = 3.15,\quad pH = 14.00 - 3.15 = 10.85$

(e) $[NaOH]_0 = [OH^-]$

$\text{number of moles of NaOH} = \dfrac{0.0136\ \text{g}}{40.00\ \text{g} \cdot \text{mol}^{-1}} = 3.40 \times 10^{-4}\ \text{mol}$

$[NaOH]_0 = \dfrac{3.40 \times 10^{-4}\ \text{mol}}{0.350\ \text{L}} = 9.71 \times 10^{-4}\ \text{mol} \cdot L^{-1} = [OH^-]$

$pOH = -\log(9.71 \times 10^{-4}) = 3.01,\quad pH = 14.00 - 3.01 = 10.99$

(f) $[HBr]_0 = [H_3O^+]$

$[H_3O^+] = \left(\dfrac{75.0\ \text{mL}}{500\ \text{mL}}\right) \times (3.5 \times 10^{-4}\ \text{mol} \cdot L^{-1}) = 5.2 \times 10^{-5}\ \text{mol} \cdot L^{-1}$

$pH = -\log(5.3 \times 10^{-5}) = 4.28,\quad pOH = 14.00 - 4.28 = 9.72$

6B.7 Because $pH = -\log[H_3O^+]$, $\log[H_3O^+] = -pH$. Taking the antilogs of both sides gives $[H_3O^+] = 10^{-pH}\ \text{mol} \cdot L^{-1}$.

(a) $[H_3O^+] = 10^{-3.3} = 5 \times 10^{-4}\ \text{mol} \cdot L^{-1}$

(b) $[H_3O^+] = 10^{-6.7}\ \text{mol} \cdot L^{-1} = 2 \times 10^{-7}\ \text{mol} \cdot L^{-1}$

(c) $[H_3O^+] = 10^{-4.4}\ \text{mol} \cdot L^{-1} = 4 \times 10^{-5}\ \text{mol} \cdot L^{-1}$

(d) $[H_3O^+] = 10^{-5.3}\ \text{mol} \cdot L^{-1} = 5 \times 10^{-6}\ \text{mol} \cdot L^{-1}$

6B.9 (a)

	$[H_3O^+]$ $(mol \cdot L^{-1})$	$[OH^-]$ $(mol \cdot L^{-1})$	pH	pOH
(i)	**1.50**	1.50×10^{-14}	0.176	13.824
(ii)	1.50×10^{-14}	**1.50**	13.824	0.176
(iii)	0.18	5.6×10^{-14}	**0.75**	13.25
(iv)	5.6×10^{-14}	0.18	13.25	**0.75**

(b) (ii) < (iv) < (iii) < (i)

6B.11 (a) (i) in diluted solution:

pH = 13.25, pOH = 0.75; $[OH^-] = 10^{-0.75} = 0.18$ mol $\cdot L^{-1}$

(ii) $[OH^-]_0 = 0.18 \times \left(\dfrac{500.0 \, mL}{5.00 \, mL} \right) = 18$ mol $\cdot L^{-1}$

(b) The reaction is $Na_2O(s) + H_2O(l) \rightarrow 2 \, NaOH(aq)$

The mass of Na_2O is: 18 mol $\cdot L^{-1} \times (0.200 \, L) \times \left(\dfrac{1 \, mol \, Na_2O}{2 \, mol \, NaOH} \right)$

$\times \left(\dfrac{61.98 \, g \, Na_2O}{1 \, mol \, Na_2O} \right) = 110.$ grams Na_2O.

6C.1 (i) $HClO_2$

(a) $HClO_2(aq) + H_2O(l) \rightleftharpoons H_3O^+(aq) + ClO_2^-(aq)$

$K_a = \dfrac{[H_3O^+][ClO_2^-]}{[HClO_2]}$

(b) $ClO_2^-(aq) + H_2O(l) \rightleftharpoons HClO_2(aq) + OH^-(aq)$

$K_b = \dfrac{[HClO_2][OH^-]}{[ClO_2^-]}$

(ii) HCN

(a) $HCN(aq) + H_2O(l) \rightleftharpoons H_3O^+(aq) + CN^-(aq)$

$$K_a = \frac{[H_3O^+][CN^-]}{[HCN]}$$

(b) $CN^-(aq) + H_2O(l) \rightleftharpoons HCN(aq) + OH^-(aq)$

$$K_b = \frac{[HCN][OH^-]}{[CN^-]}$$

(iii) C_6H_5OH

(a) $C_6H_5OH(aq) + H_2O(l) \rightleftharpoons H_3O^+(aq) + C_6H_5O^-(aq)$

$$K_a = \frac{[H_3O^+][C_6H_5O^-]}{[C_6H_5OH]}$$

(b) $C_6H_5O^-(aq) + H_2O(l) \rightleftharpoons C_6H_5OH(aq) + OH^-(aq)$

$$K_b = \frac{[C_6H_5OH][OH^-]}{[C_6H_5O^-]}$$

6C.3 $pK_{a1} = -\log K_{a1}$; therefore, after taking antilogs, $K_{a1} = 10^{-pK_{a1}}$.

	Acid	pK_{a1}	K_{a1}
(i)	H_3PO_4	2.12	7.6×10^{-3}
(ii)	H_3PO_3	2.00	1.0×10^{-2}
(iii)	H_2SeO_3	2.46	3.5×10^{-3}
(iv)	$HSeO_4^-$	1.92	1.2×10^{-2}

(b) The larger K_{a1}, the stronger the acid; therefore

$$H_2SeO_3 < H_3PO_4 < H_3PO_3 < HSeO_4^-.$$

6C.5 The conjugated base for HCOOH is: $HCOO^-$

$$pK_b = pK_w - pK_a = 14.0 - 3.75 = 10.25$$

6C.7 Decreasing pK_a will correspond to increasing acid strength because

$pK_a = -\log K_a$. The pK_a values (given in parentheses) determine the

following ordering:

$(CH_3)_2 NH_2^+$ $(14.00 - 3.27 = 10.73) <$ $^+NH_3OH$ $(14.00 - 7.97 = 6.03)$

$< HNO_2$ $(3.37) < HClO_2$ (2.00).

Remember that the pK_a for the conjugate acid of a weak base will be

given by $pK_a + pK_b = 14$.

6C.9 Decreasing pK_b will correspond to increasing base strength because

$pK_b = -\log K_b$. The pK_b values (given in parentheses) determine the

following ordering:

F^- $(14.00 - 3.45 = 10.55) < CH_3COO^-$ $(14.00 - 4.75 = 9.25)$

$< C_5H_5N$ $(8.75) \ll NH_3$ (4.75).

Remember that the pK_b for the conjugate base of a weak acid will be

given by $pK_a + pK_b = 14$.

6C.11 The larger the K_a, the stronger the corresponding acid.

2,4,6-Trichlorophenol is the stronger acid because the chlorine atoms have

a greater electron-withdrawing power than the hydrogen atoms present in

the unsubstituted phenol.

6C.13 The larger the pK_a of an acid, the stronger the corresponding conjugate

base; hence, the order is aniline < ammonia < methylamine < ethylamine.

Although we should not draw conclusions from such a small data set, we

might suggest the possibility that

(1) arylamines < ammonia < alkylamines

(2) methyl < ethyl < etc.

(Arylamines are amines in which the nitrogen of the amine is attached to a

benzene ring.)

6C.15 For oxoacids, the greater the number of highly electronegative O atoms attached to the central atom, the stronger the acid. This effect is related to the increased oxidation number of the central atom as the number of O atoms increases. Therefore, HIO_3 is the stronger acid, with the lower pK_a.

6C.17 The smaller the pK_b of a base, the stronger the base;
$pK_b(\text{morphine}) = 5.79$; $pK_b(BrO^-) = pK_w - pK_a = 14.0 - 8.69 = 5.31$
Hence, hypobromite ion is a stronger base.

6C.19 (a) HCl is the stronger acid, because its bond strength is much weaker than the bond in HF, and bond strength is the dominant factor in determining the strength of binary acids.

(b) $HClO_2$ is stronger; there is one more O atom attached to the Cl atom in $HClO_2$ than in HClO. The additional O in $HClO_2$ helps to pull the electron of the H atom out of the H—O bond. The oxidation state of Cl is higher in $HClO_2$ than in HClO.

(c) $HClO_2$ is stronger; Cl has a greater electronegativity than Br, making the H—O bond $HClO_2$ more polar than in $HBrO_2$.

(d) $HClO_4$ is stronger; Cl has a greater electronegativity than P.

(e) HNO_3 is stronger. The explanation is the same as that for part (b). HNO_3 has one more O atom.

(f) H_2CO_3 is stronger; C has greater electronegativity than Ge. See part (c).

6C.21 (a) The —CCl_3 group that is bonded to the carboxyl group, —COOH, in trichloroacetic acid, is more electron withdrawing than the —CH_3 group in acetic acid. Thus, trichloroacetic acid is the stronger acid.

(b) The —CH_3 group in acetic acid has electron-donating properties, which means that it is less electron withdrawing than the —H attached to the carboxyl group in formic acid, HCOOH. Thus, formic acid is a slightly stronger acid than acetic acid. However, it is not nearly as strong as trichloroacetic acid. The order is $CCl_3COOH \gg HCOOH > CH_3COOH$.

6D.1 (a) Concentration

(mol·L^{-1})	CH_3COOH	+ H_2O	$\leftrightarrow$ H_3O^+	+ $CH_3CO_2^-$
initial	0.20	—	0	0
change	$-x$	—	$+x$	$+x$
equilibrium	0.20 - x	—	x	x

$$K_a = 1.8 \times 10^{-5} = \frac{[H_3O^+][CH_3CO_2^-]}{[CH_3COOH]} = \frac{x^2}{0.20 - x} \approx \frac{x^2}{0.20}$$

$$x = [H_3O^+] = 1.9 \times 10^{-3} \text{ mol} \cdot L^{-1}$$

$$pH = -\log(1.9 \times 10^{-3}) = 2.72; \quad pOH = 14.00 - 2.72 = 11.28$$

$$\% \text{ deprotonation} = (\frac{1.9 \times 10^{-3}}{0.20}) \times 100 = 0.95\%$$

(b) The equilibrium table for (b) is similar to that for (a).

$$K_a = 3.0 \times 10^{-1} = \frac{[H_3O^+][CCl_3CO_2^-]}{[CCl_3COOH]} = \frac{x^2}{0.20 - x}$$

$$x^2 + 0.30\,x - 0.060 = 0; \quad x = 0.14, -0.44$$

The negative root is not possible and can be eliminated.

$$x = [H_3O^+] = 0.14 \text{ mol} \cdot L^{-1},$$

$$pH = -\log(0.14) = 0.85; \quad pOH = 14.00 - 0.85 = 13.15$$

$$\% \text{ deprotonation} = (\frac{0.14}{0.20}) \times 100 = 70\%$$

(c) The equilibrium table for (c) is similar to that for (a).

$$K_a = 1.8 \times 10^{-4} = \frac{[H_3O^+][HCO_2^-]}{[HCOOH]} = \frac{x^2}{0.20 - x} \approx \frac{x^2}{0.20}$$

$$x = [H_3O^+] = 6.0 \times 10^{-3} \text{ mol} \cdot L^{-1}$$

$pH = -\log(6.0 \times 10^{-3}) = 2.22;\ pOH = 14.00 - 2.72 = 11.78$

$\%\ \text{deprotonation} = (\dfrac{6.0 \times 10^{-3}}{0.20}) \times 100 = 3.0\%$

(d) Acidity increases when the hydrogen atoms in the methyl group of acetic acid are replaced by atoms that have a higher electronegativity, such as chlorine. In general, as the electron-withdrawing ability of the function group that attached to the carboxylic acid increases, the acidity of the carboxylic acid will increase.

6D.3 (a) $HClO_2 + H_2O \rightarrow H_3O^+ + ClO_2^-$

$[H_3O^+] = [ClO_2^-] = 10^{-pH} = 10^{-1.2} = 0.06\ mol \cdot L^{-1}$

$K_a = \dfrac{[H_3O^+][ClO_2^-]}{[HClO_2]} = \dfrac{(0.06)^2}{0.10 - 0.06} = 0.09\ (1\ sf)$

$pK_a = -\log(0.09) = 1.0$

(b) $C_3H_7NH_2 + H_2O \rightarrow C_3H_7NH_3^+ + OH^-$

$pOH = 14.00 - 11.86 = 2.14$

$[C_3H_7NH_3^+] = [OH^-] = 10^{-2.14} = 7.2 \times 10^{-3}\ mol \cdot L^{-1}$

$K_b = \dfrac{[C_3H_7NH_3^+][OH^-]}{[C_3H_7NH_2]} = \dfrac{(7.2 \times 10^{-3})^2}{0.10 - 7.2 \times 10^{-3}} = 5.6 \times 10^{-4}$

$pK_b = -\log(5.6 \times 10^{-4}) = 3.25$

6D.5 (a) Concentration

$(mol \cdot L^{-1})$	H_2O	$+$	NH_3	$\rightleftharpoons$	NH_4^+	$+$	OH^-
initial	—		0.057		0		0
change	—		$-x$		$+x$		$+x$
equilibrium	—		$0.057 - x$		x		x

$$K_b = \frac{[NH_4^+][OH^-]}{[NH_3]} = \frac{x \cdot x}{0.057 - x} \approx \frac{x^2}{0.057} = 1.8 \times 10^{-5}$$

$$x = [OH^-] = \sqrt{0.057 \times 1.8 \times 10^{-5}} = 1.0 \times 10^{-3} \text{ mol} \cdot L^{-1}$$

$$pOH = -\log(1.0 \times 10^{-3}) = 3.00, \; pH = 14.00 - 3.00 = 11.00$$

$$\text{percentage protonation} = \frac{1.0 \times 10^{-3}}{0.057} \times 100\% = 1.8\%$$

(b) Concentration

$(\text{mol} \cdot L^{-1})$	NH_2OH	$+$	H_2O	$\rightleftharpoons$	$^+NH_3OH$	$+$	OH^-
initial	0.162		—		0		0
change	$-x$		—		$+x$		$+x$
equilibrium	$0.162 - x$		—		x		x

$$K_b = 1.1 \times 10^{-8} = \frac{x^2}{0.162 - x} \approx \frac{x^2}{0.162}$$

$$x = [OH^-] = 4.2 \times 10^{-5} \text{ mol} \cdot L^{-1}$$

$$pOH = -\log(4.2 \times 10^{-5}) = 4.38, \; pH = 14.00 - 4.38 = 9.62$$

$$\text{percentage protonation} = \frac{4.2 \times 10^{-5}}{0.162} \times 100\% = 0.026\%$$

(c) Concentration

$(\text{mol} \cdot L^{-1})$	$(CH_3)_3N$	$+$	H_2O	$\rightleftharpoons$	$(CH_3)_3NH^+$	$+$	OH^-
initial	0.35		—		0		0
change	$-x$		—		$+x$		$+x$
equilibrium	$0.35 - x$		—		$+x$		$+x$

$$6.5 \times 10^{-5} = \frac{x^2}{0.35 - x}$$

Assume $x \ll 0.35$

Then $x = 4.8 \times 10^{-3} \text{ mol} \cdot L^{-1}$

$$[OH^-] = 4.8 \times 10^{-3} \text{ mol} \cdot L^{-1}$$

$$pOH = -\log(4.8 \times 10^{-3}) = 2.32, \; pH = 14.00 - 2.32 = 11.68$$

$$\text{percentage protonation} = \frac{4.8 \times 10^{-3}}{0.35} \times 100\% = 1.4\%$$

(d) $pK_b = 14.00 - pK_a = 14.00 - 8.21 = 5.79, \; K_b = 1.6 \times 10^{-6}$

codeine + $H_2O \rightleftharpoons$ codeineH$^+$ + OH$^-$

$$K_b = 1.6 \times 10^{-6} = \frac{x^2}{0.0073 - x} \approx \frac{x^2}{0.0073}$$

$x = [OH^-] = 1.1 \times 10^{-4} \text{ mol} \cdot L^{-1}$

pOH $= -\log(1.1 \times 10^{-4}) = 3.96$, pH $= 14.00 - 3.96 = 10.04$

$$\text{percentage protonation} = \frac{1.1 \times 10^{-4}}{0.0073} \times 100\% = 2.5\%$$

6D.7 (a) pH = 4.60, $[H_3O^+] = 10^{-pH} = 10^{-4.60} = 2.5 \times 10^{-5} \text{ mol} \cdot L^{-1}$

Let x = nominal concentration of HClO, then

Concentration

(mol $\cdot$ L^{-1})	HClO	+	H$_2$O	$\rightleftharpoons$	H$_3$O$^+$	+	ClO$^-$
nominal	x		—		0		0
equilibrium	$x - 2.5 \times 10^{-5}$		—		2.5×10^{-5}		2.5×10^{-5}

$$K_a = 3.0 \times 10^{-8} = \frac{(2.5 \times 10^{-5})^2}{x - 2.5 \times 10^{-5}}$$

Solve for x; $x = \dfrac{(2.5 \times 10^{-5})^2 + (2.5 \times 10^{-5})(3.0 \times 10^{-8})}{3.0 \times 10^{-8}}$

$$= 2.1 \times 10^{-2} \text{ mol} \cdot L^{-1} = 0.021 \text{ mol} \cdot L^{-1}$$

(b) pOH $= 14.00 - $ pH $= 14.00 - 10.20 = 3.80$

$[OH^-] = 10^{-pOH} = 10^{-3.80} = 1.6 \times 10^{-4}$

Let x = nominal concentration of NH$_2$NH$_2$, then

Concentration

(mol $\cdot$ L^{-1})	NH$_2$NH$_2$	+	H$_2$O	$\rightleftharpoons$	NH$_2$NH$_3^+$	+	OH$^-$
nominal	x		—		0		0
equilibrium	$x - 1.6 \times 10^{-4}$		—		1.6×10^{-4}		1.6×10^{-4}

$$K_b = 1.7 \times 10^{-6} = \frac{(1.6 \times 10^{-4})^2}{x - 1.6 \times 10^{-4}}$$

Solve for x; $x = 1.5 \times 10^{-2} \text{ mol} \cdot L^{-1}$

6D.9 Concentration

(mol·L^{-1})	C$_6$H$_5$COOH	+ H$_2$O	$\rightleftharpoons$	H$_3$O$^+$	+ C$_6$H$_5$CO$_2^-$
initial	0.110	—		0	0
change	$-x$	—		$+x$	$+x$
equilibrium	0.110 $- x$	—		x	x

$x = 0.024 \times 0.110 \text{ mol} \cdot \text{L}^{-1} = [\text{H}_3\text{O}^+] = [\text{C}_6\text{H}_5\text{CO}_2^-]$

$K_a = \dfrac{[\text{H}_3\text{O}^+][\text{C}_6\text{H}_5\text{COO}^-]}{[\text{C}_6\text{H}_5\text{COOH}]} = \dfrac{(0.024 \times 0.110)^2}{(1 - 0.024) \times 0.110} = 6.5 \times 10^{-5}$

$\text{pH} = -\log(2.6 \times 10^{-3}) = 2.58$

6D.11 (a) less than 7, $\text{NH}_4^+(\text{aq}) + \text{H}_2\text{O}(\text{l}) \rightleftharpoons \text{H}_3\text{O}^+(\text{aq}) + \text{NH}_3(\text{aq})$

(b) greater than 7, $\text{H}_2\text{O}(\text{l}) + \text{CO}_3^{2-}(\text{aq}) \rightleftharpoons \text{HCO}_3^-(\text{aq}) + \text{OH}^-(\text{aq})$

(c) greater than 7, $\text{H}_2\text{O}(\text{l}) + \text{F}^-(\text{aq}) \rightleftharpoons \text{HF}(\text{aq}) + \text{OH}^-(\text{aq})$

(d) neutral

(e) less than 7,

$\text{Al}(\text{H}_2\text{O})_6^{3+}(\text{aq}) + \text{H}_2\text{O}(\text{l}) \rightleftharpoons \text{H}_3\text{O}^+(\text{aq}) + \text{Al}(\text{H}_2\text{O})_5\text{OH}^{2+}(\text{aq})$

(f) less than 7,

$\text{Cu}(\text{H}_2\text{O})_6^{2+}(\text{aq}) + \text{H}_2\text{O}(\text{l}) \rightleftharpoons \text{H}_3\text{O}^+(\text{aq}) + \text{Cu}(\text{H}_2\text{O})_5\text{OH}^+(\text{aq})$

6D.13 The rank in increasing solution pH is:

(c) < (a) < (b) < (d)

Justification: (d) is a weak base, pH will be the highest (pH = 8.97)

(c) is a weak acid but the concentration high (pH = 2.72), which gives this

solution the lowest pH; (a) is a strong acid, but the concentration is low

(pH = $-\log(1.0 \times 10^{-5})$ = 5.00); (b) is a conjugated acid of CH$_3$NH$_2$ (K_b =

3.6×10^{-4}) with a $K_a = 2.8 \times 10^{-11}$, which is a very weak acid (pH = 5.63).

6D.15 (a) $K_a = \dfrac{K_w}{K_b} = \dfrac{1.00 \times 10^{-14}}{1.8 \times 10^{-5}} = 5.6 \times 10^{-10}$

Concentration (mol·L^{-1})	NH$_4^+$(aq)	+ H$_2$O(l)	$\rightleftharpoons$	H$_3$O$^+$(aq)	+ NH$_3$(aq)
initial	0.19	—		0	0
change	$-x$	—		$+x$	$+x$
equilibrium	$0.19 - x$	—		x	x

$K_a = \dfrac{[H_3O^+][NH_3]}{[NH_4Cl]} = 5.6 \times 10^{-10} = \dfrac{x^2}{0.19 - x} \approx \dfrac{x^2}{0.19}$

$x = 1.0 \times 10^{-5} \text{ mol·L}^{-1} = [H_3O^+]$

$pH = -\log(1.0 \times 10^{-5}) = 5.00$

(b) Concentration

(mol·L^{-1}) Al(H$_2$O)$_6^{3+}$(aq)	+ H$_2$O(l)	$\rightleftharpoons$	H$_3$O$^+$(aq)	+ Al(H$_2$O)$_5$OH^{2+}(aq)
initial 0.055	—		0	0
change $-x$	—		$+x$	$+x$
equilibrium $0.055 - x$	—		x	x

$K_a = \dfrac{[H_3O^+][Al(H_2O)_5OH^{2+}]}{[Al(H_2O)_6^{3+}]} = 1.4 \times 10^{-5} = \dfrac{x^2}{0.055 - x} \approx \dfrac{x^2}{0.055}$

$x = 8.8 \times 10^{-4} \text{ mol·L}^{-1} = [H_3O^+]$

$pH = -\log(8.8 \times 10^{-4}) = 3.06$

6D.17 (a) $K_b = \dfrac{K_w}{K_a} = \dfrac{1.00 \times 10^{-14}}{1.8 \times 10^{-5}} = 5.6 \times 10^{-10}$

Concentration

(mol·L^{-1}) CH$_3$CO$_2^-$(aq)	+ H$_2$O(l)	$\rightleftharpoons$	HCH$_3$CO$_2$(aq)	+ OH$^-$(aq)
initial 0.63	—		0	0
change $-x$	—		$+x$	$+x$
equilibrium $0.63 - x$	—		x	x

$K_b = \dfrac{[HCH_3CO_2][OH^-]}{[CH_3CO_2^-]} = 5.6 \times 10^{-10} = \dfrac{x^2}{0.63 - x} \approx \dfrac{x^2}{0.63}$

$x = 1.9 \times 10^{-5} = [OH^-], pOH = -\log(1.9 \times 10^{-5}) = 4.72$

$pH = 14.00 - pOH = 14.00 - 4.72 = 9.28$

(b) Concentration

$(mol \cdot L^{-1})$	$H_2O(l)$ +	$CN^-(aq)$	$\rightleftharpoons$	$HCN(aq)$ +	$OH^-(aq)$
initial	—	0.65		0	0
change	—	$-x$		$+x$	$+x$
equilibrium	—	$0.65 - x$		x	x

$K_b = \dfrac{K_w}{K_a} = \dfrac{1.00 \times 10^{-14}}{4.9 \times 10^{-10}} = 2.0 \times 10^{-5} = \dfrac{[HCN][OH^-]}{[CN^-]} = \dfrac{x^2}{0.65 - x} \approx \dfrac{x^2}{0.65}$

$x = [OH^-] = 3.6 \times 10^{-3} \, mol \cdot L^{-1}$

$pOH = -\log(3.6 \times 10^{-3}) = 2.44, pH = 11.56$

6D.19 Molarity of $CH_3NH_3Cl = \dfrac{15.5 \, g / (67.52 \, g \cdot mol^{-1})}{0.450 \, L} = 0.510 \, mol \times L^{-1}$

Concentration

$(mol \cdot L^{-1})$	$CH_3NH_3^+(aq)$ +	$H_2O(l)$	$\rightleftharpoons$	$H_3O^+(aq)$ +	$CH_3NH_2(aq)$
initial	0.510	—		0	0
change	$-x$	—		$+x$	$+x$
equilibrium	$0.510 - x$	—		x	x

$K_a = \dfrac{[H_3O^+][CH_3NH_2]}{[CH_3NH_3^+]} = 2.8 \times 10^{-11} = \dfrac{x^2}{0.510 - x} \approx \dfrac{x^2}{0.510}$

$x = 3.8 \times 10^{-6} \, mol \cdot L^{-1} = [H_3O^+]$

$pH = -\log(3.8 \times 10^{-6}) = 5.42$

6D.21 pH of the NaA is 10.35, $pOH = 3.65$, $[OH^-] = 10^{-3.65} = 2.2 \times 10^{-4}$

Hydrolysis reaction: $A^-(aq) + H_2O(l) \rightarrow HA(aq) + OH^-(aq)$

$K_b = \dfrac{[HA][OH^-]}{[A^-]} = \dfrac{(2.2 \times 10^{-4})^2}{(0.010)} = 4.84 \times 10^{-6}$

$K_a = K_w/K_b = 2.1 \times 10^{-9}$. The formula of the acid is HBrO.

6E.1 The initial concentrations of HSO_4^- and H_3O^+ are both 0.15 mol·L^{-1} as a result of the complete ionization of H_2SO_4 in the first step. The second ionization is incomplete.

Concentration

(mol·L^{-1})	HSO_4^-	+	H_2O	$\rightleftharpoons$	H_3O^+	+	SO_4^{2-}
initial	0.15		—		0.15		0
change	$-x$		—		$+x$		$+x$
equilibrium	$0.15 - x$		—		$0.15 + x$		x

$$K_{a2} = 1.2 \times 10^{-2} = \frac{[H_3O^+][SO_4^{2-}]}{[HSO_4^-]} = \frac{(0.15 + x)(x)}{0.15 - x}$$

$$x^2 + 0.162x - 1.8 \times 10^{-3} = 0$$

$$x = \frac{-0.162 + \sqrt{(0.162)^2 + (4)(1.8 \times 10^{-3})}}{2} = 0.0104 \text{ mol·L}^{-1}$$

$$[H_3O^+] = 0.15 + x = (0.15 + 0.0104) \text{ mol·L}^{-1} = 0.16 \text{ mol·L}^{-1}$$

$$pH = -\log(0.16) = 0.80$$

6E.3 (a) Because $K_{a2} \ll K_{a1}$, the second ionization can be ignored.

Concentration

(mol·L^{-1})	H_2CO_3	+	H_2O	$\rightleftharpoons$	H_3O^+	+	HCO_3^-
initial	0.010		—		0		0
change	$-x$		—		$+x$		$+x$
equilibrium	$0.010 - x$		—		x		x

$$K_{a1} = \frac{[H_3O^+][HCO_3^-]}{[H_2CO_3]} = \frac{x^2}{0.010 - x} \approx \frac{x^2}{0.010} = 4.3 \times 10^{-7}$$

$$x = [H_3O^+] = 6.6 \times 10^{-5} \text{ mol·L}^{-1}$$

$$pH = -\log(6.6 \times 10^{-5}) = 4.18$$

(b) Because $K_{a2} \ll K_{a1}$, the second ionization can be ignored.

Concentration

$$\begin{array}{cccccccc}
(mol \cdot L^{-1}) & (COOH)_2 & + & H_2O & \rightleftharpoons & H_3O^+ & + & (COOH)CO_2^- \\
\text{initial} & 0.10 & & — & & 0 & & 0 \\
\text{change} & -x & & — & & +x & & +x \\
\text{equilibrium} & 0.10-x & & — & & x & & x
\end{array}$$

$$K_{a1} = 5.9 \times 10^{-2} = \frac{[H_3O^+][(COOH)CO_2^-]}{[(COOH)_2]} = \frac{x^2}{0.10-x}$$

$$x^2 + 5.9 \times 10^{-2} x - 5.9 \times 10^{-3} = 0$$

$$x = \frac{-5.9 \times 10^{-2} + \sqrt{(5.9 \times 10^{-2})^2 + (4)(5.9 \times 10^{-3})}}{2} = 0.053 \ mol \cdot L^{-1}$$

$$pH = -\log(0.053) = 1.28$$

(c) Because $K_{a2} \ll K_{a1}$, the second ionization can be ignored.

Concentration $(mol \cdot L^{-1})$

$$\begin{array}{cccccccc}
 & H_2S & + & H_2O & \rightleftharpoons & H_3O^+ & + & HS^- \\
\text{equilibrium} & 0.20-x & & — & & x & & x
\end{array}$$

$$K_{a1} = 1.3 \times 10^{-7} = \frac{[H_3O^+][HS^-]}{[H_2S]} = \frac{x^2}{0.20-x} \approx \frac{x^2}{0.20}$$

$$x = [H_3O^+] = 1.6 \times 10^{-4} \ mol \cdot L^{-1}$$

$$pH = -\log(1.6 \times 10^{-4}) = 3.80$$

6E.5 (a) The pH is given by $pH = \frac{1}{2}(pK_{a1} + pK_{a2})$. From Table 6E.1, we find

$$K_{a1} = 1.5 \times 10^{-2} \quad pK_{a1} = 1.81$$
$$K_{a2} = 1.2 \times 10^{-7} \quad pK_{a2} = 6.91$$
$$pH = \frac{1}{2}(1.81 + 6.91) = 4.36$$

(b) The pH of a salt solution of a polyprotic acid is independent of the concentration of the salt; therefore, pH = 4.36.

6E.7 (a) The pH is given by $pH = \frac{1}{2}(pK_{a1} + pK_{a2})$. For the monosodium salt, the pertinent values are pK_{a1} and pK_{a2}:

$$pH = \frac{1}{2}(3.14 + 5.95) = 4.55$$

(b) For the disodium salt, the pertinent values are pK_{a2} and pK_{a3}:

$$pH = \tfrac{1}{2}(5.95 + 6.39) = 6.17$$

6E.9 The equilibrium reactions of interest are

$$H_2CO_3(aq) + H_2O(l) \rightleftharpoons H_3O^+(aq) + HCO_3^-(aq) \quad K_{a1} = 4.3 \times 10^{-7}$$
$$HCO_3^-(aq) + H_2O(l) \rightleftharpoons H_3O^+(aq) + CO_3^{2-}(aq) \quad K_{a2} = 5.6 \times 10^{-11}.$$

Because the second ionization constant is much smaller than the first, we can assume that the first step dominates:

Concentration				
(mol·L^{-1})	$H_2CO_3(aq)$ +	$H_2O(l)$	$\rightleftharpoons$ $H_3O^+(aq)$ +	$HCO_3^-(aq)$
initial	0.0456	—	0	0
change	$-x$	—	$+x$	$+x$
final	$0.0456 - x$	—	$+x$	$+x$

$$K_{a1} = \frac{[H_3O^+][HCO_3^-]}{[H_2CO_3]}$$

$$4.3 \times 10^{-7} = \frac{(x)(x)}{0.0456 - x} = \frac{x^2}{0.0456 - x}$$

Assume that $x \ll 0.0456$.

Then $x^2 = (4.3 \times 10^{-7})(0.0456)$.

$x = 1.4 \times 10^{-4}$

Because $x < 1\%$ of 0.0456, the assumption was valid.

$$x = [H_3O^+] = [HCO_3^-] = 1.4 \times 10^{-4} \text{ mol·L}^{-1}$$

This means that the concentration of H_2CO_3 is

$$0.0456 \text{ mol·L}^{-1} - 0.00014 \text{ mol·L}^{-1} = 0.0455 \text{ mol·L}^{-1}.$$

We can then use the other equilibria to determine the remaining concentrations:

$$K_{a2} = \frac{[H_3O^+][CO_3^{2-}]}{[HCO_3^-]}$$

$$5.6 \times 10^{-11} = \frac{(1.4 \times 10^{-4})[CO_3^{2-}]}{(1.4 \times 10^{-4})}$$

$$[CO_3^{2-}] = 5.6 \times 10^{-11} \text{ mol} \cdot \text{L}^{-1}$$

Because $5.6 \times 10^{-11} \ll 1.4 \times 10^{-4}$, the initial assumption that the first ionization would dominate is valid.

To calculate $[OH^-]$, we use the K_w relationship:

$$K_w = [H_3O^+][OH^-]$$

$$[OH^-] = \frac{K_w}{[H_3O^+]} = \frac{1.00 \times 10^{-14}}{1.4 \times 10^{-4}} = 7.1 \times 10^{-11} \text{ mol} \cdot \text{L}^{-1}$$

In summary, $[H_2CO_3] = 0.0455 \text{ mol} \cdot \text{L}^{-1}$, $[H_3O^+] = [HCO_3^-]$
$$= 1.4 \times 10^{-4} \text{ mol} \cdot \text{L}^{-1},$$

$$[CO_3^{2-}] = 5.6 \times 10^{-11} \text{ mol} \cdot \text{L}^{-1}, [OH^-] = 7.1 \times 10^{-11} \text{ mol} \cdot \text{L}^{-1}.$$

6E.11 The equilibrium reactions of interest are now the base forms of the carbonic acid equilibria, so K_b values should be calculated for the following changes:

$$CO_3^{2-}(aq) + H_2O(l) \rightleftharpoons HCO_3^-(aq) + OH^-(aq)$$

$$K_{b1} = \frac{K_w}{K_{a2}} = \frac{1.00 \times 10^{-14}}{5.6 \times 10^{-11}} = 1.8 \times 10^{-4}$$

$$HCO_3^-(aq) + H_2O(l) \rightleftharpoons H_2CO_3(aq) + OH^-(aq)$$

$$K_{b2} = \frac{K_w}{K_{a1}} = \frac{1.00 \times 10^{-14}}{4.3 \times 10^{-7}} = 2.3 \times 10^{-8}$$

Because the second hydrolysis constant is much smaller than the first, we can assume that the first step dominates:

Concentration

$$(\text{mol} \cdot \text{L}^{-1}) \quad CO_3^{2-}(aq) \quad + \quad H_2O(l) \quad \rightleftharpoons \quad HCO_3^-(aq) \quad + \quad OH^-(aq)$$

initial	0.0456	—	0	0
change	$-x$	—	$+x$	$+x$
final	$0.0456 - x$	—	$+x$	$+x$

$$K_{b1} = \frac{[HCO_3^-][OH^-]}{[CO_3^{2-}]}$$

$$1.8 \times 10^{-4} = \frac{(x)(x)}{0.0456 - x} = \frac{x^2}{[0.0456 - x]}$$

Assume that $x \ll 0.0456$.

Then $x^2 = (1.8 \times 10^{-4})(0.0456)$.

$x = 2.9 \times 10^{-3}$

Because $x > 5\%$ of 0.0456, the assumption was not valid and the full expression should be solved using the quadratic equation:

$$x^2 + 1.8 \times 10^{-4} x - (1.8 \times 10^{-4})(0.0456) = 0$$

Solving using the quadratic equation gives $x = 0.0028 \text{ mol} \cdot \text{L}^{-1}$.

$x = [HCO_3^-] = [OH^-] = 0.0028 \text{ mol} \cdot \text{L}^{-1}$

Therefore, $[CO_3^{2-}] = 0.0456 \text{ mol} \cdot \text{L}^{-1} - 0.0028 \text{ mol} \cdot \text{L}^{-1} = 0.0428 \text{ mol} \cdot \text{L}^{-1}$.

We can then use the other equilibria to determine the remaining concentrations:

$$K_{b2} = \frac{[H_2CO_3][OH^-]}{[HCO_3^-]}$$

$$2.3 \times 10^{-8} = \frac{[H_2CO_3](0.0028)}{(0.0028)}$$

$[H_2CO_3] = 2.3 \times 10^{-8} \text{ mol} \cdot \text{L}^{-1}$

Because $2.3 \times 10^{-8} \ll 0.0028$, the initial assumption that the first hydrolysis would dominate is valid. To calculate $[H_3O^+]$, we use the K_w relationship:

$$K_w = [H_3O^+][OH^-]$$

$$[H_3O^+] = \frac{K_w}{[OH^-]} = \frac{1.00 \times 10^{-14}}{0.0028} = 3.6 \times 10^{-12} \text{ mol} \cdot L^{-1}$$

In summary, $[H_2CO_3] = 2.3 \times 10^{-8}$ mol $\cdot$ L^{-1}, $[OH^-] = [HCO_3^{\ -}] =$

0.0028 mol $\cdot$ L^{-1}, $[CO_3^{\ 2-}] = 0.0428$ mol $\cdot$ L^{-1}, $[H_3O^+] = 3.6 \times 10^{-12}$ mol $\cdot$ L^{-1}.

6E.13 The equilibria present in solution are

$$H_2SO_3(aq) + H_2O(l) \rightleftharpoons H_3O^+(aq) + HSO_3^{\ -}(aq) \quad K_{a1} = 1.5 \times 10^{-2}$$
$$HSO_3^{\ -}(aq) + H_2O(l) \rightleftharpoons H_3O^+(aq) + SO_3^{\ 2-}(aq) \quad K_{a2} = 1.2 \times 10^{-7}.$$

The calculation of the desired concentrations follows exactly after the

method derived in Equation 15, substituting H_2SO_3 for

H_2CO_3, $HSO_3^{\ -}$ for $HCO_3^{\ -}$, and $SO_3^{\ 2-}$ for $CO_3^{\ 2-}$. First, calculate the

quantity f (at pH = 5.50 $[H_3O^+] = 10^{-5.5} = 3.2 \times 10^{-6}$ mol $\cdot$ L^{-1}):

$$f = [H_3O^+]^2 + [H_3O^+] K_{a1} + K_{a1}K_{a2}$$
$$= (3.2 \times 10^{-6})^2 + (3.2 \times 10^{-6})(1.5 \times 10^{-2}) + (1.5 \times 10^{-2})(1.2 \times 10^{-7})$$
$$= 5.0 \times 10^{-8}$$

The fractions of the species present are then given by

$$\alpha(H_2SO_3) = \frac{[H_3O^+]^2}{f} = \frac{(3.2 \times 10^{-6})^2}{5.0 \times 10^{-8}} = 2.1 \times 10^{-4}$$

$$\alpha(HSO_3^{\ -}) = \frac{[H_3O^+]K_{a1}}{f} = \frac{(3.2 \times 10^{-6})(1.5 \times 10^{-2})}{5.0 \times 10^{-8}} = 0.96$$

$$\alpha(SO_3^{\ 2-}) = \frac{K_{a1}K_{a2}}{f} = \frac{(1.5 \times 10^{-2})(1.2 \times 10^{-7})}{5.0 \times 10^{-8}} = 0.036$$

Thus, in a 0.150 mol $\cdot$ L^{-1} solution at pH 5.50, the dominant species will

be $HSO_3^{\ -}$ with a concentration of $(0.150$ mol $\cdot$ $L^{-1})(0.96) = 0.14$ mol $\cdot$ L^{-1}.

The concentration of H_2SO_3 will be:

$(2.1 \times 10^{-4})(0.150$ mol $\cdot$ $L^{-1}) = 3.2 \times 10^{-5}$ mol $\cdot$ L^{-1} and the concentration of

$SO_3^{\ 2-}$ will be $(0.036)(0.150$ mol $\cdot$ $L^{-1}) = 0.0054$ mol $\cdot$ L^{-1}.

6E.15 (a) Concentration

$$(\text{mol} \cdot \text{L}^{-1}) \qquad B(OH)_3 \quad + \quad 2\,H_2O \rightleftharpoons H_3O^+ \quad + \quad B(OH)_4^-$$

initial	1.0×10^{-4}	—	0	0
change	$-x$	—	$+x$	$+x$
equilibrium	$1.0 \times 10^{-4} - x$	—	x	x

$$K_a = 7.2 \times 10^{-10} = \frac{[H_3O^+][B(OH)_4^-]}{[B(OH)_3]} = \frac{x^2}{1.0 \times 10^{-4} - x} \approx \frac{x^2}{1.0 \times 10^{-4}}$$

$$x = [H_3O^+] = 2.7 \times 10^{-7} \text{ mol} \cdot \text{L}^{-1}$$

$$pH = -\log(2.7 \times 10^{-7}) = 6.57$$

Note: This value of $[H_3O^+]$ is not much different from the value for pure water, 1.0×10^{-7} mol $\cdot$ L^{-1}; therefore, it is at the lower limit of safely ignoring the contribution to $[H_3O^+]$ from the autoprotolysis of water. The exercise should be solved by simultaneously considering both equilibria. Concentration

$$(\text{mol} \cdot \text{L}^{-1}) \qquad B(OH)_3 \quad + \quad 2\,H_2O \quad \rightleftharpoons \quad H_3O^+ \quad + \quad B(OH)_4^-$$

equilibrium	$1.0 \times 10^{-4} - x$	—	x	y

Concentration (mol $\cdot$ L^{-1}) $2\,H_2O \rightleftharpoons H_3O^+ + OH^-$

equilibrium	—	x	z

Because there are now two contributions to $[H_3O^+]$, $[H_3O^+]$ is no longer equal to $[B(OH)_4^-]$, nor is it equal to $[OH^-]$, as in pure water. To avoid a cubic equation, x will again be ignored relative to 1.0×10^{-4} mol $\cdot$ L^{-1}. This approximation is justified by the approximate calculation above, and because K_a is very small relative to 1.0×10^{-4}. Let $a =$ initial concentration of $B(OH)_3$, then

$$K_a = 7.2 \times 10^{-10} = \frac{xy}{a - x} \approx \frac{xy}{a} \text{ or } y = \frac{aK_a}{x}$$

$$K_w = 1.0 \times 10^{-14} = xz.$$

Electroneutrality requires

$x = y + z$ or $z = x - y$; hence, $K_w = xz = x(x - y)$.

Substituting for y from above:

$$x \times \left(x - \frac{aK_a}{x} \right) = K_w$$

$$x^2 - aK_a = K_w$$

$$x^2 = K_w + aK_a$$

$$x = \sqrt{K_w + aK_a} = \sqrt{1.0 \times 10^{-14} + 1.0 \times 10^{-4} \times 7.2 \times 10^{-10}}$$

$$x = 2.9 \times 10^{-7} \text{ mol} \cdot \text{L}^{-1} = [H_3O^+]$$

$$pH = -\log(2.9 \times 10^{-7}) = 6.54$$

This value is slightly, but measurably, different from the value 6.57 obtained by ignoring the contribution to $[H_3O^+]$ from water.

(b) In this case, the second ionization can safely be ignored; $K_{a2} \ll K_{a1}$.

Concentration

(mol·L^{-1})	H_3PO_4	+	H_2O	$\rightleftharpoons$	H_3O^+	+	$H_2PO_4^-$
initial	0.015		—		0		0
change	$-x$		—		$+x$		$+x$
equilibrium	$0.015 - x$		—		x		x

$$K_{a1} = 7.6 \times 10^{-3} = \frac{x^2}{0.015 - x}$$

$$x^2 + 7.6 \times 10^{-3} x - 1.14 \times 10^{-4} = 0$$

$$x = [H_3O^+] = \frac{-7.6 \times 10^{-3} + \sqrt{(7.6 \times 10^{-3})^2 + 4.56 \times 10^{-4}}}{2}$$

$$= 7.5 \times 10^{-3} \text{ mol} \cdot \text{L}^{-1}$$

$$pH = -\log(7.5 \times 10^{-3}) = 2.12$$

(c) In this case, the second ionization can safely be ignored; $K_{a2} \ll K_{a1}$.

Concentration

(mol · L^{-1})	H_2SO_3	+	H_2O	$\rightleftharpoons$	H_3O^+	+	HSO_3^-
initial	0.1		—		0		0
change	$-x$		—		$+x$		$+x$

equilibrium $\quad 0.1 - x \qquad — \qquad\qquad x \qquad\qquad x$

$$K_{a1} = 1.5 \times 10^{-2} = \frac{x^2}{0.10 - x}$$

$$x^2 + 1.5 \times 10^{-2} x - 1.5 \times 10^{-3} = 0$$

$$x = [H_3O^+] = \frac{-1.5 \times 10^{-2} + \sqrt{(1.5 \times 10^{-2})^2 + 6.0 \times 10^{-3}}}{2} = 0.032 \text{ mol} \cdot L^{-1}$$

$$pH = -\log(0.032) = 1.49$$

6E.17 The three equilibria involved are

$$H_3PO_4(aq) \rightleftharpoons H_2PO_4^-(aq) + H_3O^+(aq), \quad K_{a_1} = 7.6 \times 10^{-3} = \frac{[H_3O^+][H_2PO_4^-]}{[H_3PO_4]}$$

$$H_2PO_4^-(aq) \rightleftharpoons HPO_4^{2-}(aq) + H_3O^+(aq), \quad K_{a_2} = 6.2 \times 10^{-8} = \frac{[H_3O^+][HPO_4^{2-}]}{[H_2PO_4^-]}$$

$$HPO_4^{2-}(aq) \rightleftharpoons PO_4^{3-}(aq) + H_3O^+(aq), \quad K_{a_3} = 2.1 \times 10^{-13} = \frac{[H_3O^+][PO_4^{3-}]}{[HPO_4^{2-}]}$$

We also know that the combined concentration of all the phosphate species is

$$[H_3PO_4] + [H_2PO_4^-] + [HPO_4^{2-}] + [PO_4^{3-}] = 1.5 \times 10^{-2} \text{ mol} \cdot L^{-1}$$

and the hydronium ion concentration is

$$[H] = 10^{-pH} = 10^{-2.25} = 5.62 \times 10^{-3} \text{ mol} \cdot L^{-1}.$$

At this point it is a matter of solving this set of simultaneous equations to obtain the concentrations of the phosphate-containing species. We start by dividing both sides of the equilibrium constant expressions above by the given hydronium ion concentration to obtain three ratios:

$$1.35 = \frac{[H_2PO_4^-]}{[H_3PO_4]}, \quad 1.10 \times 10^{-5} = \frac{[HPO_4^{2-}]}{[H_2PO_4^-]}, \quad \text{and} \quad 3.74 \times 10^{-11} = \frac{[PO_4^{3-}]}{[HPO_4^{2-}]}$$

Through rearrangement and substitution of these three ratios, we can obtain the following expressions:

$$[H_2PO_4^-] = 1.35 \cdot [H_3PO_4],$$

$$[HPO_4^{2-}] = 1.10 \times 10^{-5} \cdot [H_2PO_4^-] = 1.10 \times 10^{-5} \cdot 1.35 \cdot [H_3PO_4]$$

$$= 1.48 \times 10^{-5} \cdot [H_3PO_4], \text{ and}$$

$$[PO_4^{3-}] = 3.74 \times 10^{-11} \cdot [HPO_4^{2-}] = 3.74 \times 10^{-11} \cdot 1.48 \times 10^{-5} \cdot [H_3PO_4]$$

$$= 5.54 \times 10^{-16} \cdot [H_3PO_4]$$

Substituting these expressions back into the sum:

$$[H_3PO_4] + [H_2PO_4^-] + [HPO_4^{2-}] + [PO_4^{3-}] =$$

$$= [H_3PO_4] + (1.35 \cdot [H_3PO_4]) + (1.48 \times 10^{-5} \cdot [H_3PO_4]) + (5.54 \times 10^{-16} \cdot [H_3PO_4])$$

$$= 1.5 \times 10^{-2}$$

we find:

$$[H_3PO_4] = 6.4 \times 10^{-3} \text{ mol} \cdot L^{-1},$$

$$[H_2PO_4^-] = 1.35 \cdot [H_3PO_4] = 8.6 \times 10^{-3} \text{ mol} \cdot L^{-1},$$

$$[HPO_4^{2-}] = 1.48 \times 10^{-5} \cdot [H_3PO_4] = 9.5 \times 10^{-8} \text{ mol} \cdot L^{-1}, \text{ and}$$

$$[PO_4^{3-}] = 5.54 \times 10^{-16} \cdot [H_3PO_4] = 3.5 \times 10^{-18} \text{ mol} \cdot L^{-1}.$$

6F.1 We can use the relationship derived in the text,

$$[H_3O^+]^2 - [HA]_{initial}[H_3O^+] - K_w = 0, \text{ in which HA is any strong acid.}$$

$$[H_3O^+]^2 - (6.55 \times 10^{-7})[H_3O^+] - (1.00 \times 10^{-14}) = 0$$

Solving using the quadratic equation gives

$$[H_3O^+] = 6.70 \times 10^{-7}, \text{ pH} = 6.174.$$

This value is slightly lower than the value calculated, based on the acid concentration alone $(\text{pH} = -\log(6.55 \times 10^{-7}) = 6.184)$.

6F.3 We can use the relationship derived in the text,

$$[H_3O^+]^2 + [B]_{initial}[H_3O^+] - K_w = 0, \text{ in which B is any strong base.}$$

$$[H_3O^+]^2 + (9.78 \times 10^{-8})[H_3O^+] - (1.00 \times 10^{-14}) = 0$$

Solving using the quadratic equation gives

$$[H_3O^+] = 6.24 \times 10^{-8}, \text{ pH} = 7.205.$$

This value is higher than the value calculated, based on the base concentration alone $(pOH = -\log(9.78 \times 10^{-8}) = 7.009)$, $pH = 6.991$.

6F.5 In general, when $[H_3O^+] \leq 10^{-6}$ (that is, when the pH lies between 6 and 8), the autoprotolysis of water must be taken into account. We can use Equation (3b) in 6F.2:

$$x^3 + K_a x^2 - (K_w + K_a \cdot [HA]_{initial})x - K_w \cdot K_a = 0, \text{ where } x = [H_3O^+]$$

To solve the $[HA]_{initial}$, you substitute the values of $[H_3O^+] = 1.0 \times 10^{-6}$, $K_w = 1.00 \times 10^{-14}$, and $K_a = 1.8 \times 10^{-5}$ into this equation and then solve for $[HA]_{initial}$. $[HA]_{initial} = 1.0 \times 10^{-6}$ M.

That is, when acetic acid concentration is 1.0×10^{-6} M or higher, the autoprotolysis of water will not be taken into account.

6F.7 Before we answer whether the criteria $6 < pH < 8$ should be modified, we need to use equation 3b in 6F.2 to resolve the example 6D.4 to see whether autoprotolysis of water should be considered:

$$x^3 + K_a x^2 - (K_w + K_a \cdot [HA]_{initial})x - K_w \cdot K_a = 0, \text{ where } x = [H_3O^+]$$

$[HA]_{initial} = 0.15$ M; $Ka = K_w/K_b(ammonia) = 5.6 \times 10^{-10}$

Solve for x: $x = [H_3O^+] = 9.2 \times 10^{-6}$ M; $pH = 5.04$

It is clear that the pH does not change because the $pH < 6$ and the autoprotolysis of water can still be ignored. Therefore, the criteria $6 < pH < 8$ is a good guide for when water autoprotolysis can be ignored.

6F.9 (a) In the absence of a significant effect due to the autoprotolysis of water, the pH values of the 8.50×10^{-5} M and 7.37×10^{-6} M HCN solutions can be calculated as described earlier.

For 8.50×10^{-5} mol $\cdot$ L^{-1} :

Concentration

(mol $\cdot$ L^{-1}) $HCN(aq)$ + $H_2O(l)$ $\rightleftharpoons$ $H_3O^+(aq)$ + $CN^-(aq)$

initial	8.50×10^{-5}	—	0	0
change	$-x$	—	$+x$	$+x$
final	$8.50 \times 10^{-5} - x$	—	$+x$	$+x$

$$K_a = \frac{[H_3O^+][CN^-]}{[HCN]}$$

$$4.9 \times 10^{-10} = \frac{(x)(x)}{8.5 \times 10^{-5} - x} = \frac{x^2}{[8.5 \times 10^{-5} - x]}$$

Assume $x \ll 8.5 \times 10^{-5}$.

$$x^2 = (4.9 \times 10^{-10})(8.5 \times 10^{-5})$$

$$x = 2.0 \times 10^{-7}$$

Because $x < 1\%$ of 8.50×10^{-5}, the assumption was valid. Given this value, the pH is then calculated to be $-\log(2.0 \times 10^{-7}) = 6.69$.

For $7.37 \times 10^{-6} \ mol \cdot L^{-1}$:

Concentration

$(mol \cdot L^-)$	$HCN(aq)$	+	$H_2O(l)$	$\rightleftharpoons$	$H_3O^+(aq)$	+	$CN^-(aq)$
initial	7.37×10^{-6}		—		0		0
change	$-x$		—		$+x$		$+x$
final	$7.37 \times 10^{-6} - x$		—		$+x$		$+x$

$$K_a = \frac{[H_3O^+][CN^-]}{[HCN]}$$

$$4.9 \times 10^{-10} = \frac{(x)(x)}{7.37 \times 10^{-6} - x} = \frac{x^2}{[7.37 \times 10^{-6} - x]}$$

Assume $x \ll 7.37 \times 10^{-6}$.

$$x^2 = (4.9 \times 10^{-10})(7.37 \times 10^{-6})$$

$$x = 6.0 \times 10^{-8}$$

x is $< 1\%$ of 7.37×10^{-6}, so the assumption is still reasonable. The pH is then calculated to be $-\log(6.0 \times 10^{-8}) = 7.22$. This answer is not reasonable because we know HCN is an acid.

(b) To calculate the value, taking into account the autoprotolysis of water, we can use Equation 21:

$x^3 + K_a x^2 - (K_w + K_a \cdot [HA]_{initial})x - K_w \cdot K_a = 0$, where $x = [H_3O^+]$

To solve the expression, you substitute the values of $K_w = 1.00 \times 10^{-14}$, the initial concentration of acid, and $K_a = 4.9 \times 10^{-10}$ into this equation and then solve the expression either by trial and error or, preferably, by using a graphing calculator such as the one found on the CD accompanying this text.

Alternatively, you can use a computer program designed to solve simultaneous equations. Because the unknowns include $[H_3O^+], [OH^-], [HBrO]$, and $[BrO^-]$, you will need four equations. As seen in the text, the pertinent equations are

$$K_a = \frac{[H_3O^+][CN^-]}{[HCN]}$$

$K_w = [H_3O^+][OH^-]$

$[H_3O^+] = [OH^-] + [CN^-]$

$[HCN]_{initial} = [HCN] + [CN^-]$.

All methods should produce the same result.

For $[HCN] = 8.5 \times 10^{-5}$ mol $\cdot$ L^{-1}, the values obtained are

$[H_3O^+] = 2.3 \times 10^{-7}$ mol $\cdot$ L^{-1}, pH = 6.64 (compare with 6.69 obtained in (a))

$[CN^-] = 1.8 \times 10^{-7}$ mol $\cdot$ L^{-1}

$[HCN] \cong 8.5 \times 10^{-5}$ mol $\cdot$ L^{-1}

$[OH^-] = 4.4 \times 10^{-8}$ mol $\cdot$ L^{-1}.

Similarly, for $[HCN]_{initial} = 7.37 \times 10^{-6}$,

$[H_3O^+] = 1.2 \times 10^{-7}$ mol $\cdot$ L^{-1}, pH = 6.92 (compare with 7.22 obtained in (a))

$[CN^-] = 3.1 \times 10^{-8}$ mol $\cdot$ L^{-1}

$[HCN] \cong 7.3 \times 10^{-6}$ mol $\cdot$ L^{-1}

$[OH^-] = 8.6 \times 10^{-8}$ mol $\cdot$ L^{-1}.

Note that for the more concentrated solution, the effect of the autoprotolysis of water is smaller. Notice also that the less concentrated solution is more acidic, because of the autoprotolysis of water, than would be predicted if this effect were not operating.

6G.1 (a) When solid sodium acetate is added to an acetic acid solution, the concentration of H_3O^+ decreases because the equilibrium

$$HC_2H_3O_2(aq) + H_2O(l) \rightleftharpoons H_3O^+(aq) + C_2H_3O_2^-(aq)$$

shifts to the left to relieve the stress imposed by the increase of $[C_2H_3O_2^-]$ (Le Chatelier's principle).

(b) When HCl is added to a benzoic acid solution, the percentage of

benzoic acid that is deprotonated decreases because the equilibrium

$$C_6H_5COOH(aq) + H_2O(l) \rightleftharpoons H_3O^+(aq) + C_6H_5CO_2^-(aq)$$

shifts to the left to relieve the stress imposed by the increased $[H_3O^+]$ (Le Chatelier's principle).

(c) When solid NH_4Cl is added to an ammonia solution, the

concentration of OH^- decreases because the equilibrium

$NH_3(aq) + H_2O(l) \rightleftharpoons NH_4^+(aq) + OH^-(aq)$ shifts to the left to relieve the

stress imposed by the increased $[NH_4^+]$ (Le Chatelier's principle).

Because $[OH^-]$ decreases, $[H_3O^+]$ increases and pH decreases.

6G.3 (a) $K_a = \dfrac{[H_3O^+][A^-]}{[HA]}$; $pK_a = pH - \log \dfrac{[A^-]}{[HA]}$. If $[A^-] = [HA]$, then

$pK_a = pH$.

$pH = pK_a = 3.52$, $K_a = 3.02 \times 10^{-4}$

(b) Let $x = [\text{glycerate ion}] = [L^-]$ and $y = [H_3O^+]$

Concentration

$(\text{mol} \cdot L^{-1})$	HL(aq)	+	H$_2$O(l)	$\rightleftharpoons$	H$_3$O$^+$(aq)	+	L$^-$(aq)
initial	$2x$		—		—		x

change $\qquad$ $-y$ $\qquad$ $\qquad$ $+y$ $\qquad$ $+y$

equilibrium $\qquad$ $2x - y$ $\qquad$ $\qquad$ y $\qquad$ $y + x$

$$K_a = \frac{[H_3O^+][L^-]}{[HL]} = \frac{(y)(y+x)}{(2x-y)} \cong \frac{(y)(x)}{(2x)} = 3.02 \times 10^{-4}$$

$$y = 2(3.02 \times 10^{-4}) \cong 6.04 \times 10^{-4} \ mol \cdot L^{-1} \cong [H_3O^+]$$

$$pH = 3.22$$

6G.5 **(a)** The reaction is $HCN(aq) + H_2O(l) \rightleftharpoons H_3O^+(aq) + CN^-(aq)$.

Concentration

$(mol \cdot L^{-1})$ HCN(aq) +	H$_2$O(l)	$\rightleftharpoons$	H$_3$O$^+$(aq) +	CN$^-$(aq)
initial	0.075	—	0	0.060
change	$-x$	—	$+x$	$+x$
equilibrium	$0.075 - x$	—	x	$0.060 + x$

$$K_a = \frac{[H_3O^+][CN^-]}{[HCN]} = \frac{(x)(0.060 + x)}{(0.075 - x)} = 4.9 \times 10^{-10}$$

$$x = [[H_3O^+] \approx 6.1 \times 10^{-10} \ mol \cdot L^{-1}$$

(b) 0.30 M NaCl will have no effect:

Concentration

$(mol \cdot L^{-1})$	NH$_2$NH$_2$(aq) +	H$_2$O(l) $\rightleftharpoons$	NH$_2$NH$_3{}^+$(aq) +	OH$^-$(aq)
initial	0.20	0	0	
change	$-x$	—	$+x$	$+x$
equilibrium	$0.20 - x$	—	x	x

$$K_b = \frac{[NH_2NH_3{}^+][OH^-]}{[NH_2NH_2]}$$

$$1.7 \times 10^{-6} = \frac{[OH^-]^2}{0.20}$$

$$[OH^-] = \sqrt{3.4 \times 10^{-7}} = 5.8 \times 10^{-4} \ mol \cdot L^{-1}$$

$$[H_3O^+] = \frac{K_w}{[OH^-]} = \frac{1.00 \times 10^{-14}}{5.8 \times 10^{-4}} = 1.7 \times 10^{-11} \ mol \cdot L^{-1}$$

(c) Setup is similar to part (a).

$$K_a = 4.9 \times 10^{-10} = \frac{[H_3O^+][CN^-]}{[HCN]} = \frac{(x)(0.030 + x)}{(0.015 - x)}$$

$$4.9 \times 10^{-10} \approx \frac{[H_3O^+][0.030]}{(0.015)}$$

$$[H_3O^+] = 2.5 \times 10^{-10} \ mol \cdot L^{-1}$$

(d) When the concentrations of a weak base and its conjugate acid are equal, the pOH equals the pK_b. Therefore, the pOH of hydrazine = pK_b = 5.77, and

$$pH = 14.00 - pOH = 14.00 - 5.77 = 8.23$$

$$[H_3O^+] = 10^{-8.23} = 5.9 \times 10^{-9} \ mol \cdot L^{-1}$$

6G.7   In each case, the equilibrium involved is

$$HSO_4^-(aq) + H_2O(l) \rightleftharpoons H_3O^+(aq) + SO_4^{2-}(aq).$$

$HSO_4^-(aq)$ and $SO_4^{2-}(aq)$ are conjugate acid and base; therefore, the pH calculation is most easily performed with the Henderson–Hasselbalch equation:

$$pH = pK_a + \log\left(\frac{[base]}{[acid]}\right) = pK_a + \log\left(\frac{[SO_4^{2-}]}{[HSO_4^-]}\right)$$

(a) $pH = 1.92 + \log\left(\dfrac{0.25 \ mol \cdot L^{-1}}{0.5 \ mol \cdot L^{-1}}\right) = 1.62, \quad pOH = 14.00 - 1.62 = 12.38$

(b) $pH = 1.92 + \log\left(\dfrac{0.10 \ mol \cdot L^{-1}}{0.50 \ mol \cdot L^{-1}}\right) = 1.22, \quad pOH = 12.78$

(c) $pH = pK_a = 1.92, \quad pOH = 12.08$

See solution to Exercise 6G.3.

6G.9   (a) $HCN(aq) + H_2O(l) \rightleftharpoons H_3O^+(aq) + CN^-(aq)$

total volume = 100 mL = 0.100 L

moles of HCN = $0.0300 \ L \times 0.050 \ mol \cdot L^{-1} = 1.5 \times 10^{-3}$ mol HCN moles of NaCN = $0.0700 \text{ L} \times 0.030 \text{ mol} \cdot \text{L}^{-1} = 2.1 \times 10^{-3}$ mol NaCN

initial $[\text{HCN}]_0 = \dfrac{1.5 \times 10^{-3} \text{ mol}}{0.100 \text{ L}} = 1.5 \times 10^{-2} \text{ mol} \cdot \text{L}^{-1}$

initial $[\text{CN}^-]_0 = \dfrac{2.1 \times 10^{-3} \text{ mol}}{0.100 \text{ L}} = 2.1 \times 10^{-2} \text{ mol} \cdot \text{L}^{-1}$

Concentration $(\text{mol} \cdot \text{L}^{-1})$	HCN(aq)	+	H_2O(l)	$\rightleftharpoons$	H_3O^+(aq)	+	CN^-(aq)
initial	1.5×10^{-2}		—		0		2.1×10^{-2}
change	$-x$		—		$+x$		$+x$
equilibrium	$1.5 \times 10^{-2} - x$		—		x		$2.1 \times 10^{-2} + x$

$$K_a = \frac{[\text{H}_3\text{O}^+][\text{CN}^-]}{[\text{HCN}]} = \frac{(x)(2.1 \times 10^{-2} + x)}{(1.5 \times 10^{-2} - x)} \approx \frac{(x)(2.1 \times 10^{-2})}{(1.5 \times 10^{-2})} = 4.9 \times 10^{-10}$$

$x \approx 3.5 \times 10^{-10} \text{ mol} \cdot \text{L}^{-1} \approx [\text{H}_3\text{O}^+]$

$\text{pH} = -\log[\text{H}_3\text{O}^+] = -\log(3.5 \times 10^{-10}) = 9.46$

(b) The solution here is the same as for part (a), except for the initial concentrations:

$$[\text{HCN}]_0 = \frac{0.0400 \text{ L} \times 0.030 \text{ mol} \cdot \text{L}^{-1}}{0.100 \text{ L}} = 1.2 \times 10^{-2} \text{ mol} \cdot \text{L}^{-1}$$

$$[\text{CN}^-]_0 = \frac{0.0600 \text{ L} \times 0.050 \text{ mol} \cdot \text{L}^{-1}}{0.100 \text{ L}} = 3.0 \times 10^{-2} \text{ mol} \cdot \text{L}^{-1}$$

$$K_a = 4.9 \times 10^{-10} = \frac{(x)(3.0 \times 10^{-2})}{(1.2 \times 10^{-2})}$$

$x = [\text{H}_3\text{O}^+] = 2.0 \times 10^{-10} \text{ mol} \cdot \text{L}^{-1}$

$\text{pH} = -\log(2.0 \times 10^{-10}) = 9.71$

(c) $[\text{HCN}]_0 = [\text{NaCN}]_0$ after mixing; therefore,

$$K_a = 4.9 \times 10^{-10} = \frac{(x)[\text{NaCN}]_0}{[\text{HCN}]_0} = x = [\text{H}_3\text{O}^+]$$

$\text{pH} = \text{p}K_a = -\log(4.9 \times 10^{-10}) = 9.31$

6G.11 (a) $\quad pH = pK_a + \log\left(\dfrac{[CH_3CO_2^{-}]}{[CH_3COOH]}\right)$

$pH = pK_a + \log\left(\dfrac{0.100}{0.100}\right) = 4.75$ (initial pH)

final pH: $(0.0100 \text{ L})(0.950 \text{ mol} \cdot \text{L}^{-1}) = 9.50 \times 10^{-3}$ mol NaOH (strong

base) produces 9.50×10^{-3} mol $CH_3CO_2^{-}$ from CH_3COOH

$0.100 \text{ mol} \cdot \text{L}^{-1} \times 0.100 \text{ L} = 1.00 \times 10^{-2}$ mol CH_3COOH initially

$0.100 \text{ mol} \cdot \text{L}^{-1} \times 0.100 \text{ L} = 1.00 \times 10^{-2}$ mol $CH_3CO_2^{-}$ initially

After adding NaOH:

$[CH_3COOH] = \dfrac{(1.00 \times 10^{-2} - 9.5 \times 10^{-3}) \text{ mol}}{0.110 \text{ L}} = 5 \times 10^{-3} \text{ mol} \cdot \text{L}^{-1}$

$[CH_3CO_2^{-}] = \dfrac{(1.00 \times 10^{-2} + 9.5 \times 10^{-3}) \text{ mol}}{0.110 \text{ L}} = 1.77 \times 10^{-1} \text{ mol} \cdot \text{L}^{-1}$

$pH = 4.75 + \log\left(\dfrac{1.77 \times 10^{-1} \text{ mol} \cdot \text{L}^{-1}}{5 \times 10^{-3} \text{ mol} \cdot \text{L}^{-1}}\right) = 4.75 + 1.5 = 6.3$

$\Delta pH = 6.3 - 4.75 = 1.55$

(b) $\quad (0.0200 \text{ L})(0.100 \text{ mol} \cdot \text{L}^{-1}) = 2.00 \times 10^{-3}$ mol HNO_3 (strong acid)

produces 2.00×10^{-3} mol CH_3COOH from $CH_3CO_2^{-}$.

After adding HNO_3 [see part (a) of this exercise]:

$[CH_3COOH] = \dfrac{(1.00 \times 10^{-2} + 2.00 \times 10^{-3}) \text{ mol}}{0.120 \text{ L}} = 1.00 \times 10^{-1} \text{ mol} \cdot \text{L}^{-1}$

$[CH_3CO_2^{-}] = \dfrac{(1.00 \times 10^{-2} - 2.00 \times 10^{-3}) \text{ mol}}{0.120 \text{ L}} = 6.7 \times 10^{-2} \text{ mol} \cdot \text{L}^{-1}$

$pH = 4.75 + \log\left(\dfrac{6.7 \times 10^{-2} \text{ mol} \cdot \text{L}^{-1}}{1.00 \times 10^{-1} \text{ mol} \cdot \text{L}^{-1}}\right) = 4.75 - 0.17 = 4.58$

$\Delta pH = -0.17$

6G.13 $\left(\dfrac{0.356\ \text{g NaF}}{0.050\ \text{L}}\right)\left(\dfrac{1\ \text{mol NaF}}{41.99\ \text{g NaF}}\right)=0.17\ \text{mol}\cdot\text{L}^{-1}$

Concentration

(mol·L^{-1})	HF(aq)	+	H$_2$O(l)	⇌	H$_3$O$^+$(aq)	+	F$^-$(aq)
initial	0.40		—		0		0.17
change	$-x$		—		$+x$		$+x$
equilibrium	$0.40-x$		—		x		$0.17+x$

$$K_a=\frac{[\text{H}_3\text{O}^+][\text{F}^-]}{[\text{HF}]}=\frac{(x)(0.17+x)}{(0.40-x)}\approx\frac{(x)(0.17)}{(0.40)}=3.5\times10^{-4}$$

$x\cong 8.2\times10^{-4}\ \text{mol}\cdot\text{L}^{-1}\cong[\text{H}_3\text{O}^+]$

$\text{pH}=-\log[\text{H}_3\text{O}^+]=-\log(8.2\times10^{-4})=3.09$

change in pH $=3.09-1.93=1.16$

6G.15 In a solution containing HClO(aq) and ClO$^-$(aq), the following equilibrium occurs:

$$\text{HClO(aq)}+\text{H}_2\text{O(l)}\rightleftharpoons\text{H}_3\text{O}^+(\text{aq})+\text{ClO}^-(\text{aq})$$

The ratio [ClO$^-$]/[HClO] is related to pH, as given by the

Henderson–Hasselbalch equation: $\text{pH}=pK_a+\log\!\left(\dfrac{[\text{ClO}^-]}{[\text{HClO}]}\right)$, or

$$\log\!\left(\frac{[\text{ClO}^-]}{[\text{HClO}]}\right)=\text{pH}-pK_a=6.50-7.53=-1.03$$

$$\frac{[\text{ClO}^-]}{[\text{HClO}]}=9.3\times10^{-2}$$

6G.17 The rule of thumb we use is that the effective range of a buffer is roughly within plus or minus one pH unit of the pK_a of the acid. Therefore:

(a) $pK_a=3.08$; pH range 2–4

(b) $pK_a=4.19$; pH range 3–5

(c) $pK_{a3} = 12.68$; pH range 11.5–13.5

(d) $pK_{a2} = 7.21$; pH range 6–8

(e) $pK_b = 7.97$, $pK_a = 6.03$; pH range 5–7

6G.19 Choose a buffer system in which the conjugate acid has a pK_a close to the desired pH. Therefore:

(a) $HClO_2$ and $NaClO_2$, $pK_a = 2.00$

(b) NaH_2PO_4 and Na_2HPO_4, $pK_{a2} = 7.21$

(c) $CH_2ClCOOH$ and $NaCH_2ClCO_2$, $pK_a = 2.85$

(d) Na_2HPO_4 and Na_3PO_4, $pK_a = 12.68$

6G.21 (a) $HCO_3^-(aq) + H_2O(l) \rightarrow CO_3^{2-}(aq) + H_3O^+(aq)$

$$K_{a2} = \frac{[H_3O^+][CO_3^{2-}]}{[HCO_3^-]}, \quad pK_{a2} = 10.25$$

$$pH = pK_{a2} + \log\left(\frac{[CO_3^{2-}]}{[HCO_3^-]}\right)$$

$$\log\left(\frac{[CO_3^{2-}]}{[HCO_3^-]}\right) = pH - pK_{a2} = 11.0 - 10.25 = 0.75$$

$$\frac{[CO_3^{2-}]}{[HCO_3^-]} = 5.6$$

(b) $[CO_3^{2-}] = 5.6 \times [HCO_3^-] = 5.6 \times 0.100 \text{ mol} \cdot L^{-1} = 0.56 \text{ mol} \cdot L^{-1}$

moles of CO_3^{2-} = moles of K_2CO_3 = $0.56 \text{ mol} \cdot L^{-1} \times 1 \text{ L} = 0.56$ mol

mass of $K_2CO_3 = 0.56 \text{ mol} \times \left(\frac{138.21 \text{ g } K_2CO_3}{1 \text{ mol } K_2CO_3}\right) = 77 \text{ g } K_2CO_3$

(c) $[HCO_3^-] = \frac{[CO_3^{2-}]}{5.6} = \frac{0.100 \text{ mol} \cdot L^{-1}}{5.6} = 1.8 \times 10^{-2} \text{ mol} \cdot L^{-1}$

moles of HCO_3^- = moles of $KHCO_3$ = $1.8 \times 10^{-2} \text{ mol} \cdot L^{-1} \times 1 \text{ L}$
$$= 1.8 \times 10^{-2} \text{ mol}$$

mass $KHCO_3 = 1.8 \times 10^{-2}$ mol $\times 100.12$ g $\cdot$ mol$^{-1} = 1.8$ g $KHCO_3$

(d) $[CO_3{}^{2-}] = 5.6 \times [HCO_3{}^-]$

moles of

$HCO_3{}^- =$ moles $KHCO_3 = 0.100$ mol $\cdot$ L$^{-1} \times 0.100$ L $= 1.00 \times 10^{-2}$ mol

Because the final total volume is the same for both

$KHCO_3$ and K_2CO_3, the number of moles of K_2CO_3 required is

$5.6 \times 1.00 \times 10^{-2}$ mol $= 5.6 \times 10^{-2}$ mol.

Thus,

volume of K_2CO_3 solution $= \dfrac{5.6 \times 10^{-2}\ \text{mol}}{0.200\ \text{mol} \cdot \text{L}^{-1}} = 0.28$ L $= 2.8 \times 10^2$ mL.

6H.1 (a) The titration curve is as follows (See example 6H.1 for calculations):

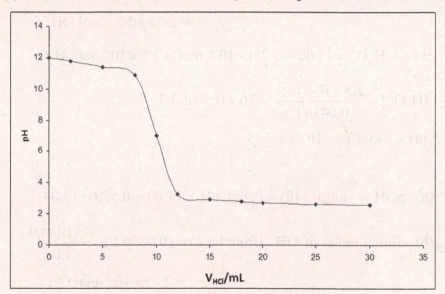

The initial pH was 12.00; the final pH was 2.54 (when 30.0 mL of HCl was added); pH at stoichiometric point was 7.00

(b) 10.0 mL; (c) 5.00 mL

6H.3 $HCl(aq) + NaOH(aq) \longrightarrow H_2O(l) + Na^+(aq) + Cl^-(aq)$

(a) $V_{HCl} = (\tfrac{1}{2})(25.0\ \text{mL})\left(\dfrac{10^{-3}\ \text{L}}{1\ \text{mL}}\right)\left(\dfrac{0.110\ \text{mol NaOH}}{1\ \text{L}}\right)$

$$\left(\frac{1 \text{ mol HCl}}{1 \text{ mol NaOH}}\right)\left(\frac{1 \text{ L HCl}}{0.150 \text{ mol HCl}}\right)$$

$$= 9.17 \times 10^{-3} \text{ L HCl}$$

(b) $2 \times 9.17 \times 10^{-3} \text{ L} = 0.0183 \text{ L}$

(c) volume $= (0.0250 + 0.0183) \text{ L} = 0.0433 \text{ L}$

$$[Na^+] = (0.0250 \text{ L})\left(\frac{0.110 \text{ mol NaOH}}{1 \text{ L}}\right)\left(\frac{1 \text{ mol Na}^+}{1 \text{ mol NaOH}}\right)\left(\frac{1}{0.0433 \text{ L}}\right)$$

$$= 0.0635 \text{ mol} \cdot \text{L}^{-1}$$

(d) number of moles of H_3O^+ (from acid) $= (0.0200 \text{ L})\left(\frac{0.150 \text{ mol}}{1 \text{ L}}\right)$

$$= 3.00 \times 10^{-3} \text{ mol } H_3O^+$$

number of moles of OH^- (from base) $= (0.0250 \text{ L})\left(\frac{0.110 \text{ mol Na}^+}{1 \text{ L}}\right)$

$$= 2.75 \times 10^{-3} \text{ mol } OH^-$$

excess $H_3O^+ = (3.00 - 2.75) \times 10^{-3} \text{ mol} = 2.5 \times 10^{-4} \text{ mol } H_3O^+$

$$[H_3O^+] = \frac{2.5 \times 10^{-4} \text{ mol}}{0.0450 \text{ L}} = 5.6 \times 10^{-3} \text{ mol} \cdot \text{L}^{-1}$$

$$pH = -\log(5.6 \times 10^{-3}) = 2.25$$

6H.5 (a) $pOH = -\log(0.110) = 0.959$, $pH = 14.00 - 0.959 = 13.04$

(b) initial moles of OH^- (from base) $= (0.0250 \text{ L})\left(\frac{0.110 \text{ mol}}{1 \text{ L}}\right)$

$$= 2.75 \times 10^{-3} \text{ mol } OH^-$$

moles of H_3O^+ added $= (0.0050 \text{ L})\left(\frac{0.150 \text{ mol}}{1 \text{ L}}\right) = 7.5 \times 10^{-4} \text{ mol } H_3O^+$

excess $OH^- = (2.75 - 0.75) \times 10^{-3} \text{ mol} = 2.00 \times 10^{-3} \text{ mol } OH^-$

$$[OH^-] = \frac{2.00 \times 10^{-3} \text{ mol}}{0.0300 \text{ L}} = 0.0667 \text{ mol} \cdot \text{L}^{-1}$$

$$pOH = -\log(0.0667) = 1.176, \quad pH = 14.00 - 1.176 = 12.82$$

(c) moles of H_3O^+ added $= 2 \times 7.5 \times 10^{-4}$ mol $= 1.50 \times 10^{-3}$ mol H_3O^+

excess $OH^- = (2.75 - 1.50) \times 10^{-3}$ mol $= 1.25 \times 10^{-3}$ mol OH^-

$[OH^-] = \dfrac{1.25 \times 10^{-3} \text{ mol}}{0.0350 \text{ L}} = 0.0357$ mol·L^{-1}

$pOH = -\log(0.0357) = 1.447$, $pH = 14.00 - 1.447 = 12.55$

(d) $pH = 7.00$

$V_{HCl} = (2.75 \times 10^{-3} \text{ mol NaOH}) \left(\dfrac{1 \text{ mol HCl}}{1 \text{ mol NaOH}} \right) \left(\dfrac{1 \text{ L HCl}}{0.150 \text{ mol HCl}} \right)$

$= 0.0183$ L

(e) $[H_3O^+] = (0.0050 \text{ L}) \left(\dfrac{0.150 \text{ mol}}{1 \text{ L}} \right) \left(\dfrac{1}{(0.0250 + 0.0183 + 0.0050) \text{ L}} \right)$

$= 0.016$ mol·L^{-1}

$pH = -\log(0.016) = 1.80$

(f) $[H_3O^+] = \left(\dfrac{0.010 \text{ L}}{0.0533 \text{ L}} \right) \left(\dfrac{0.150 \text{ mol}}{1 \text{ L}} \right) = 0.028$ mol·L^{-1}

$pH = -\log(0.028) = 1.55$

6H.7 Mass of pure NaOH $= (0.0342 \text{ L HCl}) \left(\dfrac{0.0695 \text{ mol HCl}}{1 \text{ L HCl}} \right)$

$\left(\dfrac{1 \text{ mol NaOH}}{1 \text{ mol HCl}} \right) \left(\dfrac{40.00 \text{ g NaOH}}{1 \text{ mol NaOH}} \right) \left(\dfrac{300 \text{ mL}}{25.0 \text{ mL}} \right)$

$= 1.14$ g

percent purity $= \dfrac{1.14 \text{ g}}{1.436 \text{ g}} \times 100\% = 79.4$

6H.9 At stoichiometric point of titration, the volume of required NaOH is:

$V_{NaOH} = \dfrac{25.00 \text{ mL} \times 0.120 \text{ M}}{0.0230 \text{ M}} = 130$ mL

At stoichiometric point of titration, all C_6H_5COOH are converted to

$C_6H_5COO^-$. Then the following reaction will occur to reach to equilibrium:

$$C_6H_5COO^-(aq) + H_2O(l) \rightarrow C_6H_5COOH(aq) + OH^-(aq)$$

The initial concentration of $C_6H_5COO^-$ at stoichiometric point of titration

$$= \frac{25.00 \text{ mL} \times 0.120 \text{ M}}{155 \text{ mL}} = 0.0194 \text{ M}$$

Concentration

$(mol \cdot L^{-1})$	$C_6H_5COO^-(aq)$	$+ H_2O(l)$	$\rightleftharpoons C_6H_5COOH(aq)$	$+ OH^-(aq)$
initial	0.0194	—	0	0
change	$-x$	—	$+x$	$+x$
equilibrium	$0.0194 - x$	—	x	x

$$K_b = \frac{K_w}{K_a} = \frac{1.0 \times 10^{-14}}{6.5 \times 10^{-5}} = 1.5 \times 10^{-10} = \frac{x^2}{0.0194 - x}$$

Assuming x is small because of the small K_b, $x = [OH^-] = 1.7 \times 10^{-6}$ M

pOH = 5.77;   pH = 14 − 5.77 = 8.23

6H.11 The moles of OH^- added are equivalent to the number of moles of HA

present: $(0.350 \text{ mol} \cdot L^{-1})(0.052 \text{ L}) = 0.0182 \text{ mol OH}^-$

Therefore, 0.0182 mol HA were present in solution

(a) molar mass of HA: $\frac{4.25 \text{ g}}{0.0182 \text{ mol}} = 234 \text{ g} \cdot \text{mol}^{-1}$

(b) pK_a = pH at half-way titration = 3.82

6H.13 (a) Concentration

$(mol \cdot L^{-1})$	$CH_3COOH(aq)$	$+ H_2O(l)$	$\rightleftharpoons H_3O^+(aq)$	$+ CH_3CO_2^-(aq)$
initial	0.10	—	0	0
change	$-x$	—	$+x$	$+x$
equilibrium	$0.10 - x$	—	x	x

$$K_a = \frac{[H_3O^+][CH_3CO_2^-]}{[CH_3COOH]} = 1.8 \times 10^{-5} = \frac{x^2}{0.10 - x} \approx \frac{x^2}{0.10}$$

$$x^2 = 1.8 \times 10^{-6}$$

$$x = 1.3 \times 10^{-3} \text{ mol} \cdot \text{L}^{-1} = [\text{H}_3\text{O}^+]$$

$$\text{initial pH} = -\log(1.3 \times 10^{-3}) = 2.89$$

(b)   moles of $\text{CH}_3\text{COOH} = (0.0250 \text{ L})(0.10 \text{ M})$

$$= 2.50 \times 10^{-3} \text{ mol CH}_3\text{COOH}$$

moles of $\text{NaOH} = (0.0100 \text{ L})(0.10 \text{ M}) = 1.0 \times 10^{-3} \text{ mol OH}^-$

After neutralization,

$$\frac{1.50 \times 10^{-3} \text{ mol CH}_3\text{COOH}}{0.0350 \text{ L}} = 4.29 \times 10^{-2} \text{ mol} \cdot \text{L}^{-1} \text{ CH}_3\text{COOH}$$

$$\frac{1.0 \times 10^{-3} \text{ mol CH}_3\text{CO}_2^-}{0.0350 \text{ L}} = 2.86 \times 10^{-2} \text{ mol} \cdot \text{L}^{-1} \text{ CH}_3\text{CO}_2^-$$

Then consider equilibrium, $K_a = \dfrac{[\text{H}_3\text{O}^+][\text{CH}_3\text{CO}_2]}{[\text{CH}_3\text{COOH}]}$

Concentration

$$(\text{mol} \cdot \text{L}^{-1}) \quad \text{CH}_3\text{COOH(aq)} + \text{H}_2\text{O(l)} \rightleftharpoons \text{H}_3\text{O}^+ \text{(aq)} + \text{CH}_3\text{CO}_2^- \text{(aq)}$$

initial	4.29×10^{-2}	—	0	2.86×10^{-2}
change	$-x$	—	$+x$	$+x$
equilibrium	$4.29 \times 10^{-2} - x$	—	x	$2.86 \times 10^{-2} + x$

$$1.8 \times 10^{-5} = \frac{(x)(x + 2.86 \times 10^{-2})}{(4.29 \times 10^{-2} - x)}; \text{ assume } +x \text{ and } -x \text{ negligible.}$$

$$[\text{H}_3\text{O}^+] = x \doteq 2.7 \times 10^{-5} \text{ mol} \cdot \text{L}^{-1} \text{ and pH} = -\log(2.7 \times 10^{-5}) = 4.56$$

(c) Because acid and base concentrations are equal, their volumes are equal at the stoichiometric point. Therefore, 25.0 mL NaOH is required to reach the stoichiometric point, and 12.5 mL NaOH is required to reach the half stoichiometric point.

(d) At the half stoichiometric point, pH = $pK_a = 4.75$.

(e) 25.0 mL; see part (c)

(f) The pH is that of 0.050 M NaCH_3CO_2.

Concentration

$$(\text{mol} \cdot \text{L}^{-1}) \quad \text{H}_2\text{O(l)} + \text{CH}_3\text{CO}_2^- \text{(aq)} \rightleftharpoons \text{CH}_3\text{COOH(aq)} + \text{OH}^- \text{(aq)}$$

		initial	—	0.050	0	0
		change	—	$-x$	$+x$	$+x$
		equilibrium	—	$0.050 - x$	x	x

$$K_b = \frac{K_w}{K_a} = \frac{1.00 \times 10^{-14}}{1.8 \times 10^{-5}} = 5.6 \times 10^{-10} = \frac{x^2}{0.050 - x} \approx \frac{x^2}{0.050}$$

$$x^2 = 2.8 \times 10^{-11}$$

$$x = 5.3 \times 10^{-6} \text{ mol} \cdot \text{L}^{-1} = [OH^-]$$

$$pOH = 5.28, \ pH = 14.00 - 5.28 = 8.72$$

6H.15 When a buffer solution $pH = pK_a$ (or $pOH = pK_b$), the buffer has the highest buffer capacity. Therefore, at a point when 50% of the weak base is neutralized with a strong acid, the solution has the highest buffer capacity.

6H.17 (a) $K_b = \dfrac{[NH_4^+][OH^-]}{[NH_3]} = 1.8 \times 10^{-5}$

Concentration

(mol · L^{-1})	H$_2$O(l)	+	NH$_3$(aq)	$\rightleftharpoons$	NH$_4^+$(aq)	+	OH$^-$(aq)
initial	—		0.15		0		0
change	—		$-x$		$+x$		$+x$
equilibrium	—		$0.15 - x$		x		x

$$1.8 \times 10^{-5} = \frac{x^2}{0.15 - x} \approx \frac{x^2}{0.15}$$

$$[OH^-] = x = 1.6 \times 10^{-3} \text{ mol} \cdot \text{L}^{-1}$$

$$pOH = 2.80, \ \text{initial } pH = 14.00 - 2.80 = 11.20$$

(b) initial moles of

$$NH_3 = (0.0150 \text{ L})(0.15 \text{ mol} \cdot \text{L}^{-1}) = 2.3 \times 10^{-3} \text{ mol NH}_3$$

moles of HCl $= (0.0150 \text{ L})(0.10 \text{ mol} \cdot \text{L}^{-1}) = 1.5 \times 10^{-3} \text{ mol HCl}$

$$\frac{(2.3 \times 10^{-3} - 1.5 \times 10^{-3}) \text{ mol NH}_3}{0.0300 \text{ L}} = 2.7 \times 10^{-2} \text{ mol} \cdot \text{L}^{-1} \text{ NH}_3$$

$$\frac{1.5 \times 10^{-3} \text{ mol HCl}}{0.0300 \text{ L}} = 5.0 \times 10^{-2} \text{ mol} \cdot \text{L}^{-1} \text{ HCl} \approx 5.0 \times 10^{-2} \text{ mol} \cdot \text{L}^{-1} \text{ NH}_4^+$$

Then consider the equilibrium:

Concentration

$(\text{mol} \cdot \text{L}^{-1})$	$H_2O(l)$	$+$	$NH_3(aq)$	$\rightleftharpoons$	$NH_4^+(aq)$	$+$	$OH^-(aq)$
initial	—		2.7×10^{-2}		5.0×10^{-2}		0
change	—		$-x$		$+x$		$+x$
equilibrium	—		$2.7 \times 10^{-2} - x$		$5.0 \times 10^{-2} + x$		x

$$K_b = \frac{[NH_4^+][OH^-]}{[NH_3]} = 1.8 \times 10^{-5}$$

$$= \frac{(x)(5.0 \times 10^{-2} + x)}{(2.7 \times 10^{-2} - x)}; \text{ assume that } +x \text{ and } -x \text{ are negligible}$$

$[OH^-] = x = 9.7 \times 10^{-6} \text{ mol} \cdot \text{L}^{-1}$ and pOH $= 5.01$

Therefore, pH $= 14.00 - 5.01 = 8.99$.

(c) At the stoichiometric point, moles of NH_3 = moles of HCl.

volume HCl added $= \dfrac{(0.15 \text{ mol} \cdot \text{L}^{-1} \text{ NH}_3)(0.0150 \text{ L})}{0.10 \text{ mol} \cdot \text{L}^{-1} \text{ HCl}} = 0.022 \text{ L HCl}$

Therefore, halfway to the stoichiometric point, volume HCl added

$= 22/2 = 11 \text{ mL}$.

(d) At half stoichiometric point, pOH $= pK_b$ and pOH $= 4.75$.

Therefore, pH $= 14.00 - 4.75 = 9.25$.

(e) 22 mL; see part (c)

(f) $NH_4^+(aq) + H_2O(l) \rightleftharpoons H_3O^+(aq) + NH_3(aq)$

The initial moles of NH_3 have now been converted to moles of NH_4^+ in a

$(15 + 22 = 37)$ mL volume:

$$[NH_4^+] = \frac{2.25 \times 10^{-3}\ mol}{0.037\ L} = 0.061\ mol \cdot L^{-1}$$

$$K_a = \frac{K_w}{K_b} = \frac{1.00 \times 10^{-14}}{1.8 \times 10^{-5}} = 5.6 \times 10^{-10}$$

Concentration

$(mol \cdot L^{-1})$	$NH_4^+(aq)$	$+$	$H_2O(l)$	$\rightleftharpoons$	$H_3O^+(aq)$	$+$	$NH_3(aq)$
initial	0.061		—		0		0
change	$-x$		—		$+x$		$+x$
equilibrium	$0.061 - x$		—		x		x

$$K_a = 5.6 \times 10^{-10} = \frac{x^2}{0.061 - x} \approx \frac{x^2}{0.061}$$

$$x = [H_3O^+] = 5.8 \times 10^{-6}\ mol \cdot L^{-1}$$

$$pH = -\log(5.8 \times 10^{-6}) = 5.24$$

(g) Methyl red will be appropriate for this titration.

6H.19 (a) The acid is a weak acid because the initial pH of the acid = 5.0 and the stoichiometric point occurs at pH = 10 (the anion is a base).

(b) pH = 5.0, $[H_3O^+] = 1 \times 10^{-5}\ mol \cdot L^{-1}$

(c) At halfway titration, pH = pK_a. It takes 5.0 mL base to reach halfway, the pH = 7.5 = pK_a of the acid. $K_a = 10^{-7.5} = 3 \times 10^{-8}$.

(d) $HA(aq) + H_2O(l) \rightarrow H_3O^+(aq) + A^-(aq)$

When HA self-dissociation reaches to an equilibrium:

$[H_3O^+] = [A^-] = 1 \times 10^{-5}\ mol \cdot L^{-1}$; $[HA]_{eq} = [HA]_{initial} - (1 \times 10^{-5})$

$$K_a = 3 \times 10^{-8} = \frac{[H_3O^+][A^-]}{[HA]_{eq}} = \frac{(1 \times 10^{-5})^2}{[HA]_{initial} - (1 \times 10^{-5})}$$

Since K_a is small, the dissociation level of HA will be very small.

Therefore, $[HA]_{initial} - (1 \times 10^{-5}) \approx [HA]_{initial}$.

Solve for the equation: $[HA]_{initial} = 3 \times 10^{-3}\ mol \cdot L^{-1}$

(e) It takes 10.0 mL base to reach to stoichiometric point. $[HA]_{initial} = 3.16 \times 10^{-3}\ mol \cdot L^{-1}$. If the volume of acid = 25.0 mL, then

$$[B] = \frac{[HA]V_{HA}}{V_{base}} = \frac{(3 \times 10^{-3})(25\ mL)}{(10.0\ mL)} = 8 \times 10^{-3}\ M.$$

(f) Phenolphthalein will be an appropriate indicator.

6H.21 At the stoichiometric point, the volume of solution will have doubled; therefore, the concentration of $CH_3CO_2^-$ will be 0.10 M. The equilibrium is

Concentration

$(mol \cdot L^{-1})\quad CH_3CO_2^-(aq) + H_2O(l) \rightleftharpoons CH_3CO_2H(aq) + OH^-(aq)$

initial	0.10	—	0	0
change	$-x$	—	$+x$	$+x$
equilibrium	$0.10 - x$	—	x	x

$$K_b = \frac{K_w}{K_a} = \frac{1.00 \times 10^{-14}}{1.8 \times 10^{-5}} = 5.6 \times 10^{-10}$$

$$K_b = \frac{[CH_3CO_2H][OH^-]}{[CH_3CO_2^-]} = \frac{x^2}{0.10 - x} \approx \frac{x^2}{0.10} = 5.6 \times 10^{-10}$$

$x = 7.5 \times 10^{-6}\ mol \cdot L^{-1} = [OH^-]$

$pOH = -\log(7.5 \times 10^{-6}) = 5.12,\ pH = 14.00 - 5.12 = 8.88$

From Table 6H.2, we see that this pH value lies within the range for phenolphthalein and thymol blue; the others would not.

6H.23 For both exercise 6H.9 and 6H.14: thymol blue or phenolphthalein.

6H.25 (a) To reach the first stoichiometric point, we must add enough solution to neutralize one H^+ on the H_3AsO_4. To do this, we will require

$0.0750\ L \times 0.137\ mol \cdot L^{-1} = 0.0103\ mol\ OH^-$. The volume of base required will be given by the number of moles of base required divided by the concentration of base solution:

$$\frac{0.0750\ L \times 0.137\ mol \cdot L^{-1}}{0.275\ mol \cdot L^{-1}} = 0.0374\ L\ or\ 37.4\ mL$$

(b) and (c) To reach the second stoichiometric point will require double
the amount calculated in (a), or 74.8 mL, and the third stoichiometric
point will be reached with three times the amount added in (a), or 112 mL.

6H.27 (a) The base HPO_3^{2-} is the fully deprotonated form of phosphorous acid
H_3PO_3 (the remaining H attached to P is not acidic). It will require an
equal number of moles of HNO_3 to react with HPO_3^{2-} in order to reach
the first stoichiometric point (formation of $H_2PO_3^-$). The value will be

given by $\dfrac{0.0355\ L \times 0.158\ mol \cdot L^{-1}}{0.255\ mol \cdot L^{-1}} = 0.0220\ L$ or 22.0 mL

(b) To reach the second stoichiometric point would require double the
amount of solution calculated in (a), or 44.0 mL.

6H.29 (a)

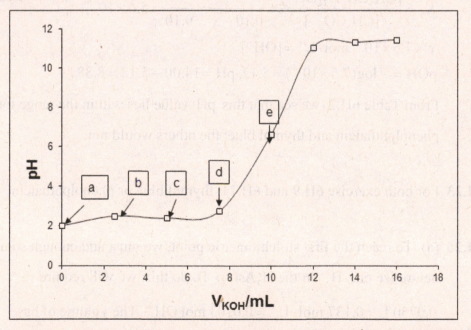

Point (a): initial point, $K_{a1} = 0.25 = [H^+]^2/(0.010-[H^+])$, $[H^+] = 0.0096$
$mol \cdot L^-$; pH = 2.02; point (b): first halfway point, pH = pK_{a1} = 0.6; point
(c): first stoichiometric point (see calculation in part c); point (d): second

halfway point, pH = pK_{a2} = 1.74; point (e) second stoichiometric point (see calculation in part c below)

Note: (i) The initial pH is higher than points b and c because we assume that the initial [H^+] from the first dissociation only (actually second dissociation contributes also); (ii) due to the close pK_a values of thiosulfuric acid, only one obvious titration curve can be observed. Two protons will be titrated at the same time for this acid.

(b) First stoichiometric point: (C) = 5.0 mL; second stoichiometric point: (E) = 10 mL

(c) First stoichiometric point: pH = $\frac{1}{2}(pK_{a_1} + pK_{a_2}) = \frac{1}{2}(0.6 + 1.74) = 1.2$;

second stoichiometric point:

$$S_2O_3^{2-} + H_2O \rightleftharpoons HS_2O_3^- + OH^-$$

$$K_{b_1} = \frac{K_W}{K_{a_2}} = 5.56 \times 10^{-13}, \ [S_2O_3^{2-}]_{initial} = (5.0 \times 0.01)/15.0 = 0.0033 \text{ M}$$

$$K_{b_1} = \frac{x^2}{0.0033 - x} = 5.56 \times 10^{-13}, \ x = [OH^-] = 4.30 \times 10^{-8} < [OH^-]_{water}$$

Therefore, pH = 7.0.

6H.31 (a) This value is calculated as described in Example 6E.2. First, we calculate the molarity of the starting phosphorous acid solution:

$\dfrac{0.122 \text{ g}}{81.99 \text{ g} \cdot \text{mol}^{-1}} \bigg/ 0.0500 \text{ L} = 0.0298 \text{ mol} \cdot \text{L}^{-1}$. We then use the first acid

dissociation of phosphorous acid as the dominant equilibrium. The

K_{a1} is 1.0×10^{-2}. Let H_2P represent the fully protonated phosphorus acid.

Concentration (mol·L^{-1})	$H_2P(aq)$	+ $H_2O(l)$	$\rightleftharpoons$ $HP^-(aq)$	+ $H_3O^+(aq)$
initial	0.0298	—	0	0
change	$-x$	—	$+x$	$+x$
final	$0.0298 - x$	—	$+x$	$+x$

$$K_a = \frac{[H_3O^+][HP^-]}{[H_2P]} = 1.0 \times 10^{-2}$$

$$1.0 \times 10^{-2} = \frac{x \cdot x}{0.0298 - x} = \frac{x^2}{0.0298 - x}$$

If we assume $x \ll 0.0298$, then the equation becomes

$$x^2 = (1.0 \times 10^{-2})(0.0298) = 2.98 \times 10^{-4}$$

$$x = 1.73 \times 10^{-2}.$$

Because this value is more than 10% of 0.0300, the full quadratic solution should be undertaken. The equation is

$$x^2 = (1.0 \times 10^{-2})(0.0298 - x) \text{ or}$$

$$x^2 + (1.0 \times 10^{-2}x) - (2.98 \times 10^{-4}) = 0.$$

Using the quadratic formula, we obtain $x = 0.013$.

pH $= 1.89$

(b) First, carry out the reaction between phosphorous acid and the strong base to completion:

$$H_2P(aq) + OH^-(aq) \longrightarrow HP^-(aq) + H_2O(l)$$

moles of $H_2P = (0.0298 \text{ mol} \cdot L^{-1})(0.0500 \text{ L}) = 1.49 \times 10^{-3} \text{ mol}$

moles of $OH^- = (0.00500 \text{ L})(0.175 \text{ mol} \cdot L^{-1}) = 8.75 \times 10^{-4} \text{ mol}$

8.75×10^{-4} mol OH^- will react completely with 1.49×10^{-3} mol H_2P

to give 8.75×10^{-4} mol HP^-, with 6.15×10^{-4} moles of H_2P remaining.

$$[H_2P] = \frac{6.15 \times 10^{-4} \text{ mol}}{0.0550 \text{ L}} = 0.0112 \text{ mol} \cdot L^{-1}$$

$$[HP^-] = \frac{8.75 \times 10^{-4} \text{ mol}}{0.0550 \text{ L}} = 0.0159 \text{ mol} \cdot L^{-1}$$

Concentration

(mol·L⁻¹)	$H_2P(aq)$	+	$H_2O(l)$	$\rightleftharpoons$	$HP^-(aq)$	+	$H_3O^+(aq)$
initial	0.0112		—		0.0159		0
change	$-x$		—		$+x$		$+x$
final	$0.0112 - x$		—		$0.0159 + x$		$+x$

The calculation is performed as in part (a):

$$1.0 \times 10^{-2} = \frac{(0.0159 + x)x}{0.0112 - x}$$

$$x = 3.75 \times 10^{-3}$$

$$pH = 2.42$$

(c) moles of $H_2P = 1.49 \times 10^{-3}$

moles of $OH^- = 8.75 \times 10^{-4} + 8.75 \times 10^{-4} = 1.75 \times 10^{-3}$

Following the reaction between H_2P and OH^-, 0 mol of H_2P remain and

2.60×10^{-4} mol OH^- remain. $[OH^-] = 4.33 \times 10^{-3}$ M

1.49×10^{-3} mol of HP^- remain, $[HP^-] = \dfrac{1.49 \times 10^{-3} \text{ mol}}{0.060 \text{ L}} = 2.48 \times 10^{-2}$ M.

Since 4.33×10^{-3} M OH^- will produce 4.33×10^{-3} M P^{2-},

the equilibrium $[HP^-] = 2.48 \times 10^{-2} - 4.33 \times 10^{-3} = 2.05 \times 10^{-2}$ M

Concentration

(mol·L⁻¹)	HP^-	+	H_2O	$\rightleftharpoons$	P^{2-}	+	H_3O^+
initial	2.05×10^{-2}		—		4.33×10^{-3}		0
change	$-x$		$-x$		$+x$		$+x$
final	$2.05 \times 10^{-2} - x$		—		$4.33 \times 10^{-3} + x$		x

The calculation is performed as in part (a):

$$K_{a2} = 2.60 \times 10^{-7} = \frac{x(4.33 \times 10^{-3} + x)}{(2.05 \times 10^{-2} - x)}$$

Assuming x is small, we find $x = 1.23 \times 10^{-6}$,

$pH = -\log(1.23 \times 10^{-6}) = 5.91$.

6H.33 (a) The reaction of the base Na_2HPO_4 with the strong acid will be taken

to completion first:

$$HPO_4^{2-}(aq) + H_3O^+(aq) \longrightarrow H_2PO_4^- + H_2O(l)$$

Initially, moles of

$HPO_4^{2-} =$ moles of $H_3O^+ = 0.0500 \text{ L} \times 0.275 \text{ mol·L}^{-1} = 0.0138$ mol

Because this reaction proceeds with no excess base or acid, we are dealing with a solution that can be viewed as being composed of $H_2PO_4^-$. The problem then becomes one of estimating the pH of this solution, which can be done from the relationship

$$pH = \tfrac{1}{2}(pK_{a1} + pK_{a2})$$

$$pH = \tfrac{1}{2}(2.12 + 7.21) = 4.66.$$

(b) This reaction proceeds as in (a), but there is more strong acid available, so the excess acid will react with $H_2PO_4^-$ to produce H_3PO_4. Addition of the first 50.0 mL of acid solution will convert all the HPO_4^{2-} into $H_2PO_4^-$. The additional 25.0 mL of the strong acid will react with $H_2PO_4^-$:

$$H_2PO_4^-(aq) + H_3O^+(aq) \longrightarrow H_3PO_4(aq) + H_2O(l)$$

0.0138 mol $H_2PO_4^-$ will react with 0.006 88 mol H_3O^+ to give 0.0069 mol H_3PO_4, with 0.069 mol $H_2PO_4^-$ in excess. The concentrations will be

$$[H_3PO_4] = [H_2PO_4^-] = \frac{0.0069 \text{ mol}}{0.125 \text{ L}} = 0.055.$$ The appropriate relationship to use is then

Concentration

$(mol \cdot L^{-1})$	$H_3PO_4(aq)$	$+$	$H_2O(l)$	$\rightleftharpoons$	$H_2PO_4^-(aq)$	$+$	$H_3O^+(aq)$
initial	0.055		—		0.055		0
change	$-x$		—		$+x$		$+x$
final	$0.055 - x$		—		$0.055 + x$		$+x$

$$K_{a1} = \frac{[H_2PO_4^-][H_3O^+]}{[H_3PO_4]}$$

Because the equilibrium constant is not small compared with 0.055, the full quadratic solution must be calculated:

$$x^2 + 0.055x = 7.6 \times 10^{-3} (0.055 - x)$$

$x^2 + 0.063x - 4.2 \times 10^{-4} = 0$

$x = 1.6 \times 10^{-3}$

$pH = -\log(1.6 \times 10^{-3}) = 2.80$

(c) The reaction of Na_2HPO_4 with strong acid goes only halfway to completion. 0.275 mol of HPO_4^{2-} will react with

$(0.025 \text{ L} \times 0.275 \text{ mol} \cdot L^{-1}) = 6.9 \times 10^{-3}$ mol HCl to produce 6.9×10^{-3} mol $H_2PO_4^-$ and leave 6.9×10^{-3} HPO_4^{2-} unreacted.

6.9×10^{-3} mol $\div$ 0.075 L = 0.092 mol $\cdot L^{-1}$

Concentration

$(mol \cdot L^{-1})$	$H_2PO_4^- (aq)$	$+ H_2O(l)$	$\rightleftharpoons$	$H_2PO_4^{2-}(aq)$	$+ H_3O^+(aq)$
intial	0.092	—		0.092	0
change	$-x$	—		$+x$	$+x$
final	$0.092 - x$	—		$0.092 + x$	$+x$

$K_{a2} = 6.2 \times 10^{-8} = \dfrac{[HPO_4^{2-}][H_3O^+]}{[H_2PO_4^-]}$

$\dfrac{[0.092 + x][H_3O^+]}{[0.092 - x]} = 6.2 \times 10^{-8}$

assuming $x \ll$ than 0.092

$x = [H_3O^+] = 6.2 \times 10^{-8}$

$pH = -\log(6.2 \times 10^{-8}) = 7.21$

6I.1   (a)   The solubility equilibrium is $AgBr(s) \rightleftharpoons Ag^+(aq) + Br^-(aq)$.

$[Ag^+] = [Br^-] = 8.8 \times 10^{-7}$ mol $\cdot L^{-1} = S =$ solubility

$K_{sp} = [Ag^+][Br^-] = (8.8 \times 10^{-7})(8.8 \times 10^{-7}) = 7.7 \times 10^{-13}$

(b)   The solubility equilibrium is $PbCrO_4(s) \rightleftharpoons Pb^{2+}(aq) + CrO_4^{2-}(aq)$.

$[Pb^{2+}] = 1.3 \times 10^{-7}$ mol $\cdot L^{-1} = S$, $[CrO_4^{2-}] = 1.3 \times 10^{-7}$ mol $\cdot L^{-1} = S$

$K_{sp} = [Pb^{2+}][CrO_4^{2-}] = (1.3 \times 10^{-7})(1.3 \times 10^{-7}) = 1.7 \times 10^{-14}$

(c) The solubility equilibrium is $Ba(OH)_2(s) \rightleftharpoons Ba^{2+}(aq) + 2\,OH^-(aq)$.

$[Ba^{2+}] = 0.11\ mol \cdot L^{-1} = S, \quad [OH^-] = 0.22\ mol \cdot L^{-1} = 2S$

$K_{sp} = [Ba^{2+}][OH^-]^2 = (0.11)(0.22)^2 = 5.3 \times 10^{-3}$

(d) The solubility equilibrium is $MgF_2(s) \rightleftharpoons Mg^{2+}(aq) + 2\,F^-(aq)$.

$[Mg^{2+}] = 1.2 \times 10^{-3}\ mol \cdot L^{-1} = S, \quad [F^-] = 2.4 \times 10^{-3}\ mol \cdot L^{-1} = 2S$

$K_{sp} = [Mg^{2+}][F^-]^2 = (1.2 \times 10^{-3})(2.4 \times 10^{-3})^2 = 6.9 \times 10^{-9}$

6I.3 $Tl_2CrO_4(s) \rightleftharpoons 2\,Tl^+(aq) + CrO_4^{\,2-}(aq)$

$[CrO_4^{\,2-}] = S = 6.3 \times 10^{-5}\ mol \cdot L^{-1}$

$[Tl^+] = 2S = 2(6.3 \times 10^{-5})\ mol \cdot L^{-1}$

$K_{sp} = [Tl^+]^2[CrO_4^{\,2-}] = (2S)^2 \times (S)$

$K_{sp} = [2(6.3 \times 10^{-5})]^2 \times (6.3 \times 10^{-5}) = 1.0 \times 10^{-12}$

6I.5 (a) Equilibrium equation: $BiI_3(s) \rightleftharpoons Bi^{3+}(aq) + 3\,I^-(aq)$

$K_{sp} = 7.71 \times 10^{-19} = [Bi][I^-]^3 = (S)(3S)^3$

$S = 1.30 \times 10^{-5}\ mol \cdot L^{-1}$

(b) Equilibrium equation: $CuCl(s) \leftrightarrow Cu^+(aq) + Cl^-(aq)$

$K_{sp} = 1.0 \times 10^{-6} = [Cu^+][Cl] = S^2$

$S = 1.0 \times 10^{-3}\ mol \cdot L^{-1}$

(c) Equilibrium equation: $CaCO_3(s) \rightleftharpoons Ca^{2+}(aq) + CO_3^{\,2-}(aq)$

$K_{sp} = [Ca^{2+}][CO_3^{\,2-}] = S \times S = S^2 = 8.7 \times 10^{-9}$

$S = 9.3 \times 10^{-5}\ mol \cdot L^{-1}$

6I.7 (a)

Concentration $(mol \cdot L^{-1})$	$AgCl(s)$	$\rightleftharpoons$	$Ag^+(aq)$	$+$	$Cl^-(aq)$
initial	—		0		0.20
change	—		$+S$		$+S$
equilibrium	—		S		$S + 0.20$

$K_{sp} = [Ag^+][Cl^-] = (S) \times (S + 0.20) = 1.6 \times 10^{-10}$

Assume S in $(S + 0.20)$ is negligible, so $0.20\,S = 1.6 \times 10^{-10}$

$S = 8.0 \times 10^{-10}$ mol$\cdot$L^{-1} = [Ag$^+$] = molar solubility of AgCl in

0.15 M NaCl

(b) Concentration (mol$\cdot$L^{-1})   Hg$_2$Cl$_2$(s) $\rightleftharpoons$ Hg$_2$$^{2+}$(aq)   +   2 Cl$^-$(aq)

initial	—	0	0.150
change	—	+S	+2S
equilibrium	—	S	0.150 + 2S

$K_{sp} = [\text{Hg}_2{}^{2+}][\text{Cl}^-]^2 = (S) \times (2S + 0.150)^2 = 2.6 \times 10^{-18}$

Assume $2S$ in $(2S + 0.150)$ is negligible, so $(0.150)^2 S = 2.6 \times 10^{-18}$.

$S = 1.2 \times 10^{-16}$ mol$\cdot$L^{-1} = [Hg$_2$$^{2+}$] = molar solubility of Hg$_2$Cl$_2$ in

0.150 M NaCl.

(c) Concentration (mol$\cdot$L^{-1})   PbCl$_2$(s) $\rightleftharpoons$ Pb^{2+}(aq)   +   2 Cl$^-$(aq)

initial	—	0	2 $\times$ 0.025 = 0.05
change	—	+S	+2S
equilibrium	—	S	2S + 0.05

$K_{sp} = [\text{Pb}^{2+}][\text{Cl}^-]^2 = S \times (2S + 0.05)^2 = 1.6 \times 10^{-5}$

S may not be negligible relative to 0.05, so the full cubic form may be required. We do it both ways:

For $S^3 + 0.20\,S^2 + (0.0025 \times 10^{-2}\,S) - (1.6 \times 10^{-5}) = 0$, the solution by

standard methods is $S = 4.6 \times 10^{-3}$ mol$\cdot$L^{-1}.

If S had been neglected, the answer would be $S = 6.4 \times 10^{-3}$.

(d) Concentration (mol$\cdot$L^{-1})   Fe(OH)$_2$(s) $\rightleftharpoons$ Fe^{2+}(aq)   +   2 OH$^-$(aq)

initial	—	2.5×10^{-3}	0
change	—	+S	+2S
equilibrium	—	2.5×10^{-3} + S	2S

$$K_{sp} = [\text{Fe}^{2+}][\text{OH}^-]^2 = (S + 2.5 \times 10^{-3}) \times (2S)^2 = 1.6 \times 10^{-14}$$

Assume S in $(S + 2.5 \times 10^{-3})$ is negligible, so

$$4S^2 \times (2.5 \times 10^{-3}) = 1.6 \times 10^{-1}.$$

$$S^2 = 1.6 \times 10^{-12}$$

$$S = 1.3 \times 10^{-6} \text{ mol} \cdot \text{L}^{-1} = \text{molar solubility of Fe(OH)}_2 \text{ in } 2.5 \times 10^{-3} \text{ M FeCl}_2.$$

6I.9 (a) pH = 7.0; $[\text{OH}^-] = 1.0 \times 10^{-7}$ mol·L^{-1}

$$\text{Al}^{3+}(\text{aq}) + 3\,\text{OH}^-(\text{aq}) \rightleftharpoons \text{Al(OH)}_3(\text{s})$$

$$[\text{Al}^{3+}][\text{OH}^-]^3 = K_{sp} = 1.0 \times 10^{-33}$$

$$S \times (10^{-7})^3 = 1.0 \times 10^{-33}$$

$$S = \frac{1.0 \times 10^{-33}}{1 \times 10^{-21}} = 1.0 \times 10^{-12} \text{ mol} \cdot \text{L}^{-1} = [\text{Al}^{3+}]$$

= molar solubility of Al(OH)$_3$ at pH = 7.0

(b) pH = 4.5; pOH = 9.5; $[\text{OH}^-] = 3.2 \times 10^{-10}$ mol·L^{-1}

$$[\text{Al}^{3+}][\text{OH}^-]^3 = K_{sp} = 1.0 \times 10^{-33}$$

$$S \times (3.2 \times 10^{-10})^3 = 1.0 \times 10^{-33}$$

$$S = \frac{1.0 \times 10^{-33}}{3.3 \times 10^{-29}} = 3.1 \times 10^{-5} \text{ mol} \cdot \text{L}^{-1} = [\text{Al}^{3+}]$$

= molar solubility of Al(OH)$_3$ at pH = 4.5

(c) pH = 7.0; $[\text{OH}^-] = 1.0 \times 10^{-7}$ mol·L^{-1}

$$\text{Zn}^{2+}(\text{aq}) + 2\,\text{OH}^-(\text{aq}) \rightleftharpoons \text{Zn(OH)}_2(\text{s})$$

$$[\text{Zn}^{2+}][\text{OH}^-]^2 = K_{sp} = 2.0 \times 10^{-17}$$

$$S \times (1.0 \times 10^{-7})^2 = 2.0 \times 10^{-17}$$

$$S = \frac{2.0 \times 10^{-17}}{1.0 \times 10^{-14}} = 2.0 \times 10^{-3} \text{ mol} \cdot \text{L}^{-1} = [\text{Zn}^{2+}]$$

= molar solubility of Zn(OH)$_2$ at pH = 7.0

(d) pH = 6.0; pOH = 8.0; $[\text{OH}^-] = 1.0 \times 10^{-8}$ mol·L^{-1}

$[Zn^{2+}][OH^-]^2 = 2.0 \times 10^{-17} = K_{sp}$

$S \times (1.0 \times 10^{-8})^2 = 2.0 \times 10^{-17}$

$S = \dfrac{2.0 \times 10^{-17}}{1.0 \times 10^{-16}} = 2.0 \times 10^{-1} = 0.20 \text{ mol} \cdot L^{-1} = [Zn^{2+}]$

= molar solubility of $Zn(OH)_2$ at pH = 6.0

6I.11 $\quad AgBr(s) \rightleftharpoons Ag^+(aq) + Br^-(aq) \qquad\qquad\qquad K_{sp} = 7.7 \times 10^{-13}$

$Ag^+(aq) + 2\,CN^-(aq) \rightleftharpoons Ag(CN)_2^-(aq) \qquad\qquad K_f = 5.6 \times 10^8$

$AgBr(s) + 2\,CN^-(aq) \rightleftharpoons Ag(CN)_2^-(aq) + Br^-(aq) \qquad K = 4.3 \times 10^{-4}$

Hence, $K = \dfrac{[Ag(CN)_2^-][Br^-]}{[CN^-]^2} = 4.3 \times 10^{-4}$

Concentration

$(mol \cdot L^{-1})$	$AgBr(s)$	+	$2\,CN^-(aq)$	$\rightleftharpoons$	$Ag(CN)_2^-(aq)$	+	$Br^-(aq)$
initial	—		0.10		0		0
change	—		$-2S$		$+S$		$+S$
equilibrium	—		$0.10 - 2S$		S		S

$\dfrac{[Ag(CN)_2^-][Br^-]}{[CN^-]^2} = \dfrac{S^2}{(0.10 - 2S)^2} = 4.3 \times 10^{-4}$

$\dfrac{S}{0.10 - 2S} = \sqrt{4.3 \times 10^{-4}} = 2.1 \times 10^{-2}$

$S = (2.1 \times 10^{-3}) - (4.2 \times 10^{-2}S)$

$1.042S = 2.1 \times 10^{-3}$

$S = 2.0 \times 10^{-3} \text{ mol} \cdot L^{-1}$ = molar solubility of AgBr

6J.1 (a) $Ag^+(aq) + Cl^-(aq) \rightleftharpoons AgCl(s)$

Concentration $(mol \cdot L^{-1})$	Ag^+	Cl^-
initial	0	1.0×10^{-5}
change	$+x$	0
equilibrium	x	1.0×10^{-5}

$$K_{sp} = [Ag^+][Cl^-] = 1.6 \times 10^{-10} = (x)(1.0 \times 10^{-5})$$

$$x = [Ag^+] = 1.6 \times 10^{-5} \text{ mol} \cdot L^{-1}$$

(b) mass $AgNO_3$

$$= \left(\frac{1.6 \times 10^{-5} \text{ mol } AgNO_3}{1 \text{ L}} \right) (0.100 \text{ L}) \left(\frac{169.88 \text{ g } AgNO_3}{1 \text{ mol } AgNO_3} \right) \left(\frac{1 \mu g}{10^{-6} \text{ g}} \right)$$

$$= 2.7 \times 10^2 \text{ } \mu g \text{ } AgNO_3$$

6J.3 (a) $Ni^{2+}(aq) + 2 OH^-(aq) \rightleftharpoons Ni(OH)_2(s)$

Concentration $(\text{mol} \cdot L^{-1})$	Ni^{2+}	OH^-
initial	0.060	0
change	0	$+x$
equilibrium	0.060	x

$$K_{sp} = [Ni^{2+}][OH^-]^2 = 6.5 \times 10^{-18} = (0.060)(x)^2$$

$$[OH^-] = x = 1.0 \times 10^{-8} \text{ mol} \cdot L^{-1}$$

$$pOH = -\log(1.0 \times 10^{-8}) = 8.00, \ pH = 14.00 - 8.00 = 6.00$$

(b) A similar setup for $[Ni^{2+}] = 0.030$ M.

gives $x = 1.5 \times 10^{-8}$

$$pOH = -\log(1.5 \times 10^{-8}) = 7.82$$

$$pH = 14.00 - 7.82 = 6.18$$

6J.5 $\left(\dfrac{1 \text{ mL}}{20 \text{ drops}} \right) \times 1 \text{ drop} = 0.05 \text{ mL} = 0.05 \times 10^{-3} \text{ L} = 5 \times 10^{-5} \text{ L}$

and $(5 \times 10^{-5} \text{ L})(0.010 \text{ mol} \cdot L^{-1}) = 5 \times 10^{-7} \text{ mol NaCl} = 5 \times 10^{-7} \text{ mol Cl}^-$

(a) $Ag^+(aq) + Cl^-(aq) \rightleftharpoons AgCl(s), \quad [Ag^+][Cl^-] = K_{sp}$

$$Q_{sp} = \left[\frac{(0.010 \text{ L})(0.0040 \text{ mol} \cdot L^{-1})}{0.010 \text{ L}} \right] \left[\frac{5 \times 10^{-7} \text{ mol}}{0.010 \text{ L}} \right] = 2 \times 10^{-7}$$

Will precipitate because $Q_{sp}(2 \times 10^{-7}) > K_{sp}(1.6 \times 10^{-10})$.

(b) $Pb^{2+}(aq) + 2 Cl^-(aq) \rightleftharpoons PbCl_2(s), \ [Pb^{2+}][Cl^-]^2 = K_{sp}$

$$Q_{sp} = \left[\frac{(0.0100 \text{ L})(0.0040 \text{ mol} \cdot \text{L}^{-1})}{0.010 \text{ L}}\right]\left[\frac{5 \times 10^{-7} \text{ mol}}{0.010 \text{ L}}\right]^2 = 1 \times 10^{-11}$$

Will not precipitate because Q_{sp} $(1 \times 10^{-11}) < K_{sp}$ (1.6×10^{-5}).

6J.7 (a) $K_{sp}[\text{Ni(OH)}_2] < K_{sp}[\text{Mg(OH)}_2] < K_{sp}[\text{Ca(OH)}_2]$

This is the order for the solubility products of these hydroxides. Thus, the order of precipitation is (first to last) Ni(OH)_2, Mg(OH)_2, Ca(OH)_2.

(b) $K_{sp}[\text{Ni(OH)}_2] = 6.5 \times 10^{-18} = [\text{Ni}^{2+}][\text{OH}^-]^2$

$$[\text{OH}^-]^2 = \frac{6.5 \times 10^{-18}}{0.0010} = 6.5 \times 10^{-15}$$

$[\text{OH}^-] = 8.1 \times 10^{-8}$

$\text{pOH} = -\log[\text{OH}^-] = 7.09$ $\text{pH} \approx 7$

$K_{sp}[\text{Mg(OH)}_2] = 1.1 \times 10^{-11} = [\text{Mg}^{2+}][\text{OH}^-]^2$

$$[\text{OH}^-] = \sqrt{\frac{1.1 \times 10^{-11}}{0.0010}} = 1.0 \times 10^{-4}$$

$\text{pOH} = -\log(1.0 \times 10^{-4}) = 4.00$ $\text{pH} = 14.00 - 4.00 = 10.00$, $\text{pH} \approx 10$

$K_{sp}[\text{Ca(OH)}^2] = 5.5 \times 10^{-6} = [\text{Ca}^{2+}][\text{OH}^-]^2$

$$[\text{OH}^-] = \sqrt{\frac{5.5 \times 10^{-6}}{0.0010}} = 7.4 \times 10^{-2}$$

$\text{pOH} = -\log(7.4 \times 10^{-2}) = 1.13$ $\text{pH} = 14.00 - 1.13 = 12.87$, $\text{pH} \approx 13$

6J.9 The K_{sp} values are: MgF_2 6.4×10^{-9}

BaF_2 1.7×10^{-6}

MgCO_3 1.0×10^{-5}

BaCO_3 8.1×10^{-9}

The difference in these numbers suggests that there is a greater solubility difference between the carbonates, and thus this anion should give a better separation. Because different numbers of ions are involved, it is instructive to convert the K_{sp} values into molar solubility. For the fluorides the reaction is:

$$MF_2(s) \rightleftharpoons M^{2+}(aq) + 2\,F^-(aq)$$

Change $+x$ $+2x$

$$K_{sp} = x(2x)^2$$

Solving this for MgF_2 gives 0.0012 M and for BaF_2 gives 0.0075 M.

For the carbonates:

$$MCO_3(s) \rightleftharpoons M^{2+}(aq) + CO_3^{2-}(aq)$$

$$+x \qquad +x$$

$$K_{sp} = x^2$$

Solving this for $MgCO_3$ gives 0.0032 M and for $BaCO_3$ gives

9.0×10^{-5} M. Clearly, the solubility difference is greatest between the two

carbonates, and CO_3^{2-} is the better choice of anion.

6J.11 $Cu(IO_3)_2$ ($K_{sp} = 1.4 \times 10^{-7}$) is more soluble than

$Pb(IO_3)_2$ ($K_{sp} = 2.6 \times 10^{-13}$), so Cu^{2+} will remain in solution until

essentially all the $Pb(IO_3)_2$ has precipitated. Thus, we expect very little

Pb^{2+} to be the left in solution by the time we reach the point at which

$Cu(IO_3)_2$ begins to precipitate.

The concentration of IO_3^- at which Cu^{2+} begins to precipitate will be

given by $\quad K_{sp} = [Cu^{2+}][IO_3^-]^2 = 1.4 \times 10^{-7} = [0.0010][IO_3^-]^2$
$$[IO_3^-] = 0.012 \ mol \cdot L^{-1}.$$

The concentration of Pb in solution when the $[IO_3^-] = 0.012 \ mol \cdot L^{-1}$ is

given by

$$K_{sp} = [Pb^{2+}][IO_3^-]^2 = 2.6 \times 10^{-13} = [Pb^{2+}][0.012]^2$$
$$[Pb^{2+}] = 1.8 \times 10^{-9} \ mol \cdot L^{-1}.$$

6J.13 $CaF_2(s) \rightleftharpoons Ca^{2+}(aq) + 2F^-(aq)$ $\qquad$ $K_{sp} = 4.0 \times 10^{-11}$

$F^-(aq) + H_2O(l) \rightleftharpoons HF(aq) + OH^-(aq)$ $K_b(F^-) = 2.9 \times 10^{-11}$

(a) Multiply the second equilibrium equation by 2 and add to the first equilibrium:

$CaF_2(s) + 2H_2O(l) \rightleftharpoons Ca^{2+}(aq) + 2HF(aq) + 2OH^-(aq)$

$K = K_{sp} \cdot K_b^2 = (4.0 \times 10^{-11})(2.9 \times 10^{-11})^2 = 3.4 \times 10^{-32}$

(b) (i) The calculation of K_{sp} is complicated by the fact that the anion of the salt is part of a weak base-acid pair. If we wish to solve the equation algebraically, then we need to consider which equilibrium is the dominant one at pH = 7.0, for which $[H_3O^+] = 1 \times 10^{-7}$.

To determine whether F^- or HF is the dominant species at this pH (if either), consider the base hydrolysis reaction:

$F^-(aq) + H_2O(l) \rightleftharpoons HF(aq) + OH^-(aq)$ $K_b = 2.9 \times 10^{-11}$

$$K_b = \frac{[HF][OH^-]}{[F^-]}$$

$$2.9 \times 10^{-11} = \frac{[HF][1 \times 10^{-7}]}{[F^-]}$$

$$\frac{[HF]}{[F^-]} = \frac{2.9 \times 10^{-11}}{1 \times 10^{-7}} = 3 \times 10^{-4}$$

Given that the ratio of HF to F^- is on the order of 10^{-4} to 1, the F^- species is still dominant. The appropriate equation to use is thus the original one for the K_{sp} of $CaF_2(s)$:

$CaF_2(s) \rightleftharpoons Ca^{2+}(aq) + 2F^-(aq)$ $\qquad$ $K_{sp} = 4.0 \times 10^{-11}$

$K_{sp} = [Ca^{2+}][F^-]^2$

$4.0 \times 10^{-11} = x(2x)^2 = 4x^3$

$x = 2.0 \times 10^{-4}$

molar solubility $= 2.0 \times 10^{-4}$ mol$\cdot$L^{-1}

(ii) At

pH = 3.0, $[H_3O^+] = 1 \times 10^{-3}$ mol$\cdot$L^{-1} and $[OH^-] = 1 \times 10^{-11}$ mol$\cdot$L^{-1}.

Under these conditions,

$$K_b = \frac{[HF][OH^-]}{[F^-]}$$

$$2.9 \times 10^{-11} = \frac{[HF][1 \times 10^{-11}]}{[F^-]}$$

$$\frac{[HF]}{[F^-]} = \frac{2.9 \times 10^{-11}}{1 \times 10^{-11}} = 3$$

$$[HF] = 3[F^-].$$

As can be seen, at pH 3.0 the amounts of F^- and HF are comparable, so the protonation of F^- to form HF cannot be ignored. The relation $2[Ca^{2+}] = [F^-] + [HF]$ is required by the mass balance as imposed by the stoichiometry of the dissolution equilibrium:

$$2[Ca^{2+}] = [F^-] + 3[F^-]$$

$$2[Ca^{2+}] = 4[F^-]$$

$$[Ca^{2+}] = 2[F^-]$$

Using this with K_{sp} relationship:

$$(2[F^-])[F^-]^2 = 4.0 \times 10^{-11}$$

$$2[F^-]^3 = 4.0 \times 10^{-11}$$

$$[F^-] = 2 \times 10^{-4}$$

$$[Ca^{2+}] = (2)(2 \times 10^{-4}) = 4 \times 10^{-4} \, mol \cdot L^{-1}$$

The solubility is about double that at pH = 7.0.

6J.15 The two salts can be distinguished by their solubility in NH_3. The equilibria that are pertinent are:

$$AgCl(s) + 2NH_3(aq) \rightleftharpoons Ag(NH_3)_2^+(aq) + Cl^-(aq)$$
$$K = K_{sp} \cdot K_f = 2.6 \times 10^{-3}$$

$$AgI(s) + 2NH_3(aq) \rightleftharpoons Ag(NH_3)_2^+(aq) + I^-(aq) \quad K = K_{sp} \cdot K_f = 2.4 \times 10^{-9}$$

For example, let's consider the solubility of these two salts in 1.00 M NH_3 solution:

For AgCl $K = \dfrac{[Ag(NH_3)_2{}^+][Cl^-]}{[NH_3]^2} = 2.6 \times 10^{-3}$

Concentration

$(mol \cdot L^{-1})$ $AgCl(s) + 2\,HN_3(aq) \rightleftharpoons Ag(NH_3)_2{}^+(aq) + Cl^-(aq)$

initial	—	1.00	0	0
change	—	$-2x$	$+x$	$+x$
final	—	$1.00 - 2x$	$+x$	$+x$

$K = \dfrac{[Ag(NH_3)_2{}^+][Cl^-]}{[NH_3]^2} = 2.6 \times 10^{-3}$

$2.6 \times 10^{-3} = \dfrac{[x][x]}{[1.00 - 2x]^2} = \dfrac{x^2}{[1.00 - 2x]^2}$

$0.051 = \dfrac{x}{1.00 - 2x}$

$x = 0.046$

0.046 mol AgCl will dissolve in 1.00 L of aqueous solution. The molar mass of AgCl is 143.32 $g \cdot mol^{-1}$; this corresponds to

$0.046 \ mol \cdot L^{-1} \times 142.32 \ g \cdot mol^{-1} = 6.5 \ g \cdot L^{-1}$.

For AgI, the same calculation gives $x = 4.9 \times 10^{-5} \ mol \cdot L^{-1}$. The molar mass of AgI is 234.77 $g \cdot mol^{-1}$, giving a solubility of

$4.9 \times 10^{-5} \ mol \cdot L^{-1} \times 234.77 \ g \cdot mol^{-1} = 0.023 \ g \cdot L^{-1}$.

Thus, we could treat a 0.10 g sample of the compound with 20.0 mL of 1.00 M NH_3. The AgCl would all dissolve, whereas practically none of the AgI would dissolve.

Note: AgI is also slightly yellow in color, whereas AgCl is white, so an initial distinction could be made based on the color of the sample.

6J.17 In order to use qualitative analyses, the sample must first be dissolved. This can be accomplished by digesting the sample with concentrated HNO_3 and then diluting the resulting solution. HCl or H_2SO_4 could not be used because some of the metal compounds formed would be insoluble, whereas all of the nitrates would dissolve. Once the sample is dissolved and diluted, an aqueous solution containing chloride ions can be introduced. This should precipitate the Ag^+ as AgCl but would leave the bismuth and nickel in solution, as long as the solution was acidic. The remaining solution can then be treated with H_2S. In acidic solution, Bi_2S_3 will precipitate but NiS will not. Once the Bi_2S_3 has been precipitated, the pH of the solution can be raised by addition of base. Once this is done, NiS should precipitate.

6K.1 (a) Cr reduced from 6+ to 3+; C oxidized from 2– to 1–;

(b) $C_2H_5OH(aq) \rightarrow C_2H_4O(aq) + 2\,H^+(aq) + 2\,e^-$;

(c) $Cr_2O_7^{2-}(aq) + 14\,H^+(aq) + 6\,e^- \rightarrow 2\,Cr^{3+}(aq) + 7\,H_2O(l)$;

(d) $8\,H^+(aq) + Cr_2O_7^{2-}(aq) + 3\,C_2H_5OH(aq) \rightarrow$
$\qquad 2\,Cr^{3+}(aq) + 3\,C_2H_4O(aq) + 7\,H_2O(l)$

6K.3 In each case, first obtain the balanced half-reactions. Multiply the oxidation and reduction half-reactions by appropriate factors that will result in the same number of electrons being present in both half-reactions. Then add the half-reactions, canceling electrons in the process, to obtain the balanced equation for the whole reaction. Check to see that the final equation is balanced.

(a) $4[Cl_2(g) + 2\,e^- \rightarrow 2\,Cl^-(aq)]$

$1[S_2O_3^{2-}(aq) + 5\,H_2O(l) \rightarrow 2\,SO_4^{2-}(aq) + 10\,H^+(aq) + 8\,e^-]$

$4\,Cl_2(g) + S_2O_3^{2-}(aq) + 5\,H_2O(l) + 8\,e^- \rightarrow$
$\qquad 8\,Cl^-(aq) + 2\,SO_4^{2-}(aq) + 10\,H^+(aq) + 8\,e^-$

$4\,Cl_2(g) + S_2O_3^{2-}(aq) + 5\,H_2O(l) \rightarrow 8\,Cl^-(aq) + 2\,SO_4^{2-}(aq) + 10\,H^+(aq)$

Cl_2 is the oxidizing agent and $S_2O_3^{2-}$ is the reducing agent.

(b) $2[MnO_4^-(aq) + 8\,H^+(aq) + 5\,e^- \rightarrow Mn^{2+}(aq) + 4\,H_2O(l)]$

$5[H_2SO_3(aq) + H_2O(l) \rightarrow HSO_4^-(aq) + 3 H^+(aq) + 2 e^-]$

$2 MnO_4^-(aq) + 16 H^+(aq) + 5 H_2SO_3(aq) + 5 H_2O(l) + 10 e^- \rightarrow$

$\qquad 2 Mn^{2+}(aq) + 8 H_2O(l) + 5 HSO_4^-(aq) + 15 H^+(aq) + 10 e^-$

$2 MnO_4^-(aq) + H^+(aq) + 5 H_2SO_3(aq) \rightarrow$

$\qquad 2 Mn^{2+}(aq) + 3 H_2O(l) + 5 HSO_4^-(aq)$

MnO_4^- is the oxidizing agent and H_2SO_3 is the reducing agent.

(c) $Cl_2(g) + 2 e^- \rightarrow 2 Cl^-(aq)$

$H_2S(aq) \rightarrow S(s) + 2 H^+(aq) + 2 e^-$

$Cl_2(g) + H_2S(aq) + 2 e^- \rightarrow 2 Cl^-(aq) + S(s) + 2 H^+(aq) + 2 e^-$

$Cl_2(g) + H_2S(aq) \rightarrow 2 Cl^-(aq) + S(s) + 2 H^+(aq)$

Cl_2 is the oxidizing agent and H_2S is the reducing agent.

(d) $Cl_2(g) + 2 e^- \rightarrow 2 Cl^-(aq)$

$2 H_2O(l) + Cl_2(g) \rightarrow 2 HOCl(aq) + 2 H^+(aq) + 2 e^-$

$2 H_2O(l) + 2 Cl_2(g) + 2 e^- \rightarrow 2 HOCl(aq) + 2 H^+(aq) + 2 Cl^-(aq) + 2 e^-$

or $H_2O(l) + Cl_2(g) \rightarrow HOCl(aq) + H^+(aq) + Cl^-(aq)$

Cl_2 is both the oxidizing and the reducing agent.

6K.5 (a) $O_3(g) \rightarrow O_2(g)$

$O_3(g) \rightarrow O_2(g) + H_2O(l)$ ⠀⠀⠀(balances O)

$2 H_2O(l) + O_3(g) \rightarrow O_2(g) + H_2O(l) + 2 OH^-(aq)$ ⠀⠀⠀(balances H)

$H_2O(l) + O_3(g) \rightarrow O_2(g) + 2 OH^-(aq)$ ⠀⠀⠀(cancels H_2O)

$H_2O(l) + O_3(g) + 2 e^- \rightarrow O_2(g) + 2 OH^-(aq)$ ⠀⠀⠀(balances charge);

$Br^-(aq) \rightarrow BrO_3^-(aq)$

$3 H_2O(l) + Br^-(aq) \rightarrow BrO_3^-(aq)$ ⠀⠀⠀(balances O)

$6 OH^-(aq) + 3 H_2O(l) + Br^-(aq) \rightarrow BrO_3^-(aq) + 6 H_2O(l)$ ⠀(balances H)

$6 OH^-(aq) + 3 H_2O(l) + Br^-(aq) \rightarrow BrO_3^-(aq) + 6 H_2O(l) + 6 e^-$

(balances charge)

Combining half-reactions yields

$$3[H_2O(l) + O_3(g) + 2\,e^- \rightarrow O_2(g) + 2\,OH^-(aq)]$$

$$6\,OH^-(aq) + 3\,H_2O(l) + Br^-(aq) \rightarrow BrO_3^-(aq) + 6\,H_2O(l) + 6\,e^-$$

$$6\,H_2O(l) + 3\,O_3(g) + 6\,OH^-(aq) + Br^-(aq) + 6\,e^- \rightarrow$$

$$3\,O_2(g) + 6\,OH^-(aq) + BrO_3^-(aq) + 6\,H_2O(l) + 6\,e^-$$

and $3\,O_3(g) + Br^-(aq) \rightarrow 3\,O_2(g) + BrO_3^-(aq)$

O_3 is the oxidizing agent and Br^- is the reducing agent.

(b) $Br_2(l) + 2\,e^- \rightarrow 2\,Br^-(aq)$ (balanced reduction half-reaction)

$Br_2(l) + 6\,H_2O(l) \rightarrow 2\,BrO_3^-(aq)$ (O balanced); then

$Br_2(l) + 6\,H_2O(l) + 12\,OH^-(aq) \rightarrow 2\,BrO_3^-(aq) + 12\,H_2O(l)$ (H balanced);

and $Br_2(l) + 12\,OH^-(aq) \rightarrow 2\,BrO_3^-(aq) + 6\,H_2O(l) + 10\,e^-$ (electrons balanced)

Combining half-reactions yields

$$5[Br_2(l) + 2\,e^- \rightarrow 2\,Br^-(aq)]$$

$$1[Br_2(l) + 12\,OH^-(aq) \rightarrow 2\,BrO_3^-(aq) + 6\,H_2O(l) + 10\,e^-]$$

$$6\,Br_2(l) + 12\,OH^-(aq) + 10\,e^- \rightarrow$$

$$10\,Br^-(aq) + 2\,BrO_3^-(aq) + 6\,H_2O(l) + 10\,e^-$$

$$6\,Br_2(l) + 12\,OH^-(aq) \rightarrow 10\,Br^-(aq) + 2\,BrO_3^-(aq) + 6\,H_2O(l)$$

Dividing by 2 gives

$$3\,Br_2(l) + 6\,OH^-(aq) \rightarrow 5\,Br^-(aq) + BrO_3^-(aq) + 3\,H_2O(l)$$

Br_2 is both the oxidizing agent and the reducing agent.

(c) $Cr^{3+}(aq) + 4\,H_2O(l) \rightarrow CrO_4^{2-}(aq)$ (O balanced); then

$Cr^{3+}(aq) + 4\,H_2O(l) + 8\,OH^-(aq) \rightarrow CrO_4^{2-}(aq) + 8\,H_2O(l)$ (H balanced); and

$Cr^{3+}(aq) + 8\,OH^-(aq) \rightarrow CrO_4^{2-}(aq) + 4\,H_2O(l) + 3\,e^-$ (charge balanced)

$MnO_2(s) \rightarrow Mn^{2+}(aq) + 2H_2O(l)$; then

$MnO_2(s) + 4\,H_2O(l) \rightarrow Mn^{2+}(aq) + 2\,H_2O(l) + 4\,OH^-(aq)$ (H balanced); and

$MnO_2(s) + 2 H_2O(l) + 2 e^- \rightarrow Mn^{2+}(aq) + 4 OH^-(aq)$ (charge balanced)

Combining half-reactions yields

$2[Cr^{3+}(aq) + 8 OH^-(aq) \rightarrow CrO_4^{2-}(aq) + 4 H_2O(l) + 3 e^-]$

$3[MnO_2(s) + 2 H_2O(l) + 2 e^- \rightarrow Mn^{2+}(aq) + 4 OH^-(aq)]$

$2 Cr^{3+}(aq) + 16 OH^-(aq) + 3 MnO_2(s) + 6 H_2O(l) + 6 e^- \rightarrow$
$\qquad 2 CrO_4^{2-}(aq) + 8 H_2O(l) + 3 Mn^{2+}(aq) + 12 OH^-(aq) + 6 e^-$

$2 Cr^{3+}(aq) + 4 OH^-(aq) + 3 MnO_2(s) \rightarrow$
$\qquad 2 CrO_4^{2-}(aq) + 2 H_2O(l) + 3 Mn^{2+}(aq)$

Cr^{3+} is the reducing agent and MnO_2 is the oxidizing agent.

(d) $3[P_4(s) + 8 OH^-(aq) \rightarrow 4 H_2PO_2^-(aq) + 4 e^-]$

$P_4(s) + 12 H_2O(l) + 12 e^- \rightarrow 4 PH_3(g) + 12 OH^-(aq)$

$4 P_4(s) + 12 H_2O(l) + 24 OH^-(aq) + 12 e^- \rightarrow$
$\qquad 12 H_2PO_2^-(aq) + 4 PH_3(g) + 12 OH^-(aq) + 12 e^-$

$4 P_4(s) + 12 H_2O(l) + 12 OH^-(aq) \rightarrow 12 H_2PO_2^-(aq) + 4 PH_3(g)$

or $P_4(s) + 3 H_2O(l) + 3 OH^-(aq) \rightarrow 3 H_2PO_2^-(aq) + PH_3(g)$

$P_4(s)$ is both the oxidizing and the reducing agent.

6K.7 $P_4S_3(aq) \rightarrow H_3PO_4(aq) + SO_4^{2-}(aq)$

For the oxidation of P_4S_3, both the P and S atoms are oxidized. The assignment of oxidation states to the P and S atoms is complicated by the presence of P—P bonds in the molecule, which leads to non-integral values. As long as we are consistent in our assignments, the end result should be the same. We will assume that S in P_4S_3 is 2 – and, therefore, loses 8 electrons on going to S^{6+} in the sulfate ion. Because P_4S_3 is a neutral molecule and, if S has an oxidation number of -2, then each phosphorus atom will have an oxidation number of $+1.5$. Phosphorus in phosphoric acid has an oxidation number of $+5$. so each P atom of P_4S_3

must lose 3.5 electrons. The total number of electrons lost is

$(4 \times 3.5) + (3 \times 8) = 38$.

$$P_4S_3(aq) \rightarrow 4\,H_3PO_4(aq) + 3\,SO_4{}^{2-}(aq) + 38\,e^-$$

We balance the charge by adding H^+ in an acidic solution:

$$P_4S_3(aq) \rightarrow 4\,H_3PO_4(aq) + 3\,SO_4{}^{2-}(aq) + 44\,H^+(aq) + 38\,e^-$$

The final balance is achieved by adding water to provide the oxygen and hydrogen atoms:

$$P_4S_3(aq) + 28\,H_2O(l) \rightarrow 4\,H_3PO_4(aq) + 3\,SO_4{}^{2-}(aq) + 44\,H^+(aq) + 38\,e^-$$

The other half-reaction is simpler.

$$NO_3{}^-(aq) \rightarrow NO(g)$$

N has an oxidation number of $+5$ in the nitrate ion and $+2$ in nitric oxide.

Each nitrogen atom gains three electrons in the course of the reaction.

$$NO_3{}^-(aq) + 3\,e^- \rightarrow NO(g)$$

Charge balance is again achieved by adding H^+:

$$NO_3{}^-(aq) + 4\,H^+(aq) + 3\,e^- \rightarrow NO(g)$$

The number of hydrogen and oxygen atoms is completed by the addition of water:

$$NO_3{}^-(aq) + 4\,H^+(aq) + 3\,e^- \rightarrow NO(aq) + 2\,H_2O(l)$$

Combining the two half-reactions gives

$$38\,[NO_3{}^-(aq) + 4\,H^+(aq) + 3\,e^- \rightarrow NO(g) + 2\,H_2O(l)]$$
$$+3\,[P_4S_3(aq) + 28\,H_2O(l) \rightarrow$$
$$4\,H_3PO_4(aq) + 3\,SO_4{}^{2-}(aq) + 44\,H^+(aq) + 38\,e^-]$$
$$3\,P_4S_3(aq) + 38\,NO_3{}^-(aq) + 20\,H^+(aq) + 8\,H_2O(l) \rightarrow$$
$$12\,H_3PO_4(aq) + 9\,SO_4{}^{2-}(aq) + 38\,NO(g)$$

6L.1 (a) $\Delta G_r^\circ = -nFE^\circ = -2(9.6485 \times 10^4\,\mathrm{C.\,mol^{-1}})(1.08\,\mathrm{J.\,C^{-1}})$

$= -2.08 \times 10^5\,\mathrm{J.\,mol^{-1}}$

(b) $\Delta G_r^\circ = -nFE^\circ = -6(9.6485 \times 10^4\,\mathrm{C.\,mol^{-1}})(-1.29\,\mathrm{J.\,C^{-1}})$

$= 7.47 \times 10^5\,\mathrm{J.\,mol^{-1}}$

6L.3 (a) $Ag^+(aq) + e^- \rightarrow Ag(s)$ $E°(\text{cathode}) = +0.80$ V

$Ni^{2+}(aq) + 2\,e^- \rightarrow Ni(s)$ $E°(\text{anode}) = -0.23$ V

Reversing the anode half-reaction yields

$Ni(s) \rightarrow Ni^{2+}(aq) + 2\,e^-$

and the cell reaction is, upon addition of the half-reactions,

$2\,Ag^+(aq) + Ni(s) \rightarrow 2\,Ag(s) + Ni^{2+}(aq)$ $E°_{\text{cell}} = +0.80$ V $- (-0.23)$ V

$= +1.03$ V

(b) $2\,H^+(aq) + 2\,e^- \rightarrow H_2(g)$ $E°(\text{anode}) = 0.00$ V

$Cl_2(g) + 2\,e^- \rightarrow 2\,Cl^-(aq)$ $E°(\text{cathode}) = +1.36$ V

Therefore, at the anode, after reversal,

$H_2(g) \rightarrow 2\,H^+(aq) + 2\,e^-$

and, the cell reaction is, upon addition of the half-reactions,

$Cl_2(g) + H_2(g) \rightarrow 2\,H^+(aq) + 2\,Cl^-(aq)$ $E°_{\text{cell}} = +1.36$ V $- 0.00$ V

$= +1.36$ V

(c) $Cu^{2+}(aq) + 2\,e^- \rightarrow Cu(s)$ $E°(\text{anode}) = +0.34$ V

$Ce^{4+}(aq) + e^- \rightarrow Ce^{3+}(aq)$ $E°(\text{cathode}) = +1.61$ V

Therefore, at the anode, after reversal,

$Cu(s) \rightarrow Cu^{2+}(aq) + 2\,e^-$

and, the cell reaction is, upon addition of the half-reactions,

$2\,Ce^{4+}(aq) + Cu(s) \rightarrow Cu^{2+}(aq) + 2\,Ce^{3+}(aq)$ $E°_{\text{cell}} = 1.61$ V $- (0.34$ V$)$

$= +1.27$ V

(d) $O_2(g) + 2\,H_2O(l) + 4\,e^- \rightarrow 4\,OH^-(aq)$ $E°(\text{cathode}) = 0.40$ V

$O_2(g) + 4\,H^+(aq) + 4\,e^- \rightarrow 2\,H_2O(l)$ $E°(\text{anode}) = 1.23$ V

Reversing the anode half-reaction yields

$2\,H_2O(l) \rightarrow O_2(g) + 4\,H^+(aq) + 4\,e^-$

and the cell reaction is, upon addition of the half-reactions,

$4\,H_2O(l) \rightarrow 4\,H^+(aq) + 4\,OH^-(aq)$ $E°_{\text{cell}} = 0.40$ V $- 1.23$ V $= -0.83$ V or, $H_2O(l) \rightarrow H^+(aq) + OH^-(aq)$

Note: This balanced equation corresponds to the cell notation given. The spontaneous process is the reverse of this reaction.

(e) $Sn^{4+}(aq) + 2\,e^- \rightarrow Sn^{2+}(aq)$ $E°(anode) = +0.15\ V$

$Hg_2Cl_2(s) + 2\,e^- \rightarrow 2\,Hg(l) + 2\,Cl^-(aq)$ $E°(cathode) = +0.27\ V$

Therefore, at the anode, after reversal,

$Sn^{2+}(aq) \rightarrow Sn^{4+}(aq) + 2\,e^-$

and the cell reaction is, upon addition of the half-reactions,

$Sn^{2+}(aq) + Hg_2Cl_2(s) \rightarrow 2\,Hg(l) + 2\,Cl^-(aq) + Sn^{4+}(aq)$

$E°_{cell} = 0.27\ V - 0.15\ V = 0.12\ V$

6L.5   (a) $Ni^{2+}(aq) + 2\,e^- \rightarrow Ni(s)$ $E°(cathode) = -0.23\ V$

$Zn^{2+}(aq) + 2\,e^- \rightarrow Zn(s)$ $E°(anode) = -0.76\ V$

Reversing the anode reaction yields

$Zn(s) \rightarrow Zn^{2+}(aq) + 2\,e^-$ (at anode); then, upon addition,

$Ni^{2+}(aq) + Zn(s) \rightarrow Ni(s) + Zn^{2+}(aq)$ (overall cell)

$E°_{cell} = -0.23\ V - (-0.76\ V) = +0.53\ V$

and $Zn(s)\,|\,Zn^{2+}(aq)\,\|\,Ni^{2+}(aq)\,|\,Ni(s)$

(b) $2[Ce^{4+}(aq) + e^- \rightarrow Ce^{3+}(aq)]$ $E°(cathode) = +1.61\ V$

$I_2(s) + 2\,e^- \rightarrow 2\,I^-(aq)$ $E°(anode) = +0.54\ V$

Reversing the anode reaction yields

$2\,I^-(aq) \rightarrow 2\,e^- + I_2(s)$ (at anode); then, upon addition,

$2\,I^-(aq) + 2\,Ce^{4+}(aq) \rightarrow 2\,Ce^{3+}(aq) + I_2(s)$ (overall cell)

$E°_{cell} = +1.61\ V - 0.54\ V = +1.07\ V$

and $Pt(s)\,|\,I^-(aq)\,|\,I_2(s)\,\|\,Ce^{4+}(aq),\,Ce^{3+}(aq)\,|\,Pt(s)$

An inert electrode such as Pt is necessary when both oxidized and reduced species are in the same solution.

(c) $Cl_2(g) + 2\,e^- \rightarrow 2\,Cl^-(aq)$ $E°(\text{cathode}) = +1.36\ V$

$2\,H^+(aq) + 2\,e^- \rightarrow H_2(g)$ $E°(\text{anode}) = 0.00\ V$

Reversing the anode reaction yields

$H_2(g) \rightarrow 2\,H^+(aq) + 2\,e^-$ (at anode); then, upon addition,

$H_2(g) + Cl_2(g) \rightarrow 2\,HCl(aq)$ (overall cell) $E°_{cell} = +1.36\ V - 0.00\ V$
$$= +1.36\ V$$

and $Pt(s)\,|\,H_2(g)\,|\,H^+(aq)\,||\,Cl^-(aq)\,|\,Cl_2(g)\,|\,Pt(s)$

An inert electrode such as Pt is necessary for gas/ion electrode reactions.

(d) $3[Au^+(aq) + e^- \rightarrow Au(s)]$ $E°(\text{cathode}) = +1.69\ V$

$Au^{3+}(aq) + 3\,e^- \rightarrow Au(s)$ $E°(\text{anode}) = +1.40\ V$

Reversing the anode reaction yields

$Au(s) \rightarrow Au^{3+}(aq) + 3\,e^-$ then, upon addition (anode),

$3\,Au^+(aq) \rightarrow$
$\quad 2\,Au(s) + Au^{3+}(aq)$ (overall cell) $E°_{cell} = +1.69\ V - 1.40\ V$
$$= +0.29\ V$$

and $Au(s)\,|\,Au^{3+}(aq)\,||\,Au^+(aq)\,|\,Au(s)$

6L.7   (a) $Ag^+(aq) + e^- \rightarrow Ag(s)$ $E°(\text{cathode}) = +0.80\ V$

$AgBr(s) + e^- \rightarrow Ag(s) + Br^-(aq)$ $E°(\text{anode}) = +0.07\ V$

Reversing the anode reaction yields

$Ag(s) + Br^-(aq) \rightarrow AgBr(s) + e^-$ then, upon addition,

$Ag^+(aq) + Br^-(aq) \rightarrow AgBr(s)$ (overall cell) $E°_{cell} = +0.80\ V - 0.07\ V$
$$= +0.73\ V$$

This is the direction of the spontaneous standard cell reaction that could be used to study the reverse of the given solubility equilibrium.   A cell diagram for this favorable process is

$Ag(s)\,|AgBr(s)\,|\,Br^-(aq)\,||\,Ag^+(aq)\,|\,Ag(s)$

(b) To conform to the notation of this chapter, the neutralization is rewritten as

$H^+(aq) + OH^- \rightarrow H_2O(l)$

$O_2(g) + 4\,H^+(aq) + 4\,e^- \rightarrow 2\,H_2O(l)$ $E°(cathode) = +1.23\ V$

$O_2(g) + 2\,H_2O(l) + 4\,e^- \rightarrow 4\,OH^-(aq)$ $E°(anode) = +0.40\ V$

Reversing the anode reaction yields

$4\,OH^-(aq) \rightarrow O_2(g) + 2\,H_2O(l) + 4\,e^-$; then, upon addition,

$4\,H^+(aq) + 4\,OH^-(aq) \rightarrow 4\,H_2O(l)$

or $H^+(aq) + OH^-(aq) \rightarrow H_2O(l)$ (overall cell) $E° = +1.23\ V - 0.40\ V$
$$= +0.83\ V$$

and $Pt(s)\,|O_2(g)\,|OH^-(aq)\,\|\,H^+(aq)\,|O_2(g)\,|Pt(s)$

(c) $Cd(OH)_2(s) + 2\,e^- \rightarrow Cd(s) + 2\,OH^-(aq)$ $E°(anode) = -0.81\ V$

$Ni(OH)_3(s) + e^- \rightarrow Ni(OH)_2(s) + OH^-(aq)$ $E°(cathode) = +0.49\ V$

Reversing the anode reaction and multiplying the cathode reaction by 2 yields

$Cd(s) + 2\,OH^-(aq) \rightarrow Cd(OH)_2(s) + 2\,e^-$

$2\,Ni(OH)_3 + 2\,e^- \rightarrow 2\,Ni(OH)_2(s) + 2\,OH^-(aq)$ then, upon addition,

$2\,Ni(OH)_2(s) + Cd(s) \rightarrow Cd(OH)_2(s) + 2\,Ni(OH)_2(s)$

overall cell $E° = +1.30\ V$

and $Cd(s)\,|\,Cd(OH)_2(s)\,|\,KOH(aq)\,\|\,Ni(OH)_3(s)\,|\,Ni(OH)_2(s)\,|\,Ni(s)$

with Ni(s) used as the conducting electrode for the cathode.

6L.9 (a) $MnO_4^-(aq) + 8\,H^+(aq) + 5\,e^- \rightarrow Mn^{2+}(aq) + 4\,H_2O(l)$ (cathode half-reaction)

$5[Fe^{2+}(aq) \rightarrow Fe^{3+}(aq) + e^-]$ (anode half-reaction)

(b) Adding the two equations yields:

$MnO_4^-(aq) + 5\,Fe^{2+}(aq) + 8\,H^+(aq) \rightarrow Mn^{2+}(aq) + 5\,Fe^{3+}(aq) + 4\,H_2O(l)$

The cell diagram is:

$Pt(s)\,|\,Fe^{3+}(aq), Fe^{2+}(aq)\,\|\,H^+(aq), MnO_4^-(aq), Mn^{2+}(aq)\,|\,Pt(s)$

6M.1 The cell, as written $Cu(s)|Cu^{2+}(aq)\|M^{2+}(aq)|M(s)$, makes the Cu/Cu^{2+} electrode the anode, because this is where oxidation is occurring; the M^{2+}/M electrode is the cathode. The calculation is

$$E° = E°(cathode) - E°(anode)$$
$$-0.689 \text{ V} = E°(cathode) - (+0.34 \text{ V})$$
$$E°(cathode) = -0.349 \text{ V}$$

6M.3 A galvanic cell has a positive potential difference; therefore, identify as cathode and anode the electrodes that make $E°$ (cell) positive upon calculating

$$E°(cell) = E°(cathode) - E°(anode)$$

There are only two possibilities: If your first guess gives a negative $E°$ (cell), switch your identification.

(a) $Cu^{2+}(aq) + 2 e^- \rightarrow Cu(s)$ $E°(cathode) = +0.34 \text{ V}$

$Cr^{3+}(aq) + e^- \rightarrow Cr^{2+}(aq)$ $E°(anode) = -0.41 \text{ V}$
$E°(cell) = +0.34 \text{ V} - (-0.41 \text{ V}) = +0.75 \text{ V}$

(b) $AgCl(s) + e^- \rightarrow Ag(s) + Cl^-(aq)$ $E°(cathode) = +0.22 \text{ V}$

$AgI(s) + e^- \rightarrow Ag(s) + I^-(aq)$ $E°(anode) = -0.15 \text{ V}$
$E°(cell) = +0.22 \text{ V} - (-0.15 \text{ V}) = +0.37 \text{ V}$

(c) $Hg_2^{2+}(aq) + 2 e^- \rightarrow 2 Hg(l)$ $E°(cathode) = +0.79 \text{ V}$

$Hg_2Cl_2(s) + 2 e^- \rightarrow 2 Hg(l) + 2 Cl^-(aq)$ $E°(anode) = +0.27 \text{ V}$
$E°(cell) = +0.79 \text{ V} - (+0.27 \text{ V}) = +0.52 \text{ V}$

(d) $Pb^{4+}(aq) + 2 e^- \rightarrow Pb^{2+}(aq)$ $E°(cathode) = +1.67 \text{ V}$

$Sn^{4+}(aq) + 2 e^- \rightarrow Sn^{2+}(aq)$ $E°(anode) = +0.15 \text{ V}$
$E°(cell) = +1.67 \text{ V} - (+0.15 \text{ V}) = +1.52 \text{ V}$

6M.5   In each case, determine the cathode and anode half-reactions corresponding to the reactions *as written*. Look up the standard reduction potentials for these half-reactions and then calculate $E^\circ_{cell} = E^\circ(cathode) - E^\circ(anode)$. If E°_{cell} is positive, the reaction is spontaneous under standard conditions.

(a) $E^\circ_{cell} = E^\circ(cathode) - E^\circ(anode) = +0.96\ V - (+0.79\ V) = +0.17\ V$.

Therefore, spontaneous galvanic cell:

$Hg(l)\,|\,Hg_2^{2+}(aq)\,\|\,NO_3^-(aq),\,H^+(aq)\,|\,NO(g)\,|\,Pt(s)$

$\Delta G^\circ_r = -nFE^\circ = -(6)(9.65 \times 10^4\ C \cdot mol^{-1})(+0.17\ J \cdot C^{-1}) = -98\ kJ \cdot mol^{-1}$

(b) $E^\circ_{cell} = E^\circ(cathode) - E^\circ(anode) = +0.92\ V - (+1.09\ V) = -0.17\ V$

Therefore, not spontaneous.

(c) $E^\circ_{cell} = E^\circ(cathode) - E^\circ(anode) = +1.33\ V - (+0.97\ V) = +0.36\ V$

Therefore, spontaneous galvanic cell.

$Pt(s)\,|\,Pu^{3+}(aq),\,Pu^{4+}(aq)\,\|\,Cr_2O_7^{2-}(aq),\,Cr^{3+}(aq),\,H^+(aq)\,|\,Pt(s)$

$\Delta G^\circ_r = -nFE^\circ = -(6)(9.65 \times 10^4\ C \cdot mol^{-1})(0.36\ J \cdot C^{-1}) = -208\ kJ \cdot mol^{-1}$

6M.7   The more negative (less positive) the standard reduction potential, the stronger is the metal as a reducing agent.

(a) $Cu < Fe < Zn < Cr$

(b) $Mg < Na < K < Li$

(c) $V < Ti < Al < U$

(d) $Au < Ag < Sn < Ni$

6M.9   The appropriate half-reactions are:

$U^{4+} + e^- \rightarrow U^{3+}$ $E^\circ = -0.61$ (A)

$U^{3+} + 3\,e^- \rightarrow U$ $E^\circ = -1.79$ (B)

(A) and (B) add to give the desired half-reaction (C):

$U^{4+} + 4\,e^- \rightarrow U$ $E^\circ = ?$ (C)

In order to calculate the potential of a *half-reaction*, we need to convert the $E°$ values into $\Delta G°$ values:

$$\Delta G°(A) = -nFE°(A) = -1F(-0.61 \text{ V})$$
$$\Delta G°(B) = -nFE°(B) = -3F(-1.79 \text{ V})$$
$$\Delta G°(C) = -nFE°(C) = -4FE°(C)$$
$$\Delta G°(C) = \Delta G°(A) + \Delta G°(B)$$
$$-4FE°(C) = -1F(-0.61 \text{ V}) + [-3F(-1.79 \text{ V})]$$

The constant F will cancel from both sides, leaving:

$$-4E°(C) = -1(-0.61 \text{ V}) - 3(-1.79 \text{ V})$$
$$E°(C) = -[0.61 \text{ V} + 5.37 \text{ V}]/4 = -1.50 \text{ V}$$

6M.11 In each case, identify the couple with the more positive reduction potential. This will be the couple at which reduction occurs, and therefore which contains the oxidizing agent. The other couple contains the reducing agent.

(a) Co^{2+}/Co $E° = -0.28$ V, Co^{2+} is the oxidizing agent (cathode)

Ti^{3+}/Ti^{2+} $E° = -0.37$ V, Ti^{2+} is the reducing agent (anode)

$Pt(s) \mid Ti^{2+}(aq), Ti^{3+}(aq) \parallel Co^{2+}(aq) \mid Co(s)$

$E°_{cell} = E°(cathode) - E°(anode) = -0.28 \text{ V} - (-0.37 \text{ V}) = +0.09 \text{ V}$

(b) U^{3+}/U $E° = -1.79$ V, U^{3+} is the oxidizing agent (cathode)

La^{3+}/La $E° = -2.52$ V, La is the reducing agent (anode)

$La(s) \mid La^{3+}(aq) \parallel U^{3+}(aq) \mid U(s)$

$E°_{cell} = -1.79 \text{ V} - (-2.52 \text{ V}) = +0.73 \text{ V}$

(c) Fe^{3+}/Fe^{2+} $E° = +0.77$ V, Fe^{3+} is the oxdizing agent (cathode)

H^+/H_2 $E° = 0.00$ V, H_2 is the reducing agent (anode)

$Pt(s) \mid H_2(g) \mid H^+(aq) \parallel Fe^{2+}(aq), Fe^{3+}(aq) \mid Pt(s)$

$E°_{cell} = +0.77 \text{ V} - 0.00 \text{ V} = +0.77 \text{ V}$

(d) $O_3/O_2, OH^-$ $E° = +1.24$ V, O_3 is the oxidizing agent (cathode)

Ag^+/Ag $E° = +0.80$ V, Ag is the reducing agent (anode)

$Ag(s) \mid Ag^+(aq) \parallel OH^-(aq) \mid O_3(g), O_2(g) \mid Pt(s)$

$E°_{cell} = +1.24 \text{ V} - 0.80 \text{ V} = +0.44 \text{ V}$

6M.13 (a) $E°(Cl_2, Cl^-) = +1.36$ V (cathode)

$E°(Br_2, Br^-) = +1.09$ V (anode)

Because $E°(Cl_2, Cl^-) > E°(Br_2, Br^-)$ the reaction favors products.

$E°_{cell} = +1.36$ V $- 1.09$ V $= +0.27$ V

$Cl_2(g)$ is the oxidizing agent.

(b) $E°(Ce^{4+}/Ce^{3+}) = +1.61$ V (anode)

$E°(MnO_4^-/Mn^{2+}) = +1.51$ V (cathode)

Because $E°(Ce^{4+}/Ce^{3+}) > E°(MnO_4^-/Mn^{2+})$, the reaction does not favor products.

(c) $E°(Pb^{4+}/Pb^{2+}) = +1.67$ V (anode)

$E°(Pb^{2+}/Pb) = -0.13$ V (cathode)

Because $E°(Pb^{4+}/Pb^{2+}) > E°(Pb^{2+}/Pb)$, the reaction does not favor products.

(d) $E°(NO_3^-/NO_2/H^+) = +0.80$ V (cathode)

$E°(Zn^{2+}/Zn) = -0.76$ V (anode)

Because $E°(NO_3^-/NO_2/H^+) > E°(Zn^{2+}/Zn)$, the reaction favors products.

$E°_{cell} = +0.80$ V $- (-0.76$ V$) = +1.56$ V

NO_3^- is the oxidizing agent.

6N.1 (a) $Ti^{2+}(aq) + 2 e^- \rightarrow Ti(s)$ $E°$(cathode) $= -1.63$ V

$Mn^{2+}(aq) + 2 e^- \rightarrow Mn(s)$ $E°$(anode) $= -1.18$ V

Note: These equations represent the cathode and anode half-reactions for the overall reaction as written. The spontaneous direction of this reaction under standard conditions is the opposite of that given.

$E°_{cell} = E°$(cathode) $- E°$(anode) $= -1.63$ V $- (-1.18$ V$) = -0.45$ V, and

$$\ln K = \frac{nFE^\circ}{RT}. \qquad \text{At } 25^\circ C \quad \ln K = \frac{nE^\circ}{0.02569 \text{ V}}.$$

$$\therefore \quad \ln K = \frac{(2)(-0.45 \text{ V})}{0.02569 \text{ V}} = -35 \quad \text{and} \quad K = 6 \times 10^{-16}.$$

(b) $\text{In}^{3+}(\text{aq}) + 2\text{ e}^- \rightarrow \text{In}^{2+}(\text{aq}) \quad E^\circ(\text{cathode}) = -0.49 \text{ V}$

$\text{U}^{4+}(\text{aq}) + \text{e}^- \rightarrow \text{U}^{3+}(\text{aq}) \qquad\qquad\qquad E^\circ(\text{anode}) = -0.61 \text{ V}$

$E^\circ_{\text{cell}} = E^\circ(\text{cathode}) - E^\circ(\text{anode}) = -0.49 \text{ V} - (-0.61 \text{ V}) = +0.12 \text{ V}$, and

$$\text{at } 25^\circ C \quad \ln K = \frac{(2)(+0.12 \text{ V})}{0.02569 \text{ V}} = +9.3.$$

$$\therefore \quad K = 1 \times 10^4.$$

6N.3 In each case, $E^\circ_{\text{cell}} = E^\circ(\text{cathode}) - E^\circ(\text{anode})$. Recall that the values for E° at the electrodes refer to the electrode potential for the half-reaction written as a reduction reaction. In balancing the cell reaction, the half-reaction at the anode is reversed. However, this does not reverse the sign of electrode potential used at the anode, because the value always refers to the reduction potential. As an approximation use: 1 bar = 1 atm = 760 Torr

(a) $2\text{ H}^+(\text{aq}, 1.0 \text{ M}) + 2\text{ e}^- \rightarrow \text{H}_2(\text{g}, 1 \text{ atm}) \quad E^\circ(\text{cathode}) = 0.00 \text{ V}$

$\text{H}_2(\text{g}, 1 \text{ atm}) \rightarrow 2\text{ H}^+(\text{aq}, 0.075 \text{ M}) + 2\text{ e}^- \quad E^\circ(\text{anode}) = 0.00 \text{ V}$

$2\text{ H}^+(\text{aq}, 1.0 \text{ M}) + \text{H}_2(\text{g}, 1 \text{ atm}) \rightarrow 2\text{ H}^+(\text{aq}, 0.075 \text{ M}) + \text{H}_2(\text{g}, 1 \text{ atm})$

$E^\circ_{\text{cell}} = 0.00 \text{ V}$

Then, $E = E^\circ - \left(\dfrac{0.025\,693 \text{ V}}{n}\right) \ln \left(\dfrac{[\text{H}^+, 0.075 \text{ M}]^2 P_{\text{H}_2}}{[\text{H}^+, 1.0 \text{ M}]^2 P_{\text{H}_2}}\right)$

$E = 0.00 \text{ V} - \left(\dfrac{0.025\,693 \text{ V}}{2}\right) \ln \left(\dfrac{(0.075 \text{ M})^2 \times 1 \text{ atm}}{(1.0 \text{ M})^2 \times 1 \text{ atm}}\right)$

$E = -0.0129 \text{ V} \ln (0.075)^2 = +0.067 \text{ V}$

(b) $\text{Ni}^{2+}(\text{aq}) + 2\text{ e}^- \rightarrow \text{Ni(s)} \quad E^\circ(\text{cathode}) = -0.23 \text{ V}$

$\text{Zn(s)} \rightarrow \text{Zn}^{2+}(\text{aq}) + 2\text{ e}^- \quad E^\circ(\text{anode}) = -0.76 \text{ V}$

$\text{Ni}^{2+}(\text{aq}) + \text{Zn(s)} \rightarrow \text{Ni(s)} + \text{Zn}^{2+}(\text{aq}) \quad E^\circ_{\text{cell}} = +0.53 \text{ V}$

Then, $E = E° - \left(\dfrac{0.025\ 693\ \text{V}}{n} \right) \ln \left(\dfrac{[\text{Zn}^{2+}]}{[\text{Ni}^{2+}]} \right)$

$E = 0.53\ \text{V} - \left(\dfrac{0.025\ 693\ \text{V}}{2} \right) \ln \left(\dfrac{0.37}{0.059} \right) = 0.53\ \text{V} - 0.02\ \text{V} = 0.51\ \text{V}$

(c) $2\ \text{H}^+(\text{aq}) + 2\ \text{e}^- \rightarrow \text{H}_2(\text{g})\ E°(\text{cathode}) = 0.00\ \text{V}$

$2\ \text{Cl}^-(\text{aq}) \rightarrow \text{Cl}_2(\text{g}) + 2\ \text{e}^-\ E°(\text{anode}) = +1.36\ \text{V}$

$2\ \text{H}^+(\text{aq}) + 2\ \text{Cl}^-(\text{aq}) \rightarrow \text{H}_2(\text{g}) + \text{Cl}_2(\text{g})\ E°_{\text{cell}} = -1.36\ \text{V}$

Then,

$E = E° - \left(\dfrac{0.025\ 693\ \text{V}}{n} \right) \ln \left(\dfrac{P_{\text{H}_2} P_{\text{Cl}_2}}{[\text{H}^+]^2 [\text{Cl}^-]^2} \right)$

$E = -1.36\ \text{V} - \left(\dfrac{0.025\ 693\ \text{V}}{2} \right) \ln \left(\dfrac{\left(\dfrac{125}{760} \right) \left(\dfrac{250}{760} \right)}{(0.85)^2\ (1.0)^2} \right)$

$E = -1.36\ \text{V} + 0.03\ \text{V}$

$\quad = -1.33\ \text{V}$

(d) $\text{Sn}^{4+}(\text{aq},\ 0.867\ \text{M}) + 2\ \text{e}^- \rightarrow \text{Sn}^{2+}(\text{aq},\ 0.55\ \text{M})$

$\quad E°(\text{cathode}) = +0.15\ \text{V}$

$\text{Sn}(\text{s}) \rightarrow \text{Sn}^{2+}(\text{aq},\ 0.277\ \text{M}) + 2\ \text{e}^-\ E°(\text{anode}) = -0.14\ \text{V}$

$\text{Sn}^{4+}(\text{aq},\ 0.867\ \text{M}) + \text{Sn}(\text{s}) \rightarrow \text{Sn}^{2+}(\text{aq},\ 0.55) + \text{Sn}^{2+}(\text{aq},\ 0.277\ \text{M})$

$E°_{\text{cell}} = 0.29\ \text{V}$

$E = E° - \left(\dfrac{0.025\ 693\ \text{V}}{2} \right) \ln \left(\dfrac{(0.55)(0.277)}{(0.867)} \right)$

$E = 0.29\ \text{V} + 0.02\ \text{V} = 0.31\ \text{V}$

6N.5 In each case, obtain the balanced equation for the cell reaction from the half-cell reactions at the electrodes, by reversing the reduction equation for the half-reaction at the anode, multiplying the half-reaction equations by an appropriate factor to balance the number of electrons, and then adding the half-reactions. Calculate $E°_{\text{cell}} = E°(\text{cathode}) - E°(\text{anode})$.

Then write the Nernst equation for the cell reaction and solve for the unknown.

(a) $Hg_2Cl_2(s) + 2\,e^- \rightarrow 2\,Hg(l) + 2\,Cl^-(aq)$ $E°(\text{cathode}) = +0.27$ V

$H_2(g) \rightarrow 2\,H^+(aq) + 2\,e^-$ $E°(\text{anode}) = 0.00$ V

$H_2(g) + Hg_2Cl_2(s) \rightarrow 2\,H^+(aq) + 2\,Hg(l) + 2\,Cl^-(aq)$ $E°_{\text{cell}} = +0.27$ V

$$E = E° - \left(\frac{0.025\ 693\ \text{V}}{n}\right) \ln\left(\frac{[H^+]^2[Cl^-]^2}{[H_2]}\right)$$

$$0.33\ \text{V} = 0.27\ \text{V} - \left(\frac{0.025\ 693\ \text{V}}{2}\right) \ln\left(\frac{[H^+]^2(1)^2}{(1)}\right)$$

$$= 0.27\ \text{V} - (0.0129\ \text{V}) \ln [H^+]^2$$

$$0.06\ \text{V} = -0.0257\ \text{V} \ln [H^+] = -0.0257\ \text{V} \times (2.303 \log [H^+])$$

$$\text{pH} = \frac{0.06\ \text{V}}{(2.303)\,(0.025\ 693\ \text{V})} = 1.0$$

(b)

$2[MnO_4^-(aq) + 8\,H^+(aq) + 5\,e^- \rightarrow$
$\quad Mn^{2+}(aq) + 4\,H_2O(l)]$ $E°(\text{cathode}) = +1.51$ V

$5[2\,Cl^-(aq) \rightarrow Cl_2(g) + 2\,e^-]$ $E°(\text{anode}) = +1.36$ V

$2\,MnO_4^-(aq) + 16\,H^+(aq) + 10\,Cl^-(aq) \rightarrow$
$\quad 5\,Cl_2(g) + 2\,Mn^{2+}(aq) + 8\,H_2O(l)$ $E°_{\text{cell}} = +0.15$ V

$$E = E° - \left(\frac{0.0257\ \text{V}}{n}\right) \ln\left(\frac{[Cl_2]^5[Mn^{2+}]^2}{[MnO_4^-]^2[H^+]^{16}[Cl^-]^{10}}\right)$$

$$-0.30 \text{ V} = +0.15 \text{ V} - \left(\frac{0.0257 \text{ V}}{10} \right) \ln \left(\frac{(1)^5 (0.10)^2}{(0.010)^2 (1 \times 10^{-4})^{16} (\text{Cl}^-)^{10}} \right)$$

$$-0.45 \text{ V} = -(0.002\,5693 \text{ V}) \ln \left(\frac{1 \times 10^{-2}}{(1 \times 10^{-4})(1 \times 10^{-64})[\text{Cl}^-]^{10}} \right)$$

$$= -0.002\,5693 \text{ V} \left[\ln (1 \times 10^{66}) + \ln \left(\frac{1}{[\text{Cl}^-]^{10}} \right) \right]$$

$$= -0.390 \text{ V} + (0.0025\,693 \text{ V}) \ln [\text{Cl}^-]^{10}$$

$$-0.0594 \text{ V} = 0.002\,5693 \text{ V} \ln[\text{Cl}^-]^{10}$$

$$= (0.025\,693 \text{ V}) \ln[\text{Cl}^-]$$

$$\ln[\text{Cl}^-] = \frac{-0.06 \text{ V}}{0.025\,693 \text{ V}} = -2$$

$$[\text{Cl}^-] = 10^{-1} \text{ mol} \cdot \text{L}^{-1}$$

6N.7 (a) $\text{Cu}^{2+} (\text{aq}, 0.010\,\text{M}) + 2 \text{ e}^- \rightarrow \text{Cu(s) (cathode)}$

$\text{Cu}^{2+} (\text{aq}, 0.0010 \text{ M}) + 2 \text{ e}^- \rightarrow \text{Cu(s) (anode)}$

$\text{Cu}^{2+} (\text{aq}, 0.010 \text{ M}) \rightarrow \text{Cu}^{2+} (\text{aq}, 0.0010 \text{ M}), n = 2$

$E°_{\text{cell}} = E°(\text{cathode}) - E°(\text{anode}) = 0 \text{ V}$

$E_{\text{cell}} = E°_{\text{cell}} - \left(\frac{RT}{nF} \right) \ln Q = - \left(\frac{0.025\,693 \text{ V}}{2} \right) \ln Q$ at 25°C

$E_{\text{cell}} = - \left(\frac{0.025\,693 \text{ V}}{2} \right) \ln \left(\frac{0.0010 \text{ M}}{0.010 \text{ M}} \right) = +0.030 \text{ V}$

(b) at pH = 3.0, $[\text{H}^+] = 1 \times 10^{-3}$ M

at pH = 4.0, $[\text{H}^+] = 1 \times 10^{-4}$ M

Cell reaction is $\text{H}^+ (\text{aq}, 1 \times 10^{-3} \text{ M}) \rightarrow \text{H}^+ (\text{aq}, 1 \times 10^{-4} \text{ M}), n = 1$

$E°_{\text{cell}} = 0 \text{ V}$ $E_{\text{cell}} = E°_{\text{cell}} - \left(\frac{RT}{nF} \right) \ln Q = - \left(\frac{0.025\,693 \text{ V}}{1} \right) \ln \left(\frac{1 \times 10^{-4}}{1 \times 10^{-3}} \right)$

$$= +6 \times 10^{-2} \text{ V}$$

6N.9 Since the reduction potential of tin(II) is negative relative to the SHE, we will assume the tin electrode to be the anode such that the standard cell potential would be positive. Then we can use the Nernst equation to solve for the hydrogen ion activity in order to calculate the pH. The cell reaction is $Sn(s) + 2 H^+ \rightarrow Sn^{2+} + H_2(g)$.

$$E = E° - \left(\frac{0.025\ 693\ V}{n}\right) \ln\left(\frac{[Sn^{2+}]p_{H_2}}{[H^+]^2}\right)$$

$$0.061\ V = 0.14\ V - \left(\frac{0.025\ 693\ V}{2}\right) \ln\left(\frac{[0.015][1]}{[H^+]^2}\right)$$

$$= 0.14\ V - (0.01284\ V)(\ln(0.015) - \ln[H^+]^2)$$

$$\frac{0.079\ V}{0.01284\ V} = \ln(0.015) - 2\ \ln[H^+]$$

$$-10.352 = 2\ \ln[H^+]$$

$$\ln[H^+] = -5.176, \quad [H^+] = 5.650 \times 10^{-3}$$

$$pH = -\log(5.650 \times 10^{-3}) = 2.25$$

6N.11 To calculate this value, we need to determine the $E°$ value for the solubility reaction: $Hg_2Cl_2(s) \rightarrow Hg_2^{2+}(aq) + 2\ Cl^-(aq)$

The equations that will add to give the net equation we want are

$Hg_2Cl_2(s) + 2\ e^- \rightarrow 2\ Hg(l) + 2\ Cl^-(aq)$ $\qquad E° = +0.27\ V$

$2\ Hg(l) \rightarrow Hg_2^{2+}(aq) + 2\ e^-$ $\qquad E° = 0.79\ V$

Adding these two equations together gives the desired net reaction, and summing the $E°$ values will give the $E°$ value for that process:

$$E° = (+0.29\ V) + (-0.79\ V) = -0.50\ V$$

$$\ln K_{sp} = \frac{nFE°}{RT} = \frac{(2)(9.65 \times 10^4\ C \cdot mol^{-1})(-0.50\ V)}{(8.314\ J \cdot K^{-1} \cdot mol^{-1})(298\ K)} = -38.95$$

$$K_{sp} = 1.2 \times 10^{-17}$$

(b) The calculated value is a factor of 10 greater than the experimentally determined (measured) value (1.3×10^{-18}). Such differences in calculated and measured values are common for very small and very large values.

6N.13 (a) $Pb^{4+}(aq) + 2\ e^- \rightarrow Pb^{2+}(aq)$ $\quad E°(cathode) = +1.67\ V$

$Sn^{2+}(aq) \rightarrow Sn^{4+}(aq) + 2\ e^-$ $\quad E°(anode) = +0.15\ V$

$$Pb^{4+}(aq) + Sn^{2+}(aq) \rightarrow Pb^{2+}(aq) + Sn^{4+}(aq) \quad E^\circ_{cell}$$
$$= 1.67 \text{ V} - (0.15 \text{ V}) = +1.52 \text{ V}$$

Then, $E = E^\circ - \left(\dfrac{0.025\ 693 \text{ V}}{n}\right) \ln Q$; $1.33 \text{ V} = 1.52 \text{ V} - \left(\dfrac{0.025\ 693 \text{ V}}{2}\right) \ln Q$

$$\ln Q = \frac{1.52 \text{ V} - 1.33 \text{ V}}{0.0129 \text{ V}} = \frac{0.19 \text{ V}}{0.0129 \text{ V}} = 15 \quad Q = 10^6$$

(b)

$$2[Cr_2O_7^{2-}(aq) + 14 \text{ H}^+(aq) + 6 \text{ e}^- \rightarrow$$
$$2 \text{ Cr}^{3+}(aq) + 7 \text{ H}_2O(l)] \quad E^\circ(\text{cathode}) = 1.33 \text{ V}$$

$$3[2 \text{ H}_2O(l) \rightarrow O_2(g) + 4 \text{ H}^+(aq) + 4 \text{ e}^-] \quad E^\circ(\text{anode}) = +1.23 \text{ V}$$

$$2 Cr_2O_7^{2-}(aq) + 16 \text{ H}^+(aq) \rightarrow 4 \text{ Cr}^{3+}(aq) + 8 \text{ H}_2O(l) + 3 O_2(g) \quad E^\circ_{cell}$$
$$= 0.10 \text{ V}$$

Then, $E = E^\circ - \left(\dfrac{0.0257 \text{ V}}{n}\right) \ln Q$; $0.10 \text{ V} = +0.10 \text{ V} - \left(\dfrac{0.0257 \text{ V}}{12}\right) \ln Q$

$$\ln Q = 0.00 \quad Q = 1.0$$

6N.15 This is a concentration cell, so $E^\circ = 0$ V. The half-reactions are:

Cathode: $Ni^{2+}(aq) + 2 \text{ e}^- \rightarrow Ni(s)$

Anode: $Ni(s) \rightarrow Ni^{2+}(aq) + 2 \text{ e}^-$

The value of $n = 2$ and the overall cell reaction is:

$Ni^{2+}(aq, \text{cathode}) \rightarrow Ni^{2+}(aq, \text{anode})$

$Q = [Ni^{2+}(aq, \text{anode})]/[Ni^{2+}(aq, \text{cathode})]$

Anode:

When hydroxide is added to the second compartment, the net ionic

equation for the reaction in that compartment is:

$$Ni^{2+}(aq) + 2 \text{ OH}^-(aq) \rightarrow Ni(OH)_2(aq)$$

Because the concentration of Ni^{2+} ions is less in the half cell with

hydroxide ion, more nickel will need to be oxidized for equilibrium to be

established. Therefore, this half-cell serves as the anode. The

concentration of Ni^{2+} ions at the anode can be obtained from the solubility

product of $Ni(OH)_2$, in Table 6I.1:

$K_{sp}\{Ni(OH)_2\} = [Ni^{2+}][OH^-]^2 = 6.5 \times 10^{-18}$

[OH$^-$] at the anode is given by the measured pH:

$pOH = 14.00 - pH = 14.00 - 11.0 = 3.0$

$[OH^-] = 10^{-3.0} = 0.01 \text{ mol.L}^{-1}$

$[Ni^{2+}]_{anode} = K_{sp}/[OH^-]^2 = 6.5 \times 10^{-18}/(0.01)^2 = 6.5 \times 10^{-12} \text{ mol.L}^{-1}$

(*Note:* we are retaining an extra significant figure until the end.)

Use the Nernst equation to find the cell potential:

$[Ni^{2+}]_{anode} = 6.5 \times 10^{-12} \text{ mol.L}^{-1}$

$[Ni^{2+}]_{cathode} = 1.0 \text{ mol.L}^{-1}$ (given)

$Q = [Ni^{2+}]_{anode}/[Ni^{2+}]_{cathode} = (6.5 \times 10^{-12})/1.0 = 6.5 \times 10^{-12}$

$n = 2$ (from the half-reactions)

$E^° = 0$ V

At 298 K, $RT/F = 0.025\ 693$ V

$E = E^° - \dfrac{RT}{nF} \ln Q = 0 \text{ V} - \dfrac{(0.025\ 693 \text{ V})}{2} \ln(6.5 \times 10^{-12}) = +0.3 \text{ V}$

6N.17 This cell uses two silver electrodes, so $E^° = 0$ and E is determined by the ratio of $[Ag^+]_{anode}$ to $[Ag^+]_{cathode}$. Since $[Ag^+]_{anode} < [Ag^+]_{cathode}$, the ratio is less than 1 and $E > 0$, so the cell can do work because $\Delta G = w_{max} = -nFE$.

$$E = -\left(\frac{0.025\ 693 \text{ V}}{n}\right) \ln\left(\frac{[Ag^+]_{anode}}{[Ag^+]_{cathode}}\right) = -\left(\frac{0.025\ 693 \text{ V}}{1}\right) \ln\left(\frac{5.0 \times 10^{-3}}{0.15}\right)$$

$$= 0.0874 \text{ V}$$

$\Delta G = w_{max} = -nFE = -(1 \text{ mol})(96\ 485 \text{ J V}^{-1}\text{mol}^{-1})(0.0874 \text{ V})$

$$= -8.4 \text{ kJ}$$

Therefore, the maximum work that the cell can perform is 8.4 kJ per mole of Ag.

6N.19 For the standard calomel electrode, $E^° = +0.27$ V. If this were set equal to 0, all other potentials would also be decreased by 0.27 V. (a) Therefore, the standard hydrogen electrode's standard reduction potential would be

0.00 V − 0.27 V or −0.27 V.   (b) The standard reduction potential for

Cu^{2+}/Cu would be 0.34 V − 0.27 V or +0.07 V.

6N.21 (a) $Fe_2O_3 \cdot H_2O$ (b) H_2O and O_2 jointly oxidize iron.   (c) Water is more highly conducting if it contains dissolved ions, so the rate of rusting is increased.

6N.23 (a) aluminum or magnesium; both are below titanium in the electrochemical series

(b) cost, availability, and toxicity of products in the environment

(c) $Cu^{2+} + 2 e^- \rightarrow Cu(s)$ $E° = +0.34$ V

$Cu^+ + e^- \rightarrow Cu(s)$ $E° = +0.52$ V

$Fe^{3+} + 3 e^- \rightarrow Fe(s)$ $E° = -0.04$ V

$Fe^{2+} + 2 e^- \rightarrow Fe(s)$ $E° = -0.44$ V

Fe could act as the anode of an electrochemical cell if Cu^{2+} or Cu^+ were present; therefore, it could be oxidized at the point of contact. Water with dissolved ions would act as the electrolyte.

6O.1 The strategy is to consider the possible competing cathode and anode reactions. At the cathode, choose the reduction reaction with the most positive (least negative) standard reduction potential ($E°$ value). At the anode, choose the oxidation reaction with the least positive (most negative) standard reduction potential ($E°$ value, as given in the table). Then calculate $E°_{cell} = E°(\text{cathode}) - E°(\text{anode})$. The negative of this value is the minimum potential that must be supplied.

(a) cathode: Ni^{2+} (aq) $+ 2 e^- \rightarrow Ni(s)$ $E° = -0.23$ V

(rather than $2 H_2O(l) + 2 e^- \rightarrow H_2(g) + 2 OH^-$ (aq) $E° = -0.83$ V)

(b) anode: $2 H_2O(l) \rightarrow O_2(g) + 4 H^+(aq) + 4 e^-$ $E° = +1.23$ V (the SO_4^{2-} ion will not oxidize)

(c) $E°_{cell} = E°(cathode) - E°(anode) = -0.23 \text{ V} - (+1.23 \text{ V}) = -1.46 \text{ V}$

Therefore, E (supplied) must be $> +1.46$ V (1.46 V is the minimum).

6O.3 In each case, compare the reduction potential of the ion with the reduction potential of water $(E° = -0.42 \text{ V})$ and choose the process with the least negative $E°$ value.

(a) $Mn^{2+}(aq) + 2 e^- \rightarrow Mn(s)$ $E° = -1.18 \text{ V}$

(b) $Al^{3+}(aq) + 3 e^- \rightarrow Al(s)$ $E° = -1.66 \text{ V}$

The reactions in (a) and (b) evolve hydrogen rather than yield a metallic deposit because water is reduced, according to

$2 H_2O(l) + 2 e^- \rightarrow H_2(g) + 2 OH^-(aq)$ $(E° = -0.42 \text{ V}, \text{ at } pH = 7)$

(c) $Ni^{2+}(aq) + 2 e^- \rightarrow Ni(s)$ $E° = -0.23 \text{ V}$

(d) $Au^{3+}(aq) + 3 e^- \rightarrow Au(s)$ $E° = +1.69 \text{ V}$

In (c) and (d) the metal ion will be reduced.

6O.5 $4500 \text{ C} \div 9.65 \times 10^4 \text{ C} \cdot \text{F}^{-1} = 0.047 \text{ F} = 0.047 \text{ mol } e^-$

(a)
  $(0.047 \div 3) \text{ mol Bi}^{3+} + 0.047 \text{ mol } e^- \rightarrow$
    $(0.047 \div 3) \text{ mol Bi, or } 0.016 \text{ mol Bi} = 3.3 \text{ g}$

(b)
  $0.047 \text{ mol H}^+ + 0.047 \text{ mol } e^- \rightarrow (0.047 \div 2) \text{ mol H}_2$

  $0.024 \text{ mol H}_2 \times 22.4 \text{ L.mol}^{-1} = 0.54 \text{ L}$

(c)
  $(0.047 \div 3) \text{ mol Co}^{3+} + 0.047 \text{ mol } e^- \rightarrow$
    $(0.047 \div 3) \text{ mol Co} = 0.016 \text{ mol Co or } 0.94 \text{ g}$

6O.7   (a) $Ag^+(aq) + e^- \rightarrow Ag(s)$

$$time = (1.50 \text{ g Ag})\left(\frac{1 \text{ mol Ag}}{107.98 \text{ g Ag}}\right)\left(\frac{1 \text{ mol } e^-}{1 \text{ mol Ag}}\right)$$

$$\left(\frac{9.65 \times 10^4 \text{ C}}{1 \text{ mol } e^-}\right)\left(\frac{1 \text{ A} \cdot \text{s}}{1 \text{ C}}\right)\left(\frac{1}{0.0136}\right) = 9.9 \times 10^4 \text{ s or 27 h}$$

(b) $Cu^{2+}(aq) + 2 e^- \rightarrow Cu(s)$

$$mass \text{ Cu} = (9.9 \times 10^4 \text{ s})(0.0136 \text{ A})\left(\frac{1 \text{ C}}{1 \text{ A} \cdot \text{s}}\right)\left(\frac{1 \text{ mol } e^-}{9.65 \times 10^4 \text{ C}}\right)$$

$$\left(\frac{0.50 \text{ mol Cu}}{1 \text{ mol } e^-}\right)\left(\frac{63.5 \text{ g Cu}}{1 \text{ mol Cu}}\right) = 0.44 \text{ g Cu}$$

6O.9   (a)
$Cr(VI) + 6e^- \rightarrow Cr(s)$

$$current = \frac{charge}{time}$$

$$= \frac{8.2 \text{ g Cr} \left(\frac{1 \text{ mol Cr}}{52.00 \text{ g Cr}}\right)\left(\frac{6 \text{ mol } e^-}{1 \text{ mol Cr}}\right)\left(\frac{9.65 \times 10^4 \text{ C}}{1 \text{ mol } e^-}\right)}{24 \text{ h} \times 3600 \text{ s} \cdot \text{h}^{-1}}$$

$$= 1.056 \text{ C} \cdot \text{s}^{-1} = 1.1 \text{ A}$$

(b)
$$Na^+ + e^- \rightarrow Na(s)$$

$$current \quad current = \frac{8.2 \text{ g Na} \left(\frac{1 \text{ mol}}{22.99 \text{ g Na}}\right)\left(\frac{1 \text{ mol } e^-}{1 \text{ mol Na}}\right)\left(\frac{9.65 \times 10^4 \text{ C}}{1 \text{ mol } e^-}\right)}{24 \text{ h} \times 3600 \text{ s} \cdot \text{h}^-}$$

$$= 0.398 \text{ C} \cdot \text{s}^{-1} = 0.40 \text{ A}$$

6O.11 $Ru^{n+}(aq) + n e^- \rightarrow Ru(s)$; solve for n

$$moles \text{ of Ru} = (0.0310 \text{ g Ru})\left(\frac{1 \text{ mol}}{101.07 \text{ g Ru}}\right) = 3.07 \times 10^{-4} \text{ mol}$$

$$\text{total charge} = (500 \text{ s})(120 \text{ mA})\left(\frac{10^{-3} \text{ A}}{1 \text{ mA}}\right)\left(\frac{1 \text{ C} \cdot \text{s}^{-1}}{1 \text{ A}}\right) = 60 \text{ C}$$

$$\text{moles of } e^- = (60 \text{ C})\left(\frac{1 \text{ mol } e^-}{96\,500 \text{ C}}\right) = 6.2 \times 10^{-4} \text{ mol } e^-$$

$$n = \frac{6.2 \times 10^{-4} \text{ mol } e^-}{3.07 \times 10^{-4} \text{ mol}} = \frac{2 \text{ mol charge}}{1 \text{ mol}}$$

Therefore, oxidation number of Ru^{2+} is +2.

6O.13 (a) $2Cu_{(aq)}^{2+} + 4e^- \rightarrow 2Cu(s)$ $E° = 0.34$ V

$$\frac{2H_2O(l) \rightarrow O_2(g) + 4H^+_{(aq)} + 4e^- \qquad\qquad E° = -0.82 \text{ V (pH7)}}{2Cu_{(aq)}^{2+} + 2H_2O(l) \rightarrow 2Cu(s) + O_2(g) + 4H^+_{(aq)} \qquad E° = -0.48 \text{ V}}$$

Based on the half-reactions $H^+_{(aq)}$ or hydronium ions are generated at the anode.

(b) $Q = nF$ and $Q = It$

$$\therefore n = \frac{It}{F} = \frac{(0.120 \text{ A})(1.08 \times 10^5 \text{s})}{9.6485 \times 10^4 \text{C. mol}^{-1}} = 0.134 \text{ mol of electrons}$$

Since 4 moles of electrons are needed in the balanced reaction to generate 4 moles of H_3O^+ there is a 1:1 ratio. Since 0.134 mol of electrons were supplied a maximum of 0.134 mol H_3O^+ can be generated.

(c)

$$[H_3O^+] = \frac{0.134 \text{ mol}}{0.2000 \text{ L}}$$
$$= 0.67 \text{ mol. L}^{-1}$$

$$pH = -\log[H_3O^+] = 0.17$$

6O.15 (a) $n_{e^-} = n_{Ag^+} = \dfrac{It}{F} = \dfrac{(3.5 \text{ A})(395.0 \text{ s})}{(96\,485 \text{ C} \cdot \text{mol}^{-1})} = 1.43 \times 10^{-2} \text{ mol Ag}$

$$1.43 \times 10^{-2} \text{ mol Ag}\left(\frac{107.87 \text{ g Ag}}{\text{mol Ag}}\right) = 1.55 \text{ g Ag}$$

$$\frac{1.55 \text{ g}}{2.69 \text{ g}} \times 100 = 57.4\% \text{ Ag}$$

(b) The mass of X is: $2.69 \text{ g} - 1.55 \text{ g} = 1.14 \text{ g X}$

Since the salt is 1:1 Ag:X, the molar mass of X is

$$\frac{1.14 \text{ g}}{1.43 \times 10^{-2} \text{ mol}} = 79.7 \text{ g} \cdot \text{mol}^{-1}$$

This molar mass is closest to bromine, so the formula is AgBr.

6.1   (a) Acid (1) is a strong acid because it totally dissociates.

(b) Acid (3) has the strongest base because it is the weakest acid.

(c) Acid (3) has the largest pK_a. Since $pK_a = -\log K_a$. The smallest K_a value will result the largest pK_a.

6.3 $T' = 30°C = 303 \text{ K}$, $T = 20°C = 293 \text{ K}$

$$\ln\left(\frac{K_a'}{K_a}\right) = \frac{\Delta H°}{R}\left(\frac{1}{T} - \frac{1}{T'}\right) = \frac{\Delta H°}{R}\left(\frac{T' - T}{TT'}\right)$$

This is the van 't Hoff equation.

$$\ln\left(\frac{1.768 \times 10^{-4}}{1.765 \times 10^{-4}}\right) = \frac{\Delta H°}{8.314 \text{ J} \cdot \text{K}^{-1} \cdot \text{mol}^{-1}}\left(\frac{303 \text{ K} - 293 \text{ K}}{293 \text{ K} \times 303 \text{ K}}\right)$$

$$\Delta H° = \frac{\ln\left(\frac{1.768 \times 10^{-4}}{1.765 \times 10^{-4}}\right)\left(8.314 \text{ J} \cdot \text{K}^{-1} \cdot \text{mol}^{-1}\right)}{\left(\frac{303 \text{ K} - 293 \text{ K}}{293 \text{ K} \times 303 \text{ K}}\right)} = 1.3 \times 10^2 \text{ J} \cdot \text{mol}^{-1}$$

6.5   (a) The reaction is: $H_2O_2 \text{ (g)} + SO_3 \text{ (g)} \rightarrow H_2SO_5\text{(g)}$

(b) H_2O_2, $\text{H}-\ddot{\text{O}}-\ddot{\text{O}}-\text{H}$

SO_3,

H_2SO_5,

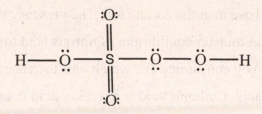

(c) H_2O_2 = Lewis base; SO_3 = Lewis acid

6.7 (a) We begin by finding the empirical formula of the compound:

$$C: \frac{0.942 \text{ g CO}_2}{44.011 \text{ g} \cdot \text{mol}^{-1}} = 0.0214 \text{ mol CO}_2 \therefore 0.214 \text{ mol C}$$

$$(0.214 \text{ mol C})(12.011 \text{ g} \cdot \text{mol}^{1}) = 0.257 \text{ g C}$$

$$H: \frac{0.0964 \text{ g H}_2\text{O}}{18.0158 \text{ g} \cdot \text{mol}^{-1}} = 0.00535 \text{ mol H}_2\text{O} \therefore 0.0107 \text{ mol H}$$

$$(0.0107 \text{ mol H})(1.008 \text{ g} \cdot \text{mol}^{-1}) = 0.0108 \text{ g H}$$

$$Na: \frac{0.246 \text{ g H}_2\text{O}}{22.99 \text{ g} \cdot \text{mol}^{-1}} = 0.0107 \text{ mol Na}$$

$$(0.0107 \text{ mol Na})(22.99 \text{ g} \cdot \text{mol}^{-1}) = 0.0246 \text{ g Na}$$

O: mass of O = 1.200 g − 0.257 g − 0.0108 g − 0.0246 g = 0.686 g of O

$$\frac{0.686 \text{ g O}}{16.00 \text{ g} \cdot \text{mol}^{-1}} = 0.0429 \text{ mol O}$$

Dividing through by 0.0107 moles, we find the empirical formula to be: C_2HNaO_4.

A molar mass of 112.02 g·mol^{-1} indicates that this is also the molecular formula.

(b)

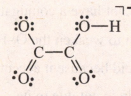

(c) The dissolved substance is sodium oxalate, it is capable of gaining or losing a proton and, therefore, amphiprotic. $\text{pH} = \frac{1}{2}\left(\text{p}K_{a_1} + \text{p}K_{a2}\right) = 2.7$.

6.9   (a) Nitrous acid will act like a strong acid because its conjugate base, the nitrite ion, is a weaker base than the acetate ion. The presence of acetate ions will shift the proton transfer equilibrium of nitrous acid toward the products (NO_2^- and H^+) by consuming H^+, which will increase the apparent K_a of nitrous acid. Carbonic acid is a weaker acid than acetic acid, so no shift in equilibrium occurs. (b) Ammonia will act like a strong base because its conjugate acid, the ammonium ion, is a weaker acid than acetic acid. The presence of acetic acid will shift the proton transfer equilibrium of ammonia with water toward the products (NH_4^+ and OH^-) by consuming OH^-, which will increase the apparent K_b of ammonia.

6.11 $$K = \frac{[NO_2^-][NH_4^+]}{[HNO_2][NH_3]} = \frac{K_a(HNO_2) \times K_b(NH_3)}{K_w}$$

$$= \frac{(4.3 \times 10^{-4})(1.8 \times 10^{-5})}{1.0 \times 10^{-14}} = 7.7 \times 10^5$$

This means that the reaction is in favor of the products.

6.13   (a) The Lewis structure of boric acid is

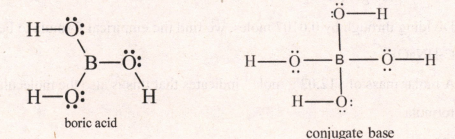

boric acid   conjugate base

No. $B(OH)_3$ is a very weak acid because it does not have a conjugate system to delocalize the electrons on the oxygen to weaken the O–H bond. (b) In that reaction, boric acid acts as a Lewis acid because it accepts electron pairs (OH^-) from water. But, its acidity is not due to its dissociation.

6.15 We wish to calculate K_a for the reaction

$$HF(aq) + H_2O(l) \rightleftharpoons H_3O^+(aq) + F^-(aq).$$

This equation is equivalent to

$$HF(aq) \rightleftharpoons H^+(aq) + F^-(aq).$$

This latter writing of the expression is simpler for the purpose of the thermodynamic calculations.

The $\Delta G°$ value for this reaction is easily calculated from the free energies given in the appendix:

$$= (-278.79 \text{ kJ} \cdot \text{mol}^{-1}) - (-296.82 \text{ kJ} \cdot \text{mol}^{-1}) = 18.03 \text{ kJ} \cdot \text{mol}^{-1}$$
$$= -RT \ln K$$
$$K = e^{-\Delta G°/RT}$$
$$K = e^{-(18030 \text{ J} \cdot \text{mol}^{-1})/[8.314 \text{ J} \cdot \text{K}^{-1} \cdot \text{mol}^{-1})(298 \text{ K})]} = 6.9 \times 10^{-4}$$

6.17 (a) $D_2O + D_2O \leftrightarrow D_3O^+ + OD^-$

(b) $K_{hw} = [D_3O^+][OD^-] = 1.35 \times 10^{-15}$; $pK_{hw} = -\log K_{hw} = 14.870$

(c) $[D_3O^+] = [OD^-] = \sqrt{1.35 \times 10^{-15}} = 3.67 \times 10^{-8} \text{ mol} \cdot \text{L}^{-1}$

(d) $pD = -\log(3.67 \times 10^{-8}) = 7.435 = pOD$

(e) $pD + pOD = pK_{hw} = 14.870$

6.19 (a) The H^+ in lactic acid can interact with HbO_2^- to produce HHb and release more oxygen molecules, which means that the concentration of HbO_2^- will be lower in the tissues.

(b) Since more Hb^- will return to the lung, more oxygen will be bound to Hb^- to produce HbO_2^-, which will increase the concentration of HbO_2^-.

6.21 (a) Two protons can be accepted (one on each N);

(b)

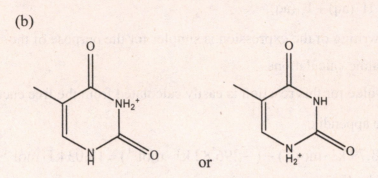

or

(c) Each of the two nitro- groups will show amphiprotic behavior in
aqueous solution (can either accept a proton and donate a proton).

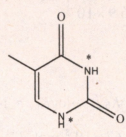

6.23 The relation to use is $pH = pK_a + \log \frac{[\text{Base form}]}{[\text{Acid form}]}$. The K_a value for acetic

acid is 1.8×10^{-5} and $pK_a = 4.74$. Because we are adding acid, the pH will

fall upon the addition, and we want the final pH to be no more than 0.20 pH

units different from the initial pH, or 4.54.

$$4.54 = 4.74 + \log \frac{[\text{Base form}]}{[\text{Acid form}]}$$

$$-0.20 = \log \frac{[\text{Base form}]}{[\text{Acid form}]}$$

$$\frac{[\text{Base form}]}{[\text{Acid form}]} = 0.63$$

We want the concentration of the base form to be 0.63 times that of the acid

form. We do not know the initial number of moles of base or acid forms, but we know that the two amounts were equal. Let C = initial number of moles of acetic acid and the initial number of moles of sodium acetate. The number of moles of H_3O^+ to be added (in the form of HCl(aq)) is

$0.001\ 00\ L \times 6.00\ mol \cdot L^{-1} = 0.006\ 00\ mol$. The total final volume will be 0.1010 L.

$$\frac{\dfrac{C - 0.006\ 00}{0.1010\ L}}{\dfrac{C + 0.006\ 00}{0.1010\ L}} = 0.63$$

$$\frac{C - 0.006\ 00}{C + 0.006\ 00} = 0.63$$

$$C - 0.006\ 00 = 0.63(C + 0.006\ 00)$$

$$C - 0.006\ 00 = 0.63\ C + 0.003\ 78$$

$$0.37\ C = 0.009\ 78$$

$$C = 0.026$$

The initial buffer solution must contain at least 0.026 mol acetic acid and 0.026 mol sodium acetate. The concentration of the initial solution will then be $0.026\ mol \div 0.100\ L = 0.260\ M$ in both acetic acid and sodium acetate.

6.25 (a) $pH = pK_{a1} + \log\dfrac{[HOOCCH_2CO_2^-]}{[HOOCCH_2COOH]}$;

when $[HOOCCH_2COOH] = [HOOCCH_2COO^-]$, $pH = pK_{a1} = 2.8$

(b) Similar to (a), when $[HOOCCH_2COO^-] = [^-OOCCH_2COO^-]$, $pH = pK_{a2} = 5.7$

(a) At $pH = 4.2$, $HOOCCH_2COO^-$ is the predominant species. The reason is that between $pH > 2.8$ and $pH < 5.7$, $HOOCCH_2COO^-$ is the major species in the solution from the results in (a) and (b).

6.27 In addition to the reaction corresponding to the dissolution of PbF_2(s):

$$PbF_2(s) \leftrightarrow Pb^{2+}(aq) + 2\ F^-(aq) \quad K = 3.7 \times 10^{-8}$$

The buffer will provide a source of $H_3O^+(aq)$ ions, which will allow the reaction

$$H_3O^+(aq) + F^- \leftrightarrow HF(aq) + H_2O(l) \quad K = 1/(K_a(HF)) = 1/(3.5 \times 10^{-4}) = 2.86 \times 10^3.$$

These two coupled reactions give two equilibrium expressions that must be simultaneously satisfied:

$$[F^-]^2[Pb^{2+}] = 3.7 \times 10^{-8} \quad \text{and} \quad \frac{[HF]}{[H_3O^+][F^-]} = 2.86 \times 10^{-3}$$

Given that all fluoride ions come from $PbF_2(s)$ and wind up as either $F^-(aq)$ or $HF(aq)$, and that for every 1 mole of $Pb^{2+}(aq)$ generated 2 moles of $F^-(aq)$ are also produced, we can write a third equation that relates the concentration of the fluoride-containing species to the concentration of dissolved barium:

$$[Pb^{2+}] = \tfrac{1}{2}([F^-] + [HF])$$

In the end, the concentration of $Pb^{2+}(aq)$ will be equal to the solubility of $PbF_2(s)$. To determine the equilibrium concentration of $Pb^{2+}(aq)$, we first determine $[H_3O^+]$, which is fixed by the buffer system, and then use the three simultaneous equations above to solve for $[Pb^{2+}]_{eq}$.

The buffer determines the equilibrium concentration of $H_3O^+(aq)$. The initial concentrations of $H_3O^+(aq)$ and $NaCH_3CO_2(aq)$ are

$$[H_3O^+]_i = \frac{(0.055\ L)(0.15\ mol \cdot L^{-1})}{0.10\ L} = 0.0825\ M \quad \text{and}$$

$$[CH_3CO_2^-]_i = \frac{(0.045\ L)(0.65\ mol \cdot L^{-1})}{0.10\ L} = 0.293\ M.$$

To determine their equilibrium concentrations we solve using the familiar method:

Concentration

(mol·L^{-1}) CH$_3$COOH(aq) + H$_2$O(l) $\rightleftharpoons$ H$_3$O$^+$(aq) + CH$_3$CO$_2^-$(aq)

initial	0	—	0.0825	0.293
change	+x	—	$-x$	$-x$
equilibrium	x	—	0.0825 $- x$	0.292 $- x$

$$K_a = 1.8 \times 10^{-5} = \frac{[H_3O^+][CH_3CO_2^-]}{[CH_3COOH]} = \frac{(0.0825 - x)(0.292 - x)}{(x)}$$

Rearranging this expression we obtain $0.024072 - 0.3745\, x + x^2 = 0$.

Using the quadratic formula, we find $x = 0.082493$, and $[H_3O^+] = 7.1 \times 10^{-6}$ M. With this equilibrium concentration of H$_3$O$^+$(aq), we revisit the three simultaneous equations from above, namely

$$[F^-]^2[Pb] = 3.7 \times 10^{-8};$$

$$[Pb^{2+}] = \tfrac{1}{2}([F^-] + [HF]); \text{ and}$$

$$\frac{[HF]}{[H_3O^+][F^-]} = 2.86 \times 10^{-3}.$$

Due to the presence of the buffer, $[H_3O^+] = 7.1 \times 10^{-6}$, and this last equation simplifies to

$$\frac{[HF]}{[F^-]} = 2.03 \times 10^{-2}.$$

Rearranging these three simultaneous equations we find that

$$[Pb^{2+}] = \frac{3.7 \times 10^{-8}}{[F^-]^2}, \quad [HF] = [F^-] \times (2.03 \times 10^{-2}), \quad \text{and}$$

$$[Pb^{2+}] = \tfrac{1}{2}([F^-] + [HF])$$

$$\frac{3.7 \times 10^{-8}}{[F^-]^2} = \tfrac{1}{2}\left[[F^-] + \left([F^-] \times (2.03 \times 10^{-2})\right)\right]. \quad \text{Solving this expression for } [F^-]:$$

$$[F^-] = \sqrt[3]{\frac{3.7 \times 10^{-8}}{0.5203}} = 4.14 \times 10^{-3}.$$

The equilibrium concentration of Pb^{2+}(aq) is then:

$$[Pb^{2+}] = \frac{3.7 \times 10^{-8}}{[F^-]^2} = \frac{3.7 \times 10^{-8}}{(4.14 \times 10^{-3})^2} = 2.16 \times 10^{-3} \text{ M}$$

Therefore, the solubility of PbF$_2$(s) is 2.2×10^{-3} M

6.29 (a) $CO_3^{2-}(aq) + 2 Ag^+(aq) \rightarrow Ag_2CO_3(s)$

Moles of Ag^+ = $(0.0362 \text{ L } Ag^+)(0.110 \text{ mol/L})$ = 3.98×10^{-3} mol Ag^+

Moles of CO_3^{2-} = 3.98×10^{-3} mol $Ag^+ \times \dfrac{1 \text{ mol } CO_3^{2-}}{2 \text{ mol } Ag^+}$

= 1.99×10^{-3} mol CO_3^{2-}

$[CO_3^{2-}] = \dfrac{1.99 \times 10^{-3} \text{ mol } CO_3^{2-}}{0.0250 \text{ L}} = 0.0796 \text{ mol} \cdot L^{-1} \, CO_3^{2-}$

(b) You can use excel to make the titration plot. Before stoichiometric point, $[CO_3^{2-}]$ is dominant, and the concentration of Ag^+ is very low.

$[CO_3^{2-}] = \dfrac{(V_{CO_3^{2-}} M_{CO_3^{2-}}) - (V_{Ag^+} M_{Ag^+})/2}{V_{Total}}$, and $[Ag^+] = \sqrt{\dfrac{K_{sp}}{[CO_3^{2-}]}}$;

At stoichiometric point, $[Ag^+] = \sqrt[3]{\dfrac{K_{sp}}{4}}$; After stoichiometric point,

$[Ag^+]$ is dominant, and the extra $[Ag^+] = \dfrac{(V_{Ag^+} M_{Ag^+})/2 - (V_{CO_3^{2-}} M_{CO_3^{2-}})}{V_{Total}}$.

The K_{sp} value of Ag_2CO_3 is 6.2×10^{-12}. The plot is as follows:

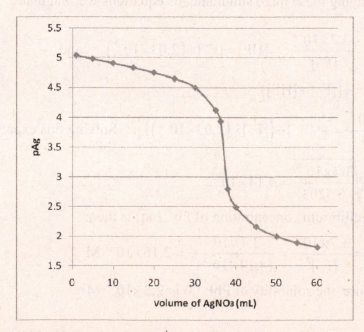

At stoichiometric point, $pAg^+ = 3.94$

6.31 For friend #1: the reaction between $CaCO_3$ with HCl is:

$$2\,HCl + CaCO_3 \rightarrow CaCl_2 + CO_2 + H_2O$$

The total mole of stomach acid (HCl) = 0.100 L × 0.10 M = 0.010 mole

$$\text{Moles of } CaCO_3 \text{ in 2 tablets} = \frac{0.750\ g}{100.09\ g \cdot mol^{-1}} \times 2 = 0.0150 \text{ mole}$$

Since 1 mole $CaCO_3$ will neutralize 2 mole of HCl, 0.0150 mole

$CaCO_3$ will neutralize all of the HCl in the stomach. If assuming all CO_2

will be escaped from stomach, the pH in the stomach will be 7.00 in the

first person (in reality, the pH will be < 7 because some of the CO_2 will

remain in the stomach).

For friend #2: the reaction of MgO with HCl is:

$$MgO(s) + 2HCl(aq) \rightarrow MgCl_2(aq) + H_2O(l)$$

$$\text{Total moles of MgO} = \frac{3 \times 0.400\ g}{40.305\ g \cdot mol^{-1}} = 2.98 \times 10^{-2} \text{ mole}$$

Since 1 mole MgO will neutralize 2 mole of HCl, 0.010 mole HCl will

consume 0.00500 mole MgO. Therefore, the moles of MgO left over are:

$2.98 \times 10^{-2} - 0.00500 = 0.0248$ mole MgO

MgO will react with H_2O producing OH^- (100% dissociation):

$$MgO(s) + H_2O(l) \rightarrow Mg^{2+}(aq) + 2\,OH^-(aq)$$

$[OH^-] = (2 \times 0.0248 \text{ mole})/(3 \times 4.93 \times 10^{-3}\ L + 0.100\ L) = 0.432$ M

Note: 1 US tsp = 4.93 mL;

pOH = 0.36; pH = 13. 64

The pH in friend #2's stomach is much higher than that of friend #1.

6.33 The K_{sp} value for PbF_2 obtained from Table 11.5 is 3.7×10^{-8}. Using this

value, the $\Delta G°$ of the dissolution reaction can be obtained from

$\Delta G° = -RT \ln K$

$\Delta G° = -(8.314 \text{ J} \cdot \text{K}^{-1} \cdot \text{mol}^{-1})(298.2 \text{ K}) \ln(3.7 \times 10^{-8})$

$\Delta G° = +42.43 \text{ kJ} \cdot \text{mol}^{-1}$

From the Appendices we find that $\Delta G°_f(\text{F}^-, \text{aq}) = -278.79 \text{ kJ} \cdot \text{mol}^{-1}$ and

$\Delta G°_f(\text{Pb}^{2+}, \text{aq}) = -24.43 \text{ kJ} \cdot \text{mol}^{-1}$.

$\Delta G° = +42.43 \text{ kJ} \cdot \text{mol}^{-1} = \Delta G°_f(\text{Pb}^{2+}, \text{aq}) + \Delta G°_f(\text{F}^-, \text{aq}) - \Delta G°_f(\text{PbF}_2, \text{s})$

$\quad +42.43 \text{ kJ} \cdot \text{mol}^{-1} = (-24.43 \text{ kJ} \cdot \text{mol}^{-1}) + (-278.79 \text{ kJ} \cdot \text{mol}^{-1})$

$- \Delta G°_f(\text{PbF}_2, \text{s})$

$\Delta G°_f(\text{PbF}_2, \text{s}) = -345.65 \text{ kJ} \cdot \text{mol}^{-1}$

6.35   (a) pH = pKa + log ([base]/[acid]) = 4.75 + log(0.300/0.200) = 4.93

(b) $6.0 = 4.75 + \log\{(0.300 \text{ L} \times 0.300 \text{ M} + x)/(0.300 \text{ L} \times 0.200 \text{ M} - x)\}$

$\quad\quad = 4.75 + \log\{(0.0900 \text{ mol} + x)/(0.0600 \text{ mol} - x)\}$

$\log\{(0.0900 \text{ mol} + x)/(0.0600 \text{ mol} - x)\} = 1.25;$

$x = \text{mol of OH}^- = 0.0520 \text{ mol} = \text{mole of NaOH}$

mass of NaOH = (0.0520 mol) × (40.01 g/mol) = 2.08 g NaOH

6.37   Let novocaine = N; $\text{N(aq)} + \text{H}_2\text{O(l)} \rightleftharpoons \text{HN}^+(\text{aq}) + \text{OH}^-(\text{aq})$.

$K_b = \dfrac{[\text{HN}^+][\text{OH}^-]}{[\text{N}]}$

$pK_a = pK_w - pK_b = 14.00 - 5.05 = 8.95$

$\text{pH} = pK_a + \log\left(\dfrac{[\text{N}]}{[\text{HN}^+]}\right)$

$\log\left(\dfrac{[\text{N}]}{[\text{HN}^+]}\right) = \text{pH} - pK_a = 7.4 - 8.95 = -1.55$

Therefore, the ratio of the concentrations of novocaine and its conjugate

acid is $[\text{N}]/[\text{HN}^+] = 10^{-1.55} = 2.8 \times 10^{-2}$.

6.39   (a) The amount of CO_2 present at equilibrium may be found using the

equilibrium expression for the reaction of interest:

$$H_3O^+(aq) + HCO_3^- \rightleftharpoons 2 H_2O(l) + CO_2(aq)$$

$$K = 7.9 \times 10^{-7} = \frac{[CO_2]}{[H_3O^+][HCO_3^-]}$$

Solving for $[CO_2]$:

$[CO_2] = (7.9 \times 10^{-7})[H_3O^+][HCO_3^-]$

Given: $[H_3O^+] = 10^{-6.1} = 8 \times 10^{-7}$ M, and $[HCO_3^-] = 5.5 \ \mu mol \cdot L^{-1} = 5.5 \times 10^{-6}$ M

$[CO_2] = (7.9 \times 10^{-7})(8 \times 10^{-7})(5.5 \times 10^{-6}) = 3 \times 10^{-18} \ mol \cdot L^{-1}$

In 1.0 L of solution there will be 3×10^{-18} mol of $CO_2(aq)$.

(b) Adding 0.65×10^{-6} mol of $H_3O^+(aq)$ to the equilibrium system in (a)

will give an initial $[H_3O^+]$ of 1.44×10^{-6} M. To determine the equilibrium

concentration of $H_3O^+(aq)$ we set up the familiar problem:

Concentration

$(mol \cdot L^{-1})$	$H_3O^+(aq)$	$+$	$HCO_3^-(l)$	$\rightleftharpoons$	$2 H_2O(l)$	$+$	$CO_2(aq)$
initial	1.44×10^{-6}		5.5×10^{-6}		—		3.45×10^{-18}
change	$-x$		$-x$		—		$+x$
equilibrium	$1.44 \times 10^{-6} - x$		$5.5 \times 10^{-6} - x$		—		$3.45 \times 10^{-18} + x$

$$K = 7.9 \times 10^{-7} = \frac{3.45 \times 10^{-18} + x}{\left(1.44 \times 10^{-6} - x\right)\left(5.5 \times 10^{-6} - x\right)}$$

Rearranging to obtain a polynomial in x:

$x = 2.8 \times 10^{-18}$

Giving: $[H_3O^+] = 1.4 \times 10^{-6}$ and pH = 5.8,

$\Delta pH = 5.8 - 6.1 = -0.3$

6.41 $K_{sp}(CaF_2) = 4.0 \times 10^{-11}$

$Q(CaF_2) = (2 \times 10^{-4})(5 \times 10^{-5})^2 = 5 \times 10^{-13}$

Since $Q(CaF_2) < K_{sp}(CaF_2)$, there will be no CaF_2 precipitation at this

condition.

6.43 (a) E_{cell}

$E_{cell}°$ does not decrease with time, whereas E_{cell} does.

(b) Both

$E_{cell}°$ and E_{cell} are temperature dependent.

(c) Neither

As long as the redox equation is correctly balanced it makes no difference to $E_{cell}°$ and E_{cell} what the stoichiometric coefficients are.

(d) $E_{cell}°$

Since $E_{cell}° = \frac{RT}{nF} \ln K$.

(e) E

Since E is dependent on concentrations.

6.45 Al, Zn, Fe, Co, Ni, Cu, Ag, Au

6.47 $2[Zn^{2+}(aq) + 2\,e^- \rightarrow Zn(s)]$ $E°(\text{cathode}) = -0.76$ V

$M(s) \rightarrow M^{4+}(aq) + 4\,e^-$ $E°(\text{anode}) = x$

$M(s) + 2\,Zn^{2+}(aq) \rightarrow 2\,Zn(s) + M^{4+}(aq)$ $E°_{cell} = 0.16$ V

$E°_{cell} = E°(\text{cathode}) - E°(\text{anode})$

$+0.16\text{ V} = -0.76\text{ V} - (x)$

$x = -0.92\text{ V} = E°(M^{4+}/M)$

6.49 The strategy is to find the $E°$ value for the solubility reaction and then find appropriate half-reactions that add to give that solubility reaction. One of these half-reactions is our unknown, the other is obtained from the Appendix:

$Cu(IO_3)_2(s) + 2\,e^- \rightarrow Cu(s) + 2\,IO_3^-(aq)$ $E° = ?$ (A)

$Cu(s) \quad\quad \rightarrow Cu^{2+}(aq) + 2\,e^-$ $E° = -0.34$ V   (B)

$Cu(IO_3)_2(s) \quad\quad \rightarrow Cu^{2+}(aq) + 2\,IO_3^-(aq)$ $E° = \dfrac{RT \ln K_{sp}}{nF}$ (C)

$$E° = \frac{RT \ln K_{sp}}{nF}$$

$$= \frac{(8.314 \text{ J} \cdot \text{K}^{-1} \cdot \text{mol}^{-1})(298.2 \text{ K}) \ln (1.4 \times 10^{-7})}{2(9.65 \times 10^4 \text{ C} \cdot \text{mol}^{-1})}$$

$$= -0.20 \text{ V}$$

$$-0.20 \text{ V} = E°(A) + (-0.34 \text{ V})$$

$$E°(A) = +0.14 \text{ V}$$

6.51 A negatively charged electrolyte flows from the cathode to the anode.

6.53 (a) Reduction takes place at the electrode with the higher concentration, which would be the chromium electrode in contact with the 1.0 M $CrCl_3$.

Cathode (reduction) 1.0 M $CrCl_3$

Anode (oxidation) 0.0010 M $CrCl_3$

(b) Further dilution of the anode concentration would increase the cell potential.

(c) Adding 100 mL 1.0 M NaOH(aq) to the cathode compartment results in the formation of insoluble $(Cr(OH)_3$ which decreases the Cr^{3+} concentration and therefore decreases the cell potential.

(d) Increasing the mass of electrodes has no effect on the cell potential, however, the life-span of larger electrodes are longer.

6.55 (a) $M_{Ag^+}V_{Ag^+} = M_{I^-}V_{I^-}$

$$M_{Ag^+} = \frac{M_{I^-}V_{I^-}}{V_{Ag^+}} = \frac{(0.015 \text{ M})(16.7 \text{ mL})}{(25.0 \text{ mL})}$$

$$= 1.0 \times 10^{-2} \text{ M}$$

(b) We can find $[Ag^+]$ by using the Nernst equation appropriately,

$E = E° - \left(\dfrac{0.025\ 693 \text{ V}}{n}\right) \ln \left(\dfrac{1}{[Ag^+]}\right)$. Since the standard reduction

potential of silver(I) is +0.80 V, it will be the reduction half reaction versus the SHE, so $[Ag^+]$ appears in the denominator of Q. In addition, $n = 1$ and $E° = 0.080\ V$.

$$0.325\ V = 0.80\ V - \left(\frac{0.025\ 693\ V}{1}\right) \ln\left(\frac{1}{[Ag^+]}\right)$$

$$-0.475\ V = (-2.567 \times 10^{-2}\ V)(\ln 1 - \ln[Ag^+])$$

$$-18.50 = \ln[Ag^+]$$

$$[Ag^+] = 9.23 \times 10^{-9}\ M$$

Recalling that $K_{sp} = [Ag^+][I^-]$, and assuming $[Ag^+] = [I^-]$ at the stoichiometric point of the titration,

$$K_{sp} = [Ag^+][I^-] = (9.23 \times 10^{-9})^2 = 8.5 \times 10^{-17}$$

6.57 $F_2(g) + 2\ e^- \rightarrow 2\ F^-(aq)$ $E°(\text{cathode}) = +2.87\ V$

$2\ HF(aq) \rightarrow F_2(g) + 2\ H^+(aq) + 2\ e^-$ $E°(\text{anode}) = +3.03\ V$

$2\ HF(aq) \rightarrow 2\ H^+(aq) + 2\ F^-(aq)$

$E°_{cell} = E°(\text{cathode}) - E°(\text{anode}) = +2.87\ V - (+3.03\ V) = -0.16\ V$

For the above reaction, $K = \dfrac{[H^+]^2[F^-]^2}{[HF]^2}$ and $\ln K = \dfrac{nFE°}{RT}$

At $25°C = \dfrac{nE°}{0.025\ 69\ V} = \dfrac{(2)(-0.16\ V)}{0.025\ 69\ V} = -12$

$\ln K = -12$

$K = 6.14 \times 10^{-6}$

$K_a = \sqrt{K} = \sqrt{6.14 \times 10^{-6}} = 2.5 \times 10^{-3}$

6.59 (a) $\Delta G_{r1}° = -nFE_1° = \Delta H_{r1}° - T_1\Delta S_{r1}°$ where r_1 represents the reaction at T_1.

$\Delta G_{r2}° = -nFE_2° = \Delta H_{r2}° - T_2\Delta S_{r2}°$ where r_2 represents the reaction at T_2.

On subtracting the first reaction from the second, we obtain

$-nFE_2° + nFE_1° = \Delta H_{r2}° - T_2\Delta S_{r2}° - [\Delta H_{r1}° - T_1\Delta S_{r1}°]$.

Since $\Delta H_{r1}° = \Delta H_{r2}°$ we obtain,

$nFE_1° - nFE_2° = -T_2\Delta S_{r2}° + T_1\Delta S_{r1}°$

which can be rewritten as $-nFE_1^\circ + nFE_2^\circ = + T_2 \Delta S_{r2}^\circ - T_1 \Delta S_{r1}^\circ$.

Since $\Delta S_{r1}^\circ = \Delta S_{r2}^\circ = \Delta S_r^\circ$ we obtain,

$-nFE_1^\circ + nFE_2^\circ = \Delta S_r^\circ (T_2 - T_1)$

$nFE_2^\circ = nFE_1^\circ + \Delta S_r^\circ (T_2 - T_1)$

$E_2^\circ = E_1^\circ + \Delta S_r^\circ (T_2 - T_1)/nF$

(b) The redox reaction is $2H_2(g) + O_2(g) \rightarrow 2H_2O(l)$.

Using the standard potentials at 25°C to calculate E_1°,

$$O_2(g) + 4H^+(aq) + 4e^- \rightarrow 2H_2O(l) \quad E^\circ = +1.23 \text{ V}$$

$$\underline{2H_2(g) \rightarrow 4H^+(aq) + 4e^- \quad\quad E^\circ - 0 \text{ V}}$$

$$2H_2(g) + O_2(g) \rightarrow 2H_2O(l) \quad E_1^\circ = +1.23 \text{ V}$$

$\Delta S_r^\circ = \sum S_{m,\text{products}}^\circ - \sum S_{m,\text{reactants}}^\circ$

$= 2(69.91 \text{ J·K}^{-1}\text{·mol}^{-1}) - [2(130.68 \text{ J·K}^{-1}\text{·mol}^{-1}) +$

$(205.14 \text{ J·K}^{-1}\text{·mol}^{-1})]$

$= 139.82 \text{ J·K}^{-1}\text{·mol}^{-1} - 466.50 \text{ J·K}^{-1}\text{·mol}^{-1}$

$= -326.68 \text{ J·K}^{-1}\text{·mol}^{-1}$

$E_{r, 80°C}^\circ = E_{r, 25°C}^\circ + \Delta E_r^\circ$

$\qquad = E_{r, 25°C}^\circ + \dfrac{\Delta S_r^\circ}{nF} \Delta T$

$= +1.23 \text{ V} + \dfrac{-326.68 \text{ J·K}^{-1}\text{·mol}^{-1}}{4(9.6485 \times 10^4 \text{J·V}^{-1}\text{·mol}^{-1})}(353.15 \text{ K} - 298.15 \text{ K})$

$= +1.23 \text{ V} + \dfrac{-326.68 \text{ J·K}^{-1}\text{·mol}^{-1}}{3.8594 \times 10^5 \text{ J·V}^{-1}\text{·mol}^{-1}}(55.0 \text{ K})$

$= +1.23 \text{ V} - 4.65 \times 10^{-2} \text{ V}$

$= +1.18 \text{ V}$

6.61 The wording of this exercise suggests that K^+ ions participate in an electrolyte concentration cell reaction. Therefore, $E_{\text{cell}}^\circ = 0.00$ V, because the two half cells would be identical under standard conditions. Then,

$$E = E° - \left(\frac{0.0257 \text{ V}}{n}\right) \ln \left(\frac{[K_{out}^+]}{[K_{in}^+]}\right) = 0.00 \text{ V} - \left(\frac{0.0257 \text{ V}}{1}\right) \ln \left(\frac{1}{30}\right)$$

$$= +0.09 \text{ V}$$

and $E = 0.00 \text{ V} - \left(\frac{0.0257 \text{ V}}{1}\right) \ln \left(\frac{1}{20}\right) = +0.08 \text{ V}$

The range of potentials is 0.08 V to 0.09 V.

6.63 Buffer system. $HA \rightarrow H^+ + A^-$

$$Q = \frac{(H^+)(A^-)}{(HA)}$$

Note: (H^+), as opposed to $[H^+]$, indicates a nonequilibrium molarity.

Because in a buffer system $(A) \approx (HA)$, we can write

$$Q = (H^+)$$

$$E_{cell} = E°_{cell} - \frac{RT}{nF} \ln (H^+)$$

$$0.060 \text{ V} = E°_{cell} - \left(\frac{0.025\ 693}{1}\right)(2.303)(\log(H^+))$$

Because $\log(H^+) = -[-\log(H^+)] = -pH$, we have

$$0.060 \text{ V} = E°_{cell} - 0.0592 \times (-pH)$$

$$0.060 \text{ V} = E°_{cell} + 0.0592 \times pH$$

$$0.060 \text{ V} = E°_{cell} + 0.0592 \times 9.40$$

$$0.060 \text{ V} = E°_{cell} + 0.556 \text{ V}$$

$$E° = 0.060 \text{ V} - 0.556 \text{ V} = -0.496 \text{ V}$$

Similarly, $0.22 \text{ V} = -0.496 \text{ V} + 0.0592 \text{ V} \times pH$

$$pH = \frac{0.22 \text{ V} + 0.496 \text{ V}}{0.0592 \text{ V}} = 12$$

6.65 Using $\Delta G° = -nFE°$ and $\Delta G° = -RT \ln K$, one obtains the relationship

$$E° = \frac{RT}{nF} \ln K.$$

Must have $E^o = 0$ when pH = 7.

pH = 7 when concentration of H^+ and OH^- are equal to 1.0×10^{-7} mol.L^{-1}.

When $K = 1$, then $E^o = 0$.

Therefore, at pH = 7: $K = \dfrac{[H^+]}{[OH^-]} = \dfrac{1.0 \times 10^{-7} \text{ mol.L}^{-1}}{1.0 \times 10^{-7} \text{ mol.L}^{-1}} = 1$ and $E^o = 0$

At pH = 1: $E^o = \dfrac{RT}{nF} \ln K = 0.025693 \text{ V} \ln \dfrac{1.0 \text{ mol.L}^{-1}}{1.0 \times 10^{-14} \text{ mol.L}^{-1}} = +0.828$ V

At pH = 14: $E^o = \dfrac{RT}{nF} \ln K = 0.025693 \text{ V} \ln \dfrac{1.0 \times 10^{-14} \text{ mol.L}^{-1}}{1.0 \text{ mol.L}^{-1}}$

$-= -0.828 \text{V}$

6.67 (a) The more dilute electrolyte in a concentration cell is always the anode. The anode generates the electrons (in this case $Ag(s) \rightarrow Ag^+(aq) + 1e^-$) and the concentration of $Ag^+(aq)$ increases.

In all chemical reactions the species that increases in concentration is the product. Therefore, in the Nernst equation, for a concentration cell, the more dilute electrolyte is always the product.

$$E = E^\circ - \dfrac{RT}{nF} \ln \dfrac{[Ag^+]_{anode}}{[Ag^+]_{cathode}}$$

A plot of cell voltage, E, versus $\ln [Ag^+]_{anode}$ would be a linear increase with positive slope because a larger difference in concentration results in a larger cell voltage. In a concentration cell the larger the concentration difference between the anode and cathode the higher the cell voltage. In other words, a decreasing value for the numerator, $[Ag^+]_{anode}$, makes the $\ln$ term increasingly negative and therefore the term, $-\dfrac{RT}{nF} \ln \dfrac{[Ag^+]_{anode}}{[Ag^+]_{cathode}}$, increasingly positive.

(b) $E = E^\circ - \dfrac{RT}{nF} \ln Q = E^\circ - \dfrac{0.025693 \text{ V}}{n} \ln Q$

For this redox reaction $n = 1$.

The slope is 0.025693 V, and this value corresponds to the terms $\frac{RT}{nF}$.

What is being done here is a rearrangement of the Nernst equation to

$$E = -\frac{0.025693\,V}{n} \ln Q + E°,\ \text{resulting in a straight line plot of the form}$$

$y = mx + c$ where the slope, m, is $\frac{RT}{nF}$ and the y-intercept, c, is $E°$.

(c) y-intercept is $E°$, and for all concentration cells $E° = 0$

6.69 The strategy for working this problem is to create a set of equations that will add to the desired equilibrium reaction:

$$HClO(aq) + H_2O(l) \rightarrow H_3O^+(aq) + ClO^-(aq)$$

From the book Appendix, we find

$$2\,HClO + 2\,H^+ + 2\,e^- \rightarrow Cl_2 + 2\,H_2O \quad E° = +1.63V \qquad (1)$$

$$ClO^- + H_2O + 2\,e^- \rightarrow Cl^- + 2\,OH^- \quad E° = +0.89V \qquad (2)$$

On examination of these equations, it is clear that we will also need a half-reaction that, when combined with the two above, will eliminate Cl_2 and Cl^-. The obvious choice is

$$Cl_2 + 2\,e^- \rightarrow 2\,Cl^- \quad E° = +1.36V \qquad (3)$$

We combine these by adding (1) and (3) and twice the reverse reaction of (2):

$$2\,HClO + 2\,H^+ + 2\,e^- \rightarrow Cl_2 + 2\,H_2O$$
$$2(Cl^- + 2\,OH^- \rightarrow ClO^- + H_2O + 2\,e^-)$$
$$Cl_2 + 2\,e^- \rightarrow 2\,Cl^-$$

$$\overline{2\,HClO + 2\,H^+ + 4\,OH^- \rightarrow 2\,ClO^- + 4\,H_2O}$$

Caution: We must be careful here in adding the $E°$ values—we have created essentially a new half-reaction by summing these reactions, which requires that we convert to ΔG values. Whenever one sums more than two half-reactions, it is necessary to convert to the ΔG values using $\Delta G° = nFE°$, in order to work the problem:

$$2\,HClO + 2\,H^+ + 2\,e^- \rightarrow Cl_2 + 2\,H_2O$$

$$\Delta G^\circ = -2(9.65 \times 10^4\ C \cdot mol^{-1})(+1.63\ V) = -315\ kJ \cdot mol^{-1}$$

$$ClO^- + H_2O + 2\,e^- \rightarrow Cl^- + 2\,OH^-$$

$$\Delta G^\circ = -2(9.65 \times 10^4\ C \cdot mol^{-1})(+0.89\ V) = -172\ kJ \cdot mol^{-1}$$

$$Cl_2^- + 2\,e^- \rightarrow 2\,Cl^-$$

$$\Delta G^\circ = -2(9.65 \times 10^4\ C \cdot mol^{-1})(+1.36\ V) = -262\ kJ \cdot mol^{-1}$$

For $2\,HClO + 2\,H^+ + 4\,OH^- \rightarrow 2\,ClO^- + 4\,H_2O$

$$\Delta G^\circ = -315\ kJ + 2(+172\ kJ) - 262\ kJ = -233\ kJ \cdot mol^{-1}$$

We now see that we will need to eliminate OH^- from the left side of the equation. This can be done in one of two ways: we can use the K_W value for the autoprotolysis of water or, equivalently, we can use appropriate half-reactions that sum to the autoprotolysis of water. The appropriate half-reactions are

$$2\,H_2O \rightarrow O_2 + 4\,H^+ + 4\,e^- \quad E^\circ = -1.23\ V$$

$$O_2 + 2\,H_2O + 4\,e^- \rightarrow 4\,OH^- \quad E^\circ = +0.40\ V$$

These sum to give

$$4\,H_2O \rightarrow 4\,H^+ + 4\,OH^- \quad E^\circ = -0.83\ V$$

This is a $4\,e^-$ reaction. Alternatively, one can write the $1\,e^-$ process that will have the same E° value.

$$H_2O \rightarrow H^+ + OH^- \quad E^\circ = -0.83\ V$$

$$\Delta G^\circ = -(1)(9.65 \times 10^4\ C \cdot mol^{-1})(-0.83\ V) = +80\ kJ \cdot mol^{-1}$$

$$2\,HClO + 2\,H^+ + 4\,OH^- \rightarrow 2\,ClO^- + 4\,H_2O \quad \Delta G^\circ = -233\ kJ \cdot mol^{-1}$$

$$4(H_2O \rightarrow H^+ + OH^-) \qquad 4(\Delta G^\circ = +80\ kJ \cdot mol^{-1})$$

$$2\,HClO \rightarrow 2\,H^+ + 2\,ClO^- \quad \Delta G^\circ = -233\ kJ \cdot mol^{-1} + 4(+80\ kJ \cdot mol^{-1})$$

$$= +87\ kJ \cdot mol^{-1}$$

The desired reaction is half of this, for which $\Delta G^\circ = +44\ kJ \cdot mol^{-1}$.

Using $\Delta G^\circ = -RT \ln K$, we obtain $K = 1.9 \times 10^{-8}$, which is in reasonable agreement for this type of calculation with the value of 3.0×10^{-8} given in Table 6C.1.

6.71 In order to determine the current applied, we need to find the number of moles of electrons transferred. The electrolysis of water to produce gaseous oxygen and hydrogen,

$2 H_2O(l) \rightarrow O_2(g) + 2 H_2(g)$, transfers 4 moles of electrons for each mole of oxygen gas produced: $4 OH^-(aq) \rightarrow O_2(g) + 2 H_2O(l) + 4 e^-$. We can determine the number of moles of oxygen from its volume, partial pressure, and temperature.

$$n_{O_2} = \frac{p_{O_2}V}{RT} = \frac{(p_{tot} - p_{H_2O})V}{RT}$$

$$= \frac{(722 \text{ Torr} - 19.83 \text{ Torr})(25.0 \text{ mL})}{(0.08206 \text{ L} \cdot \text{atm} \cdot \text{K}^{-1} \cdot \text{mol}^{-1})(295 \text{ K})} \cdot \frac{1 \text{ L}}{1000 \text{ mL}} \cdot \frac{1 \text{ atm}}{760 \text{ Torr}}$$

$$= 9.542 \times 10^{-4} \text{ mol } O_2 \text{ produced}$$

$$n_{e^-} = n_{O_2} \times \frac{4 \text{ mol e}^-}{\text{mol } O_2} = 9.542 \times 10^{-4} \text{ mol } O_2 \times \frac{4 \text{ mol e}^-}{\text{mol } O_2}$$

$$= 3.817 \times 10^{-3} \text{ mol e}^-$$

$$It = nF$$

$$I = \frac{nF}{t} = \frac{(3.817 \times 10^{-3} \text{ mol e}^-)}{(30.0 \text{ min})(60 \text{ s} \cdot \text{min}^{-1})} \cdot \left(96\ 485 \frac{C}{\text{mol e}^-}\right)$$

$$= 0.205 \text{ A}$$

6.73 (a) Assuming alkaline conditions:

oxidation/anode: $6 OH^-(aq) + 2 Al(s) \rightarrow 2 Al(OH)_3(aq) + 6 e^-$

reduction/cathode: $3 H_2O(l) + 1\frac{1}{2} O_2(g) + 6 e^- \rightarrow 6 OH^-(aq)$

overall: $3 H_2O(l) + 2 Al(s) + 1\frac{1}{2} O_2(g) \rightarrow 2 Al(OH)_3(aq)$

(b) $E°_{cell} = E°_{cathode} - E°_{anode}$

$$= (+0.40 \text{ V}) - (-1.66 \text{ V})$$

$$= +2.06 \text{ V}$$

6.75 (a) $Pb_{(s)} + HSO^-_{4(aq)} \rightarrow PbSO_4(s) + H^+_{(aq)} + 2e^-$

$$\frac{PbO_2(s) + 3\,H^+_{(aq)} + HSO_4^-{}_{(aq)} + 2e^- \rightarrow PbSO_4(s) + 2\,H_2O(l)}{Pb(s) + PbO_2(s) + 2\,HSO_4^-{}_{(aq)} + 2H^+_{(aq)} \rightarrow 2PbSO_4(s) + 2\,H_2O(l)}$$

(b) pH increases

Amount of $PbO_2(s)$ decreases.

Total amount of lead in the battery stays the same.

6.77 (a) The electrolyte is KOH(aq)/HgO(s), which will have the consistency of a moist paste.

(b) The oxidizing agent is HgO(s).

(c) $HgO(s) + Zn(s) \rightarrow Hg(l) + ZnO(s)$

6.79 Assuming all the energy comes from reduction of oxygen focuses attention on this half reaction:

$$O_2(g) + 4\,H^+(aq) + 4\,e^- \rightarrow 2\,H_2O(l) \qquad E° = +1.23\ V$$

Body conditions are far from standard state values, so the actual value of E would be reduced by about 0.5 V if we take pH, p_{O_2} and T into account.

However, we are only estimating an average current to one significant digit, so $E = 1.23 \pm 0.5\ V \approx 1\ V$ is adequate. With these approximations in mind, we can calculate the current.

$$It = nF = \frac{\Delta G}{-E} \quad \text{or}$$

$$I = \frac{\Delta G}{-Et} = \frac{(-10 \times 10^6\ J)}{-(1\ V)(24\ h)(3600\ s \cdot h^{-1})} \cdot \frac{1\ V \cdot C}{1\ J} = 115\ A \approx 100\ A$$

6.81 (a) (i) $2H^+(aq) + 2e^- \rightarrow H_2(g)\ E° = 0$

$$E^* = E° - \frac{RT}{nF}\ln Q$$

$$E^* = E^\circ - \frac{0.025693\,V}{2}\ln\frac{[1]}{\left[10^{-7}\right]^2}$$

$$E^* = 0 - 0.41\text{ V} = -0.41\text{ V}$$

(ii) $NO_3^-(aq) + 4H^+(aq) + 3e^- \rightarrow NO(g) + 2H_2O$ $\quad E^\circ = +0.96\,V$

$$E^* = E^\circ - \frac{RT}{nF}\ln Q$$

$$E^* = E^\circ - \frac{0.025693\,V}{3}\ln\frac{[1][1]^2}{[1]\left[10^{-7}\right]^4}$$

$$E^* = +0.96\text{ V} - 0.55\text{ V} = +0.41\text{ V}$$

(b) $NAD^+(aq) + H^+(aq) + 2e^- \rightarrow NADH(aq)$ $\quad E^\circ = -0.099\text{ V}$

$$E^* = E^\circ - \frac{RT}{nF}\ln Q$$

$$E^\oplus = E^\circ - \frac{0.025693\,V}{n}\ln\frac{\left[NADH\right]}{\left[NAD^+\right]\left[H^+\right]}$$

$$E^* = E^\circ - \frac{0.025693\,V}{2}\ln\frac{[1]}{[1]\left[10^{-7}\right]}$$

$$E^* = -0.099\,V - \frac{0.025693\,V}{2}\ln 10^7$$

$$E^* = -0.099\text{ V} - 0.207\text{ V} = -0.31\text{ V}$$

(c)

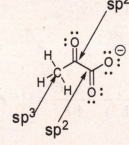

(d)

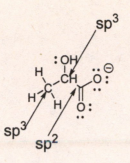

(e)

$$NADH(aq) \rightarrow NAD^+(aq) + H^+(aq) + 2e^- \qquad E^* = +0.31\,V$$

$$pyruvate + 2H^+(aq) + 2e^- \rightarrow lactate \qquad E^* = -0.190\,V$$

$$NADH(aq) + pyruvate + H^+(aq) \rightarrow NAD^+(aq) + lactate \quad E^* = +0.12\ V$$

$$E^{\oplus} = E^\circ - \frac{RT}{nF}\ln Q$$

$$E^\circ = E^{\oplus} + \frac{RT}{nF}\ln Q$$

$$E^\circ = E^{\oplus} + \frac{0.025693\,V}{n}\ln\frac{\left[lactate\right]\left[NAD^+(aq)\right]}{\left[NADH(aq)\right]\left[pyruvate\right]\left[H^+(aq)\right]}$$

$$E^\circ = +0.12\,V + \frac{0.025693\,V}{2}\ln\frac{[1][1]}{[1][1]\left[10^{-7}\right]}$$

$$E^\circ = +0.12\,V + 0.0128465\,V\ \ln 10^7$$

$$E^\circ = +0.12\,V + 0.207\ V$$

$$E^\circ = +0.33\,V$$

(f) $\Delta G_r^\circ = -nFE^\circ$

$$\Delta G_r^\circ = -2(9.64853 \times 10^4\,C.mol^{-1})(+0.33\,V)$$

$$\Delta G_r^\circ = -63.7\ kJ.mol^{-1} = -64\ kJ.mol^{-1}$$

(g) $\Delta G_r^\circ = -RT\ln K$ and $\Delta G_r^\circ = -nFE^\circ$

Therefore, $\ln K = \dfrac{nFE^\circ}{RT}$; $K = e^{\frac{nFE^\circ}{RT}}$

$$K = e\left(\frac{(+63700\ J\cdot mol^{-1})}{(8.314\,J.K^{-1}\cdot mol^{-1})(298.15\,K)}\right)$$

$$K = e^{25.69}; \qquad K = +1.43 \times 10^{11}$$

FOCUS 7

KINETICS

7A.1 (a) $\text{rate}(N_2) = \text{rate}(H_2) \times \left(\dfrac{1 \text{ mol } N_2}{3 \text{ mol } H_2}\right) = \dfrac{1}{3} \times \text{rate}(H_2)$

(b) $\text{rate}(NH_3) = \text{rate}(H_2) \times \left(\dfrac{2 \text{ mol } NH_3}{3 \text{ mol } H_2}\right) = \dfrac{2}{3} \times \text{rate}(H_2)$

(c) $\text{rate}(NH_3) = \text{rate}(N_2) \times \left(\dfrac{2 \text{ mol } NH_3}{1 \text{ mol } N_2}\right) = 2 \times \text{rate}(N_2)$

7A.3 (a) Rate of oxygen consumption $= 0.44 \times 3 = 1.3 \text{ mol} \cdot L^{-1} \cdot s^{-1}$

(b) Rate of water formation $= 0.44 \times 2 = 0.88 \text{ mol} \cdot L^{-1} \cdot s^{-1}$

7A.5 (a) and (c)

Note that the curves for the $[I_2]$ and $[H_2]$ are identical and only the $[I_2]$ curve is shown.

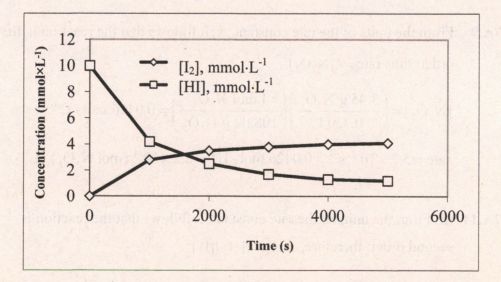

(b) The rates at individual points are given by the slopes of the lines tangent to the points in question. If these are determined graphically, there may be some variation from the numbers given below.

time, s	rate, mmol $\cdot$ L^{-1} $\cdot$ s^{-1}
0	0.0060
1000	0.003
2000	0.000 98
3000	0.000 61
4000	0.000 40
5000	0.000 31

7A.7 For A $\longrightarrow$ products, rate = (mol A) $\cdot$ L^{-1} $\cdot$ s^{-1}

(a) $[(\text{mol A}) \cdot L^{-1} \cdot s^{-1}] = k_0[A]^0 = k_0$, so units of k_0 are (mol A) $\cdot$ L^{-1} $\cdot$ s^{-1}

(same as the units for the rate, in this case)

(b) $[(\text{mol A}) \cdot L^{-1} \cdot s^{-1}] = k_1[A]$, so units of k_1 are $\dfrac{(\text{mol A}) \cdot L^{-1} \cdot s^{-1}}{(\text{mol A}) \cdot L^{-1}} = s^{-1}$

(c) $[(\text{mol A}) \cdot L^{-1} \cdot s^{-1}] = k_1[A]^2$, so units of k_1 are $\dfrac{(\text{mol A}) \cdot L^{-1} \cdot s^{-1}}{\left[(\text{mol A}) \cdot L^{-1}\right]^2}$

$$= L \cdot (\text{mol A})^{-1} \cdot s^{-1}$$

7A.9 From the units of the rate constant, k, it follows that the reaction is first order, thus rate $= k[N_2O_5]$.

$$[N_2O_5] = \left(\frac{3.45 \text{ g } N_2O_5}{0.750 \text{ L}}\right)\left(\frac{1 \text{ mol } N_2O_5}{108.02 \text{ g } N_2O_5}\right) = 0.0426 \text{ mol} \cdot L^{-1}$$

rate $= 5.2 \times 10^{-3} \text{ s}^{-1} \times 0.0426 \text{ mol} \cdot L^{-1} = 2.2 \times 10^{-4} \text{ (mol } N_2O_5) \cdot L^{-1} \cdot s^{-1}$

7A.11 (a) From the units of the rate constant, it follows that the reaction is second order; therefore, rate $= k[H_2][I_2]$

$$= (0.063 \text{ L} \cdot \text{mol}^{-1} \cdot \text{s}^{-1}) \left(\frac{0.52 \text{ g H}_2}{0.750 \text{ L}} \right) \left(\frac{1 \text{ mol H}_2}{2.016 \text{ g H}_2} \right) \left(\frac{0.19 \text{ g I}_2}{0.750 \text{ L}} \right) \left(\frac{1 \text{ mol I}_2}{253.8 \text{ g I}_2} \right)$$

$$= 2.2 \times 10^{-5} \text{ mol} \cdot \text{L}^{-1} \cdot \text{s}^{-1}$$

(b) $rate(new) = k \times 2 \times [H_2]_{initial}[I_2] = 2 \times rate(initial)$, so, by a factor of 2

7A.13 Because the rate increased in direct proportion to the concentrations of both reactants, the rate is first order in both reactants.

$$rate = k[CH_3Br][OH^-]$$

7A.15 (a) Use experiments 1 and 4 to show that [C] is independent of the rate. Use experiments 2 and 4 to solve for the order with respect to A:

$$\frac{20^a}{10} = \frac{4}{2} \qquad\qquad 2^a = 2 \qquad\qquad \text{Therefore } a = 1.$$

Use experiments 2 and 3 to solve for the order with respect to B:

$$\frac{200^b}{100} = \frac{16}{4} \qquad\qquad 2^b = 4 \qquad\qquad \text{Therefore } b = 2. \quad \text{Overall order} = 3$$

(b) $rate = k[A][B]^2$

(c) Experiment 1: $2.0 \text{ mmol L}^{-1} \text{ s}^{-1} = k(10 \text{ mmol L}^{-1})(100 \text{ mmol L}^{-1})^2$

$$k = 2.0 \times 10^{-5} \text{ L}^2 \text{ mmol}^{-2} \text{ s}^{-1}$$

(d) $rate = k[A][B]^2$

$$= (2.0 \times 10^{-5} \text{ L}^2 \text{ mmol}^{-2} \text{ s}^{-1})(4.62 \text{ mmol L}^{-1})(0.177 \text{ mmol L}^{-1})^2$$

$$= 2.9 \times 10^{-6} \text{ mmol L}^{-1} \text{ s}^{-1}$$

7A.17 (a) Doubling the concentration of A (experiments 1 and 2) doubles the rate; therefore, the reaction is first order in A. Increasing the concentration of B by the ratio 3.02/1.25 (experiments 2 and 3) increases the rate by $(3.02/1.25)^2$; hence, the reaction is second order in B. Tripling the concentration of C (experiments 3 and 4) increases the rate by $3^2 = 9$; thus, the reaction is second order in C. Therefore, $rate = k[A][B]^2[C]^2$.

(b) overall order $= 5$

(c) $k = \dfrac{\text{rate}}{[A][B]^2[C]^2}$

Using the data from experiment 4, we get

$$k = \left(\dfrac{0.457\ \text{mol}}{L \cdot s}\right)\left(\dfrac{L}{1.25 \times 10^{-3}\ \text{mol}}\right)\left(\dfrac{L}{3.02 \times 10^{-3}\ \text{mol}}\right)^2\left(\dfrac{L}{3.75 \times 10^{-3}\ \text{mol}}\right)^2$$

$$= 2.85 \times 10^{12}\ L^4 \cdot \text{mol}^{-4} \cdot s^{-1}$$

From experiment 3, we get

$$k = \left(\dfrac{5.08 \times 10^{-2}\ \text{mol}}{L \cdot s}\right)\left(\dfrac{L}{1.25 \times 10^{-3}\ \text{mol}}\right)\left(\dfrac{L}{3.02 \times 10^{-3}\ \text{mol}}\right)^2$$

$$\times \left(\dfrac{L}{1.25 \times 10^{-3}\ \text{mol}}\right)^2$$

$$= 2.85 \times 10^{12}\ L^4 \cdot \text{mol}^{-4} \cdot s^{-1}\ \text{(Checks!)}$$

(d) $\text{rate} = \left(\dfrac{2.85 \times 10^{12}\ L^4}{\text{mol}^4 \cdot s}\right)\left(\dfrac{3.01 \times 10^{-3}\ \text{mol}}{L}\right)\left(\dfrac{1.00 \times 10^{-3}\ \text{mol}}{L}\right)^2$

$$\times \left(\dfrac{1.15 \times 10^{-3}\ \text{mol}}{L}\right)^2$$

$$= 1.13 \times 10^{-2}\ \text{mol} \cdot L^{-1} \cdot s^{-1}$$

7B.1 $\quad \ln \dfrac{[A]_t}{[A]_0} = -kt$

$$\ln \dfrac{[A]_t}{[A]_0} = -(7.6 \times 10^{-3}\ \text{min}^{-1})(300\ \text{min})$$

$$\dfrac{[A]_t}{[A]_0} = e^{-2.28}$$

$[A]_t = 0.10[A]_0 \qquad$ 10% of the initial drug concentration

remains in the body.

Mass of drug remaining in the body, $A_t = (0.10)(20\ \text{mg}) = 2.0\ \text{mg}$

7B.3 (a) $k = \dfrac{0.693}{t_{1/2}} = \dfrac{0.693}{1000\ \text{s}} = 6.93 \times 10^{-4}\ s^{-1}$

(b) We use $\ln\left(\dfrac{[A]_0}{[A]_t}\right) = kt$ and solve for k.

$$k = \frac{\ln([A]_0/[A]_t)}{t} = \frac{\ln\left(\dfrac{0.67\ \text{mol}\cdot\text{L}^{-1}}{0.53\ \text{mol}\cdot\text{L}^{-1}}\right)}{25\ \text{s}} = 9.4\times10^{-3}\ \text{s}^{-1}$$

(c) $[A]_t = \left(\dfrac{0.153\ \text{mol A}}{\text{L}}\right) - \left[\left(\dfrac{2\ \text{mol A}}{1\ \text{mol B}}\right)\left(\dfrac{0.034\ \text{mol B}}{\text{L}}\right)\right]$

$\qquad = 0.085\ (\text{mol A})\cdot\text{L}^{-1}$

$$k = \frac{\ln\left(\dfrac{0.153\ \text{mol}\cdot\text{L}^{-1}}{0.085\ \text{mol}\cdot\text{L}^{-1}}\right)}{115\ \text{s}} = 5.1\times10^{-3}\ \text{s}^{-1}$$

7B.5 (a) $t_{1/2} = \dfrac{0.693}{k} = \left(\dfrac{0.693\ \text{s}}{3.7\times10^{-5}}\right)\left(\dfrac{1\ \text{min}}{60\ \text{s}}\right)\left(\dfrac{1\ \text{h}}{60\ \text{min}}\right) = 5.2\ \text{h}$

(b) $[A]_t = [A]_0\, e^{-kt}$

$t = 3.5\ \text{h} \times 3600\ \text{s}\cdot\text{h}^{-1} = 1.3\times10^4\ \text{s}$

$[N_2O_5] = 0.0567\ \text{mol}\cdot\text{L}^{-1} \times e^{-(3.7\times10^{-5}\ \text{s}^{-1})(1.3\times10^4\ \text{s})} = 3.5\times10^{-2}\ \text{mol}\cdot\text{L}^{-1}$

(c) Solve for t from $\ln\left(\dfrac{[A]_0}{[A]_t}\right) = kt$, which gives

$$t = \frac{\ln\left(\dfrac{[A]_0}{[A]_t}\right)}{k} = \frac{\ln\left(\dfrac{[N_2O_5]_0}{[N_2O_5]_t}\right)}{k} = \frac{\ln\left(\dfrac{0.0567}{0.0135}\right)}{3.7\times10^{-5}\ \text{s}^{-1}} = 3.9\times10^4\ \text{s}$$

$\qquad = (3.9\times10^4\ \text{s})\left(\dfrac{1\ \text{min}}{60\ \text{s}}\right) = 6.5\times10^2\ \text{min}$

7B.7 (a) $\dfrac{[A]}{[A]_0} = \dfrac{1}{8} = \left(\dfrac{1}{2}\right)^3$ so the time elapsed is 3 half-lives.

$t = 3 \times 355\ \text{s} = 1065\ \text{s}$

(b) $\dfrac{[A]}{[A]_0} = \dfrac{1}{4} = \left(\dfrac{1}{2}\right)^2$; so the time elapsed is 2 half-lives.

$t = 2 \times 355 \, \text{s} = 710 \, \text{s}$

(c) Because 15% is not a multiple of $\frac{1}{2}$, we cannot work directly from the half-life. But $k = 0.693/t_{1/2}$

so $k = \dfrac{0.693}{355 \, \text{s}} = 1.95 \times 10^{-3} \, \text{s}^{-1}$

$t = \dfrac{\ln\left(\dfrac{[A]_0}{[A]_t}\right)}{k} = \dfrac{\ln\left(\dfrac{1}{0.15}\right)}{1.95 \times 10^{-3} \, \text{s}^{-1}} = 9.7 \times 10^2 \, \text{s}$

(d) $t = \dfrac{\ln\dfrac{[A]_0}{\frac{1}{9}[A]_0}}{k} = \dfrac{\ln 9}{1.95 \times 10^{-3} \, \text{s}^{-1}} = 1.1 \times 10^3 \, \text{s}$

7B.9 (a) We first calculate the concentration of A at 3.0 min.

$[A]_t = [A]_0 - \left(\dfrac{1 \, \text{mol A}}{3 \, \text{mol B}}\right) \times [B]_t$

$= 0.015 \, \text{mol} \cdot \text{L}^{-1} - \left(\dfrac{1 \, \text{mol A}}{3 \, \text{mol B}}\right) \times 0.018 \, (\text{mol B}) \cdot \text{L}^{-1}$

$= 0.009 \, \text{mol} \cdot \text{L}^{-1}$

The rate constant is then determined from the first-order integrated rate law.

$k = \dfrac{\ln\left(\dfrac{[A]_0}{[A]_t}\right)}{t} = \dfrac{\ln\left(\dfrac{0.015}{0.009}\right)}{3.0 \, \text{min}} = 0.17 \, \text{min}^{-1}$

(b) $[A]_t = 0.015 \, \text{mol} \cdot \text{L}^{-1} - \left(\dfrac{1 \, \text{mol A}}{3 \, \text{mol B}}\right) \times 0.030 \, (\text{mol B}) \cdot \text{L}^{-1}$

$= 0.005 \, \text{mol} \cdot \text{L}^{-1}$

$t = \dfrac{\ln\left(\dfrac{[A]_0}{[A]_t}\right)}{k} = \dfrac{\ln\left(\dfrac{0.015}{0.005}\right)}{0.17 \, \text{min}^{-1}} = 6.5 \, \text{min}$

additional time $= 6.5 \, \text{min} - 3.0 \, \text{min} = 3.5 \, \text{min}$

7B.11 (a) $\dfrac{1}{[A]_t} = \dfrac{1 + [A]_0\, kt}{[A]_0} = \dfrac{1}{[A]_0} + kt$

Thus, if the reaction is second order, a plot of 1/[HI] against time should give a straight line of slope k. As can be seen from the graph, the data fit the equation for a second-order reaction quite well. The slope is determined by a least squares fit of the data by the graphing program.

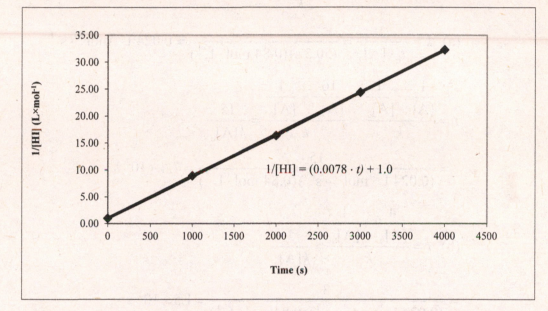

$$1/[HI] = (0.0078 \cdot t) + 1.0$$

(b) (i). The rate constant for the rate law for the loss of HI is simply the slope of the best fit line, $7.8 \times 10^{-3}\ \text{L} \cdot \text{mol}^{-1} \cdot \text{s}^{-1}$. (ii). Since two moles of HI are consumed per mole of reaction, the rate constant for the unique rate law is half the slope or $3.9 \times 10^{-3}\ \text{L} \cdot \text{mol}^{-1} \cdot \text{s}^{-1}$.

7B.13 Obtain an expression for the half-life of a second-order reaction.

$$[A]_t = \dfrac{[A]_0}{1 + [A]_0 kt}$$

$$\dfrac{[A]_{t_{1/2}}}{[A]_0} = \dfrac{1}{2} = \dfrac{1}{1 + [A]_0 kt_{1/2}}$$

Therefore, $1 + [A]_0 kt_{1/2} = 2$, or $[A]_0 kt_{1/2} = 1$, or

$$t_{1/2} = \frac{1}{k[A]_0} \quad \text{and} \quad k = \frac{1}{t_{1/2}[A]_0}$$

It is also convenient to solve for t. Take reciprocals:

$$\frac{1}{[A]_t} = \frac{1}{[A]_0} + kt$$

Giving
$$t = \frac{\dfrac{1}{[A]_t} - \dfrac{1}{[A]_0}}{k}$$

(a) $k = \dfrac{1}{t_{1/2}[A]_0} = \dfrac{1}{(50.5 \text{ s})(0.84 \text{ mol} \cdot L^{-1})} = 0.024 \text{ L} \cdot mol^{-1} \cdot s^{-1}$

$$t = \frac{\dfrac{1}{[A]} - \dfrac{1}{[A]_0}}{k} = \frac{\dfrac{16}{[A]_0} - \dfrac{1}{[A]_0}}{k} = \frac{15}{k[A]_0}$$

$$= \frac{15}{(0.024 \text{ L} \cdot mol^{-1} \cdot s^{-1})(0.84 \text{ mol} \cdot L^{-1})} = 7.4 \times 10^2 \text{ s}$$

(b) $t = \dfrac{\dfrac{4}{[A]_0} - \dfrac{1}{[A]_0}}{k} = \dfrac{3}{k[A]_0}$

$$= \frac{3}{(0.024 \text{ L} \cdot mol^{-1} \cdot s^{-1})(0.84 \text{ mol} \cdot L^{-1})} = 1.5 \times 10^2 \text{ s}$$

(c) $t = \dfrac{\dfrac{5}{[A]_0} - \dfrac{1}{[A]_0}}{k} = \dfrac{4}{k[A]_0}$

$$= \frac{4}{(0.024 \text{ L} \cdot mol^{-1} \cdot s^{-1})(0.84 \text{ mol} \cdot L^{-1})} = 2.0 \times 10^2 \text{ s}$$

7B.15 (a) $\quad t_{1/2} = \dfrac{0.693}{k} = \dfrac{0.693}{2.81 \times 10^{-3} \text{ min}^{-1}} = 247 \text{ min}$

(b)
$$t = \frac{\ln\left(\dfrac{[SO_2Cl_2]_0}{[SO_2Cl_2]_t}\right)}{k} = \frac{\ln 10}{2.81 \times 10^{-3} \text{ min}^{-1}} = 819 \text{ min}$$

(c)
$$[A]_t = [A]_0 \, e^{-kt}$$

Because the vessel is sealed, masses and concentrations are proportional, and we write

$$(\text{mass left})_t = (\text{mass})_0 \, e^{-kt}$$

$$= 14.0 \text{ g} \times e^{-(2.81 \times 10^{-3} \text{ min}^{-1} \times 60 \text{ min} \cdot \text{h}^{-1} \times 1.5 \text{ h})}$$

$$= 10.9 \text{ g}$$

Note: Knowledge of the volume of the vessel is not required.

However, we could have converted mass to concentration, solved for the new concentration at 1.5 hour, and finally converted back to the new (remaining) mass. But this is not necessary.

7B.17 (a)
$$t = \frac{\dfrac{1}{[A]} - \dfrac{1}{[A]_0}}{k} = \frac{\dfrac{1 \text{ L}}{0.080 \text{ mol}} - \dfrac{1 \text{ L}}{0.10 \text{ mol}}}{0.015 \text{ L} \cdot \text{mol}^{-1} \cdot \text{min}^{-1}} = 1.7 \times 10^2 \text{ min}$$

(b)
$$[A] = \frac{0.15 \text{ mol A}}{L} - \left[\left(\frac{0.19 \text{ mol B}}{L}\right)\left(\frac{1 \text{ mol A}}{2 \text{ mol B}}\right)\right]$$

$$= 0.055(\text{mol A}) \cdot L^{-1} = 0.37[A]_0$$

$$t = \frac{\dfrac{1}{[A]_t} - \dfrac{1}{[A]_0}}{k}$$

$$= \frac{\dfrac{1}{0.055 \text{ mol} \cdot L^{-1}} - \dfrac{1}{0.15 \text{ mol} \cdot L^{-1}}}{0.0035 \text{ L} \cdot \text{mol}^{-1} \cdot \text{min}^{-1}}$$

$$= 3.3 \times 10^3 \text{ min}$$

7B.19 $rate = -\dfrac{1}{a}\dfrac{d[A]}{dt} = k[A]$

$\dfrac{d[A]}{[A]} = -ak\,dt$

$\displaystyle\int_{[A]_0}^{[A]_t} \dfrac{d[A]}{[A]} = \ln\dfrac{[A]_t}{[A]_0} = -akt,$ and $[A]_t = [A]_0\exp(-akt)$

For the half-life. At $t_{1/2}$, $[A]_t = \frac{1}{2}[A]_0$. Therefore:

$\ln\dfrac{[A]_0}{[A]_t} = \ln 2 = akt_{1/2},$ and

$t_{1/2} = \dfrac{\ln 2}{ak}$

7B.21 Given: $\dfrac{d[A]}{dt} = -k[A]^3$, we can derive an expression for the amount of time needed for the inital concentration of A, $[A]_0$, to decrease by 1/2. Begin by obtaining the integrated rate law for a third-order reaction by separation of variables:

$\displaystyle\int_{[A]_0}^{[A]_t} [A]^{-3}d[A] = \int_0^t -k\,dt = -\dfrac{1}{2}\Big[[A]_t^{-2} - [A]_0^{-2}\Big] = -kt$

To obtain an expression for the half-life, let $[A]_t = \frac{1}{2}[A]_0$ and $t = t_{1/2}$:

$-\dfrac{1}{2}\Big[(\tfrac{1}{2}[A]_0)^{-2} - [A]_0^{-2}\Big] = -kt_{1/2}$

solving for the half-life:

$t_{1/2} = \dfrac{3}{2k[A]_0^2}$

7C.1 (a) Rate $= k[NO]^2$ bimolecular

(b) Rate $= k[Cl_2]$ unimolecular

7C.3 $2AC + B \rightarrow A_2B + 2C$

Intermediate is AB.

7C.5 (a) $2HBr + NO_2 \rightarrow NO + H_2O + Br_2$

(b) Step 1 : rate $= k_1[HBr][NO_2]$; bimolecular

Step 2 : rate $= k_2[HBr][HOBr]$; bimolecular

(c) HOBr

7C.7 The first elementary reaction is the rate-controlling step, because it is the slow step. The second elementary reaction is fast and does not affect the overall reaction order, which is second order as a result of the fact that the rate-controlling step is bimolecular.

rate $= k[NO][Br_2]$

7C.9 If mechanism (I) were correct, the rate law would be rate $= k_r[NO_2][CO]$. But this expression does not agree with the experimental result and can be eliminated as a possibility. Mechanism (II) has rate $= k_r[NO_2]^2$ from the slow step. Step 2 does not influence the overall rate, but it is necessary to achieve the correct overall reaction; thus this mechanism agrees with the experimental data. Mechanism (III) is not correct, which can be seen from the rate expression for the slow step, rate $= k_r[NO_3][CO]$. [CO] cannot be eliminated from this expression to yield the experimental result, which does not contain [CO].

7C.11 (a) True

(b) False. At equilibrium, the *rates* of the forward and reverse reactions are equal, *not the rate constants*.

(c) False. Increasing the concentration of a reactant causes the rate to increase by providing more reacting molecules. It does not affect the rate constant of the reaction.

7C.13 The overall rate of formation of A is rate $= -k[A] + k'[B]$. The first term accounts for the forward reaction and is negative as this reaction reduces [A]. The second term, which is positive, accounts for the back reaction which increases [A]. Given the 1:1 stoichiometry of the reaction, if no B was present at the beginning of the reaction, [A] and [B] at any time are related by the equation: $[A] + [B] = [A]_o$ where $[A]_o$ is the initial concentration of A. Therefore, the rate law may be written:

$$\frac{d[A]}{dt} = -k[A] + k'([A]_o - [A]) = -(k + k')[A] + k'[A]_o$$

The solution of this first-order differential equation is:

$$[A] = \frac{k' + ke^{-(k'+k)t}}{k' + k}[A]_o$$

As $t \to \infty$ the exponential term in the numerator goes to zero and the concentrations reach their equilibrium values given by:

$$[A]_{eq} = \frac{k'[A]_o}{k' + k} \quad \text{and} \quad [B]_{eq} = [A]_o - [A]_\infty = \frac{k[A]_o}{k + k'}$$

taking the ratio of products over reactants we see that:

$$\frac{[B]_{eq}}{[A]_{eq}} = \frac{k}{k'} = K \quad \text{where } K \text{ is the equilibrium constant for the reaction.}$$

7D.1   We use $\ln\left(\dfrac{k'}{k}\right) = \dfrac{E_a}{R}\left(\dfrac{1}{T} - \dfrac{1}{T'}\right) = \dfrac{E_a}{R}\left(\dfrac{T' - T}{T'T}\right)$

$$\ln\left(\frac{k'}{k}\right) = \ln\left(\frac{0.87 \text{ s}^{-1}}{0.76 \text{ s}^{-1}}\right)$$

$$= \left(\frac{E_a}{8.31 \times 10^{-3} \text{ kJ} \cdot \text{K}^{-1} \cdot \text{mol}^{-1}}\right)\left(\frac{1030 \text{ K} - 1000 \text{ K}}{1030 \text{ K} \times 1000 \text{ K}}\right)$$

$$E_a = \frac{(8.31 \times 10^{-3} \text{ kJ} \cdot \text{K}^{-1} \cdot \text{mol}^{-1})(1000 \text{ K})(1030 \text{ K})}{(1030 \text{ K} - 1000 \text{ K})} \ln\left(\frac{0.87 \text{ s}^{-1}}{0.76 \text{ s}^{-1}}\right)$$

$$= 39 \text{ kJ} \cdot \text{mol}^{-1}$$

7D.3

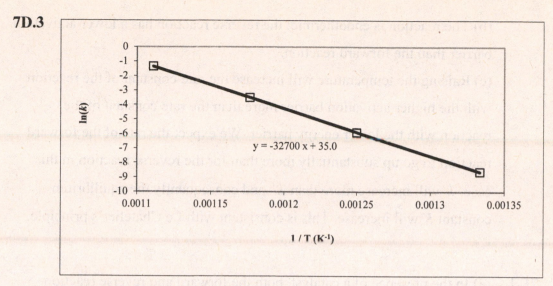

(a) Given the Arrhenius equation, $\ln k = \ln A - E_a / RT$, we see that the slope of the best fit line to the data $(-3.27 \times 10^4 \, K)$ is E_a / R and the y intercept (35.0) is $\ln A$. Therefore,

$$E_a = (3.27 \times 10^4 \, K)(8.31 \times 10^{-3} \, kJ \cdot mol^{-1} \cdot K^{-1}) = 2.72 \times 10^2 \, kJ \cdot mol^{-1}.$$

(b) At 600 °C (or 873 K), the rate constant is:

$$\ln(k) = \left(-3.27 \times 10^4 \, K\right)\frac{1}{873 \, K} + 35.0 = -2.46$$

$$k = 0.088$$

$$k = 8.8 \times 10^{-2} \, s^{-1}$$

7D.5 $k' = $ rate constant at $T' = 37 \, °C = 3130 \, K$

$$\ln\left(\frac{k'}{k}\right) = \frac{38 \, kJ \cdot mol^{-1}}{\left(0.08314 \, kJ \cdot K^{-1} \cdot mol^{-1}\right)}\left(\frac{1}{298 \, K} - \frac{1}{310 \, K}\right) - 0.59$$

$$\frac{k'}{k} = 1.8, \; k' = 1.8 \times 1.5 \times 10^{10} \, L \cdot mol^{-1} \cdot s^{-1} = 2.7 \times 10^{10} \, L \cdot mol^{-1} \cdot s^{-1}$$

7D.7 (a) The equilibrium constant will be given by the ratio of the rate constant of the forward reaction to the rate constant of the reverse reaction:

$$K = \frac{k}{k'} = \frac{265 \, L \cdot mol^{-1} \cdot min^{-1}}{392 \, L \cdot mol^{-1} \cdot min^{-1}} = 0.676$$

(b) The reaction is endothermic: the reverse reaction has a lower activation barrier than the forward reaction.

(c) Raising the temperature will increase the rate constant of the reaction with the higher activation barrier more than the rate constant of the reaction with the lower energy barrier. We expect the rate of the forward reaction to go up substantially more than for the reverse reaction in this case. k will increase more than k' and consequently the equilibrium constant K will increase. This is consistent with Le Chatelier's principle.

7E.1 (a) In the presence of a catalyst, both the forward and reverse reaction rates will increase.

(b) A catalyst will not affect the value of ΔH_r° for the reaction. Catalysts only lower the activation energy of the reaction, increasing the rate of a reaction, but not the enthalpy of the reaction. Remember not to confuse thermodynamics with kinetics.

7E.3 (a) cat = catalyzed, uncat = uncatalyzed $E_{a,cat} = \frac{75}{125} E_{a,uncat} = 0.60 E_{a,uncat}$

$$\frac{\text{rate(cat)}}{\text{rate(uncat)}} = \frac{k_{cat}}{k_{uncat}} = \frac{Ae^{-E_{a,cat}/RT}}{Ae^{-E_{a,uncat}/RT}} = \frac{e^{-(0.60)E_{a,uncat}/RT}}{e^{-E_{a,uncat}/RT}}$$

$$= e^{(-0.60+1.00)E_{a,uncat}/RT} = e^{(0.40)E_{a,uncat}/RT}$$

$$= e^{[(0.40)(125\ \text{kJ·mol}^{-1})/(8.314\times10^{-3}\ \text{kJ·K}^{-1}\text{·mol}^{-1} \times 298\ \text{K})]} = 6\times10^{8}$$

(b) The last step of the calculation in (a) is repeated with $T = 350$ K.

$e^{[(0.40)(125\ \text{kJ·mol}^{-1})/(8.314\times10^{-3}\ \text{kJ·K}^{-1}\text{·mol}^{-1} \times 350\ \text{K})]} = 3\times10^{7}$

The rate enhancement is lower at higher temperatures.

7E.5 The overall reaction is $RCN + H_2O \longrightarrow RCONH_2$.

$RC(=N^-)OH$ and $RC(=NH)OH$ are intermediates.

The hydroxide ion serves as a catalyst for the reaction.

7E.7 (a) False. A catalyst increases the rate of both the forward and reverse reactions by providing a completely different pathway.

(b) True, although a catalyst may be poisoned and lose activity.

(c) False. There is a completely different pathway provided for the reaction in the presence of a catalyst.

(d) False. The position of the equilibrium is unaffected by the presence of a catalyst.

7E.9 (a) To obtain the Michaelis–Menten rate equation, we will begin by employing the steady-state approximation, setting the rate of change in the concentration of the ES intermediate equal to zero:

$$\frac{d[ES]}{dt} = k_1[E][S] - k'_1[ES] - k_2[ES] = 0.$$

Rearranging gives: $[E][S] = \left(\dfrac{k_2 + k'_1}{k_1}\right)[ES] = K_M[ES].$

The total bound and unbound enzyme concentration, $[E]_0$, is given by:

$[E]_0 = [E] + [ES]$, and, therefore, $[E] = [ES] - [E]_0$.

Substituting this expression for $[E]$ into the preceding equation, we obtain:

$([ES] - [E]_0)[S] = K_M[ES].$

Rearranging to obtain $[ES]$ gives: $[ES] = \dfrac{[E]_0[S]}{K_M + [S]}.$

From the mechanism, the rate of appearance of the product is given by *rate* $= k_2[ES]$. Substituting the preceding equation for $[ES]$, we obtain:

$$\text{Rate} = \frac{k_2[E]_0[S]}{K_M + [S]},$$

the Michaelis–Menten rate equation, which can be rearranged to obtain:

$$\frac{1}{\text{rate}} = \frac{K_M}{k_2[E]_0[S]} + \frac{1}{k_2[E]_0}.$$

If one plots $\dfrac{1}{\text{rate}}$ versus $\dfrac{1}{[S]}$, the slope will be $\dfrac{K_M}{k_2[E]_0}$ and the y intercept

will be $\dfrac{1}{k_2[E]_0}$.

(b)

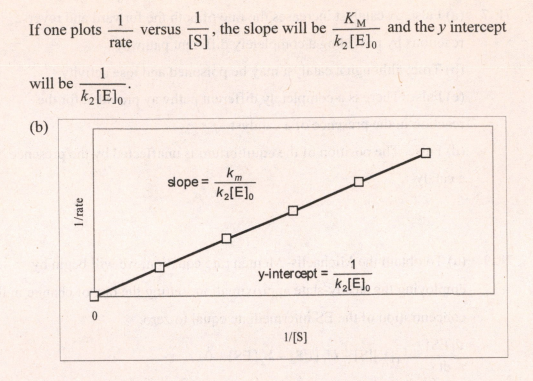

7.1 (a) $CH_3CH = CHCH_2Br$ because the positive charge on the reactive

intermediate is on a primary carbon atom. This also agrees with the result

that at high temperatures this is the predominant product as there is

sufficient energy to overcome the larger activation energy.

(b) Kinetic control predominates at low temperatures. The reaction

pathway with the lower activation energy will predominate at low

temperatures because the lower activation energy barrier results in a larger

rate constant and therefore a faster reaction.

7.3 Given $K = 326$ and $k_{attach} = 7.4 \times 10^7$ L.mol^{-1}.s^{-1}

Since $K = \dfrac{k_{forward}}{k_{reverse}} = \dfrac{k_{attach}}{k_{loss}}$

Then $k_{loss} = \dfrac{k_{attach}}{K} = \dfrac{7.4 \times 10^7 \text{ L. mol}^{-1}.\text{s}^{-1}}{326} = 2.3 \times 10^5$ L.mol^{-1}.s^{-1}

7.5 (a) $t_{1/2} = 5$ s since the flask contains 12 atoms at $t = 0$ and 6 atoms

at $t = 5$ s.

(b) Since $t_{1/2} = 5 \text{ s} = \dfrac{0.693}{k}$ then $k = 0.139 \text{ s}^{-1}$

First order, $\ln[A] = \ln[A]_0 - kt$

Then at 8 s, $\ln[A] = \ln 12 - (0.139 \text{ s}^{-1})(8 \text{ s}) = 1.37$

$\qquad\qquad A = 3.94 \approx 4$ There should be four molecules at t = 8 s.

7.7   (a) $CH_3CHO \rightarrow CH_3 + CHO$ unimolecular

$\qquad\qquad [CH_3\ldots\ldots\ldots\ldots\ldots CHO]^{\ddagger}$

      (b) $2I \rightarrow I_2$ termolecular

$\qquad\qquad [I\ldots\ldots.I\ldots\ldots..Ar]^{\ddagger}$

      (c) $O_2 + NO \rightarrow NO_2 + O$ bimolecular

$\qquad\qquad [\,O\ldots\ldots.O\ldots\ldots..NO]^{\ddagger}$

7.9   The anticipated rate for mechanism (i) is: rate $= k[C_{12}H_{22}O_{11}]$, while the expected rate for mechanism (ii) is: rate $= k[C_{12}H_{22}O_{11}][H_2O]$. The rate for mechanism (ii) will be pseudo-first-order in dilute solutions of sucrose because the concentration of water will not change. Therefore, in dilute solutions kinetic data can not be used to distinguish between the two mechanisms. However, in a highly concentrated solution of sucrose, the concentration of water will change during the course of the reaction. As a result, if mechanism (ii) is correct the kinetics will display a first-order dependence on the concentration of H_2O while mechanism (i) predicts that the rate of the reaction is independent of $[H_2O]$.

7.11   (a) The objective is to reproduce the observed rate law. If step 2 is the slow step, if step 1 is a rapid equilibrium, and if step 3 is fast also, then our proposed rate law will be rate $= k_2[N_2O_2][H_2]$. Consider the equilibrium of Step 1: $k_1[NO]^2 = k'_1[N_2O_2]$

$$[N_2O_2] = \frac{k_1}{k'_1}[NO]^2$$

Substituting in our proposed rate law, we have

$$\text{rate} = k_2\frac{k_1}{k'_1}[NO]^2[H_2] = k[NO]^2[H_2] \text{ where } k = k_2\frac{k_1}{k'_1}$$

The assumptions made above reproduce the observed rate law; therefore, step 2 is the slow step.

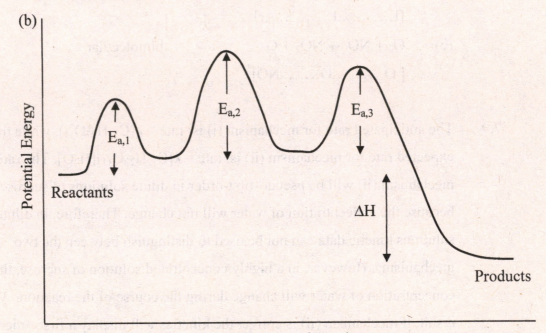

(b)

Note: The dips that represent the formation of the intermediate N_2O_2 and N_2O will not be at the same energy, but we have no information to determine which should be lower.

7.13 To get an expression for $t_{1/2}$ in terms of n, we need to evaluate an integral such as:

$$\int_{[A]_0}^{[A]} \frac{d[A]}{[A]^n} = -k\int_0^t dt = -kt$$

$$\frac{1}{n-1}\left(\frac{1}{[A]^{n-1}} - \frac{1}{[A]_0^{n-1}}\right) = kt$$

An expression for $t_{1/2}$ is then:

$$\frac{1}{n-1}\left(\frac{2^{n-1}}{[A]_0^{n-1}} - \frac{1}{[A]_0^{n-1}}\right) = kt_{1/2}$$

$$\frac{1}{n-1}\left(\frac{2^{n-1}-1}{[A]_0^{n-1}}\right) = kt_{1/2}$$

An expression for $t_{3/4}$ could be found by setting up $[A] = \frac{3}{4}[A]_0$

$$\frac{1}{n-1}\left(\frac{4^{n-1}}{3^{n-1}[A]_0^{n-1}} - \frac{1}{[A]_0^{n-1}}\right) = kt_{3/4}$$

$$\frac{1}{n-1}\left(\frac{\left(\frac{4}{3}\right)^{n-1}-1}{[A]_0^{n-1}}\right) = kt_{3/4}$$

The ratio $t_{1/2}/t_{3/4}$ is then

$$t_{1/2}/t_{3/4} = \left(\frac{2^{n-1}-1}{(4/3)^{n-1}-1}\right)$$

7.15 $$\text{rate} = \frac{k_2[E]_0[S]}{K_m + [S]}$$

$$1.21 \times 10^{-3} \text{ mol. L}^{-1}.\text{ s}^{-1} = \frac{k_2[E]_0(0.156 \text{ mol. L}^{-1})}{(0.038 \text{ mol. L}^{-1}) + (0.156 \text{ mol. L}^{-1})}$$

$$1.5 \times 10^{-3} \text{ mol. L}^{-1}.\text{ s}^{-1} = k_2[E]_0 = \text{max rate}$$

7.17 (a) three steps

(b) first step

(c) third step

(d) two

(e) none

7.19 For a third-order reaction,

$$t_{1/2} \propto \frac{1}{[A_0]^2} \text{ or } t_{1/2} = \frac{\text{constant}}{[A_0]^2}$$

(a) The time necessary for the concentration to fall to one-half of the initial concentration is one half-life:

$$\text{first half-life} = t_1 = t_{1/2} = \frac{\text{constant}}{[A_0]^2}$$

(b) This time, $t_{1/4}$, is two half-lives, but because of different starting concentrations, the half-lives are not the same:

$$\text{second half-life} = t_2 = \frac{\text{constant}}{(\frac{1}{2}[A_0])^2} = \frac{4(\text{constant})}{[A_0]^2} = 4t_1$$

$$\text{total time} = t_1 + t_2 = t_1 + 4t_1 = 5t_1 = t_{1/4}$$

(c) This time, $t_{1/16}$, is four half-lives; again, the half-lives are not the same:

$$\text{third half-life} = t_3 = \frac{\text{constant}}{(\frac{1}{4}[A_0])^2} = \frac{16(\text{constant})}{[A_0]^2} = 16t_1$$

$$\text{fourth half-life} = t_4 = \frac{\text{constant}}{(\frac{1}{8}[A_0])^2} = \frac{64(\text{constant})}{[A_0]^2} = 64t_1$$

$$\text{total time} = t_1 + t_2 + t_3 + t_4 = t_1 + 4t_1 + 16t_1 + 64t_1 = 85t_1 = t_{1/16}$$

If t_1 is known, the times $t_{1/4}$ and $t_{1/16}$ can be calculated easily.

7.21 The following plots are linear: (b)   (c)   (d)   (f)   (g)

7.23 (a) The overall reaction is: $ClO^- + I^- \rightarrow IO^- + Cl^-$

(b) The rate law will be based upon the slow step of the reaction:

$$\text{rate} = k_2[HOCl][I^-]$$

Even though HOCl is a stable species because it is an intermediate in the reaction as written, technically we should not leave the rate law in this form. The concentration of HOCl can be expressed in terms of the reactants and products using the fast equilibrium approach:

$$K = \frac{k_1}{k_1{'}} = \frac{[HOCl][OH^-]}{[OCl^-]}$$

$$[HOCl] = \frac{k_1}{k_1{'}} \frac{[OCl^-]}{[OH^-]}$$

$$\text{rate} = \frac{k_2 k_1}{k_1{'}} \frac{[OCl^-][I^-]}{[OH^-]}$$

(c) An examination of the rate law shows that the rate is dependent upon the concentration of OH^-. As the pH increases, the concentration of OH^- increases and the rate decreases.

(d) If the reaction is carried out in an organic solvent, then H_2O is no longer the solvent and its concentration must be included in calculating the equilibrium concentration of HOCl:

$$K = \frac{k_1}{k_1{'}} = \frac{[HOCl][OH^-]}{[OCl^-][H_2O]}$$

$$[HOCl] = \frac{k_1}{k_1{'}} \frac{[OCl^-][H_2O]}{[OH^-]}$$

$$\text{rate} = \frac{k_2 k_1}{k_1{'}} \frac{[OCl^-][I^-][H_2O]}{[OH^-]}$$

The rate of reaction will then show a dependence upon the concentration of water, which will be obscured when the reaction is carried out with water as the solvent.

7.25   Frogs are poikilotherms which means that they need to be able to function over a wide range of body core temperatures. This adaptation in the rhodopsin in frog eyes allows their vision to be constant despite temperature fluctuation.

7.27 Use equation for half-life of a first order reaction.

$$t_{1/2} = \frac{\ln 2}{k}$$

$$k = \frac{\ln 2}{t_{1/2}}$$

$$k = 0.154 \text{ h}^{-1}$$

$$\ln\left(\frac{[A]_0}{[A]_t}\right) = kt$$

$$\frac{[A]}{[A]_0} = e^{-kt}$$

After two hours this ratio is,

$$\frac{[A]}{[A]_0} = e^{-(0.154 \text{ h}^{-1})(2 \text{ h})} = 0.735$$

The mass of phenobarbital that remains after 2 hours is $0.735 \times 150 \text{ mg} = 110 \text{ mg}$. To restore the initial amount 40 mg must be re-injected.

7.29 (a) ClO is the reaction intermediate; Cl is the catalyst.

(b) Cl, ClO, O, O_2

(c) Step 1 and step 2 are propagating.

(d) $Cl + Cl \longrightarrow Cl_2$

7.31 We use $\ln\left(\dfrac{k'}{k}\right) = \dfrac{E_a}{R} = \left(\dfrac{1}{T} - \dfrac{1}{T'}\right) = \dfrac{E_a}{R}\left(\dfrac{T' - T}{TT'}\right)$

k' = rate constant at 700°C, $T' = (700 + 273) \text{ K} = 973 \text{ K}$

$$\ln\left(\frac{k'}{k}\right) = \left(\frac{315 \text{ kJ} \cdot \text{mol}^{-1}}{8.314 \times 10^{-3} \text{ kJ} \cdot \text{mol}^{-1}}\right)\left(\frac{973 \text{ K} - 1073 \text{ K}}{973 \text{ K} \times 1073 \text{ K}}\right)$$

$$= -3.63; \quad \frac{k'}{k} = 0.026$$

$$k' = 0.026 \times 9.7 \times 10^{10} \text{ L} \cdot \text{mol}^{-1} \cdot \text{s}^{-1} = 2.5 \times 10^{9} \text{ L} \cdot \text{mol}^{-1} \cdot \text{s}^{-1}$$

FOCUS 8
THE MAIN-GROUP ELEMENTS

8A.1 (a) nitrogen; (b) potassium; (c) gallium; (d) iodine

8A.3 (a) sulfur; (b) selenium; (c) sodium; (d) oxygen

8A.5 tellurium < selenium < oxygen

8A.7 (a) chlorine; (b) Chlorine has a higher effective nuclear charge.

8A.9 (a) bromide ion; (b) Bromide has larger size.

8A.11 (a) KCl because the ionic radius of K^+ is larger than that of Li^+.

      (b) K—O. The higher charge on Ca^{2+} makes its ionic radius much smaller than that of K^+.

8A.13 $2 K(s) + H_2(g) \rightarrow 2 KH(s)$

8A.15 (a) saline; (b) molecular; (c) molecular; (d) metallic

8A.17 (a) acidic; (b) amphoteric; (c) acidic; (d) basic

8A.19 (a) CO_2; (b) B_2O_3

8B.1 In the majority of its reactions, hydrogen acts as a reducing agent; that is, $H_2(g) \longrightarrow 2 H^+(aq) + 2 e^-$, $E° = 0$ V. In these reactions, hydrogen resembles Group 1 elements, such as Na and K. Hydrogen also has a similar electron affinity to elements in Group 1. The electron affinity of H is $+73 \, \text{kJ} \cdot \text{mol}^{-1}$, which is similar to that of Li, which is $60 \, \text{kJ} \cdot \text{mol}^{-1}$.

8B.3 (a) $C_2H_2(g) + H_2(g) \longrightarrow H_2C=CH_2(g)$
Oxidation number of C in $C_2H_2 = -1$; of C in $H_2C=CH_2 = -2$, carbon has been reduced.

(b) $CO(g) + H_2O(g) \longrightarrow CO_2(g) + H_2(g)$

(c) $BaH_2(s) + 2 \, H_2O(l) \longrightarrow Ba(OH)_2 + 2 \, H_2(g)$

8B.5 (a) $H_2(g) + Cl_2(g) \xrightarrow{\text{light}} 2 \, HCl(g)$

(b) $H_2(g) + 2 \, Na(l) \xrightarrow{\Delta} 2 \, NaH(s)$

(c) $P_4(s) + 6 \, H_2(g) \longrightarrow 4 \, PH_3(g)$

(d) $2 \, Cu(s) + H_2(g) \longrightarrow 2 \, CuH(s)$

8B.7 The reason for this trend is mainly due to the trend of the electronegativity of the central atom (N<O<F), which makes F more partially negative to attract H from another HF atom. Spatial hindrance may also contribute to this trend. There are three H atoms bonding to N, two to O, and there is only one H atom bond to F.

8C.1 Lithium is the only Group 1 element that reacts directly with nitrogen to form lithium nitride:

$6 \, Li(s) + N_2(g) \xrightarrow{\Delta} 2 \, Li_3N(s)$

Lithium reacts with oxygen to form mainly the oxide:

$4 \, Li(s) + O_2(g) \rightarrow 2 \, Li_2O(s)$

The other members of the group form mainly the peroxide or superoxide. Lithium exhibits the diagonal relationship that is common to many first members of a group. Li is similar in many of its compounds to the compounds of Mg. This behavior is related to the small ionic radius of

Li^+, 58 pm, which is closer to the ionic radius of Mg^{2+}, 72 pm, but substantially less than that of Na^+, 102 pm.

8C.3 (a) ns^1

(b) Reducing agents give up one or more electrons. It is relatively easy to remove the one valence electron of the alkali metals because they all have low first ionization energies. Loss of one electron is favorable because the resulting cation has the electron configuration of a noble gas. Alkali metal ions are strongly hydrated; the stability generated by solvation makes them unreactive toward chemical reducing agents and therefore makes the ionic form highly favorable.

8C.5 (a) $4\,Na(s) + O_2(g) \longrightarrow 2\,Na_2O(s)$

(b) $6\,Li(s) + N_2(g) \xrightarrow{\Delta} 2\,Li_3N(s)$

(c) $2\,Na(s) + 2\,H_2O(l) \longrightarrow 2\,NaOH(aq) + H_2(g)$

(d) $4\,KO_2(s) + 2\,H_2O(g) \longrightarrow 4\,KOH(s) + 3\,O_2(g)$

8D.1 $Mg(s) + 2\,H_2O(l) \longrightarrow Mg(OH)_2 + H_2(g)$

8D.3 (a) $CaO(s) + H_2O(l) \longrightarrow Ca(OH)_2(s)$

(b) $\Delta G°_r = \Delta G°_f(Ca(OH)_2, s) - [\Delta G°_f(CaO, s) + \Delta G°_f(H_2O, l)]$

$\quad\quad = -898.49\;kJ \cdot mol^{-1} - [(-604.03\;kJ \cdot mol^{-1}) + (-237.13\;kJ \cdot mol^{-1})]$

$\quad\quad = -57.33\;kJ \cdot mol^{-1}$

8D.5 (a) $Mg(OH)_2(s) + 2\,HCl(aq) \longrightarrow MgCl_2(aq) + 2\,H_2O(l)$

(b) $Ca(s) + 2\,H_2O(l) \longrightarrow Ca(OH)_2(aq) + H_2(g)$

(c) $BaCO_3(s) \xrightarrow{\Delta} BaO(s) + CO_2(g)$

8D.7   (a) $:\overset{\cdot\cdot}{\underset{\cdot\cdot}{Cl}}\!-\!Be\!-\!\overset{\cdot\cdot}{\underset{\cdot\cdot}{Cl}}:$

(b) 180°   (c) *sp*   (d) $MgCl_2$ is ionic; $BeCl_2$ is a molecular compound, thus they will have different structures.

8E.1   The overall equation for the electrolytic reduction in the Hall process is

$$4\,Al^{3+}(melt) + 6\,O^{2-}(melt) + 3\,C(s, gr) \longrightarrow 4\,Al(s) + 3\,CO_2(g)$$

8E.3   (a) $B_2O_3(s) + 3\,Mg(l) \overset{\Delta}{\longrightarrow} 2\,B(s) + 3\,MgO(s)$

(b) $2\,Al(s) + 3\,Cl_2(g) \longrightarrow 2\,AlCl_3(s)$

(c) $4\,Al(s) + 3\,O_2(g) \longrightarrow 2\,Al_2O_3(s)$

8E.5   Since there are only 22 valence electrons, or 11 electron pairs, it is not possible to draw a good conventional Lewis structure for tetraborane, B_4H_{10}, that includes four B-H-B bridges. For the suggested structure given on the left below, each bridging H and each four-coordinate B would have a formal charge of −1 while each of the six terminal H atoms and each three-coordinate B would have a formal charge of 0. The total formal charge adds up to −6 in this case even though the molecule is neutral. However, if the bridges are viewed as 3-center 2-electron bonds, then every atom can be assigned a formal charge of 0.

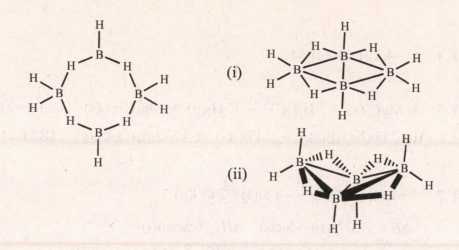

The second set of drawings, above right, shows the actual structure of tetraborane (i) in a traditional Lewis presentation, and (ii) in a stereo drawing (bold bonds are coming out toward you, dashed bonds are going back and away from you).   See: http://pubs.acs.org/doi/abs/10.1021/ba-1961-0032.ch009

8E.7 The Lewis structure of $GaBr_4^-$ is:

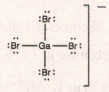

The shape of the $GaBr_4^-$ is tetrahedral.

8F.1 Silicon occurs widely in the Earth's crust in the form of silicates in rocks and as silicon dioxide in sand. It is obtained from quartzite, a form of quartz (SiO_2), by the following processes:

(1) reduction in an electric arc furnace

$$SiO_2(s) + 2\,C(s) \longrightarrow Si(s, crude) + 2\,CO(g)$$

(2) purification of the crude product in two steps

$$Si(s, crude) + 2\,Cl_2(g) \longrightarrow SiCl_4(l)$$

followed by reduction with hydrogen to the pure element

$$SiCl_4(l) + 2\,H_2(g) \longrightarrow Si(s, pure) + 4\,HCl(g)$$

8F.3   (a) +4; (b) +4; (c) +4

8F.5   (a) $MgC_2(s) + 2 H_2O(l) \rightarrow C_2H_2(g) + Mg(OH)_2(s)$ (acid-base)

(b) $2 Pb(NO_3)_2(s) \rightarrow 2 PbO(s) + 4 NO_2(g) + O_2(g)$ (Redox)

8F.7 $SiO_2(s) + 2 C(s) \longrightarrow Si(s) + 2 CO(g)$

$\Delta H°_r = \Delta H°_f(\text{products}) - \Delta H°_f(\text{reactants})$

$= [(2)(-110.53 \text{ kJ} \cdot \text{mol}^{-1})] - [-910.94 \text{ kJ} \cdot \text{mol}^{-1}]$

$= +689.88 \text{ kJ} \cdot \text{mol}^{-1}$

$\Delta S°_r = S°(\text{products}) - S°(\text{reactants})$

$= [18.83 \text{ J} \cdot \text{K}^{-1} \cdot \text{mol}^{-1} + (2)(197.67 \text{ J} \cdot \text{K}^{-1} \cdot \text{mol}^{-1})]$

$- [41.84 \text{ J} \cdot \text{K}^{-1} \cdot \text{mol}^{-1} + (2)(5.740 \text{ J} \cdot \text{K}^{-1} \cdot \text{mol}^{-1})]$

$= +360.85 \text{ J} \cdot \text{K}^{-1} \cdot \text{mol}^{-1}$

$\Delta G°_r = \Delta H°_r - T\Delta S°_r$

$= 689.88 \text{ kJ} \cdot \text{mol}^{-1} - (298.15 \text{ K})(360.85 \text{ J} \cdot \text{K}^{-1} \cdot \text{mol}^{-1})/1000 \text{ J} \cdot \text{kJ}^{-1}$

$= +582.29 \text{ kJ} \cdot \text{mol}^{-1}$

The temperature at which the equilibrium constant becomes greater than 1 is the temperature at which $\Delta G°_r = -RT \ln K = 0$ because $\ln 1 = 0$. Above this temperature, the equilibrium constant is greater than 1. $\Delta G°_r = 0$ when $T\Delta S°_r = \Delta H°_r$, or

$$T = \frac{\Delta H°_r}{\Delta S°_r} = \frac{+689.88 \times 10^3 \text{ J} \cdot \text{mol}^{-1}}{+360.85 \text{ J} \cdot \text{K}^{-1} \cdot \text{mol}^{-1}} = 1912 \text{ K}.$$

8G.1   −3 $NH_3, Li_3N, LiNH_2, NH_2^-$

−2 H_2NNH_2

−1 N_2H_2, NH_2OH

0   N_2

+1 N_2O, N_2F_2

+2   NO

+3 NF_3, NO_2^-, NO^+

+4 NO_2, N_2O_4

+5 HNO_3, NO_3^-, NO_2F

8G.3 $CO(NH_2)_2(aq) + 2 H_2O(l) \longrightarrow (NH_4)_2CO_3(aq)$

mass of $(NH_4)_2CO_3 = (4.0 \text{ kg urea}) \left(\dfrac{10^3 \text{ g}}{1 \text{ kg}} \right) \left(\dfrac{1 \text{ mol}}{60.06 \ (NH_4)_2CO_3} \right)$

$\left(\dfrac{1 \text{ mol } (NH_4)_2CO_3}{1 \text{ mol urea}} \right) \left(\dfrac{96.09 \text{ g } (NH_4)_2CO_3}{1 \text{ mol } (NH_4)_2CO_3} \right)$

$= 6.4 \times 10^3 \text{ g (or 6.4 kg)} (NH_4)_2CO_3$

8G.5   (a) One mole of $N_2(g)$ occupies 22.4 L at STP. For the reaction

$Pb(N_3)_2 \longrightarrow Pb + 3 N_2$, the volume of $N_2(g)$ produced is

$1.5 \text{ g } Pb(N_3)_2 \left(\dfrac{1 \text{ mol } Pb(N_3)_2}{291.25 \text{ g } Pb(N_3)_2} \right) \left(\dfrac{3 \text{ mol } N_2}{1 \text{ mol } Pb(N_3)_2} \right) \left(\dfrac{22.4 \text{ L } N_2}{1 \text{ mol } N_2} \right)$

$= 0.35 \text{ L } N_2(g)$

(b) $Hg(N_3)_2$ would produce a larger volume, because its molar mass is

less. Note that molar mass occurs in the denominator in this calculation.

(c) Metal azides are good explosives because the azide ion is

thermodynamically unstable with respect to the production of $N_2(g)$. This

is because the $N \equiv N$ triple bond is so strong and also because the

production of a gas is favored entropically.

8G.7 $N_2O / H_2N_2O_2$; $N_2O(g) + H_2O(l) \longrightarrow H_2N_2O_2(aq)$

N_2O_3 / HNO_2; $N_2O_3(g) + H_2O(l) \longrightarrow 2 HNO_2(aq)$

N_2O_5 / HNO_3; $N_2O_5(g) + H_2O(l) \longrightarrow 2 HNO_3(aq)$

8G.9 Ammonia (NH_3) can undergo hydrogen bonding with itself while NF_3 can not.

8H.1 (a) $4\,Li(s) + O_2(g) \xrightarrow{\Delta} 2\,Li_2O(s)$

(b) $2\,Na(s) + 2\,H_2O(l) \longrightarrow 2\,NaOH(aq) + H_2(g)$

(c) $2\,F_2(g) + 2\,H_2O(l) \longrightarrow 4\,HF(aq) + O_2(g)$

(d) $2\,H_2O(l) \longrightarrow O_2(g) + 4\,H^+(aq) + 4\,e^-$

8H.3 (a) $2\,H_2S(g) + 3\,O_2(g) \xrightarrow{\Delta} 2\,SO_2(g) + 2\,H_2O(g)$

(b) $CaO(s) + H_2O(l) \longrightarrow Ca(OH)_2(aq)$

(c) $2\,H_2S(g) + SO_2(g) \xrightarrow{300°C,\ Al_2O_3} 3\,S(s) + 2\,H_2O(l)$

8H.5 Both water and ammonia have four groups attached to their central atom and therefore both possess a tetrahedral electronic (or VSEPR) geometry. However, H_2O has two unshared electron pairs while NH_3 only has one, producing a larger dipole moment for H_2O.

8H.7 (a) Hydrogen peroxide has the following Lewis structure:

$$H - \overset{\displaystyle ..}{\underset{\displaystyle ..}{O}} - \overset{\displaystyle ..}{\underset{\displaystyle ..}{O}} - H$$

Each O in H_2O_2 is an AX_2E_2 structure; therefore, the bond angle is predicted to be $< 109.5°$. In actuality, it is ..

(i) –(iv), The reduction potential of H_2O_2 is $+1.78$ V in acidic solution. It should, therefore, be able to oxidize any ion that has a reduction potential that is less than $+1.78$ V. For the ions listed, Cu^+ and Mn^{2+} will be oxidized. It would require an input of 1.98 V to oxidize Ag^+ to Ag^{2+} and 2.87 V to oxidize F^-.

8H.9 $O_2^{2-} + H_2O \longrightarrow HO_2^- + OH^-$ essentially complete

$$HO_2^- + H_2O \rightleftharpoons H_2O_2 + OH^- \quad K_b = \frac{K_w}{K_{a_1}}$$

$$K_{a_1} = 1.8 \times 10^{-12} \quad K_b = \frac{1.00 \times 10^{-14}}{1.8 \times 10^{-12}} = 5.6 \times 10^{-3}$$

Because this K_b is relatively small, we can assume that essentially all the

OH^- is formed in the first ionization; therefore,

$$[OH^-] = \left(\frac{2.00 \text{ g Na}_2O_2}{0.200 \text{ L}} \right) \left(\frac{1 \text{ mol Na}_2O_2}{77.98 \text{ g Na}_2O_2} \right) \left(\frac{1 \text{ mol OH}^-}{1 \text{ mol Na}_2O_2} \right)$$

$$= 0.128 \text{ mol} \cdot L^{-1}$$

$$pOH = -\log(0.128) = 0.893 \quad pH = 14.00 - 0.893 = 13.11$$

If we do not ignore the second ionization, then the additional contribution

to $[OH^-]$ can be approximately calculated as follows:

$$K_b = \frac{[H_2O_2][OH^-]}{[HO_2^-]} = \frac{x(0.128 + x)}{(0.128 - x)} = 5.6 \times 10^{-3}$$

To a first approximation, $x = 5.6 \times 10^{-3} \text{ mol} \cdot L^{-1}$

To a second approximation, $x = \dfrac{K_b(0.128 - 0.0056)}{(0.128 + 0.0056)} = 0.005$

Then $[OH^-] = 0.128 + 0.005 = 0.133$; $pOH = -\log(0.133) = 0.876$; and

$pH = 13.12$. The difference between calculations is slight.

8H.11 The weaker the H—X bond, the stronger the acid. H_2Te has the weakest

bond; H_2O, the strongest. Therefore, the acid strengths are

$H_2Te > H_2Se > H_2S > H_2O$

8I.1 (a) $HIO(aq)$ $H = +1, O = -2$; therefore, $I = +1$

(b) ClO_2 $O = -2$; therefore, $Cl = +4$

(c) Cl_2O_7 $O = -2$; therefore, $Cl = +14/2 = +7$

(d) $NaIO_3$ $Na = +1, O = -2$; therefore, $I = +5$

8I.3 (a) $4 KClO_3(l) \xrightarrow{\Delta} 3 KClO_4(s) + KCl(s)$

(b) $Br_2(l) + H_2O(l) \longrightarrow HBrO(aq) + HBr(aq)$

(c) $NaCl(s) + H_2SO_4(aq) \longrightarrow NaHSO_4(aq) + HCl(g)$

(d) (a) and (b) are redox reactions. In (a), Cl is both oxidized and reduced. In (b), Br is both oxidized and reduced. (c) is a Brønsted acid-base reaction; H_2SO_4 is the acid, and Cl^- the base.

8I.5 (a) $HClO < HClO_2 < HClO_3 < HClO_4$ ($HClO_4$ is the strongest; HClO the weakest)

(b) The oxidation number of Cl increases from HClO to $HClO_4$. In $HClO_4$, chlorine has its highest oxidation number of +7, so $HClO_4$ will be the strongest oxidizing agent.

8I.7 $:\overset{..}{\underset{..}{Cl}} - \overset{..}{\underset{..}{O}} - \overset{..}{\underset{..}{Cl}}:$, AX_2E_2, angular, about 109°. The actual bond angle is 110.9°.

8I.9 (a) ICl_5 is too sterically hindered: it has too many large chlorine atoms present on the central iodine. (b) IF_2 has an odd number of electrons and as a result is a radical and highly reactive. (c) $ClBr_3$ is sterically too crowded (too many large atoms on a small central atom).

8I.11 (a) We can write the following general chemical equation for the process described:

$$x XeF_2(g) + I_2(g) \longrightarrow 2 IF_x(s) + x Xe(g)$$

Since the pressure is directly proportional to the moles of each gas present and since we only need to find the combining ratio of iodine to fluorine to determine the chemical formula of IF_x, we can use pressure in lieu of moles for the purposes of this problem. Since all of the XeF_2 is gone at reactions end, and the reaction started with 3.6 atm of XeF_2 to start, there must be 3.6 atm of Xe present in the vessel at the end of the reaction. Therefore P (total at start) = 7.2 atm = P (XeF_2 start) + P (I_2 start); solving this gives us P (I_2 start) = 3.6 atm. At the reactions end, P (total at end) = 6.0 atm = P (Xe end) + P (I_2 end); solving this gives P (I_2 end) = 2.4atm. This means that 1.2 atm of I_2 was used in the formation of IF_x. From this we can solve for x as follows:

$$1.2 \text{ atm } I_2 \left(\frac{x \text{ atm XeF}_2}{1 \text{ atm } I_2} \right) = 3.6 \text{ mol XeF}_2; x = 3$$

Therefore the chemical formula is IF_3.

(b) $3 XeF_2(g) + I_2(g) \longrightarrow 2 IF_3(s) + 3 Xe(g)$

8I.13 $Cl_2(g) + 2 e^- \longrightarrow 2 Cl^-(aq) \ E° = +1.36 \text{ V}$

$MnO_4^-(aq) + 8 H^+(aq) + 5 e^- \longrightarrow Mn^{2+}(aq) + 4 H_2O(l) \ E° = +1.51 \text{ V}$

$E°_{cell} = (1.36 - 1.51) \text{ V} = -0.15 \text{ V}$

Because $E°_{cell}$ is negative, $Cl_2(g)$ will not oxidize Mn^{2+} to form the permanganate ion in an acidic solution.

8I.15 $F^-(aq) + PbCl_2(s) \longrightarrow Cl^-(aq) + PbClF(s)$

$$\text{molarity of F}^- \text{ ions} = \left(\frac{0.765 \text{ g PbClF}}{0.0250 \text{ L}} \right) \left(\frac{1 \text{ mol PbClF}}{261.64 \text{ g PbClF}} \right) \left(\frac{1 \text{ mol F}^-}{1 \text{ mol PbClF}} \right)$$

$$= 0.117 \text{ mol} \cdot \text{L}^{-1}$$

8J.1 Helium occurs as a component of natural gases found under rock formations in certain locations, especially some in Texas. Argon is obtained by distillation of liquid air.

8J.3 (a) KrF_2:F = −1; therefore, Kr = +2

(b) XeF_6:F = -1; therefore, Xe = +6

(c) KrF_4:F = -1; therefore, Kr = +4

(d) XeO_4^{2-}:O = -2, N_{ox}(Xe) - 8 = -2; therefore, N_{ox}(Xe) = +6

8J.5 $XeF_4(aq) + 4 H^+(aq) + 4 e^- \longrightarrow Xe(g) + 4 HF(aq)$

8J.7   Because H_4XeO_6 has more highly electronegative O atoms bonded to Xe, we predict that H_4XeO_6 is more acidic than H_2XeO_4.

8.1   (a)

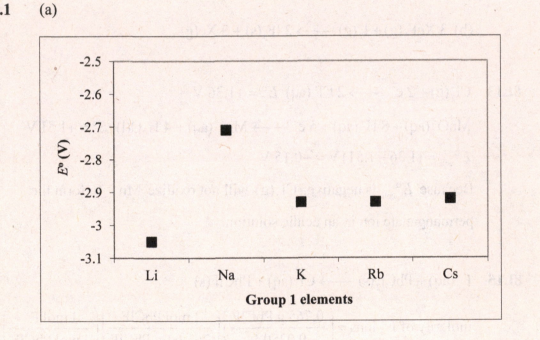

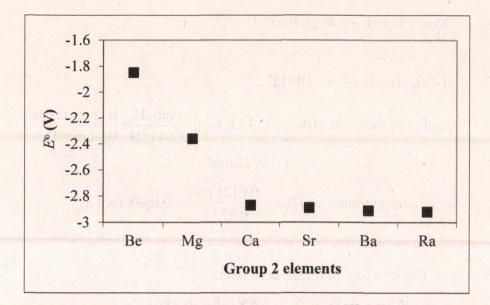

(b) For both groups, the trend in standard potentials with increasing atomic number is overall downward (they become more negative), but lithium is anomalous. This overall downward trend makes sense because we expect that it is easier to remove electrons that are farther away from the nuclei. However, because there are several factors that influence ease of removal, the trend is not smooth. The potentials are a net composite of the free energies of sublimation of solids, dissociation of gaseous molecules, ionization enthalpies, and enthalpies of hydration of gaseous ions. The origin of the anomalously strong reducing power of Li is the strongly exothermic energy of hydration of the very small Li^+ ion, which favors the ionization of the element in aqueous solution.

8.3 (a) True. (b) False; the oxides of carbon are acidic, while those of tin and lead are basic. (c) True.

8.5 In the majority of its reactions, hydrogen acts as a reducing agent. Examples are $2H_2(g) + O_2(g) \longrightarrow 2H_2O(l)$ and various ore reduction processes, such as $NiO(s) + H_2(g) \xrightarrow{\Delta} Ni(s) + H_2O(g)$. With highly electropositive elements, such as the alkali earth metals, $H_2(g)$ acts as an oxidizing agent and forms metal hydrides, for example,

$$2K(s) + H_2(g) \longrightarrow 2KH(s).$$

8.7 $H_2(g) + Br_2(l) \longrightarrow 2\,HBr(g)$

number of moles of HBr $= (0.135\ \text{L H}_2)\left(\dfrac{1\ \text{mol H}_2}{22.4\ \text{L H}_2}\right)\left(\dfrac{2\ \text{mol HBr}}{1\ \text{mol H}_2}\right)$

$$= 0.0121\ \text{mol}$$

molar concentration of HBr $= \dfrac{0.0121\ \text{mol}}{0.225\ \text{L}} = 0.0538\ \text{mol}\cdot\text{L}^{-1}$

8.9   (a) The structure of the azide anion is

$$\left[\ddot{\text{:N}}\!=\!\text{N}\!=\!\ddot{\text{N:}}\right]^-$$
   AX_2, linear 180°

(b) HCl, HBr, and HI are all strong acids. For HF, $K_a = 3.5 \times 10^{-4}$, so HF is slightly more acidic than HN_3. The small size of the azide ion suggests that the H—N bond in HN_3 is similar in strength to that of the H—F bond, so it is expected to be a weak acid.

(c) ionic: NaN_3, $Pb(N_3)_2$, AgN_3, etc.

covalent: HN_3, $B(N_3)_3$, FN_3, etc.

8.11   (a) Diborane, B_2H_6, and $Al_2Cl_6(g)$ have the same basic structure in the way in which the atoms are arranged in space. (b) The bonding between the boron atoms and the bridging hydrogen atoms is electron deficient. There are three atoms and only two electrons to hold them together in a 3-center-2-electron bond. The bonding in Al_2Cl_6 is conventional in that all the bonds involve two atoms and two electrons. Here, the lone pair of a Cl atom is donated to an adjacent Al. (c) The hybridization is sp^3 at the B and Al atoms. (d) The molecules are not planar. The Group 13 element and the terminal atoms to which it is bound lie in a plane that is perpendicular to the plane that contains the main group element and the bridging atoms. Bond angles in the ring are expected to be approximately

90°, while the angle between the terminal hydrogens and the Group 13 element is expected to be greater than 109.5°.

8.13 (a) The ionization energy of a molecule is the amount of energy required to strip one electron out of a gaseous molecule; it can be defined as follows:

$$\text{Molecule (g)} \longrightarrow \text{Molecule}^+\text{(g)} + 1e^-\text{(g)}.$$

(b) SiI_4 has more electrons than $SiCl_4$; I is also less electronegative and much bigger (and therefore more polarizable) than Cl. As a result it should be easier to remove an electron from SiI_4 meaning that $SiCl_4$ should have the higher ionization energy.

8.15 (a) 10 electron species: NH_3 and H_3O^+; 15 electron species: NO and O_2^+; 22 electron species: N_2O^+ and NO_2^+. (i) Strongest Lewis acids: H_3O^+, NO and NO_2^+; (ii) Strongest oxidizing agents: H_3O^+, NO and NO_2^+

8.17 The structure of thiosulfuric acid and sulfuric acid are:

Due to the replacement of one of the doubly bounded oxygens in sulfuric acid with a sulfur, it should be expected that an aqueous solution of thiosulfuric acid should be slightly less acidic (due to the lower electronegativity of S); in addition, the boiling point should also be expected to be slightly lower (due to reduced hydrogen bonding).

8.19 The solubility of the ionic halides is determined by a variety of factors, especially the lattice enthalpy and enthalpy of hydration. There is a delicate balance between the two factors, with the lattice enthalpy usually being the determining one. Lattice enthalpies decrease from chloride to iodide, so water molecules can more readily separate the ions in the latter. Less ionic halides, such as the silver halides, generally have a much lower solubility, and the trend in solubility is the reverse of the more ionic halides. For the less ionic halides, the covalent character of the bond allows the ion pairs to persist in water. The ions are not easily hydrated, making them less soluble. The polarizability of the halide ions, and thus, the covalency of their bonding, increases down the group.

8.21   (a) The molecular orbital diagram for NO^+ should have the oxygen orbitals slightly lower in energy than the nitrogen orbitals, because oxygen is more electronegative. This will cause the bonding to be more ionic than in either N_2 or O_2. There is an ambiguity, however, in that the MO diagram could be similar to either that of N_2 or that of O_2. Refer to Figures 2G.7 and 2G.8 where you will see that the σ_{2p} and π_{2p} have different relative energies. There are consequently two possibilities for the orbital energy diagram:

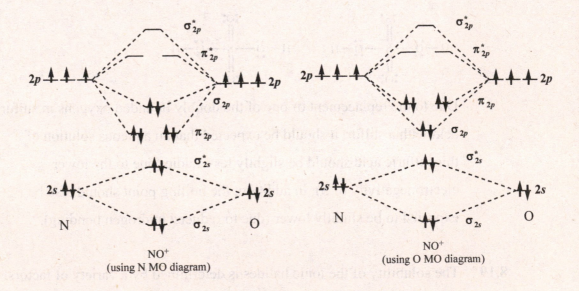

(b) The two orbital diagrams predict the same bond order (3) and same magnetic properties (diamagnetic), and so these properties cannot be used to determine which diagram is the correct one. That must be determined by more complex spectroscopic measurements.

8.23 (a) Use ΔH_f° to calculate the ΔH_r° (values can be found in Appendix 2):

$$\Delta H_r^\circ = [\Delta H_f^\circ(CO) + 2\Delta H_f^\circ(H_2)] - [\Delta H_f^\circ(CH_3OH)]$$
$$= [-110.53 + 0] - [-238.86] \text{ kJ}$$
$$= +128.33 \text{ kJ}$$

(b) $\Delta H_c^\circ(CH_3OH) = -726 \text{ kJ}$

(c) The equation for the combustion of two moles of hydrogen gas is:

$$2H_2(g) + O_2(g) \longrightarrow 2H_2O(g); \text{ this is simply } 2\Delta H_f^\circ \text{ for } H_2O(g). \text{ So}$$

$$2\Delta H_c^\circ(H_2) = 2\Delta H_f^\circ(H_2O(g)) = -483.64 \text{ kJ}$$

(d) The direct combustion of one mole of methanol will produce more heat (-726 kJ) than decomposing methanol and subsequently combusting the hydrogen gas formed (for this process, ΔH_r° will equal the heat needed to form the H₂ gas plus the heat released when the H₂ gas is combusted:

$$\Delta H_r^\circ = (+128.33 \text{ kJ}) + (-483.64 \text{ kJ}) = -355.31 \text{ kJ})$$

8.25 Species (a), (b), (c), and (d) can all function as greenhouse gases, while (e) cannot. Any molecule other than a homonuclear diatomic can exhibit a changing dipole moment as it vibrates with certain vibrational modes. Since argon is monoatomic it has no covalent bonds, no vibrational modes, and no dipole moment.

8.27 $CH_3OH(l) + \frac{3}{2} O_2(g) \longrightarrow CO_2(g) + 2 H_2O(l)$

$$\text{number of kg of } CO_2 = 1.00 \text{ L } CH_3OH \left(\frac{1000 \text{ mL}}{1 \text{ L}} \right) \left(\frac{0.791 \text{ g } CH_3OH}{1 \text{ mL } CH_3OH} \right)$$

$$\times \left(\frac{1 \text{ mol } CH_3OH}{32.04 \text{ g } CH_3OH} \right) \left(\frac{1 \text{ mol } CO_2}{1 \text{ mol } CH_3OH} \right) \left(\frac{44.02 \text{ g } CO_2}{1 \text{ mol } CO_2} \right) \left(\frac{1 \text{ kg}}{1000 \text{ g}} \right)$$

$$= 1.09 \text{ kg } CO_2$$

This mass of carbon dioxide is about half the amount generated by combusting an equivalent volume of octane (2.16 kg per liter).

However, we also need to consider how much energy is produced per liter of fuel and how the mass of carbon dioxide produced compares for a given amount of energy produced. Standard enthalpies of combustion given in Appendix 2 are $\Delta H_c^\circ = -5471 \text{ kJ} \cdot \text{mol}^{-1}$ for octane and

$\Delta H_c^\circ = -726 \text{ kJ} \cdot \text{mol}^{-1}$ for methanol.

$$\text{energy per L methanol} = 1.00 \text{ L } CH_3OH \left(\frac{1000 \text{ mL}}{1 \text{ L}}\right)\left(\frac{0.791 \text{ g } CH_3OH}{1 \text{ mL } CH_3OH}\right)$$

$$\times \left(\frac{1 \text{ mole } CH_3OH}{32.04 \text{ g } CH_3OH}\right)\left(\frac{726 \text{ kJ}}{1 \text{ mole } CH_3OH}\right)$$

$$= 1.79 \times 10^4 \text{ kJ}$$

$$\text{energy per L octane} = 1.00 \text{ L } C_8H_{18} \left(\frac{1000 \text{ mL}}{1 \text{ L}}\right)\left(\frac{0.703 \text{ g } C_8H_{18}}{1 \text{ mL } C_8H_{18}}\right)$$

$$\times \left(\frac{1 \text{ mole } C_8H_{18}}{114.22 \text{ g } C_8H_{18}}\right)\left(\frac{5471 \text{ kJ}}{1 \text{ mole } C_8H_{18}}\right)$$

$$= 3.37 \times 10^4 \text{ kJ}$$

So the combustion of octane produces almost twice as much energy per liter as methanol (octane/methanol=1.88).

For an equivalent amount of combustion energy, methanol produces $1.88 \text{ L} \times 1.09 \text{ kg } CO_2 \cdot \text{L}^{-1} = 2.05 \text{ kg } CO_2$, which is still slightly less than octane. (However, it requires that the vehicle carry about 90% more fuel by volume, 1.9 vs. 1 L, and more than twice as much fuel by mass, 1.5 kg vs. 0.7 kg.)

8.29 Ion-ion forces are among the strongest intermolecular interactions. Therefore, the interactions in (a) are the strongest shown. Hydrogen bonding is stronger than a dipole-dipole interaction indicating that the interactions in (c) are stronger than those shown in (d).

8.31 (a) C: sp^2; B: sp^2; N: sp^2

(b) A tube with a diameter of about 1.3 nm will have a circumference of $2\pi r$ or πd. Thus, the circumference of a 1.3×10^{-9} m diameter nanotube will be $(1.3 \times 10^{-9} \text{ m})(\pi) = 4.1 \times 10^{-9}$ m.

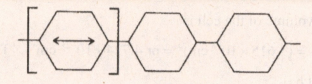

To calculate the number of C_6 rings that will be strung together, we need to calculate the distance across the C_6 ring as shown by the arrow.

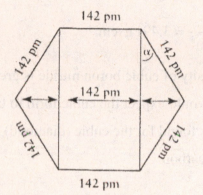

The total distance from one carbon to the opposite carbon on the ring will be given by $d = 142 \text{ pm} + 2\,(142 \text{ pm} \times \sin 30°) = 284 \text{ pm}$.

One repeat unit will be one benzene ring plus one C—C bond,

$284 \text{ pm} + 142 \text{ pm} = 426 \text{ pm}$ or 4.26×10^{-10} m. There are thus about

$4.1 \times 10^{-9} \div 4.26 \times 10^{-10} = 10$ units strung together around a nanotube that has a diameter of 1.3 nm. Since each unit contains two hexagons there will be approximately 20 hexagons per circumference.

(c) In C_{60} the carbon atoms are sp^2-hybridized and are nearly planar. However, the curvature of the molecule introduces some strain at the carbon atoms so that there is some tendency for some of the carbon atoms to undergo conversion to sp^3 hybridization. However, to make every carbon sp^3-hybridized would introduce much more strain on the carbon cage and, after a certain point, further addition of hydrogen becomes unfavorable.

(d) The spherical structures require the formation of five-member rings (see structure **3**, C_{60}). Boron nitride cannot form these rings because they would require high-energy boron-boron or nitrogen-nitrogen bonds.

(e) The unit cell described will contain a total of four B atoms and four N atoms. The volume of the cell is

$(361.5 \text{ pm})^3 = (3.615 \times 10^{-8} \text{ cm})^3 = \text{or } 4.724 \times 10^{-23} \text{ cm}^{-3}$. The mass in the unit cell will be

$$(4 \times 10.81 \text{ g} \cdot \text{mol}^{-1} + 4 \times 14.01 \text{ g} \cdot \text{mol}^{-1}) \div 6.022 \times 10^{23} \text{ mol}^{-1}$$
$$= 1.649 \times 10^{-22} \text{ g.}$$

$$d = \frac{1.649 \times 10^{-22} \text{ g}}{4.724 \times 10^{-23} \text{ cm}^3} = 3.491 \text{ g} \cdot \text{cm}^{-3}$$

(f) Because the density of cubic boron nitride is greater than that of hexagonal BN, we would expect the cubic form to be favored at high pressures, exactly as found for the cubic (diamond) and hexagonal (graphite) forms of carbon.

FOCUS 9
THE d-BLOCK ELEMENTS

9A.1 Elements at the left of the *d* block tend to have strongly negative standard potentials; this can be attributed to their lower ionization potentials which means they would rather undergo oxidation than reduction.

9A.3 (a) MnO_4^- (b) MoO_4^{2-}

9A.5 (a) Fe; (b) Cu; (c) Pt; (d) Pd; (e) Ta

9A.7 With zinc, the last of the 3d electrons is added; this closes the $n=3$ shell thereby significantly lowering the energy of the 3d orbitals. With the next element in the periodic table (gallium) is added an electron into the 4p orbitals which has little or no effect on the shielding of the 3d electrons. The proton that is added to the nucleus increases the Z (and the Z_{eff}); this causes the 3d orbitals to be drawn down closer to the nucleus, thus lowering their energy even more.

9A.9 Hg is much more dense than Cd, because the shrinkage in atomic radius that occurs between $Z = 58$ and $Z = 71$ (the lanthanide contraction) causes the atoms following the rare earths to be smaller than might have been expected for their atomic masses and atomic numbers. Zn and Cd have densities that are not too dissimilar, because the radius of Cd is subject only to a smaller *d*-block contraction.

9A.11 (a) Proceeding down a group in the *d* block (for example, from Cr to Mo to W), there is an increasing probability of finding the elements in higher oxidation states. That is, higher oxidation states become more stable on going down a group.

(b) The trend for the *p*-block elements is reversed. Because of the inert pair effect, the higher oxidation states tend to be less stable as one descends a group.

9A.13 In MO_3, M has an oxidation number of +6. Of these three elements, the +6 oxidation state is most stable for Cr. See Fig. 9A.7.

9B.1 (a) Ti(s), $MgCl_2$(s)

$$TiCl_4(g) + 2\,Mg(l) \xrightarrow{\Delta} Ti(s) + 2\,MgCl_2(s)$$

(b) Co^{2+}(aq), HCO_3^-(aq), NO_3^-(aq)

$$CoCO_3(s) + HNO_3(aq) \rightarrow Co^{2+}(aq) + HCO_3^-(aq) + NO_3^-(aq)$$

(c) V(s), CaO(s)

$$V_2O_5(s) + 5\,Ca(l) \xrightarrow{\Delta} 2\,V(s) + 5\,CaO(s)$$

9B.3 (a) titanium(IV) oxide, TiO_2

(b) iron(III) oxide, Fe_2O_3

(c) manganese(IV) oxide, MnO_2

(d) iron(II) chromite, $FeCr_2O_4$

9B.5 (a) Titanium has an oxidation state of +4.
(b) Zinc has an oxidation state of +2.

9B.7 (a) CO
(b) In Zones D & C,

$$3\,Fe_2O_3(s) + CO(g) \longrightarrow 2\,Fe_3O_4(s) + CO_2(g)$$

$$Fe_3O_4(s) + CO(g) \longrightarrow 3\,FeO(s) + CO_2(g)$$

These reactions combine to give

$$Fe_2O_3(s) + CO(g) \longrightarrow 2\,FeO(s) + CO_2(g)$$

In Zone B,

$$Fe_2O_3(s) + 3\,CO(g) \longrightarrow 2\,Fe(s) + 3\,CO_2(g)$$

$$FeO(s) + CO(g) \longrightarrow Fe(s) + CO_2(g)$$

(c) carbon

9B.9   (a) $V_2O_5(s) + 2\,H_3O^+(aq) \longrightarrow 2\,VO_2^+(aq) + 3\,H_2O(l)$

(b) $V_2O_5(s) + 6\,OH^-(aq) \longrightarrow 2\,VO_4^{3-}(aq) + 3\,H_2O(l)$

9B.11 Even though all three Group 1B/11 metal atoms have the valence shell electron configuration $(n-1)d^{10}ns^1$, Cu is more reactive than Ag or Au. Metals ordinarily lose one or more electrons to form cations when they react with some other species. As the value of n increases, d and f electrons become less effective at shielding the outermost, highest energy electron(s) from the attractive charge of the nucleus. This higher effective nuclear charge makes it more difficult to oxidize the metal atom or ion. So, for example, Cu^{2+} exists in many common compounds (and can be formed by Cu^+ disproportionation in water) while Ag^{2+} does not. Furthermore, the valence electron orbital energies for most common Lewis bases match the orbitals of Cu and its cations more closely and would interact with them more favorably to form products.

9B.13 (a) $Cr_2O_7^{2-} + 14\,H^+ + 6\,e^- \longrightarrow 2\,Cr^{2+} + 7\,H_2O$ $E° = +1.33\text{ V}$

$\qquad 2\,Br^- \longrightarrow Br_2 + 2\,e^-$ $\qquad\qquad\qquad\qquad\qquad\qquad E° = -1.09\text{ V}$

$\qquad\qquad\qquad\qquad\qquad E°(\text{overall}) = +0.99\text{ V}$

Therefore, Br^- will be oxidized to Br_2.

(b) $Cr_2O_7^{2-} + 14\,H^+ + 6\,e^- \longrightarrow 2\,Cr^{2+} + 7\,H_2O$ $E° = +1.33\text{ V}$

$\qquad Ag^+ \longrightarrow Ag^{2+} + 1\,e^-$ $\qquad\qquad\qquad\qquad\qquad\qquad E° = -1.98\text{ V}$

$\qquad\qquad\qquad\qquad\qquad E°(\text{overall}) = -0.65\text{ V}$

Therefore, no reaction will occur.

9B.15 (a) Cr^{3+} ions in water form the complex $[Cr(OH_2)_6]^{3+}(aq)$, which

behaves as a Brønsted acid:

$$[Cr(OH_2)_6]^{3+}(aq) + H_2O(l) \longrightarrow [Cr(OH_2)_5OH]^{2+}(aq) + H_3O^+(aq)$$

(b) The gelatinous precipitate is the hydroxide $Cr(OH)_3$. The precipitate

dissolves as the $Cr(OH)_4^-$ complex ion is formed:

$$Cr^{3+}(aq) + 3 OH^-(aq) \longrightarrow Cr(OH)_3(s)$$
$$Cr(OH)_3(s) + OH^-(aq) \longrightarrow Cr(OH)_4^-(aq)$$

9C.1 The general equation for determining the oxidation number of metal in a

complex ion is:

(# metal atoms)(oxidation number of the metal) +

$$\sum (\text{# each ligand})(\text{charge of each ligand}) = \text{charge of the ion}$$

for (a)-(d), let x = the oxidation number to be determined

(a) hexacyanoferrate(II) ion

$$1(x) + 6(-1) = -4$$
$$x = -4 - (-6) = +2$$

(b) hexaamminecobalt(III) ion

$$1(x) + 6(0) = +3$$
$$x = +3$$

(c) aquapentacyanocobaltate(III) ion

$$1(x) + 5(-1) + 1(0) = -2$$
$$x = -2 - (-5) = +3$$

(d) pentaamminesulfatocobalt(III) ion
$$1(x) + 1(-2) + 5(0) = +1$$
$$x = +1 - (-2) = +3$$

9C.3 (a) $K_3[Cr(CN)_6]$

(b) $[Co(NH_3)_5(SO_4)]Cl$

(c) $[Co(NH_3)_4(H_2O)_2]Br_3$

(d) $Na[Fe(OH_2)_2(C_2O_4)_2]$

9C.5 (a) The molecule $HN(CH_2CH_2NH_2)_2$ has three nitrogen atoms, each with a lone pair of electrons that may be used for bonding to a metal center. The molecule can thus function as a tridentate ligand.

(b) The CO_3^{2-} ion can bind to a metal ion through either one or two oxygen atoms. It may, therefore, serve as a mono- or bidentate ligand.

(c) H_2O is always a monodenate ligand.

(d) The oxalate ion can bind through two oxygen atoms and is usually a bidentate ligand.

9C.7 As shown below, only the molecule (b) can function as a chelating ligand. The two amine groups in (a) and (c) are arranged so that they would not be able to coordinate simultaneously to the same metal center. It is possible for each of the amine groups in (a) and (c) to coordinate to two different metal centers, however. This is not classified as chelating. When a single ligand binds to two different metal centers, it is known as a *bridging* ligand.

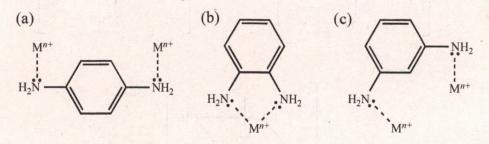

9C.9 (a) 4 (b) 2 (c) 6 (en is bidentate) (d) 6 (EDTA is hexadentate)

9C.11 (a) structural isomers, linkage isomers

(b) structural isomers, ionization isomers

(c) structural isomers, linkage isomers

(d) structural isomers, ionization isomers

9C.13 (a) yes

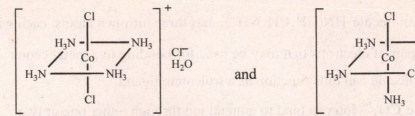 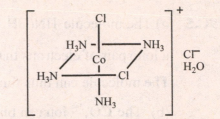

trans-tetraamminedichloridocobalt(III)   *cis*-tetraamminedichloridocobalt(III)
chloride monohydrate   chloride monohydrate

(b) no

(c) yes

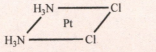   and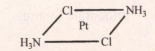

cis-diamminedichloridoplatinum(II)   *trans*-diamminedichloridoplatinum(II)

9C.15 Since nitro can bind as NO_2 or ONO, two linkage isomers are possible, as
well as one ionization isomer:

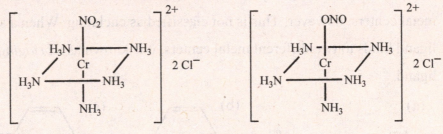

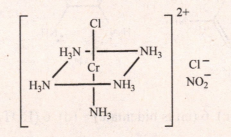

9C.17 (a)

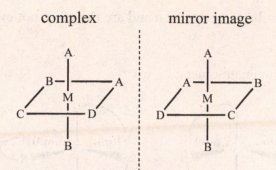

No rotation will make the complex and its mirror image match; therefore, it is chiral.

(b)

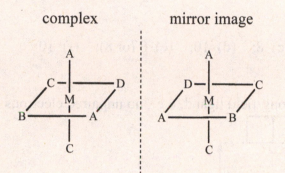

Two 90° rotations shows that the complex and its mirror image are superimposable:

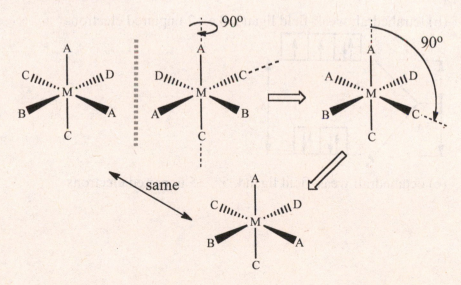

As a result, the complex is not chiral; in fact, the complex and its mirror image are identical (the same species) and are therefore not even isomers.

9C.19 $[Co(en)_3]^{3+}$ can form enantiomers:

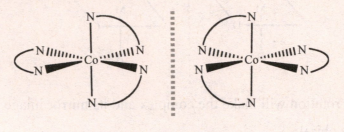

nonsuperimposable mirror images
(enantiomers)

9D.1   (a) 2;   (b) 5;   (c) 8;   (d) 10;   (e) 0 (or 8);   (f) 10

9D.3   (a) octahedral; strong-field ligand, 6 e⁻, no unpaired electrons

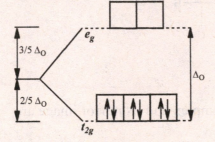

(b) tetrahedral: weak-field ligand, 8 e⁻, 2 unpaired electrons

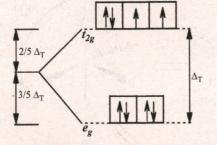

(c) octahedral: weak-field ligand, 5 e⁻, 5 unpaired electrons

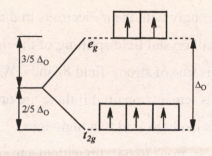

(d) octahedral: strong-field ligand, 5 e⁻, one unpaired electron

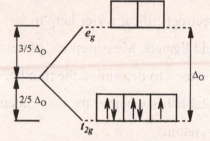

9D.5

$$\Delta_O = \frac{hc}{\lambda} = \frac{(6.626 \times 10^{-34} \text{ J} \cdot \text{s})(2.998 \times 10^8 \text{ m} \cdot \text{s}^{-1})}{560 \times 10^{-9} \text{ m}} = 3.55 \times 10^{-19} \text{ J}$$

Multiplication by Avogadro's number to get to per mole and converting J to kJ gives $\Delta_O = 214$ kJ·mol⁻¹.

9D.7 (a) $[Co(en)_3]^{3+}$; 6 e⁻; no unpaired electrons.

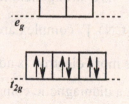

(b) $[Mn(CN)_6]^{3-}$; 4 e⁻; two unpaired electrons.

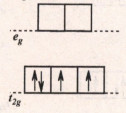

9D.9   Weak-field ligands do not interact strongly with the *d*-electrons in the metal ion, so they produce only a small crystal field splitting of the *d*-electron energy states. The opposite is true of strong-field ligands. With weak-field ligands, unpaired electrons remain unpaired if there are unfilled orbitals; hence, a weak-field ligand is likely to lead to a high-spin complex. Strong-field ligands cause electrons to pair up with electrons in lower energy orbitals. A strong-field ligand is likely to lead to a low-spin complex. Ligands arranged in the spectrochemical series help to distinguish strong-field and weak-field ligands. Measurement of magnetic susceptibility (paramagnetism) can be used to determine the number of unpaired electrons, which, in turn, establishes whether the associated ligand is weak-field or strong-field in nature.

9D.11   Use Beer's law $(A=\varepsilon bc)$ to determine the concentration:

$$c = \frac{A}{b\varepsilon} = \frac{0.262}{(175 \text{ L} \cdot \text{mol}^{-1} \cdot \text{cm}^{-1})(1.00 \text{ cm})}$$

$$c = 1.50 \times 10^{-3} \ M$$

9D.13   In both cases we are looking at the difference between an $Fe^{3+}(d^5)$ and an $Fe^{2+}(d^6)$. Since CN^- is a strong-field ligand we expect that all the *d* electrons in the $[Fe(CN)_6]^{3-}$ complex are in the lower energy t_{2g} orbitals. Upon reduction, one more electron is added, resulting in all electrons being paired, giving a diamagnetic complex:

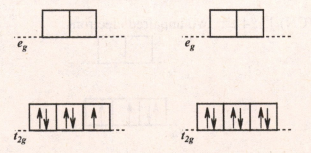

On the other hand, Cl^- is a weak-field ligand; here we expect the *d* electrons to be distributed in both the t_{2g} and e_g the orbitals. Upon reduction, one more electron is added, resulting in only one electron being paired, leading to a complex that is still paramagnetic:

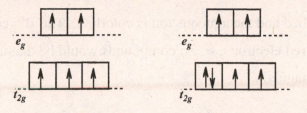

9D.15 Because F^- is a weak-field ligand and is a strong-field ligand, the splitting between levels is less in (a) than in (b). Therefore, (a) will absorb light of longer wavelength than will (b) and consequently will display a shorter wavelength color. Blue light is shorter in wavelength than yellow light, so (a) $[CoF_6]^{3-}$ is blue and (b) $[Co(en)_3]^{3+}$ is yellow.

9D.17 $$E_{photon} = \left(\frac{209\ kJ}{mol}\right)\left(\frac{1\ mol}{6.02 \times 10^{23}\ photons}\right) = 3.47 \times 10^{-22}\ kJ \cdot photon^{-1}$$

$$v = \frac{E}{h} = \frac{3.47 \times 10^{-19}\ J}{6.626 \times 10^{-34}\ J \cdot s} = 5.24 \times 10^{14}\ s^{-1}$$

$$\lambda = \frac{c}{v} = \left(\frac{3.00 \times 10^8\ m \cdot s^{-1}}{5.24 \times 10^{14}\ s^{-1}}\right)\left(\frac{10^9\ nm}{1\ m}\right) = 573\ nm$$

This wavelength is in the yellow region of the visible spectrum. Since the complex absorbs yellow light, it transmits the complement, or purple (violet).

9D.19 (a) There are no unpaired electrons in this complex (low spin, d^6).

(b) Both NH_3 and CN^- are strong field ligands, with cyanide being slightly stronger than ammonia (slightly larger splitting in $[Co(CN)_6]^{3-}$ than $[Co(NH_3)_6]^{3+}$). Exchange of the CN^- ligands for NH_3 should lead to the wavelength of the absorbed radiation to be slightly longer.

9D.21 In Zn^{2+}, the 3d-orbitals are filled (d^{10}). Therefore, there can be no electronic transitions between the t and e levels; hence, no visible light is absorbed and the aqueous ion is colorless. The d^{10} configuration has no unpaired electrons, so Zn compounds would be diamagnetic, not paramagnetic.

9D.23 (a) (i) $\Delta_O = \dfrac{hc}{\lambda} = \dfrac{(6.626 \times 10^{-34} \text{ J} \cdot \text{s}^{-1})(2.998 \times 10^8 \text{ m} \cdot \text{s}^{-1})}{740 \times 10^{-9} \text{ m}} = 2.68 \times 10^{-19} \text{ J}$

(ii) $\Delta_O = \dfrac{hc}{\lambda} = \dfrac{(6.626 \times 10^{-34} \text{ J} \cdot \text{s}^{-1})(2.998 \times 10^8 \text{ m} \cdot \text{s}^{-1})}{460 \times 10^{-9} \text{ m}} = 4.32 \times 10^{-19} \text{ J}$

(iii) $\Delta_O = \dfrac{hc}{\lambda} = \dfrac{(6.626 \times 10^{-34} \text{ J} \cdot \text{s}^{-1})(2.998 \times 10^8 \text{ m} \cdot \text{s}^{-1})}{575 \times 10^{-9} \text{m}} = 3.45 \times 10^{-19} \text{ J}$

Converting J to kJ and multiplication by 6.022×10^{23} gives $kJ \cdot mol^{-1}$.

(i) $2.68 \times 10^{-19} \text{ J}(10^{-3} \text{ kJ} \cdot \text{J}^{-1})(6.022 \times 10^{23} \text{ mol}^{-1}) = 161 \text{ kJ} \cdot \text{mol}^{-1}$

(ii) $4.32 \times 10^{-19} \text{ J}(10^{-3} \text{ kJ} \cdot \text{J}^{-1})(6.022 \times 10^{23} \text{ mol}^{-1}) = 260 \text{ kJ} \cdot \text{mol}^{-1}$

(iii) $3.45 \times 10^{-19} \text{ J}(10^{-3} \text{ kJ} \cdot \text{J}^{-1})(6.022 \times 10^{23} \text{ mol}^{-1}) = 208 \text{ kJ} \cdot \text{mol}^{-1}$

(b) $Cl^- < H_2O < NH_3$ (spectrochemical series)

9D.25 The e_g set, which comprises the $d_{x^2-y^2}$ and d_{z^2} orbitals.

9D.27 (a) (i) The CN^- ion is a π-acid ligand accepting electrons into the empty π^* orbital created by the C—N multiple bond. (ii) The Cl^- ion has extra lone pairs in addition to the one that is used to form the σ-bond to the metal, and so it can act as a π-base, donating electrons in a p-orbital to an empty d-orbital on the metal. (iii) H_2O, like Cl^- also has an "extra" lone pair of electrons that can be donated to a metal center, making it a weak π-base; (iv) en is neither a π-acid nor a π-base, because it does not have any empty π-type antibonding orbitals nor does it have any extra lone pairs of electrons to donate.

(b) $Cl^- < H_2O < en < CN^-$. Note that the spectrochemical series orders the ligands as π-bases $< \sigma$-bond only ligands $< \pi$-acceptors.

9D.29 Nonbonding or slightly antibonding. In a complex that forms only σ-bonds, the t_{2g} set of orbitals is nonbonding. If the ligands can function as weak π-donors (those close to the middle of the spectrochemical series, such as H_2O), the t_{2g} set becomes slightly antibonding by interacting with the filled p-orbitals on the ligands.

9D.31 Antibonding. The e_g set of orbitals on an octahedral metal ion are always antibonding because of interactions with ligand orbitals that form the σ-bonds. This is true regardless of whether the ligands are π-acceptors, π-donors, or neither.

9D.33 Water has two lone pairs of electrons. Once one of these is used to form the σ-bond to the metal ion, the second may be used to form a π-bond. This causes the t_{2g} set of orbitals to move up in energy, making Δ_O smaller; therefore, water is a weak-field ligand. Ammonia does not have this extra lone pair of electrons and consequently cannot function as a π-donor ligand.

9.1

$$Cr(OH)_3(s) + 3\,e^- \longrightarrow Cr(s) + 3\,OH^- \qquad E° = -1.34\ V$$

$$Cr(s) \longrightarrow Cr^{3+} + 3\,e^- \qquad E° = +0.74\ V$$

$$Cr(OH)_3(s) \longrightarrow Cr^{3+} + 3\,OH^- \qquad E° = -0.60\ V$$

$$\Delta G° = -nFE° = -RT \ln K$$

$$\ln K = \frac{nFE°}{RT} = \frac{(3)(96\ 485\ J \cdot V^{-1} \cdot mol^{-1})(-0.60\ V)}{(8.314\ J \cdot K^{-1} \cdot mol^{-1})(298\ K)} = -70.1$$

$$K_{sp} = e^{-70.1} = 3.6 \times 10^{-31}$$

9.3 (a) [PtBrCl(NH₃)₂]

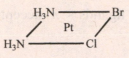 and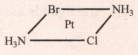

cis-
diamminebromidochloridoplatinum(II)

trans-
diamminebromidochloridoplatinum(II)

(b) If the compound were tetrahedral, there would be only one compound, not two.

9.5 (a) The first, $[Ni(SO_4)(en)_2]Cl_2$, will give a precipitate of AgCl when $AgNO_3$ is added; the second will not. (b) The second, $[NiCl_2(en)_2]I_2$, will show free I_2 when mildly oxidized with, for example, Br_2, but the first will not.

9.7 (a) $[MnCl_6]^{4-}$; 5 e⁻, Cl⁻ is a weak-field ligand

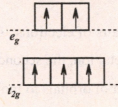

$[Mn(CN)_6]^{4-}$; 5 e⁻, CN⁻ is a strong-field ligand

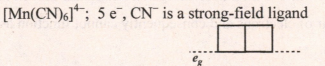

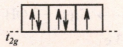

(b) $[MnCl_6]^{4-}$: five; $[Mn(CN)_6]^{4-}$: one.

(c) Complexes with weak-field ligands absorb longer wavelength light, therefore, $[MnCl_6]^{4-}$ absorbs longer wavelengths.

9.9 If the prismatic structure were the true structural form of $[CoCl_2(NH_3)_4]$ one would expect to get four possible isomers, not the two seen $(X = NH_3)$:

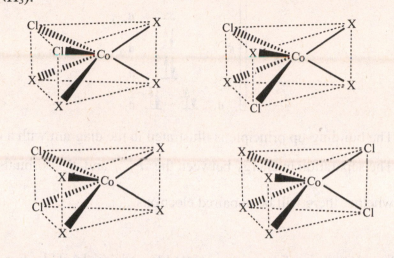

9.11 The correct structure for $[Co(NH_3)_6]Cl_3$ consists of four ions in aqueous solution $(Co(NH_3)_6^{3+}$ and three Cl^- ions). The chloride ions can be easily precipitated as AgCl. This would not be possible if they were bonded to the other (NH_3) ligands. If the structure were $Co(NH_3—NH_3—Cl)_3$, VSEPR theory would predict that the Co^{3+} ion would have a trigonal planar ligand arrangement. The splitting of the d-orbital energies would not be the same as the octahedral arrangement and would lead to different spectroscopic and magnetic properties inconsistent with the experimental evidence. In addition, neither optical nor geometrical isomers would be observed.

9.13 See Figure 9D.1. Remove (mentally) the ligands along the $\pm z$ axes. From the shape of the atomic orbitals and their orientation with respect to the x, y, and z axes, it is clear that the $d_{x^2-y^2}$-orbital will have the greatest overlap, therefore repulsion, from the ligands in the xy plane. The d_{xy}-orbital will have the next strongest repulsion from these ligands, d_{z^2} next, and finally, the weakest repulsion will be with d_{xz} and d_{yz}. Therefore, the energy-level diagram will be (schematically):

The building-up principle is illustrated in the diagram with a d^8 ion (Pd^{2+}). The separation in energy between the $d_{x^2-y^2}$ and d_{xy}- orbitals determines whether there will be unpaired electrons.

9.15   The electron configuration expected for Ni^{2+} is $[Ar]3d^8$. For this complex to have no unpaired electrons it would have to be square planar in its electronic geometry, as both the octahedral and tetrahedral geometries require a d^8 species to have two unpaired electrons. Square planar does not:

octahedral d^8 complex   tetrahedral d^8 complex

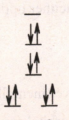

square planar d^8 complex

9.17   AgCl will only form from Cl^- found outside the coordination sphere.

(a) the amount of Cl^- outside the coordination sphere in $CoCl_3 \cdot 5NH_3 \cdot H_2O$ is:

$$0.43 \text{ g AgCl} \left(\frac{1 \text{ mol AgCl}}{143.35 \text{ g AgCl}} \right) \left(\frac{1 \text{ mol Cl}^-}{1 \text{ mol AgCl}} \right) = 0.0030 \text{ mol Cl}^- \text{; since you}$$

started with 0.0010 mol of the salt the ratio of salt to Cl$^-$ outside the

coordination sphere is 1:3. The correct formula and name for this

compound is $[Co(NH_3)_5OH_2]Cl_3$, pentaammineaquacobalt (III) chloride.

(b) the amount of Cl$^-$ outside the coordination sphere in $CoCl_3 \cdot 5NH_3$ is:

$$0.29 \text{ g AgCl} \left(\frac{1 \text{ mol AgCl}}{143.35 \text{ g AgCl}} \right) \left(\frac{1 \text{ mol Cl}^-}{1 \text{ mol AgCl}} \right) = 0.0020 \text{ mol Cl}^- \text{; since you}$$

started with 0.0010 mol of the salt the ratio of salt to Cl$^-$ outside the

coordination sphere is 1:2. The correct formula and name for this

compound is $[CoCl(NH_3)_5]Cl_2$, pentaamminechloridocobalt (III) chloride.

9.19   (a) square planar

(b) +1

(c) There are two possible isomers. Neither the cis nor the trans form is
optically active.

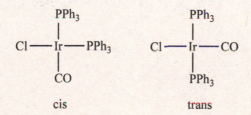

9.21   The equations for the reactions of Fe^{2+} with MnO_4^- or $Cr_2O_7^{2-}$ are:

$$MnO_4^-(aq) + 5 Fe^{2+}(aq) + 8 H^+(aq) \longrightarrow 5 Fe^{3+}(aq) + Mn^{2+}(aq) + 4 H_2O(l)$$
$$Cr_2O_7^{2-}(aq) + 6 Fe^{2+}(aq) + 14 H^+(aq) \longrightarrow 6 Fe^{3+}(aq) + 2 Cr^{3+}(aq) + 7 H_2O(l)$$

The amount of Fe^{2+} determined using 0.0210 M $Cr_2O_7^{2-}$ is:

$$25.20 \text{ mL} \left(\frac{1 \text{ L}}{1000 \text{ mL}} \right) \left(\frac{0.0210 \text{ mol } Cr_2O_7^{2-}}{1 \text{ L}} \right) \left(\frac{6 \text{ mol } Fe^{2+}}{1 \text{ mol } Cr_2O_7^{2-}} \right)$$

$$= 3.175 \times 10^{-3} \text{ mol } Fe^{2+}.$$

Based on this, the amount of 0.0420 M MnO_4^- needed to do this determination will be:

$$3.175 \times 10^{-3} \text{ mol Fe}^{2+} \left(\frac{1 \text{ mol MnO}_4^-}{5 \text{ mol Fe}^{2+}} \right) \left(\frac{1 \text{ L}}{0.0420 \text{ mol MnO}_4^-} \right) \left(\frac{1000 \text{ mL}}{1 \text{ L}} \right)$$

$$= 15.12 \text{ mL}.$$

9.23 The green color suggests chromium or copper but the blue solution formed when the mineral is dissolved in sulfuric acid points to copper as the metal ($CuSO_4$ is blue). The colorless gas that forms either upon heating or on treatment with acid is carbon dioxide; this is proven when the gas is bubbled into limewater (the cloudiness in the solution is due to calcium carbonate formation). CO_2 is formed when carbonates are either heated or treated with acid. The mineral is most likely basic copper carbonate ($Cu_2CO_3(OH)_2$).

9.25 (a) +3 is the most likely oxidation state of vanadium

(b) $2 V_2O_5(s) + 8 C(s) \longrightarrow V_4C_3(s) + 5 CO_2(g)$

FOCUS 10
NUCLEAR CHEMISTRY

10A.1 We assume that all the change in energy goes into the energy of the γ ray emitted. Then, in each case,

$$\nu = \frac{\Delta E}{h}, \qquad \lambda = \frac{c}{\nu}$$

$$\text{energy of } 1\,\text{MeV} = \left(\frac{10^6\ \text{eV}}{1\,\text{MeV}} \right)\left(\frac{1.602 \times 10^{-19}\ \text{J}}{1\,\text{eV}} \right)$$

$$= 1.602 \times 10^{-13}\ \text{J} \cdot \text{MeV}^{-1}$$

(a) $\Delta E = (1.33\,\text{MeV})\left(\dfrac{1.602 \times 10^{-13}\ \text{J}}{1\,\text{MeV}} \right) = 2.1307 \times 10^{-13}\ \text{J}$

$$\nu = \frac{\Delta E}{h} = \frac{2.1307 \times 10^{-13}\ \text{J}}{6.626 \times 10^{-34}\ \text{J} \cdot \text{s}} = 3.22 \times 10^{20}\ \text{s}^{-1} = 3.22 \times 10^{20}\ \text{Hz}$$

$$\lambda = \frac{c}{\nu} = \frac{3.00 \times 10^8\ \text{m} \cdot \text{s}^{-1}}{3.22 \times 10^{20}\ \text{s}^{-1}} = 9.32 \times 10^{-13}\ \text{m}$$

(b) $\Delta E = (1.64\,\text{MeV})\left(\dfrac{1.602 \times 10^{-13}\ \text{J}}{1\,\text{MeV}} \right) = 2.628 \times 10^{-13}\ \text{J}$

$$\nu = \frac{\Delta E}{h} = \frac{2.628 \times 10^{-13}\ \text{J}}{6.626 \times 10^{-34}\ \text{J} \cdot \text{s}} = 3.97 \times 10^{20}\ \text{s}^{-1} = 3.97 \times 10^{20}\ \text{Hz}$$

$$\lambda = \frac{3.00 \times 10^8\ \text{m} \cdot \text{s}^{-1}}{3.97 \times 10^{20}\ \text{s}^{-1}} = 7.56 \times 10^{-13}\ \text{m}$$

(c) $\Delta E = (1.10\,\text{MeV})\left(\dfrac{1.602 \times 10^{-13}\ \text{J}}{1\,\text{MeV}} \right) = 1.76 \times 10^{-13}\ \text{J}$

$$\nu = \frac{\Delta E}{h} = \frac{1.76 \times 10^{-13}\ \text{J}}{6.626 \times 10^{-34}\ \text{J} \cdot \text{s}} = 2.66 \times 10^{20}\ \text{s}^{-1} = 2.66 \times 10^{20}\ \text{Hz}$$

$$\lambda = \frac{c}{\nu} = \frac{3.00 \times 10^8\ \text{m} \cdot \text{s}^{-1}}{2.65 \times 10^{20}\ \text{s}^{-1}} = 1.13 \times 10^{-12}\ \text{m}$$

10A.3 Potassium-40 has 21 neutrons; $^{39}_{18}$Ar and $^{41}_{20}$Ca are isotones of K-40.

10A.5 (a) $^{12}_{5}$B $\longrightarrow$ $^{0}_{-1}$e + $^{12}_{6}$C

(b) $^{214}_{83}$Bi $\longrightarrow$ $^{4}_{2}\alpha$ + $^{210}_{81}$Tl

(c) $^{98}_{43}$Tc $\longrightarrow$ $^{0}_{-1}$e + $^{98}_{44}$Ru

(d) $^{266}_{88}$Ra $\longrightarrow$ $^{4}_{2}\alpha$ + $^{262}_{86}$Rn

10A.7 (a) $^{109}_{49}$In $\longrightarrow$ $^{0}_{+1}$e + $^{109}_{48}$Cd

(b) $^{23}_{12}$Mg $\longrightarrow$ $^{0}_{+1}$e + $^{23}_{11}$Na

(c) $^{202}_{82}$Pb + $^{0}_{-1}$e $\longrightarrow$ $^{202}_{81}$Tl

(d) $^{76}_{33}$As + $^{0}_{-1}$e $\longrightarrow$ $^{76}_{32}$Ge

10A.9 (a) $^{24}_{11}$Na $\rightarrow$ $^{24}_{12}$Mg + $^{0}_{-1}$e; a β particle is emitted.

(b) $^{128}_{50}$Sn $\rightarrow$ $^{128}_{51}$Sb + $^{0}_{-1}$e; a β particle is emitted.

(c) $^{140}_{57}$La $\rightarrow$ $^{140}_{56}$Ba + $^{0}_{+1}$e; a positron (β^{+}) is emitted.

(d) $^{228}_{90}$Th $\rightarrow$ $^{224}_{88}$Ra + $^{4}_{2}\alpha$; an α particle is emitted.

10A.11 (a) $^{11}_{5}$B + $^{4}_{2}\alpha$ $\longrightarrow$ $2\,^{1}_{0}$n + $^{13}_{7}$N

(b) $^{35}_{17}$Cl + $^{2}_{1}$D $\longrightarrow$ $^{1}_{0}$n + $^{36}_{18}$Ar

(c) $^{96}_{42}$Mo + $^{2}_{1}$D $\longrightarrow$ $^{1}_{0}$n + $^{97}_{43}$Tc

(d) $^{45}_{21}$Sc + $^{1}_{0}$n $\longrightarrow$ $^{4}_{2}\alpha$ + $^{42}_{19}$K

10A.13 (a) $A/Z = 68/29 = 2.34 > (A/Z)_{based}$; hence, $^{68}_{29}$Cu

is neutron rich, and β decay is most likely. $^{68}_{29}$Cu $\longrightarrow$ $^{0}_{-1}$e + $^{68}_{30}$Zn

(b) $A/Z = 103/48 = 2.15 < (A/Z)_{based}$; therefore,

$^{103}_{48}$Cd is proton rich, and β^{+} decay is most likely. $^{103}_{48}$Cd $\longrightarrow$ $^{0}_{+1}$e + $^{103}_{47}$Ag

10A.15 α $\quad ^{235}_{92}\text{U} \longrightarrow ^{4}_{2}\alpha + ^{231}_{90}\text{Th}$

β $\quad ^{231}_{90}\text{Th} \longrightarrow ^{0}_{-1}\text{e} + ^{231}_{91}\text{Pa}$

α $\quad ^{231}_{91}\text{Pa} \longrightarrow ^{4}_{2}\alpha + ^{227}_{89}\text{Ac}$

β $\quad ^{227}_{89}\text{Ac} \longrightarrow ^{0}_{-1}\text{e} + ^{227}_{90}\text{Th}$

α $\quad ^{227}_{90}\text{Th} \longrightarrow ^{4}_{2}\alpha + ^{223}_{88}\text{Ra}$

α $\quad ^{223}_{88}\text{Ra} \longrightarrow ^{4}_{2}\alpha + ^{219}_{86}\text{Rn}$

α $\quad ^{219}_{86}\text{Rn} \longrightarrow ^{4}_{2}\alpha + ^{215}_{84}\text{Po}$

β $\quad ^{215}_{84}\text{Po} \longrightarrow ^{0}_{-1}\text{e} + ^{215}_{85}\text{At}$

α $\quad ^{215}_{85}\text{At} \longrightarrow ^{4}_{2}\alpha + ^{211}_{83}\text{Bi}$

β $\quad ^{211}_{83}\text{Bi} \longrightarrow ^{0}_{-1}\text{e} + ^{211}_{84}\text{Po}$

α $\quad ^{211}_{84}\text{Po} \longrightarrow ^{4}_{2}\alpha + ^{207}_{82}\text{Pb}$

10A.17 To determine the charge and mass of the unknown particle, it helps to write $^{1}_{1}\text{p}$ and $^{1}_{0}\text{n}$ for the proton and neutron, respectively; and $^{0}_{-1}\text{e}$ and $^{0}_{+1}\text{e}$ for the β particle and positron, respectively.

(a) $^{14}_{7}\text{N} + ^{4}_{2}\alpha \longrightarrow ^{17}_{8}\text{O} + ^{1}_{1}\text{p}$

(b) $^{248}_{96}\text{Cm} + ^{1}_{0}\text{n} \longrightarrow ^{249}_{97}\text{Bk} + ^{0}_{-1}\text{e}$

(c) $^{243}_{95}\text{Am} + ^{1}_{0}\text{n} \longrightarrow ^{244}_{96}\text{Cm} + ^{0}_{-1}\text{e} + \gamma$

(d) $^{13}_{6}\text{C} + ^{1}_{0}\text{n} \longrightarrow ^{14}_{6}\text{C} + \gamma$

10A.19 (a) $^{20}_{10}\text{Ne} + ^{4}_{2}\alpha \longrightarrow ^{8}_{4}\text{Be} + ^{16}_{8}\text{O}$

(b) $^{20}_{10}\text{Ne} + ^{20}_{10}\text{Ne} \longrightarrow ^{16}_{8}\text{O} + ^{24}_{12}\text{Mg}$

(c) $^{44}_{20}\text{Ca} + ^{4}_{2}\alpha \longrightarrow \gamma + ^{48}_{22}\text{Ti}$

(d) $^{27}_{13}\text{Al} + ^{2}_{1}\text{H} \longrightarrow ^{1}_{1}\text{p} + ^{28}_{13}\text{Al}$

10A.21 For iron-56 we would expect the following nuclear reaction to occur:

$$^{56}_{26}\text{Fe} + 6\ ^{1}_{0}n \longrightarrow\ ^{62}_{26}\text{Fe} \longrightarrow\ ^{62}_{32}\text{Ge} + 6\ ^{0}_{-1}e$$

The isotope iron-56 will be converted to germanium-62.

10A.23 In each case, identify the unknown particle by performing a mass and charge balance as you did in the solutions to Exercises 10A.5 and 10A.7. Then write the complete nuclear equation.

(a) $^{14}_{7}\text{N} + ^{4}_{2}\alpha \longrightarrow\ ^{17}_{8}\text{O} + ^{1}_{1}\text{p}$

(b) $^{239}_{94}\text{Pu} + 2\ ^{1}_{0}n \longrightarrow\ ^{241}_{95}\text{Am} + ^{0}_{-1}e$

10A.25 (a) unbihexium, Ubh   (b) untrihexium, Uth   (c) binilnilium, Bnn

10B.1 $k = \dfrac{\ln 2}{t_{1/2}}$

(a) $k = \dfrac{\ln 2}{12.3\ \text{a}} = 5.64 \times 10^{-2}\ \text{a}^{-1}$

(b) $k = \dfrac{\ln 2}{0.84\ \text{s}} = 0.83\ \text{s}^{-1}$

(c) $k = \dfrac{\ln 2}{10.0\ \text{min}} = 0.0693\ \text{min}^{-1}$

10B.3 We know that initial activity $= N_0 = 2150$ dpm and final activity $= N = 1324$ dpm at $t = 6.0$ h. Therefore,

$$\ln\left(\frac{N}{N_0}\right) = -kt \text{ and } k = \frac{\ln 2}{t_{1/2}}$$

So $\ln\left(\dfrac{N}{N_0}\right) = -\dfrac{t\ln 2}{t_{1/2}}$.

$$\ln\left(\frac{1342}{2150}\right) = -\frac{(6.0\ \text{h})\ln 2}{t_{1/2}} = -0.48$$

$$t_{1/2} = -\frac{(6.0\ \text{h})\ln 2}{-0.48};\ t_{1/2} = 8.8\ \text{h}.$$

10B.5 In each case, $k = \dfrac{\ln 2}{t_{1/2}}$, $N = N_0 e^{-kt}$, $\dfrac{N}{N_0} = e^{-kt}$, and the percentage

remaining $= 100\% \times (N/N_0)$.

(a) $k = \dfrac{\ln 2}{5.73 \times 10^3 \text{ a}} = 1.21 \times 10^{-4} \text{ a}^{-1}$

percentage remaining $= 100\% \times e^{-(1.21 \times 10^{-4} \text{ a}^{-1} \times 3000 \text{ a})} = 69.6\%$

(b) $k = \dfrac{\ln 2}{12.3 \text{ a}} = 0.0563 \text{ a}^{-1}$

percentage remaining $= 100\% \times e^{-(0.0563 \text{ a}^{-1} \times 12.0 \text{ a})} = 50.9\%$

10B.7 Fraction of K-40 remaining $= \dfrac{N}{N_0} = \dfrac{3}{5}$;

$t_{1/2} = 1.26 \times 10^9 \text{ a}$, $k = \dfrac{\ln 2}{1.26 \times 10^9 \text{ a}} = 5.50 \times 10^{-10} \text{ a}^{-1}$

$\dfrac{3}{5} = e^{-kt} = e^{-(5.50 \times 10^{-10} \text{ a}^{-1})x}$

$x = 9.29 \times 10^8 \text{ a}$

10B.9 Let dis = disintegrations

activity from "old" sample $= \dfrac{1500 \text{ dis}/0.250 \text{ g}}{10.0 \text{ h}} = 600 \text{ dis} \cdot \text{g}^{-1} \cdot \text{h}^{-1}$

activity from current sample $= 921 \text{ dis} \cdot \text{g}^{-1} \cdot \text{h}^{-1}$

$k = \dfrac{\ln 2}{t_{1/2}} = \dfrac{\ln 2}{5.73 \times 10^3 \text{ a}} = 1.21 \times 10^{-4} \text{ a}^{-1}$

old activity $\propto N$, current activity $\propto N_0$

$\dfrac{\text{old activity}}{\text{current activity}} = \dfrac{N}{N_0} = e^{-kt}$, $\dfrac{N_0}{N} = e^{kt}$, $\ln\left(\dfrac{N_0}{N}\right) = kt$

Solve for t (which is the age):

$t = \dfrac{\ln\left(\dfrac{N_0}{N}\right)}{k} = \dfrac{\ln\left(\dfrac{921}{600}\right)}{1.21 \times 10^{-4} \text{ a}^{-1}} = 3.54 \times 10^3 \text{ a}$

10B.11 $k = \dfrac{\ln 2}{t_{1/2}} = \dfrac{\ln 2}{109 \text{ min}} = 6.36 \times 10^{-3} \text{ min}^{-1}$

$N = N_0 e^{-kt}$ and $\dfrac{N}{N_0} = e^{-kt}$

Taking natural log of both sides gives

$\ln\left(\dfrac{N}{N_0}\right) = -kt$

Let $N_0 = 100\%$ and $N = 10\%$; we can then write

$\ln\left(\dfrac{10\%}{100\%}\right) = -kt$

Solving for t gives

$t = -\dfrac{1}{k}\ln\left(\dfrac{10\%}{100\%}\right) = -\left(\dfrac{1}{6.36 \times 10^{-3} \text{ min}^{-1}}\right)\ln(0.10) = 400 \text{ min } (1 \text{ sf})$

10B.13 $k = \dfrac{\ln 2}{t_{1/2}} = \dfrac{\ln 2}{5.27 \text{ a}} = 0.131 \text{ a}^{-1}$

$\dfrac{N}{N_0} = e^{-kt} = e^{-(0.131 \text{ a}^{-1})(2.50 \text{ a})} = \dfrac{0.266 \text{ g}}{N_0}$

$N_0 = 0.370 \text{ g}$

$\dfrac{0.370 \text{ g}}{1.40 \text{ g}} \times 100 = 26.4\%$

10B.15 Since radioactive decay follows first-order kinetics, the rate of loss of X is

$$\dfrac{d[X]}{dt} = -k_1[X] \qquad \text{(Eq. 1)}$$

Y, which is an intermediate, is lost in the first reaction but formed in the second one, so its rate equation can be expressed as

$$\dfrac{d[Y]}{dt} = k_1[X] - k_2[Y] \qquad \text{(Eq. 2)}$$

Z is the final product of the two consecutive reactions so its rate law is

$$\dfrac{d[Z]}{dt} = k_2[Y] \qquad \text{(Eq. 3)}$$

As discussed in Chapter 14, the integrated form of equation (1) is

$$[X] = [X]_0 e^{-k_1 t} \qquad \text{(Eq. 4)}$$

Substituting this expression into the rate law for Y and rearranging gives

$$\frac{d[Y]}{dt} + k_2[Y] = k_1[X]_0 e^{-k_1 t} \qquad \text{(Eq. 5)}$$

This linear first-order differential equation has the solution

$$[Y] = \frac{k_1}{k_2 - k_1}(e^{-k_1 t} - e^{-k_2 t})[X]_0 \text{ when } [Y]_0 = 0 \qquad \text{(Eq. 6)}$$

Since $[X] + [Y] + [Z] = [X]_0$ at all times, $[Z] = [X]_0 - ([X] + [Y])$, or

$$[Z] = [X]_0 - \left([X]_0 e^{-k_1 t} + \frac{k_1}{k_2 - k_1}(e^{-k_1 t} - e^{-k_2 t})[X]_0 \right) = [X]_0 \left(1 + \frac{k_1 e^{-k_2 t} - k_2 e^{-k_1 t}}{k_2 - k_1}\right) \qquad \text{(Eq. 7)}$$

The values of the rate constants can be found from the half-lives:

$$k_1 = \frac{\ln 2}{27.4 \text{ d}} = 0.0253 \text{ d}^{-1} \qquad k_2 = \frac{\ln 2}{18.7 \text{ d}} = 0.0371 \text{ d}^{-1}$$

Using these constants and assuming $[X]_0 = 2.00$ g, equations (4), (6) and (7)
are graphed below.

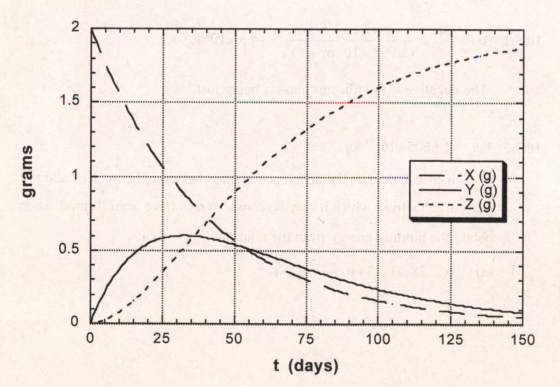

10B.17 If isotopically enriched water, such as $H_2^{18}O$, is used in the reaction, the label can be followed. Once the products are separated, a suitable technique, such as vibrational spectroscopy or mass spectrometry, can be used to determine whether the product has incorporated the ^{18}O. For example, if the methanol ends up with the O atom from the water molecules, then its molar mass would be $34\ g \cdot mol^{-1}$, rather than $32\ g \cdot mol^{-1}$ found for methanol with elements present at their natural isotopic abundance.

10B.19 From Table 10B.3, the half life for tritium (^3H) is 12.3 a. Therefore

$$k = \frac{\ln 2}{12.3\ a} = 5.64 \times 10^{-2}\ a^{-1}.$$ From the problem, $[A]_t = 8.3\%$ and $[A]_0 = 100\%$. Using the integrated rate law we get:

$$\ln\left(\frac{8.3}{100}\right) = -2.49 = -(5.64 \times 10^{-2}\ a^{-1})t;\ \text{solving for } t \text{ we get } t = 44 \text{ years.}$$

10C.1 $\Delta m = \dfrac{\Delta E}{c^2} = \dfrac{-3.9 \times 10^{26}\ J \cdot s^{-1}}{(2.998 \times 10^8\ m \cdot s^{-1})^2} = -4.3 \times 10^9\ kg \cdot s^{-1}$

The negative sign indicates mass is being lost.

10C.3 $1\ m_u = 1.6605 \times 10^{-27}\ kg$

In each case, calculate the difference in mass between the nucleus and the free particles from which it may be considered to have been formed. Then obtain the binding energy from the relation $E_{bind} = \Delta mc^2$.

(a) $^{62}_{28}Ni$: $28\ ^1H + 34\ n \longrightarrow\ ^{62}_{28}Ni$

$$\Delta m = 61.928346 \, m_u - (28 \times 1.0078 \, m_u + 34 \times 1.0087 \, m_u) = -0.586 \, m_u$$

$$\Delta m = (-0.586 \, m_u)\left(\frac{1.6605 \times 10^{-27} \, kg}{1 \, m_u}\right) = -9.73 \times 10^{-28} \, kg$$

$$E_{bind} = \left|-9.73 \times 10^{-28} \, kg\right| \times (2.998 \times 10^8 \, m \cdot s^{-1})^2$$

$$= 8.74 \times 10^{-11} \, kg \cdot m^2 \cdot s^{-2} = 8.74 \times 10^{-11} \, J$$

$$E_{bind} / \text{nucleon} = \frac{8.74 \times 10^{-11} \, J}{62 \, \text{nucleons}} = 1.41 \times 10^{-12} \, J \cdot \text{nucleon}^{-1}$$

(b) $^{239}_{94}Pu$: $94 \, ^1H + 145 \, n \longrightarrow \, ^{239}_{94}Pu$

$$\Delta m = 239.0522 \, m_u - (94 \times 1.0078 \, m_u + 145 \times 1.0087 \, m_u) = -1.94 \, m_u$$

$$\Delta m = (-1.94 \, m_u)\left(\frac{1.6605 \times 10^{-27} \, kg}{1 \, m_u}\right) = -3.23 \times 10^{-27} \, kg$$

$$E_{bind} = \left|-3.23 \times 10^{-27} \, kg\right| \times (2.998 \times 10^8 \, m \cdot s^{-1})^2$$

$$= 2.90 \times 10^{-10} \, kg \cdot m^2 \cdot s^{-2} = 2.90 \times 10^{-10} \, J$$

$$E_{bind} / \text{nucleon} = \frac{2.90 \times 10^{-10} \, J}{239 \, \text{nucleons}} = 1.21 \times 10^{-12} \, J \cdot \text{nucleon}^{-1}$$

(c) ^2_1H: $^1H + n \longrightarrow \, ^2_1H$

$$\Delta m = 2.0141 \, m_u - (1.0078 \, m_u + 1.0087 \, m_u) = -0.0024 \, m_u$$

$$\Delta m = (-0.0024 \, m_u)\left(\frac{1.6605 \times 10^{-27} \, kg}{1 \, m_u}\right) = -4.0 \times 10^{-30} \, kg$$

$$E_{bind} = \left|-4.0 \times 10^{-30} \, kg\right| \times (2.998 \times 10^8 \, m \cdot s^{-1})^2$$

$$= 3.6 \times 10^{-13} \, kg \cdot m^2 \cdot s^{-2} = 3.6 \times 10^{-13} \, J$$

$$E_{bind} / \text{nucleon} = \frac{3.6 \times 10^{-13} \, J}{2 \, \text{nucleons}} = 1.8 \times 10^{-13} \, J \cdot \text{nucleon}^{-1}$$

(d) ^3_1H: $^1H + 2 \, n \longrightarrow \, ^3_1H$

$$\Delta m = 3.01605 \, m_u - (1.0078 \, m_u + 2 \times 1.0087 \, m_u) = -0.0092 \, m_u$$

$$\Delta m = (-0.0092 \, m_u)\left(\frac{1.6605 \times 10^{-27} \, kg}{1 \, m_u}\right) = -1.5 \times 10^{-29} \, kg$$

$$E_{bind} = \left|-1.5 \times 10^{-29} \, kg\right| \times (2.998 \times 10^8 \, m \cdot s^{-1})^2$$

$$= 1.3 \times 10^{-12} \, kg \cdot m^2 \cdot s^{-2} = 1.3 \times 10^{-12} \, J$$

$$E_{bind} / \text{nucleon} = \frac{1.3 \times 10^{-12} \, J}{3 \, \text{nucleons}} = 4.3 \times 10^{-13} \, J \cdot \text{nucleon}^{-1}$$

(e) ^{62}Ni is the most stable, because it has the largest binding energy per nucleon.

10C.5 In each case, we first determine the change in mass, Δm = (mass of products) – (mass of reactants). We then calculate the energy released from $\Delta E = (\Delta m)c^2$.

(a) $D + D \longrightarrow {}^3He + n$

$2.0141\, m_u + 2.0141\, m_u \rightarrow 3.0160\, m_u + 1.0087\, m_u$

$4.0282\, m_u \rightarrow 4.0247\, m_u$

$\Delta m = (-0.0035\, m_u)\left(\dfrac{1.661\times10^{-27}\ kg}{1\, m_u}\right) = -5.8\times10^{-30}\ kg$

$\Delta E = \Delta mc^2 = (-5.8\times10^{-30}\ kg)(2.998\times10^8\ m\cdot s^{-1})^2 = -5.2\times10^{-13}\ J$

$\left(\dfrac{-5.2\times10^{-13}\ J}{4.0282\, m_u}\right)\left(\dfrac{1\, m_u}{1.661\times10^{-24}\ g}\right) = -7.8\times10^{10}\ J\cdot g^{-1}$

(b) $^3He + D \longrightarrow {}^4He + {}_1^1H$

$3.0160\, m_u + 2.0141\, m_u \rightarrow 4.0026\, m_u + 1.0078\, m_u$

$5.0301\, m_u \rightarrow 5.0104\, m_u$

$\Delta m = (-0.0197\, m_u)\left(\dfrac{1.661\times10^{-27}\ kg}{1\, m_u}\right) = -3.27\times10^{-29}\ kg$

$\Delta E = \Delta mc^2 = -(3.27\times10^{-29}\ kg)(2.998\times10^8\ m\cdot s^{-1})^2 = -2.94\times10^{-12}\ J$

$\left(\dfrac{-2.94\times10^{-12}\ J}{5.0301\, m_u}\right)\left(\dfrac{1\, m_u}{1.661\times10^{-24}\ g}\right) = -3.52\times10^{11}\ J\cdot g^{-1}$

(c) $^7Li + {}_1^1H \longrightarrow 2\ {}^4He$

$7.0160\, m_u + 1.0078\, m_u \rightarrow 2(4.0026\, m_u)$

$8.0238\, m_u \rightarrow 8.0052\, m_u$

$\Delta m = (-0.0186\, m_u)\left(\dfrac{1.661\times10^{-27}\ kg}{1\, m_u}\right) = -3.09\times10^{-29}\ kg$

$\Delta E = \Delta mc^2 = (-3.09\times10^{-29}\ kg)(2.998\times10^8\ m\cdot s^{-1})^2 = -2.78\times10^{-12}\ J$

$\left(\dfrac{-2.78\times10^{-12}\ J}{8.0238\, m_u}\right)\left(\dfrac{1\, m_u}{1.661\times10^{-24}\ g}\right) = -2.09\times10^{11}\ J\cdot g^{-1}$

(d) $D + T \longrightarrow {}^4He + n$

$$2.0141\, m_u + 3.0160\, m_u \rightarrow 4.0026\, m_u + 1.0087\, m_u$$

$$5.0301\, m_u \rightarrow 5.0113\, m_u$$

$$\Delta m = (-0.0188\, m_u)\left(\frac{1.661 \times 10^{-27}\ \text{kg}}{1\, m_u}\right) = -3.12 \times 10^{-29}\ \text{kg}$$

$$\Delta E = \Delta mc^2 = (-3.12 \times 10^{-29}\ \text{kg})(2.998 \times 10^8\ \text{m} \cdot \text{s}^{-1})^2 = -2.81 \times 10^{-12}\ \text{J}$$

$$\left(\frac{-2.81 \times 10^{-12}\ \text{J}}{5.0301\, m_u}\right)\left(\frac{1\, m_u}{1.661 \times 10^{-24}\ \text{g}}\right) = -3.36 \times 10^{11}\ \text{J} \cdot \text{g}^{-1}$$

10C.7 (a) $^{24}_{11}\text{Na} \rightarrow {}^{24}_{12}\text{Mg} + {}^{0}_{-1}\text{e}$

(b) mass $(^{24}_{11}\text{Na}) = 23.990\,96\, m_u$ and mass $(^{24}_{12}\text{Mg}) = 23.985\,04\, m_u$

The mass of the electron does not need to be explicitly included in the calculation because it is already included in the mass of Mg.

$$\Delta m = \text{mass}\,(^{24}_{12}\text{Mg}) - \text{mass}\,(^{24}_{11}\text{Na}) = 23.985\,04\, m_u - 23.990\,96\, m_u$$

$$= -5.92 \times 10^{-3}\, m_u$$

$$\Delta m\ (\text{in kg}) = -5.92 \times 10^{-3}\, m_u \times 1.661 \times 10^{-27}\ \text{kg}\ m_u^{-1} = -9.83 \times 10^{-30}\ \text{kg}$$

$$\Delta E = \Delta mc^2 = -(9.83 \times 10^{-30}\ \text{kg})(2.998 \times 10^8\ \text{m} \cdot \text{s}^{-1})^2 = -8.85 \times 10^{-13}\ \text{J}$$

(c) $\Delta E\ (\text{per nucleon}) = \dfrac{-8.85 \times 10^{-13}\ \text{J}}{24\ \text{nucleons}} = -3.69 \times 10^{-14}\ \text{J} \cdot \text{nucleon}^{-1}$

This simple calculation works because the number of nucleons is the same on both sides of the equation.

10C.9 (a) Begin by determining the mass difference:

$$235.04\, m_u + 1.0087\, m_u \longrightarrow 138.91\, m_u + 93.93\, m_u + 3(1.0087\, m_u)$$

$$236.0487\, m_u \longrightarrow 235.8661\, m_u$$

$$\Delta m = (-0.1826\, m_u)\left(\frac{1.661 \times 10^{-27}\ \text{kg}}{1\, m_u}\right) = -3.032 \times 10^{-28}\ \text{kg}$$

The per-atom energy associated with this process is:

$$\Delta E = \Delta mc^2 = (-3.032 \times 10^{-28}\ \text{kg})(2.998 \times 10^8\ \text{m} \cdot \text{s}^{-1})^2$$

$$= -2.725 \times 10^{-11}\ \text{J} \cdot \text{atom}^{-1}$$

For 5.0 grams of U-235:

$$(5.0 \text{ g U-235})\left(\frac{1 \text{ mol U-235}}{235.04 \text{ g U-235}}\right)\left(\frac{6.022 \times 10^{23} \text{ atoms}}{1 \text{ mol}}\right)\left(\frac{-2.725 \times 10^{-11} \text{ J}}{1 \text{ atom U-235}}\right)$$

$= -3.5 \times 10^{11} \text{ J or } 3.5 \times 10^8 \text{ kJ of energy released.}$

(b) The combustion of coal is $C(s) + O_2(g) \longrightarrow CO_2(g)$; this is also the enthalpy of formation for $CO_2(g)$. This value for this process is $-393.51 \text{ kJ} \cdot \text{mol}^{-1}$ (Appendix 2). The amount of coal that would have to be burned to release the same amount of energy as 5.0 g of U-235 is:

$$(3.5 \times 10^8 \text{ kJ})\left(\frac{1 \text{ mol coal}}{393.51 \text{ kJ}}\right)\left(\frac{12.01 \text{ g coal}}{1 \text{ mol coal}}\right)\left(\frac{1 \text{ kg}}{10^3 \text{ g}}\right) = 1.1 \times 10^4 \text{ kg of coal.}$$

10.1 (a) False; the dose equivalent is either equal to or larger than the actual dose, because of the Q factor.

(b) False; $1.8 \times 10^8 \text{ Bq} = 0.003 \text{ Ci}$ which is much smaller than 10 Ci.

(c) True.

10.3 (a) $1 \text{ Ci} = 3.7 \times 10^{10}$ decays per second (dps)

For 4 pCi, the decays per minute (dpm) is

$$4 \text{ pCi} = 4 \times 10^{-12} \text{ Ci} \times 3.7 \times 10^{10} \text{ dps} \times \left(\frac{60 \text{ s}}{1 \text{ min}}\right)$$

$$= 9 \text{ dpm}$$

(b) $\text{volume(L)} = (2.0 \times 3.0 \times 2.5) \text{ m}^3 \times \left(\frac{10^3 \text{ L}}{1 \text{ m}^3}\right) = 1.5 \times 10^4 \text{ L}$

$$\text{number of decays} = (1.5 \times 10^4 \text{ L})\left(\frac{4 \text{ pCi}}{1 \text{ L}}\right)\left(\frac{9 \text{ decays} \cdot \text{min}^{-1}}{4 \text{ pCi}}\right)(5.0 \text{ min})$$

$$= 7 \times 10^5 \text{ decays}$$

10.5 $N_0 = \text{number of } {}^{222}\text{Rn atoms} = 2.0 \times 10^{-5} \text{ mol} \times 6.0 \times 10^{23} \text{ atoms} \cdot \text{mol}^{-1}$

$$= 1.2 \times 10^{19} \text{ atoms}$$

$$k = \frac{\ln 2}{t_{1/2}} = \frac{\ln 2}{3.82 \text{ d}} = 0.181 \text{ d}^{-1}$$

(a) rate of decay $= k \times N = \left(\dfrac{0.181}{d}\right)\left(\dfrac{1\,d}{8.64 \times 10^4\,s}\right)(1.2 \times 10^{19}\,\text{atoms})$

$$= 2.52 \times 10^{13}\,\text{atoms} \cdot s^{-1}\ (\text{dps or Bq})$$

initial activity $= (2.52 \times 10^{13}\,\text{Bq})\left(\dfrac{1\,Ci}{3.7 \times 10^{10}\,Bq}\right)\left(\dfrac{1\,pCi}{10^{-12}\,Ci}\right)$

$$\times \left(\dfrac{1}{2000\,m^3}\right)\left(\dfrac{1\,m^3}{10^3\,L}\right)$$

$$= 3.4 \times 10^8\ \text{pCi} \cdot L^{-1}$$

(b) $N = N_0 e^{-kt} = 1.2 \times 10^{19}\ \text{atoms} \times e^{-0.181\,d^{-1} \times 1\,d} = 1.0 \times 10^{19}\ \text{atoms}$

(c) $\ln\left(\dfrac{\text{activity}}{\text{initial acitivity}}\right) = -kt$

$t = -\left(\dfrac{1}{k}\right)\ln\left(\dfrac{\text{activity}}{\text{initial activity}}\right) = -\left(\dfrac{1}{0.181\,d^{-1}}\right)\ln\left(\dfrac{4}{3.4 \times 10^8}\right)$

$$= 1 \times 10^2\ \text{days}$$

10.7 $k = \dfrac{\ln 2}{4.5 \times 10^9\,a} = 1.5 \times 10^{-10}\,a^{-1}$

$t\,(= \text{age}) = -\left(\dfrac{1}{k}\right)\ln\left(\dfrac{N}{N_0}\right)$

$\dfrac{N}{N_0} = \dfrac{\text{mass of }^{238}U}{\text{initial mass of }^{238}U} = \dfrac{1}{1 + \dfrac{\text{mass of }^{206}Pb}{\text{mass of }^{238}U}}$

(a) $\dfrac{N}{N_0} = \dfrac{1}{1 + 1.00} = \dfrac{1}{2.00}$, therefore age $= t_{1/2} = 4.5\ \text{Ga}$

(b) $\dfrac{N}{N_0} = \dfrac{1}{1 + \dfrac{1}{1.25}} = 0.556$

$t\,(= \text{age}) = -\dfrac{\ln(0.556)}{\left(\dfrac{\ln 2}{4.5\ \text{Ga}}\right)} = 3.8\ \text{Ga}$

10.9 (a) $^{98}_{42}\text{Mo} + ^{1}_{0}\text{n} \longrightarrow ^{99}_{42}\text{Mo} \rightarrow ^{99m}_{43}\text{Tc} + ^{0}_{-1}\text{e}$

(b) Tc-99m ($N/Z = 1.30$ vs 1.36 for Mo-99)

10.11 Radon-222 decays to polonium-218 by alpha emission with a half-life of 3.824 days.

$$^{222}_{86}\text{Rn} \longrightarrow ^{218}_{84}\text{Po} + ^{4}_{2}\alpha \qquad\qquad t_{1/2} = 3.824 \text{ d}$$

An alpha particle is the nucleus of ^{4}He. Assuming that (1) the alpha particles are captured inside the container, (2) they behave as an ideal gas and (3) the temperature is constant at 298 K, we can find the volume of the container by calculating the number of moles of $^{4}_{2}\alpha$ formed in 23 days and then applying the ideal gas law.

$$n_{He} = n_{Po} = n_{initial\ Rn} - n_{final\ Rn} \qquad n_{final\ Rn} \propto N_{15\ days}$$

$$k = \frac{\ln 2}{t_{1/2}} = \frac{\ln 2}{3.824 \text{ d}} = 0.181 \text{ d}^{-1}$$

$$N_{23\ days} = N_0 e^{-kt} = (2.5 \text{ g}) e^{-(0.181\ \text{d}^{-1})(23\ \text{d})} = 0.039 \text{ g}$$

$$n_{final\ Rn} = (0.039 \text{ g }^{222}\text{Rn})\left(\frac{1 \text{ mol }^{222}\text{Rn}}{222.0175 \text{ g }^{222}\text{Rn}}\right)$$

$$= 1.8 \times 10^{-4} \text{ mol }^{222}\text{Rn}$$

$$n_{initial\ Rn} = (2.5 \text{ g }^{222}\text{Rn})\left(\frac{1 \text{ mol }^{222}\text{Rn}}{222.0175 \text{ g }^{222}\text{Rn}}\right)$$

$$= 1.13 \times 10^{-2} \text{ mol }^{222}\text{Rn}$$

$$n_{He} = n_{initial\ Rn} - n_{final\ Rn} = 1.13 \times 10^{-2} \text{ mol} - 1.8 \times 10^{-4} \text{ mol}$$

$$= 1.11 \times 10^{-2} \text{ mol}$$

$$PV = nRT$$

$$V = \frac{nRT}{P} = \frac{(1.11 \times 10^{-2} \text{mol})(0.08206 \text{ L} \cdot \text{atm} \cdot \text{K}^{-1} \cdot \text{mol}^{-1})(298 \text{ K})}{1.00 \text{ atm}}$$

$$= 0.27 \text{ L}$$

10.13 The average time to reach 8000 disintegrations or counts for the sample is 22.25 min; therefore the average disintegrations or counts per minute for the sample is 359.6 dpm.

Average time to reach 500 background disintegrations or counts = 5.11 min; therefore the average disintegrations or counts per minute for the background is 97.8 dpm.

(a) Average level of radioactivity in the sample = 359.6 dpm − 97.8 dpm = 261.8 dpm = 262 dpm (3 sf).

(b) $(262 \text{ dpm})\left(\dfrac{1 \text{ dps}}{60 \text{ dpm}}\right)\left(\dfrac{1 \text{ Ci}}{3.7 \times 10^{10} \text{ dps}}\right) = 1.18 \times 10^{-10} \text{ Ci}$

$1.18 \times 10^{-10} \text{ Ci} \times \left(\dfrac{1 \,\mu\text{Ci}}{1 \times 10^{-6} \text{ Ci}}\right) = 1.18 \times 10^{-4} \,\mu\text{Ci}$

10.15 (a) Radioactive substances that emit γ radiation are most effective for diagnosis because they are the least destructive of the types of radiation listed. Additionally, γ rays pass easily through body tissues and can be counted, whereas α and β particles are stopped by the body tissues.

(b) α particles tend to be best for this application because they cause the most destruction.

(c) and (d) ^{131}I, 8d (used to image the thyroid); ^{67}Ga, 78 h (used most often as the citrate complex); ^{99m}Tc, 6 h (used for various body tissues by varying the ligands attached to the Tc atom).

10.17 (a) activity $= (17.3 \text{ Ci})\left(\dfrac{3.7 \times 10^{10} \text{ Bq}}{1 \text{ Ci}}\right)$

$= 6.4 \times 10^{11} \text{ Bq} = 6.4 \times 10^{11} \text{ nuclei} \cdot \text{s}^{-1}$

$N = (2.0 \times 10^{-3} \text{ g})\left(\dfrac{1 \, m_u}{1.661 \times 10^{-24} \text{ g}}\right)\left(\dfrac{1 \text{ nucleus}}{24 \, m_u}\right) = 5.0 \times 10^{19} \text{ nuclei}$

$k = \dfrac{\text{activity}}{N} = \dfrac{6.4 \times 10^{11} \text{ nuclei} \cdot \text{s}^{-1}}{5.0 \times 10^{19} \text{ nuclei}} = 1.3 \times 10^{-8} \text{ s}^{-1} = 1.1 \times 10^{-3} \text{ d}^{-1}$

$t_{1/2} = \dfrac{\ln 2}{k} = \dfrac{\ln 2}{1.3 \times 10^{-8} \text{ s}^{-1}} = 5.3 \times 10^7 \text{ s} = 1.5 \times 10^4 \text{ h} = 617 \text{ d}$

(b) $m = m_0 e^{-kt} = 2.0 \text{ mg} \times e^{-(1.11 \times 10^{-3} \text{ d}^{-1})(2.0 \text{ d})} = 2.0 \text{ mg}$

10.19 Nuclei that are positron emitters are proton-rich and lie below the band of stability.

(a) O stable at 16, so ^{18}O is neutron rich, will emit an electron rather than a positron; not good for PET. $^{18}_{8}O \longrightarrow {}^{18}_{9}F + {}^{0}_{-1}e$

(b) N stable at 14, so ^{13}N is proton-rich, will emit a positron; good for PET. $^{13}_{7}N \longrightarrow {}^{13}_{6}C + {}^{0}_{1}e$

(c) C stable at 12, so ^{11}C is proton rich, will emit a positron; good for PET. $^{11}_{6}C \longrightarrow {}^{11}_{5}B + {}^{0}_{1}e$

(d) F stable at 19, so ^{20}F is neutron rich, will emit an electron rather than a positron; not good for PET. $^{20}_{9}F \longrightarrow {}^{20}_{10}Ne + {}^{0}_{-1}e$

(e) O stable at 16, so ^{15}O is proton-rich, will emit a positron; good for PET. $^{15}_{8}O \longrightarrow {}^{15}_{7}N + {}^{0}_{1}e$

10.21 To determine the effective half-life we need to determine the effective rate constant, k_E. This constant is equal to the sum of the biological rate constant (k_B) and the radioactive decay rate constant (k_R), both of which can be obtained from the respective half-lives:

$$k_E = k_B + k_R = \frac{\ln 2}{65 \text{ d}} + \frac{\ln 2}{12.8 \text{ d}} = 6.5 \times 10^{-2} \text{ d}^{-1}$$

$$t_{1/2} \text{ (effective)} = \frac{\ln 2}{6.5 \times 10^{-2} \text{ d}^{-1}} = 11 \text{ d}$$

FOCUS 11
ORGANIC CHEMISTRY

11A.1 (a) alkyne

(b) alkane

(c) alkene

(d) alkene and alkyne

(e) alkene

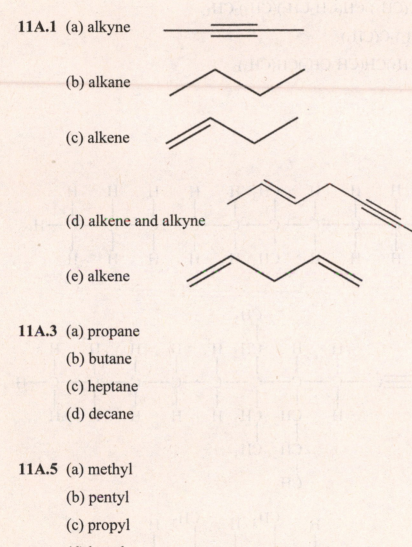

11A.3 (a) propane
(b) butane
(c) heptane
(d) decane

11A.5 (a) methyl
(b) pentyl
(c) propyl
(d) hexyl

11A.7 (a) propane
(b) ethane
(c) pentane
(d) 2,3-dimethylbutane

11A.9 (a) 4-methyl-2-pentene

(b) 2,2,3-trimethylpentane

11A.11(a) $CH_2=CHCH(CH_3)CH_2CH_3$

(b) $CH_3CH_2C(CH_3)_2CH(CH_2CH_3)(CH_2)_2CH_3$

(c) $HC\equiv C(CH_2)_2C(CH_3)_3$

(d) $CH_3CH(CH_3)CH(CH_2CH_3)CH(CH_3)_2$

11A.13

(a)

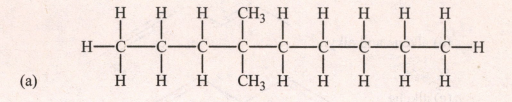

(b)

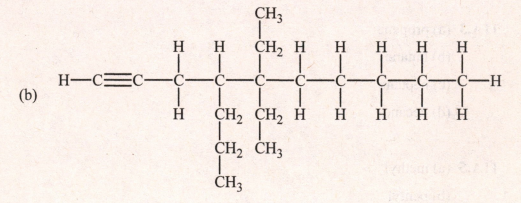

(c)

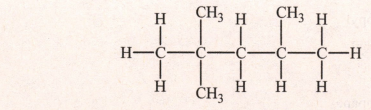

(d)

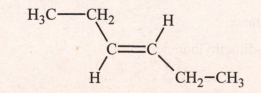

11A.15(a)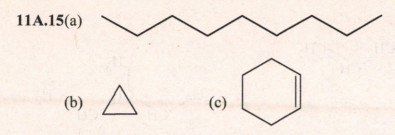

(b) 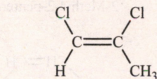   (c)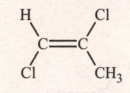

11A.17(a) four σ-type single bonds

(b) two σ-type single bonds and one double bond with a σ- and a π-bond

(c) one σ-type single bond and one triple bond with a σ-bond and two π-bonds

11A.19

cis-1,2-Dichloropropene   *trans*-1,2-Dichloropropene

cis-1,2-Dichloropropene is polar, although *trans*-1,2-Dichloropropene is slightly polar also.

11A.21(a) hexenes:

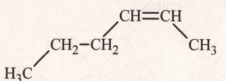

1-Hexene   cis-2-Hexene

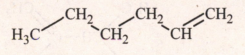

trans-2-Hexene   *cis*-3-Hexene

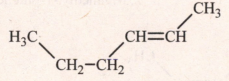

trans-3-Hexene

pentenes:

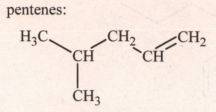
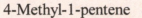
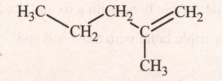
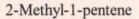

4-Methyl-1-pentene

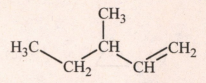
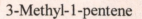
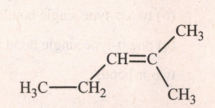

3-Methyl-1-pentene

2-Methyl-1-pentene

2-Methyl-2-pentene

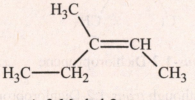
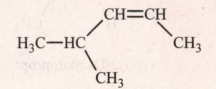

cis-3-Methyl-2-pentene

(+ trans isomer)

cis-4-Methyl-2-pentene

(+ trans isomer)

butenes:

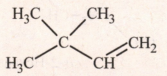
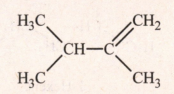

3,3-Dimethyl-1-butene

2,3-Dimethyl-1-butene

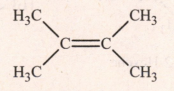
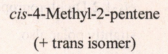

2,3-Dimethyl-2-butene

(b) cyclic molecules:

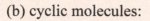

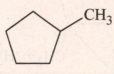

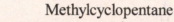

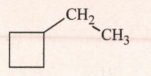

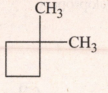

Cyclohexane

Methylcyclopentane

ethylcyclobutane

1,1-dimethylcyclobutane

The following structures are drawn to emphasize the stereochemistry

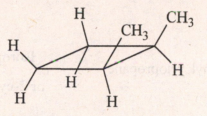

cis-1,2-Dimethylcyclobutane

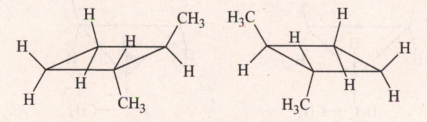

trans-1,2-Dimethylcyclobutane

(nonsuperimposable mirror images

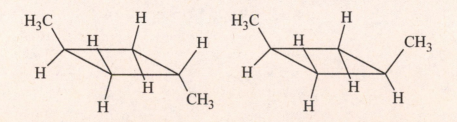

trans-1,3-Dimethylcyclobutane   *cis*-1,3-Dimethylcyclobutane

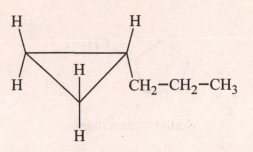

Propylcyclopropane

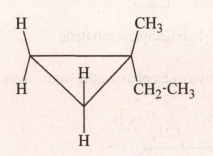

1-Ethyl-1-methylcyclopropane

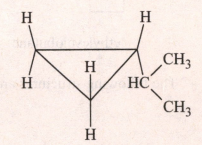

Isoropylcyclopropane

or 2-cyclopropylpropane

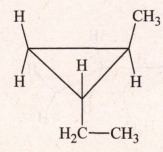 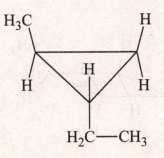

trans-1-Ethyl-2-methylcyclopropane

(nonsuperimposable mirror images)

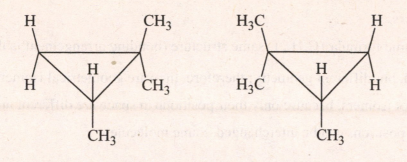

1,1,2-Trimethylcyclopropane

(nonsuperimposable mirror images)

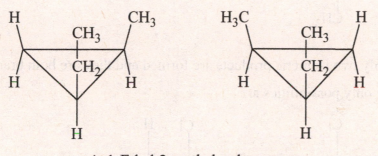

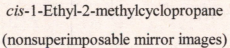

cis-1-Ethyl-2-methylcyclopropane

(nonsuperimposable mirror images)

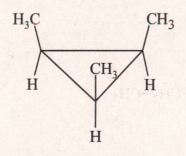

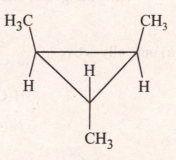

1,2,3-Trimethylcyclopropane   1,2,3-Trimethylcyclopropane

(all cis isomer)   (cis-trans isomer)

11A.23(a) Butane is C_4H_{10}, cyclobutane is C_4H_8. Because they have different

formulas, they are not isomers.

(b) Same formula, but different structures; therefore, they are structural isomers.

(c) Same formula (C_5H_{10}), same structure (bonding arrangement is the same), but different geometry; therefore, they are geometrical isomers.

(d) Not isomers, because only their positions in space are different and these positions can be interchanged. Same molecule.

11A.25(a)

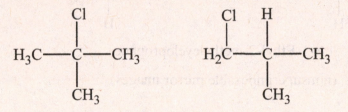

(b) If only two isomeric products are formed and they are both branched, then the only possibilities are

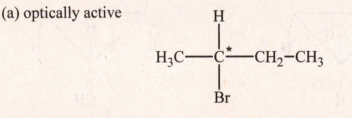

11A.27   An * designates a chiral carbon.

(a) optically active

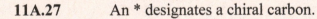

(b) not optically active

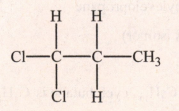

(c) optically active

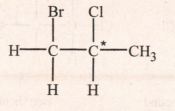

(d) optically active

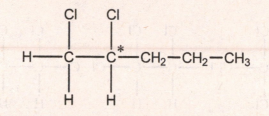

11B.1 The balanced equations are

$$C_3H_8(g) + 5\,O_2(g) \longrightarrow 3\,CO_2(g) + 4\,H_2O(l)$$

$$C_4H_{10}(g) + 13/2\,O_2(g) \longrightarrow 4\,CO_2(g) + 5\,H_2O(l)$$

$$C_5H_{12}(g) + 8\,O_2(g) \longrightarrow 5\,CO_2(g) + 6\,H_2O(l)$$

The enthalpies of combustion that correspond to these reactions are listed in Appendix 2:

Compound	(a) Enthalpy of combustion $kJ \cdot mol^{-1}$	(b) Heat released per g $kJ \cdot g^{-1}$
Propane	− 2220	50.3
Butane	− 2878	49.5
Pentane	− 3537	49.0

The molar enthalpy of combustion increases with molar mass as might be expected, because the number of moles of CO_2 and H_2O formed will increase as the number of carbon and hydrogen atoms in the compounds increases. The heat released per gram of these hydrocarbons is essentially the same because the H to C ratio is similar in the three hydrocarbons.

11B.3 There are nine possible products:

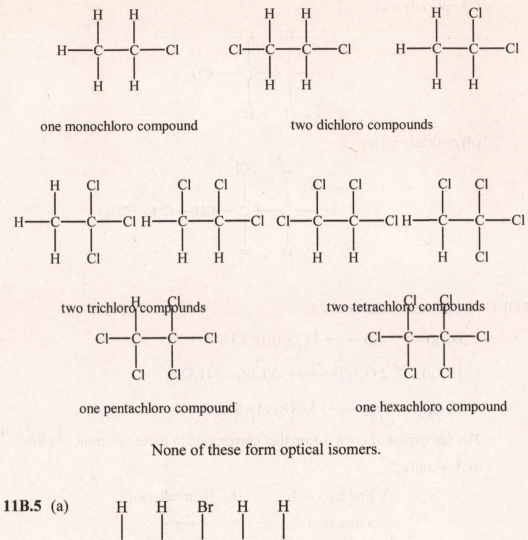

one monochloro compound   two dichloro compounds

two trichloro compounds   two tetrachloro compounds

one pentachloro compound   one hexachloro compound

None of these form optical isomers.

11B.5 (a)

3-Bromopentane

2-Bromopentane

(b) addition reaction

11B.7 (a) $C_6H_{11}Br + NaOCH_2CH_3 \rightarrow C_6H_{10} + NaBr + HOCH_2CH_3$

(b)

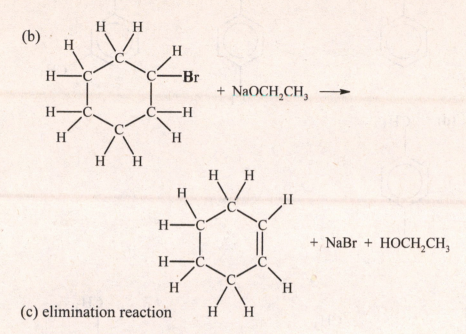

(c) elimination reaction

11B.9 $C_2H_4 + X_2 \longrightarrow C_2H_4X_2$

We will break one X—X bond and form two C—X bonds.

Using bond enthalpies:

Halogen	Cl	Br	I
X—X bond breakage (kJ·mol^{-1})	+242	+193	+151
C—X bond formation (kJ·mol^{-1})	−2(338)	−2(276)	−2(238)
Total (kJ·mol^{-1})	−434	−359	−325

The reaction is less exothermic as the halogen becomes heavier. In general, the reactivity, and also the danger associated with use of the halogens in reactions, decreases as one descends the periodic table.

11C.1 (a) 1-ethyl-3-methylbenzene

(b) pentamethylbenzene (1,2,3,4,5-pentamethylbenzene is also correct, but, because there is only one possible pentamethylbenzene, the use of the numbers is not necessary)

11C.3 (a)

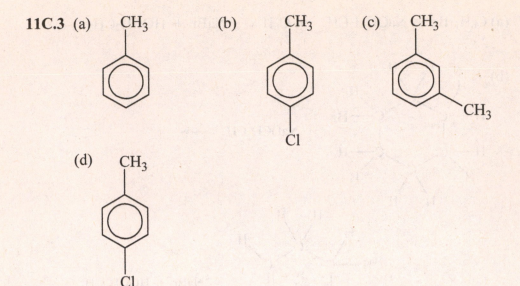

(b)

(c)

(d)

11C.5

1,3-Dichloro-2-methylbenzene   1,3-Dichloro-5-methylbenzene

1,3-Dichloro-4-methylbenzene   1,4-Dichloro-2-methylbenzene

1,2-Dichloro-3-methylbenzene   1,2-Dichloro-4-methylbenzene

(b) All of these molecules are at least slightly polar.

11C.7

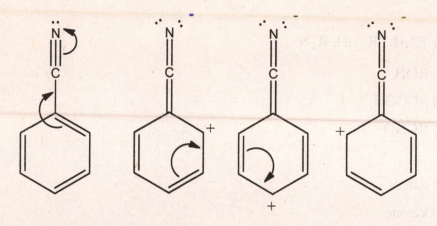

Electrophiles tend to avoid the ortho and para positions that develop slight + charges in the resonance forms.

11C.9 Two compounds can be produced. Resonance makes positions 1, 4, 6, and 9 equivalent. It also makes positions 2, 3, 7, and 8 equivalent. Positions 5 and 10 are equivalent but have no H atom.

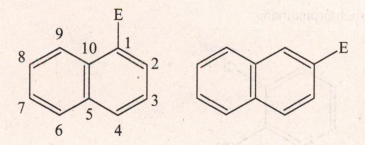

11C.11(a) C_9H_8, aromatic hydrocarbon

(b) C_6H_{14}, alkane

(c) C_7H_{10}, alkene

(d) C_6H_{12}, alkane

11C.13 (a) $C_{11}H_{24}$, alkane

(b) C_9H_{12}, alkene

(c) C_8H_{16}, alkane

(d) $C_{16}H_{12}$, alkene

11D.1 (a) RNH_2, R_2NH, R_3N

(b) ROH

(c) RCOOH

(d) RCHO

11D.3 (a) ether

(b) ketone

(c) amine

(d) ester

11D.5 (a) 2-chloropropane

(b) 2,4-dichloro-4-methylhexane

(c) 1,1,1,-triiodoethane

(d) dichloromethane

11D.7 (a) 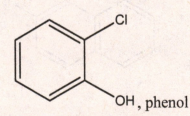, phenol

(b) $CH_3CH(CH_3)CH(OH)CH_2CH_3$, secondary alcohol

(c) $CH_3CH_2CH(CH_3)CH_2CH(CH_3)CH_2OH$, primary alcohol

(d) $CH_3C(CH_3)(OH)CH_2CH_3$, tertiary alcohol

11D.9 (a) $CH_3OCH_2CH_2CH_3$

(b) $CH_3CH_2CH_2CH_2OCH_2CH_3$

(c) CH₃CH₂CH₂OCH₂CH₂CH₃

11D.11(a) butyl propyl ether

(b) methyl phenyl ether

(c) pentyl propyl ether

11D.13(a) aldehyde, ethanol

(b) ketone, propanone

(c) ketone, 3-pentanone

11D.15(a)

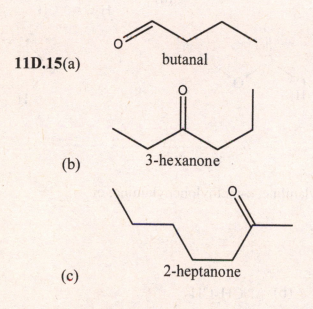

butanal

(b)

3-hexanone

(c)

2-heptanone

11D.17(a) ethanoic acid

(b) butanoic acid

(c) 2-aminoethanoic acid

11D.19 (a)

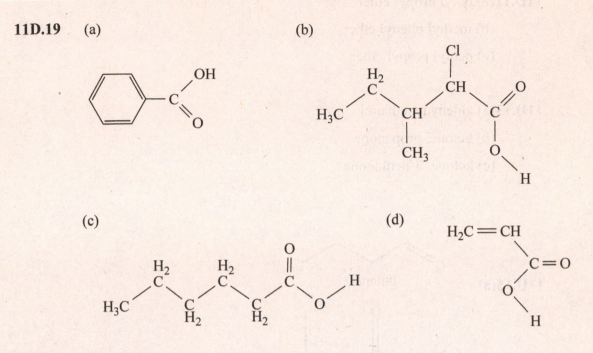

(b)

(c)

(d)

11D.21(a) methylamine

(b) diethylamine

(c) *o*-methylaniline, 2-methylaniline, *o*-methylphenylamine, or

1-amino-2-methylbenzene

11D.23

(a)

(b)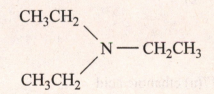

(c)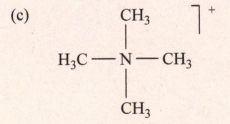

11D.25 Only (a) and (c) may function as nucleophiles, because they have lone pairs of electrons that will be attracted to a positively charged carbon center. Even though the oxygen atoms in CO_2 have lone pairs they are not readily available to act as nucleophiles. SiH_4 has no lone pairs and therefore cannot act as a nucleophile.

11D.27 (a) ethanol

Use an oxidant that is not too strong to avoid overoxidation and the formation of carboxylic acids. For example, the salt pyridinium chlorochromate (PCC), $C_5H_5NH[CrO_3Cl]$.

(b) 2-octanol

An oxidizing agent such as acidified sodium dichromate, $Na_2Cr_2O_7$.

The salt pyridinium chlorochromate (PCC), $C_5H_5NH[CrO_3Cl]$, is also suitable.

(c) 5-methyl-1-octanol

Use an oxidant that is not too strong to avoid overoxidation and the formation of carboxylic acids. For example, the salt pyridinium chlorochromate (PCC), $C_5H_5NH[CrO_3Cl]$.

11D.29

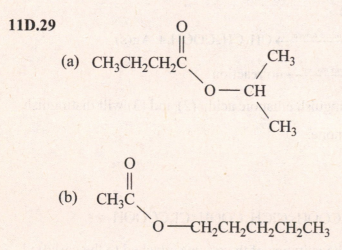

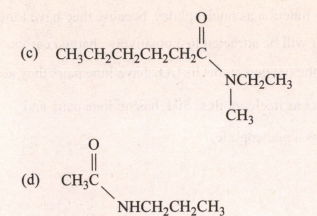

(c) $CH_3CH_2CH_2CH_2CH_2C$ with $=O$ and NCH_2CH_3, CH_3

(d) CH_3C with $=O$ and $NHCH_2CH_2CH_3$

11D.31 (a) addition

(b) condensation

(c) addition

(d) addition

(e) condensation

11D.33 The following procedures can be used:

(1) Dissolve the compounds in water and use an acid-base indicator to look for a color change.

(2) $CH_3CH_2CHO \xrightarrow{\text{Tollens reagent}} CH_3CH_2COOH + Ag(s)$

(3) $CH_3COCH_3 \xrightarrow{\text{Tollens reagent}}$ no reaction

Procedure (1) will distinguish ethanoic acid; (2) and (3) will distinguish propanal from 2-propanone.

11D.35 $CH_3CH_2COOH < CH_3COOH < ClCH_2COOH < Cl_3CCOOH$

The greater the electronegativities of the groups attached to the carboxyl group, the stronger the acid. Propanoic acid is less acidic than acetic acid because alkyl groups are more electron donating.

11E.1 (a) $-CH_2-C(CH_3)_2-CH_2-C(CH_3)_2-CH_2-C(CH_3)_2-$

(b)

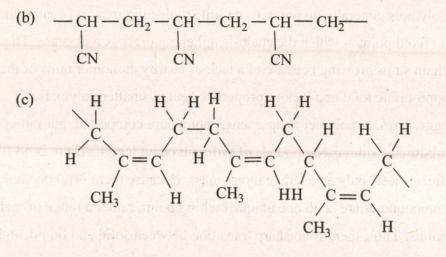

(c)

cis version

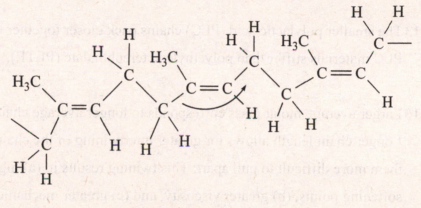

trans version

11E.3 $HOOCC_6H_{12}COOH$

$NH_2C_6H_4NH_2$

11E.5 (a) $CHCl=CH_2$

(b) $CFCl=CF_2$

11E.7 (a) $-OCCONH(CH_2)_4NHCOCONH(CH_2)_4NH-$

(b) $-OC-CH(CH_3)-NH-OC-CH(CH_3)-NH-$

11E.9 block copolymer

11E.11 Polymers generally do not have definite molecular masses because there is no fixed point at which the chain-lengthening process will cease. The chain stops growing because of a lack of nearby monomer units of the appropriate kind or a lack of properly oriented smaller polymeric aggregates. A polymer is, in a sense, not a pure compound, bur rather a mixture of similar compounds of different chain length. There is no fixed molar mass, only an average molar mass. Because there is no one unique compound, there is no one unique melting point, rather a range of melting points. Thus, there is no sharp transition between solid and liquid, and we say the solid softens rather than melts.

11E.13 The smaller polylactic acid (PLC) chains pack closer together making PLC materials stiffer than polyethylene terephthalate (PETE).

11E.15 Larger average molar mass corresponds to longer average chain length. Longer chain length allows for greater intertwining of the chains, making them more difficult to pull apart. This twining results in (a) higher softening points, (b) greater viscosity, and (c) greater mechanical strength.

11E.17 Highly linear, unbranched chains allow for maximum interaction between chains. The greater the intermolecular contact between chains, the stronger the forces between them, and the greater the strength of the material.

11E.19 (a)

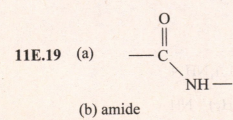

(b) amide

(c) condensation

11E.21 Side groups that contain hydroxyl, carbonyl, amino, and sulfide groups are all potentially capable of participating in hydrogen bonding that could contribute to the tertiary structure of the protein. Thus, serine, threonine, tyrosine, aspartic acid, glutamic acid, lysine, arginine, histidine, asparagine, and glutamine satisfy the criteria. Proline and tryptophan generally do not contribute through hydrogen bonding, because they are typically found in hydrophobic regions of proteins.

11E.23

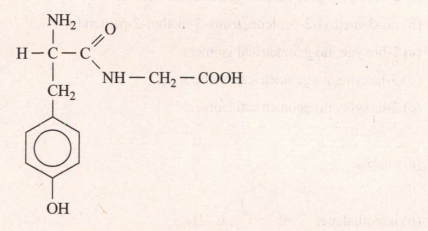

11E.25 (a) The functional groups are alcohols and aldehydes.

(b) The chiral carbon atoms are marked with asterisks.

$$OHC - \overset{\overset{\displaystyle H}{|}}{\underset{\underset{\displaystyle OH}{|}}{C}} *- \overset{\overset{\displaystyle H}{|}}{\underset{\underset{\displaystyle OH}{|}}{C}} *- \overset{\overset{\displaystyle OH}{|}}{\underset{\underset{\displaystyle H}{|}}{C}} *- \overset{\overset{\displaystyle OH}{|}}{\underset{\underset{\displaystyle H}{|}}{C}} *- CH_2OH$$

11E.27 (a) GTACTCAAT

(b) ACTTAACGT

11.1 The difference can be traced to the weaker London forces that exist in branched molecules. Atoms in neighboring branched molecules cannot lie as close together as they can in unbranched isomers.

11.3 (a) substitution, $CH_4 + Cl_2 \rightarrow CH_3Cl + HCl$

(b) addition $CH_2=CH_2 + Br_2 \rightarrow CH_2Br-CH_2Br$

11.5 Water is not used as the nonpolar reactants will not readily dissolve in a highly polar solvent like water. Also, the ethoxide ion reacts with water.

11.7 (a) 2-methyl-1-propene, no geometrical isomers

(b) *cis*-3-methyl-2-pentene, *trans*-3-methyl-2-pentene

(c) 1-hexyne, no geometrical isomers

(d) 3-hexyne, no geometrical isomers

(e) 2-hexyne, no geometrical isomers

11.9 (a) $C_{10}H_{18}$

(b) naphthalene, 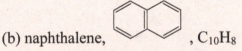 , $C_{10}H_8$

(c) Yes. Cis and trans forms (relative to the C-C bond common to the two six-membered rings) are possible.

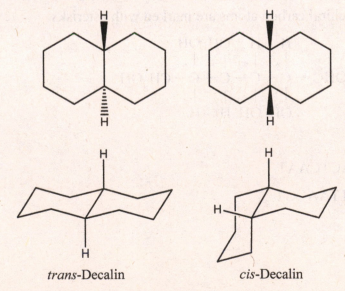

trans-Decalin   *cis*-Decalin

11.11 (a)

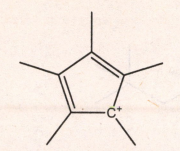

(b) five resonance structures (positive charge on any one of the five carbon atoms)

(c) four π-electrons

11.13 (a) 4-methyl-3-propylheptane

The longest chain has eight carbon atoms in it. The systematic name of the compound is 4-ethyl-5-methyloctane.

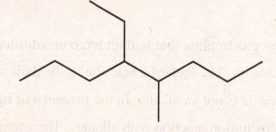

(b) 4,6-dimethyloctane

The compound name is almost correct, but the numbering scheme with the lowest numbers would be 3,5-dimethyloctane.

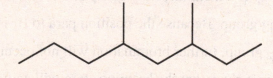

(c) 2,2-dimethyl-4-propylhexane

The longest carbon chain in the molecule is seven carbon atoms long. The systematic name is 2,2-dimethyl-4-propylheptane.

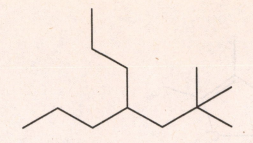

(d) 2,2-dimethyl-3-ethylhexane.

The name is essentially correct except that ethyl should be listed first The systematic name is 3-ethyl-2,2-dimethylhexane.

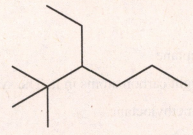

11.15 Bromine is an electrophile that will undergo an addition reaction with alkenes in the dark. The lack of a reaction in the dark with Br_2 indicates that the molecule is not an alkene. In the presence of light, bromine will undergo a substitution reaction with alkanes. Therefore, the molecule is most likely an alkane and the only alkane with the molecular formula C_3H_6 is cyclopropane.

11.17 The NO_2 group is a meta-directing group and the Br atom is an ortho, para-directing group. Because the position para to Br is already substituted with the NO_2 group, further bromination will not occur there. The resonance forms show that the bromine atom will activate the position ortho to it as expected. The NO_2 group will deactivate the group ortho to itself, thus in essence enhancing the reactivity of the position meta to the NO_2 group. This position is ortho to the Br atom, so the effects of the Br and NO_2 groups reinforce each other. Bromination is thus expected to occur as shown:

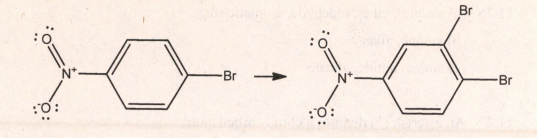

11.19 (a) and (b)

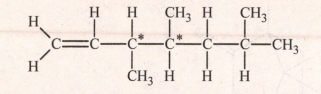

(c) No, there are no cis/trans isomers for this molecule.

11.21 C_8H_{10} will have an absorption maximum at a longer wavelength. Molecular orbital theory predicts that in conjugated hydrocarbons (molecules that contain a chain of carbon atoms with alternating single and double bonds) electrons become delocalized and are free to move up and down the chain of carbon atoms. Such electrons may be described using the one-dimensional "particle in a box" model. According to this model, as the box to which electrons are confined lengthens, the quantized energy states available to the electrons get closer together. As a result, the energy needed to excite an electron from the ground state to the next higher state is lower for electrons confined to longer boxes. Therefore, lower energy photons, that is, photons with longer wavelengths, will be absorbed by the C_8H_{10} molecule because it provides a longer "box" than C_6H_8.

11.23 (a) $C_5H_5N_5O$

(b) $C_6H_{12}O_6$

(c) $C_3H_7NO_2$

11.25 (a) alcohol, ether, aldehyde, aromatic ring

(b) ketone, alkene

(c) amine, amide, alkene

11.27 An asterisk (*) denotes a chiral carbon atom.

(a)

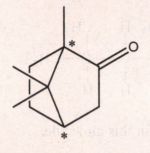

(b)

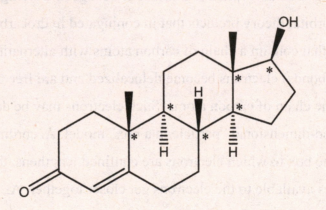

11.29

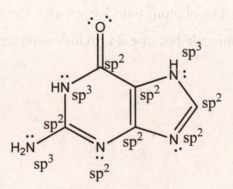

11.31 (a)

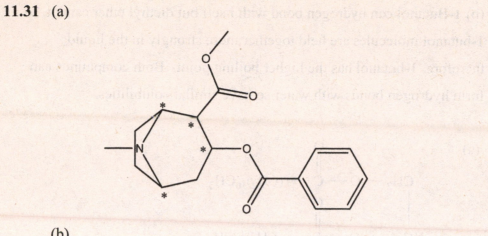

(b)

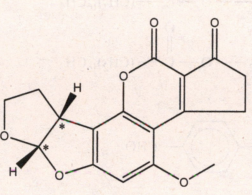

11.33 (a)

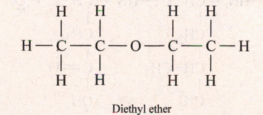

Diethyl ether

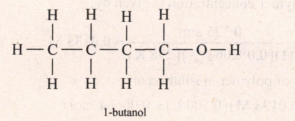

1-butanol

(b) 1-Butanol can hydrogen bond with itself but diethyl ether cannot, so 1-butanol molecules are held together more strongly in the liquid; therefore, 1-butanol has the higher boiling point. Both compounds can form hydrogen bonds with water so have similar solubilities.

11.35 (a)

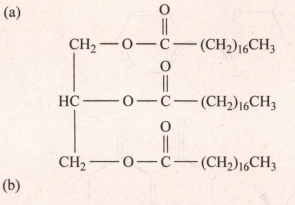

(b)

11.37

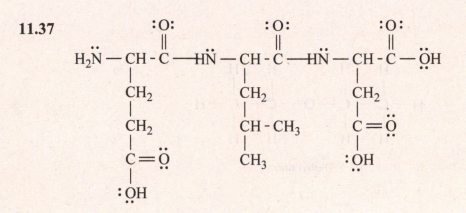

11.39 (a) The polymer concentration is given by:

$$c = \frac{\Pi}{iRT} = \frac{0.325 \text{ atm}}{(1)(0.08206 \frac{\text{L atm}}{\text{K mol}})(298 \text{ K})} = 0.0133 \text{ M}$$

The moles of polymer in solution are:

$$n_{\text{polymer}} = (0.0133 \text{ M})(0.500 \text{ L}) = 0.00664 \text{ mol}$$

and the molar mass of the polymer is:

$$\text{M.M.} = \frac{47.7 \text{ g}}{0.00664 \text{ mol}} = 7180 \frac{\text{g}}{\text{mol}}$$

(b) If a monomer in the polymer is $-CH_2CH(CN)-$, then the molar mass of a monomer is 53.06 g/mol and the average number of monomers in a polymer is: $(7180 \text{g/mol})/(53.06 \text{ g/mol}) = 135$ monomers .

(c) The pressure of $H_2O(g)$ above the mixture is given by:

$$P_{H_2O} = X_{H_2O} \cdot P_{H_2O, \text{ pure}}$$

in 100 mL, $n_{H_2O} = \dfrac{(100 \text{ mL})\left(1.00 \frac{\text{g}}{\text{mL}}\right)}{18.02 \ \frac{\text{g}}{\text{mol}}} = 5.55$ mol

and, $n_{\text{polymer}} = (0.100 \text{ L})(0.01328 \frac{\text{mol}}{\text{L}}) = 0.00133$ mol.

Therefore,

$$X_{H_2O} = \frac{5.55 \text{ mol}}{5.55 \text{ mol} + 0.00133 \text{ mol}} = 0.9998 \quad \text{and}$$

$$P_{H_2O} = (0.9998)(23.76 \text{ Torr}) = 23.75 \text{ Torr}$$

(d) Measuring the change in osmotic pressure proves to be a better method in this case. The osmotic pressure developed by the resulting polymer solution is readily measured while the change in partial pressure of $H_2O(g)$ changes by less than 0.1 % upon addition of the polymer.

11.41 (a) Primary structure is the sequence of amino acids along a protein chain. Secondary structure is the conformation of the protein, or the manner in which the chain is coiled or layered, as a result of interactions between amide and carboxyl groups. Tertiary structure is the shape into which sections of the proteins twist and intertwine, as a result of interactions between side groups of the amino acids in the protein. If the protein consists of several polypeptide units, then the manner in which the units stick together is the quaternary structure.

(b) The primary structure is held together by covalent bonds. Intermolecular forces provide the major stabilizing force of the secondary structure. The tertiary structure is maintained by a combination of London forces, hydrogen bonding, and sometimes ion-ion interactions. The same forces are responsible for the quaternary structure.

11.43 (a) $^+H_3NCH_2COOH(aq) + H_2O(l) \rightarrow {}^+H_3NCH_2COO^-(aq) + H_3O^+(aq)$

$^+H_3NCH_2COO^-(aq) + H_2O(l) \rightarrow H_2NCH_2COO^-(aq) + H_3O^+(aq)$

(b) $pK_{a1} = 2.35$ $pK_{a2} = 9.78$

$pH = 2$, $^+H_3NCH_2COOH$

$pH = 5$, $^+H_3NCH_2COO^-$

$pH = 12$, $H_2NCH_2COO^-$

11.45 Condensation polymerization involves the loss of a small molecule, often water or HCl, when monomers are combined. Dacron is more linear than the polymer obtained from benzene-1,2-dicarboxylic acid and ethylene glycol, so Dacron can be more readily spun into yarn.

11.47 (a) $CH_3CH_2CH_3(g) + 5O_2(g) \rightarrow 3CO_2(g) + 4H_2O(g)$

$CHCH + 2\frac{1}{2}O_2(g) \rightarrow 2CO_2(g) + H_2O(g)$

(b) propane, 44.10 g.mol^{-1}

$\Delta H_c° = \Sigma\Delta H_f°(\text{products}) - \Sigma\Delta H_f°(\text{reactants})$

$= [3(-393.51) + 4(-241.82)] - [-103.85]$

$= -1180.53$ kJ.mol^{-1} $- 967.28$ kJ.mol^{-1} $+ 103.85$ kJ.mol^{-1}

$= -2043.96$ kJ.mol^{-1}

$\dfrac{-2043.96 \text{ kJ.mol}^{-1}}{44.10 \text{ g.mol}^{-1}} = -46.35$ kJ.g^{-1}

ethyne, 26.04 g.mol^{-1}

$\Delta H_c° = [2(-393.51) + (-241.81)] - [+226.73]$

$= -787.02$ kJ.mol^{-1} $- 241.82$ kJ.mol^{-1} $- 226.73$ kJ.mol^{-1}

$= -1255.57$ kJ.mol^{-1}

$\dfrac{-1255.57 \text{ kJ.mol}^{-1}}{26.04 \text{ g.mol}^{-1}} = -48.22$ kJ.g^{-1}

(c) More heat is released per gram of ethyne resulting in a hotter flame.

11.49 (a)

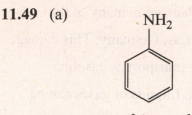

(b)   (i) sp^3 (ii) sp^3

(c) Each N atom carries one lone pair of electrons.

(d) Yes, the N atoms help to carry the current because the unhybridized *p*-orbital on each N atom is part of the extended π conjugation (delocalized π-bonds) that allows electrons to move freely along the polymer.

11.51

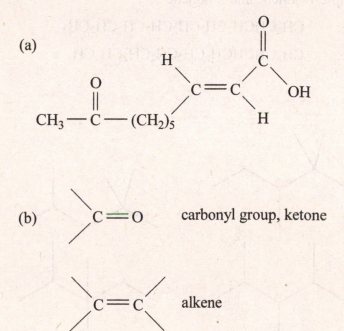

(b) $C\!=\!O$ carbonyl group, ketone

        $C\!=\!C$ alkene

        carboxylic acid

11.53 Coal is not a pure substance and, as a result, does not burn cleanly. Some types of coal produce considerable amounts of sulfur and nitrogen oxides, which contribute to air pollution. The burning of high-sulfur coal contributed very much to the environmental damage in many of the eastern European nations, such as the former East Germany. This damage persists to this day. Coal is also not as easy to transport as gasoline because it is a solid rather than a liquid or gas. Liquids or gases can be placed in fuel tanks and pipelines.

11.55 Since each compound has the same number of carbon atoms n must be nine.

C_9H_{20} $CH_3CH_2CH_2CH_2CH_2CH_2CH_2CH_2CH_3$ nonane

C_9H_{18} Straight chain with no branches and one double bond

Double bond could be located at any position on the chain

For example 1-nonene and 2-nonene

1-nonene: $CH_2CHCH_2CH_2CH_2CH_2CH_2CH_2CH_3$

2-nonene: $CH_3CHCHCH_2CH_2CH_2CH_2CH_2CH_3$

11.57 (a) (b)

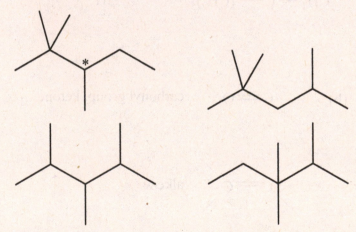

11.59 (a) Elimination

(b) $CH_3CH_2CHBrCH_3$ is 2-bromobutane; $CH_3CH_2O^-$ is ethoxide;

$CH_3CH=CHCH_3$ is 2-butene; CH_3CH_2OH is ethanol; Br^- is bromide

(c) Neither, it is a base.

(d) We will assume that 2-butene is the only desired product.

(Although the production of ethanol is useful but it would have to be separated from the other products.)

Mass of desired product is: 56.11 g.mol^{-1}

Mass of all reactants consumed:

137.2 g.mol^{-1} + 45.06 g.mol^{-1} = 182.26 g.mol^{-1}

atom economy = (mass of desired product obtained)/(mass of all reactants consumed) × 100% = 30.8 %

(e) Mass of desired product is: 56.11 g.mol^{-1}

Mass of all reactants consumed:

137.2 g.mol^{-1} + 31.03 g.mol^{-1} = 168.23 g.mol^{-1}

atom economy = (mass of desired product obtained)/(mass of all reactants consumed) × 100% = 33.4 %

(f) Mass of desired product is: 56.11 g.mol^{-1}

Mass of all reactants consumed:

137.2 g.mol^{-1} + 47.1 g.mol^{-1} = 184.3 g.mol^{-1}

atom economy = (mass of desired product obtained)/(mass of all reactants consumed) × 100% = 30.4 %

(g) The lowest mass of waste is synthesis (e).

The highest mass of waste is synthesis (f).

(h) 50.0 g of 2-bromobutane, at 100% yield should produce 0.364 moles of 2-butene (which is 20.4 g of 2-butene). The respective experimental yield for the three reactions are 79.4, 75.5, and 64.2%.

The respective experimental atom economy for the three reactions are 24.4, 25.1, and 19.5%.

(i) Based on the highest mass of 2-butene produced (and the highest percentage yield) synthesis (a) looks the best. However, based on the experimental atom economy, synthesis (e) looks the best.

MAJOR TECHNIQUES

MT1.1 The C—H bond, because the effective mass is lower and the C—H bond is stiffer than the C—Cl bond.

MT1.3

$$\tilde{\nu} = \frac{9.75 \times 10^8 \cdot s^{-1}}{3.00 \times 10^{10}\ cm \cdot s^{-1}} = 3.25 \times 10^{-2}\ cm^{-1}$$

MT1.5 This question can be answered by examining the equation that relates the reduced mass μ to the vibrational frequency: $\nu = \frac{1}{2\pi}\sqrt{\frac{k}{\mu}}$

We will assume that the force constant k is essentially the same for the Fe—H and Fe—D bonds and set up the proportionality between the frequencies of the two vibrations:

$$\frac{\nu_{Fe-D}}{\nu_{Fe-H}} = \frac{\frac{1}{2\pi}\sqrt{\frac{k}{\mu_{Fe-D}}}}{\frac{1}{2\pi}\sqrt{\frac{k}{\mu_{Fe-H}}}}$$

$$\nu_{Fe-D} = (\nu_{Fe-H})\sqrt{\frac{\mu_{Fe-H}}{\mu_{Fe-D}}}$$

$$= (1950\ cm^{-1})\sqrt{\frac{m_{Fe}m_H/m_{Fe}+m_H}{m_{Fe}m_D/m_{Fe}+m_D}}$$

$$= (1950\ cm^{-1})\sqrt{\frac{(55.85)(1.01)/55.85+1.01}{(55.85)(2.01)/55.85+2.01}}$$

$$= (1950\ cm^{-1})\sqrt{0.5113}$$

$$= 1394 \text{ cm}^{-1}$$

Note that we have also assumed the average mass for Fe to be 55.85 g.mol^{-1}. It would be more correct to use the mass of the particular isotope of Fe bonded to the H atom. Because that is not given, the average value has been used. The change in frequency due to the use of different isotopes of iron is very small compared with the change in frequency due to the substitution of D for H, because the percentage change is much greater in the latter case. The mass essentially doubles upon replacing H with D; however, only a small percentage change is observed on going from one isotope of iron to another.

MT2.1 Dyes get their color from the absorption of visible light. This absorption cannot take place unless there are bonding and antibonding orbitals available that have the correct energy spacing in order to absorb this visible light. The presence of a number of multiple bonds in molecules may lead to a delocalized set of orbitals. The more delocalized orbitals there are, the smaller the HOMO/LUMO gap and therefore the molecule absorbs light in the visible range. It is these delocalized orbitals that absorb light in the visible region, giving rise to the desired color.

MT2.3 (a) and (d) have the possibility of n-to-π^* transitions because these molecules possess both an atom with a lone pair of electrons (on O in HCOOH and on N in HCN) and a π-bond to that atom. The other molecules have either a lone pair or a π-bond, but not both.

MT3.1 (a) Based on equation: $2d \sin\theta = \lambda$, $d = \left(\dfrac{\lambda}{2\sin\theta}\right) = \left(\dfrac{152\,\text{pm}}{2\sin 12.1}\right) = 363$ pm

(b) Based on the same equation as (a) $\sin\theta = \left(\dfrac{\lambda}{2(2d)}\right) = \left(\dfrac{152}{2(726)}\right) = 0.1047$

$\theta = 6.01^{\text{o}}$

MT3.3 The lattice layers from which constructive x-ray diffraction occurs are parallel. First draw perpendicular lines from the point of intersection of the top x-ray with the lattice plane to the lower x-ray for both the incident and diffracted rays. The x-rays are in phase and parallel at point A. If we want them to be still in phase and parallel when they exit the crystal, then they must still be in phase when they reach point C, and this is only possible with an integral number of wave-lengths. The total extra distance traveled, $A \rightarrow B \rightarrow C$, is equal to $2x$. From the diagram, we can see that the angle A-D-B must also be equal to θ. The angles θ and α sum to 90°, as do the angles α and A-D-B. We can then write $\sin\theta = \dfrac{x}{d}$ and $x = d \sin\theta$. The total distance traveled is $2x = 2d \sin\theta$. So, for the two x-rays to be in phase as they exit the crystal, $2d \sin\theta$ must be equal to an integral number of wave-lengths.

MT3.5

$$d = \frac{\lambda}{2\sin\theta} = \frac{71.0\,\text{pm}}{2\sin 7.23°} = 282\,\text{pm}$$

MT3.7 The answer to this problem is obtained from Bragg's law: $\lambda = 2d \sin\theta$ where λ is the wavelength of radiation, d is the interplanar spacing, and θ is the angle of incidence of the x-ray beam. Here $d = 401.8$ pm and $\lambda = 71.07$ pm. $71.07\,\text{pm} = 2(401.8\,\text{pm})\sin\theta$; $\theta = 5.074°$

MT3.9 $3/2\,kT = 1/2\,mv^2$; $\lambda = h/mv$; $3kT = h^2/m\lambda^2$; $T = 633$ K

MT4.1 $HOOCCH(NH_2)CH_2OH$ is the more polar amino acid. Therefore, it will be retained in the column the longest.

MT4.3 Compound (b). The stationary phase is more polar than the liquid phase, so the more polar compound of (a) and (b) should be attracted more strongly to the stationary phase and should remain on the column longer. Because compound (a) has two carboxylic acid units (COOH), it will be more polar and will be eluted after compound (b).

MT5.1 For a molecule such as 1,2-dichloro-4-ethylbenzene, $C_6H_3Cl_2(CH_2CH_3)$, 175.04 u, it is relatively easy to lose heavy atoms such as chlorine and groups of atoms such as methyl and ethyl fragments. Molecules can also lose hydrogen atoms. In mass spectrometry, P is used to represent the *parent ion*, which is the ion formed from the molecule without fragmentation. Fragments are then represented as $P - x$, where x is the particular fragment lost from the parent ion to give the observed mass. Because the mass spectrum will measure the masses of individual molecules, the mass of carbon used will be 12.00 u (by definition) because the large majority of the molecules will have all ^{12}C. The mass of H is 1.0078 u. Some representative peaks that may be present are listed below.

Fragment formula	Relation to parent ion	Mass, u
$C_6H_3{}^{35}Cl_2(CH_2CH_3)$	P	174.00
$C_6H_3{}^{35}Cl{}^{37}Cl(CH_2CH_3)$	P	176.00
$C_6H_3{}^{37}Cl_2(CH_2CH_3)$	P	177.99
$C_6H_3{}^{35}Cl(CH_2CH_3)$	P-Cl	139.03
$C_6H_3{}^{37}Cl(CH_2CH_3)$	P-Cl	141.03
$C_6H_3{}^{35}Cl_2(CH_2)$	P-CH$_3$	158.98

$C_6H_3{}^{35}Cl{}^{37}Cl(CH_2)$	P-CH_3	160.97
$C_6H_3{}^{37}Cl_2(CH_2)$	P-CH_3	162.97
$C_6H_3{}^{35}Cl_2$	P-CH_2CH_3	144.96
$C_6H_3{}^{35}Cl{}^{37}Cl$	P-CH_2CH_3	146.96
$C_6H_3{}^{37}Cl_2$	P-CH_2CH_3	148.96
$C_6H_3{}^{35}Cl$	P-CH_2CH_3-Cl	109.99
$C_6H_3{}^{37}Cl$	P-CH_2CH_3-Cl	111.99
Etc.		

MT5.3 The presence of one bromine atom will produce in the ions that contain Br companion peaks that are separated by 2 u. Any fragment that contains Br will show this "doublet" in which the peaks are nearly but not exactly equal in intensity. Thus, seeing a mass spectrum of a compound that is known to have Br or that was involved in a reaction in which Br could have been added or substituted with such doublets, is almost a sure sign that Br is present in the compound. It is also fairly easy to detect Br atoms in the mass spectrum at 79 and 81 u, confirming their presence. If more than one Br atom is present, then a more complicated pattern is observed for the presence of the two isotopes. The possible combinations for a molecule of unknown formula with two Br atoms is

$^{79}Br^{79}Br$, $^{79}Br^{81}Br$, $^{81}Br^{79}Br$, and $^{81}Br^{81}Br$. Thus, a set of three peaks (the two possibilities $^{79}Br^{81}Br$ and $^{81}Br^{79}Br$ have identical masses) will be generated that differ in mass by two units. The center peak, which is produced by the $^{79}Br^{81}Br$ and $^{81}Br^{79}Br$ combinations, will have twice the intensity of the outer two peaks, because statistically there are twice as many combinations that produce this mass. All modern mass spectrometers have spectral simulation programs that can readily calculate and print our the relative isotopic distribution pattern expected for any compound formulation, so that it is possible to easily match the expected pattern for a particular ion with the experimental result.

MT6.1 Two peaks are observed with relative overall intensities 3 : 1. The larger peak is due to the three methyl protons and is split into two lines with equal intensities. The smaller peak is due to the proton on the carbonyl carbon atom and is split into four lines with relative intensities 1 : 3 : 3 : 1.

MT6.3 The peaks in the spectrum can be assigned on the basis of the integrations and the coupling to other peaks. The hydrogen atoms of the CH_3 unit of the ethyl group will have an intensity of 3 and will be split into a triplet by the two protons on the CH_2 unit. This peak is found at $\delta \approx 1.2$. The CH_2 unit will have an intensity of 2 and will be split into a quartet by the three protons on the methyl group. This peak is found at $\delta = 4.1$. The CH_2 group that is part of the butyl function will have an intensity of two but will appear as a singlet because there are no protons on adjacent carbon atoms. This is the signal found at $\delta = 2.1$. The remaining CH_3 groups are equivalent and also will not show coupling. They can be attributed to the signal at $\delta = 1.0$. Notice that the peak that is most downfield is the one for the CH_2 group attached directly to the electronegative oxygen atom, and that the second most downfield peak is the one attached to the carbonyl group.

MT6.5 If one considers the reaction, the products should be those arising from substitution of hydrogen atoms on the propane by chlorine atoms. We would expect them to form a chloropropane or perhaps a dichloropropane. Remember that in the halogenation of alkanes, substitution becomes more difficult as more halogen atoms are introduced. If we then consider the NMR spectrum, we see that there is one large peak that is neighboring a single proton, because it is split into a doublet. There is also a weaker feature, corresponding most likely to one proton, which is split into a septet. This indicates that the proton is neighboring six equivalent protons. The structure that is consistent with this spectrum is 2-chloropropane.

MT6.7 (a) ^{13}C

(b) 1.11%